3

BRITISH SOCIETY FOR DEVELOPMENTAL
BIOLOGY SYMPOSIUM

CONTENTS

PREFACE

By M. BALLS and F. S. BILLETT

The 24th meeting of the British Society for Developmental Biology, which was held at the University of Bristol on 25–27 July 1972, marked a significant step in the Society's history as, for the first time, the papers presented were to be published. In deciding that the topic for the symposium should be *The Cell Cycle in Development and Differentiation* we were aware that the relationship between cell division, growth and cell differentiation had long been recognised by developmental biologists. However, whilst earlier workers tended to stress the incompatibility between the processes of growth and differentiation (see Needham, 1942, p. 505), it has more recently been suggested that DNA synthesis and mitosis are essential precursors to cell differentiation (Ebert & Kaighn, 1966; Malamud, 1971). Possibly duplication of the chromatin, involving the synthesis of new histone and non-histone nuclear proteins and their association with the replicated DNA, provides the opportunity for reprogramming the genetic material. Thus, just as the events in the G_1 phase and the control of G_1 to S transition are of great importance in controlling the rate of cell proliferation, events in the G_1 and S phases may be important in controlling the differentiation and the range of potential activities of the daughter cells produced by the subsequent mitosis and cytokinesis. Cell populations showing variation in the length of G_2 are also known (Tobey, Petersen & Anderson, 1971), and the significance of this phase should not be underrated. Increasing awareness of the importance of the cell cycle in normal and neoplastic development is also reflected in two recent symposia edited by Cameron, Padilla & Zimmerman (1971) and Baserga (1971).

In the opening papers of this symposium, Professor Mitchison develops the concept of morphogenesis and differentiation *within* the cell cycle, and Dr Steel reviews the methods employed in the analysis of the intermitotic period and discusses some of the problems likely to be encountered by those using these methods with complex developing systems.

Since Mitchison's recent book (1971) adequately discussed present knowledge of the cell cycle in prokaryotes and in synchronous cell populations of lower eukaryotes and higher cells in-vitro, these topics were omitted from the Bristol symposium, and the next group of papers is concerned with the cell cycle and morphogenesis in the acellular eukaryotes. Dr Ord describes elegant experiments involving the combination of *Amoeba* nuclei and cytoplasms from different stages of the cell cycle. Dr Ammermann

[vii]

discusses the cell cycle, differentiation and function in the micronuclei and macronuclei of the ciliate *Stylonychia*. Dr John and his colleagues show how enzyme synthesis, mitochondria formation and cell wall production are related to the cell cycle in *Chlorella*, and Dr Grant's paper on *Physarum* indicates the advantages of working with an organism whose nuclei have a high degree of natural synchrony.

The Society is always keen to involve its botanically-oriented members in its meetings, and papers concerned with the higher plants are grouped in the next section of the volume. Dr Bennett stresses the importance of the mass of DNA in determining the duration of meiosis, and Dr Barlow discusses the factors which determine the pattern of mitosis in root meristems, including the relationship between the cap and the cells of the quiescent centre. Dr Lyndon considers the relationship between changes in the durations of the cell cycle phases and morphogenesis in the shoot apex, and Drs Yeoman and Aitchison describe changes in the activities of various enzymes during the cell cycle in Jerusalem artichoke cells in-vitro.

The largest group of papers in the symposium deals with the cell cycle during early development. Professor Giudice discusses the significance of the slow maturation of ribosomal RNA and the synthesis of histones during sea urchin oogenesis in relation to the modified cell cycle which is so characteristic of the early stages of so many embryos. The relationship between oogenesis and embryogenesis is also discussed by Drs van den Biggelaar and Boon-Niermeijer in their account of cell division synchrony in early molluscan development, where the duration and timing of the cell cycle relates not only to the subsequent fate of cells but also to their previous history as part of the egg cytoplasm. Dr Hamilton shows how the natural synchrony of the initial cleavage stages of amphibian development can be used in analysing the sensitivity of the various phases of the cell cycle to irradiation. Dr Chibon's work shows that, for various tissues of *Pleurodeles* larvae, the lengths of the cell cycle phases are temperature-dependent. However, unlike the other phases, G_1 lengthens as the temperature increases and decreases as the temperature falls, which may be a mechanism for reducing the effects of temperature fluctuations on the total cell generation time. Dr Rudkin discusses the curtailed cell cycles of the polytene cells of dipteran insects, where chromosomal, nuclear and cell division are bypassed, and endoreduplication results. Dr Graham considers the cell cycles of the inner cell mass cells of mouse embryos and polyploidisation in trophoblast cells, and Dr Snow sounds a warning to developmental biologists using radioisotope techniques, in showing that embryonic cells at certain stages may be particularly susceptible to radiation damage.

In choosing a final group of speakers for the symposium we decided to stress the extrinsic control of the cell cycle with particular reference to erythropoiesis and lymphocyte activation. Dr Harrison and his colleagues show that erythropoietin induces both proliferation in proerythroblasts and an increased rate of haemoglobinisation in foetal liver cells in-vitro, and Drs Cole and Tarbutt discuss genetic defects that upset the regulation of red cell formation in foetal mice. Drs Hardy and Ling's review of the mitotic activation of lymphocytes suggests activation via a cytoplasmic mechanism, rather than a direct action of mitotic stimulants on the genome. Drs Harris and Olsen's paper is also controversial, as they suggest that the increase in the number of reacting lymphocytes following specific stimulation is too great for their production by proliferation alone, and that the transfer of DNA between lymphocytes might be involved. Dr Rytömaa ends the volume with a discussion of the chalone theory with respect to the control of granulopoiesis in-vivo, in-vitro and in neoplasia.

In addition to the main papers, the volume contains a small number of shorter papers presented by members of the Society during the meeting. We have placed these with the major contributions in the groups of broadly related topics indicated above, although to preserve the unity of the theme we have not resorted to sub-headings. Obviously in a symposium of limited duration it is not possible to deal with all aspects of a topic. The emphasis on this occasion was on the cell cycle as a unifying concept, or perhaps less grandly as a simple theme to provide the opportunity for workers using widely diverse systems to meet and discuss a problem of common interest. At this stage it seems both relevant and useful to bring together cell cycle data from several diverse eukaryote systems. The acceptance of the cell cycle as a common theme, however, does not imply the acceptance of a concept of a simple, single cycle of events common to all eukaryote cells. It is as well to recall, for instance, that the DNA of all the chromosomes does not all replicate at the same time, and that in the early development of the sea urchin both DNA replication and RNA transcription occur at the same time. One single cycle encompasses a number of interrelated and occasionally overlapping cycles of DNA replication and RNA transcription which, as has been emphasised in this symposium, themselves relate to cyclic behaviour in the synthesis of proteins. Provided the complexity of the system is remembered, a simple theme does not become a restrictive dogma. The events which occur in the cell during that marvellously misnamed 'resting stage' of the nucleus await further investigation, but we are confident that this volume reports significant progress and will provide further inspiration.

We are grateful to the authors for the high quality of their presentations,

and to the staff of the Cambridge University Press for their collaboration
in the production of this volume, which we hope will be the first of a series
based on symposia organised by the British Society for Developmental
Biology. A special word of thanks is due to Dr Elizabeth Deuchar who acted
as Local Secretary and, together with her colleagues in the Department of
Anatomy at Bristol, did so much to make the meeting itself an enjoyable
and memorable event.

REFERENCES

BASERGA, R., ed. (1971). *The Cell Cycle and Cancer*, 481 + xii pp. New York: Marcel Dekker.

CAMERON, I. L., PADILLA, G. M. & ZIMMERMANN, A. M., eds. (1971). *Developmental Aspects of the Cell Cycle*, 387 + xii pp. New York & London: Academic Press.

EBERT, J. D. & KAIGHN, M. E. (1966). The keys to change: factors regulating differentiation. In *Major Problems in Developmental Biology*, ed. M. Locke, pp. 29–84. New York & London: Academic Press.

MALAMUD, D. (1971). Differentiation and the cell cycle. In *The Cell Cycle and Cancer*, ed. R. Baserga, pp. 132–41. New York: Marcel Dekker.

MITCHISON, J. M. (1971). *The Biology of the Cell Cycle*, 313 + v pp. London: Cambridge University Press.

NEEDHAM, J. (1942). *Biochemistry and Morphogenesis*, 785 + xvi pp. London: Cambridge University Press.

TOBEY, R. A., PETERSEN, D. F. & ANDERSON, E. C. (1971). Biochemistry of G_2 and mitosis. In *The Cell Cycle and Cancer*, ed. R. Baserga, pp. 309–53. New York: Marcel Dekker.

DIFFERENTIATION IN THE CELL CYCLE

By J. M. MITCHISON

Department of Zoology, University of Edinburgh,
West Mains Road, Edinburgh EH9 3JT

It is an honour to be asked to introduce the first Symposium of the British Society for Developmental Biology. It is also a particular pleasure for me that the subject concerns the cell cycle, since this field has not been well represented in Britain until recent years. Of the eighty-eight contributors to three important books on the cycle in the sixties, only seven were from this country (Zeuthen, 1964; Cameron & Padilla, 1966; Padilla, Whitson & Cameron, 1969). This is now changing, and there is no doubt that one of the main reasons is the growing realisation that there is a close connection between the cell cycle and development.

In the early days of embryology, one of the main subjects of interest was the overall growth of a developing organism. This pattern could be subdivided into the growth of particular regions or organs, and their differential growth or 'allometry' explained the changes in shape during development (Thompson, 1917; LeGros Clark & Medawar, 1945). At this level of analysis, the growth of an organ, as well as that of an embryo, was a smooth continuous process without obvious discontinuities. Today, we can probe more deeply into the growth process and see that it is composed of discontinuous or periodic events at the cellular and molecular levels. A cell runs through the cell cycle and divides – essentially a discontinuous process in which one cell changes sharply into two daughter cells. Whether or not this occurs, and, if it does, at what rate, is an important component of morphogenesis and differentiation, and a central theme in the Symposium. At a deeper level still, that of molecules, there is a growing body of evidence that differentiation involves periodic gene expression made manifest at the appropriate time and place in the synthesis of special molecules which are characteristic of a particular tissue or which may act as 'morphogens' and control events in other cells. My thesis in this short review is that we can find the same processes of morphogenesis and periodic gene expression in the miniature system of the cell cycle and that differentiation can be found and investigated as properly here as in the whole developing organism.

[1]

MORPHOGENESIS IN THE CYCLE

For many years, the eukaryotic cell cycle has been divided into two periods. One of these is interphase which occupies about 95% of the cycle and during which all the components of the cell are, in principle, doubled. The second period is mitosis which is essentially a separation process in which the products of interphase are physically split into two with particular care being given to an accurate splitting of the chromosomes. There can be little argument about the presence of morphogenesis and differentiation during mitosis. The cell undergoes profound changes of structure with the condensation of chromosomes, the disappearance of the nucleolus and the nuclear membrane, the formation of the mitotic apparatus and the separation first of the chromosomes and then of the daughter cells. The only difference between this type of differentiation and that of a multicellular organ or organism is that the latter involves many cells and there is almost certainly some cell-to-cell communication, whereas the former happens within the confines of a single cell.

A more subtle question is whether there is morphogenesis and differentiation during interphase. Here we need to split the two processes. Morphogenesis is the production of form, and this certainly happens during interphase. Organelles of complex structure, such as mitochondria, chloroplasts, membranes and cilia are fashioned during interphase. How they multiply in numbers is less clear. Some mitochondria and some chloroplasts appear to divide in two, but the divisions of these organelles are not necessarily synchronous either within the population of organelles or with cell division.* Centrioles show an intriguing but mysterious morphogenesis (Pitelka, 1969). They do not divide: instead, a new centriole forms near (but not touching) one end of a mature centriole and then grows outwards at right angles. In mammalian cells, this process starts at the beginning of the S period and is complete by G_2, but the separation of this pair of young and old centrioles does not take place until the following G_1 (Robbins, Jentzch & Micali, 1968). Note here that this separation is out of phase with nuclear and cell separation. Another example of de-novo morphogenesis during interphase is the development of the oral apparatus in Ciliates. In *Tetrahymena*, this starts half a cycle before the apparatus comes into use as the mouth of one of the daughter cells. Nor is morphogenesis restricted to the cellular organelles. In cells with rigid walls, the form of the cell itself has to be generated by directed growth of the wall.

The presence of differentiation during interphase is another matter. A

* References for this and many other general statements in this article can be found in Mitchison (1971).

dictionary definition of differentiation is 'to make different in the process of development'. This can be interpreted in various ways, but to me it implies, at the structural level, a change in structure with time and not merely an increase. On the whole, the evidence is against major changes in cell structure during interphase (e.g. Robbins & Scharff, 1966; Erlandson & De Harven, 1971). A cell late in the cycle usually looks the same as a cell early in the cycle except that it is larger. Recently, however, there have been descriptions of changes during interphase in the details of nuclear structure (Flickinger, 1967; González & Nardone, 1968; Erlandson & De Harven, 1971) and in the density of 70 Å particles on the cell membrane and of nuclear pores on the nuclear membrane of mammalian cells (Scott, Carter & Kidwell, 1971). It may be that other structural changes will emerge when fine detail is followed through the cycle.

CHEMICAL DIFFERENTIATION

Differentiation is mostly discussed at the structural level, but a case can be made for considering it also at the chemical level. Take a growing cell with many chemical components. If these components all increase at the same rate, there will be no change in their relative proportions and so no change in the chemical composition of the cell. This is growth without differentiation. But if some of these components change their rates of increase or even cease to increase at all, there will be changes in their proportions and in the total chemical composition of the cell. This is what I call 'chemical differentiation' and it may or may not be coupled with overall growth. My justification is that it follows the dictionary definition given above and also that the same fundamental process is involved as in structural differentiation – the ordered control of gene expression.

Does chemical differentiation occur during the cell cycle? The first positive answer can be given for that important cell component, DNA. For the last twenty years, evidence has accumulated that DNA synthesis is periodic in almost all eukaryotic cells, and periodic synthesis implies chemical differentiation. We do not know as yet why DNA synthesis starts at the end of G_1, but there are various clues. Cell fusion and nuclear transplant experiments show that one factor is likely to be an initiator present in the cytoplasm. Another factor is that protein synthesis is needed to initiate and continue DNA replication, since an inhibition of protein synthesis also inhibits DNA synthesis. A third factor is that the state of the chromatin is important since heterochromatic regions of the chromosomes start replication later than the euchromatic regions. Recent work on eukaryotic DNA has uncovered much more heterogeneity at the molecular level (repetitive and satellite DNA) than was previously suspected, and this has been mirrored in variations of the pattern of replication. In mammalian cells, GC-rich sequences are preferentially replicated at the start of the S period and AT-rich sequences at the end. There is also an interesting case in *Physarum* where the ribosomal genes appear as a separate nuclear satellite and are replicated throughout most of the cycle (Zellweger, Ryser & Braun, 1972) whereas the rest of the nuclear DNA is only replicated in the first few hours. The whole position is complex but, in view of the concentration of effort in this field, it is not unreasonable to expect an elegant simplification to emerge.

In contrast to DNA synthesis, most of the work on the bulk properties of cells such as dry mass or total protein show patterns of continuous synthesis, and so also does total RNA (apart from a block on its synthesis during mitosis). These patterns suggest growth without chemical differentiation or change in composition. Total cellular protein, however, is composed of many individual protein species and, when these are followed through the cycle, there is evidence that a substantial proportion are only synthesised during restricted periods of the cycle. The experiments here fall into two broad classes. One of them is enzyme assays on synchronous cultures, a subject which will be left until the next section. The other smaller class is where individual proteins or restricted groups have been measured through the cycle. Most work has been done on nucleo-histones, and there is a large body of evidence both from histochemical studies and from bulk measurements that histones are only accumulated in the nucleus during the S period. It is also probable that they are only synthesised during that period but the data here are less complete. In addition, some but not all of these histones are only phosphorylated during the S period in mammalian cells (Shepherd, Noland & Hardin, 1971; Balhorn *et al.* 1972). Periodic synthesis is not restricted to the basic proteins of the nucleus. Stein & Borun (1972) have found an increased rate of synthesis and accumulation of the tightly-bound acidic chromosomal protein during late G_1 in HeLa cells, though there is some synthesis throughout the cycle. A series of proteins can also be found in mouse cells which bind to DNA but are not histones (Salas & Green, 1971). These show periodic synthesis since one group is synthesised during the S period and another, which might be involved in the initiation of DNA synthesis, stops being synthesised at the end of G_1. A good case of periodic synthesis of cytoplasmic proteins is the immunoglobins which are only synthesised from late G_1 until nearly the end of S both in human lymphoid cell lines (Buell & Fahey, 1969) and in mouse myeloma cells (Byars & Kidson, 1970). Finally, a case of periodic synthesis of another type of macromolecule is chitin in budding yeast (Cabib & Farkas, 1971).

ENZYME SYNTHESIS

The largest body of evidence in favour of periodic synthesis comes from measurements of enzyme activity in synchronous cultures. This has become a popular field in the cell cycle work of the last few years, and there are several reviews (Donachie & Masters, 1969; Halvorson, Carter & Tauro, 1971; Mitchison, 1971). All I shall do here is to give a short summary together with some of the most recent results.

There are over a hundred cases in which enzymes have been followed through the cycle of cells ranging from bacteria to mammalian cells. A substantial minority of enzymes show what is apparently continuous synthesis through the cycle, whereas the majority show periodic synthesis restricted to some part of the cycle. If the enzyme is stable, the pattern of synthesis is step-wise with G_1, S and G_2 periods, as with DNA. If the enzyme is unstable, the pattern shows a peak during the period of synthesis followed by a fall as the enzyme is degraded or inactivated. This is only a rough classification, and there are some enzymes which do not fit these categories.

A natural question to ask about these step and peak enzymes is whether their periods of synthesis are associated with the S period. Not surprisingly, this does seem to be the case with the enzymes concerned with DNA synthesis which usually start their periods of synthesis at or near the beginning of the S period. With other enzymes, however, there is no clear association and, at any rate in budding and fission yeast where there is most evidence, the synthesis periods appear to be spread throughout the cycle. What is more, it has been shown in bacteria, yeast and mammalian cells that the patterns of step enzymes continue when DNA synthesis has been blocked by an inhibitor. There is therefore no immediate connection between DNA replication and the control mechanisms of periodic enzyme synthesis.

Two theories have been put forward about the control of synthesis of these enzymes. One of these, which can be called 'oscillatory repression', has been put forward by Donachie, Goodwin, Masters and Pardee in a series of papers. It is primarily concerned with biosynthetic enzymes which are controlled by end-product repression. A system of this kind has negative feedback and, with the appropriate constants, will oscillate. As, for example, the amount of an amino acid rises in the pool, so the synthesis of its biosynthetic enzymes will be repressed, and vice versa. The enzyme is under 'autogenous' control in these intermediate states of repression. The theory predicts that there should be continuous synthesis rather than steps if the enzyme is fully repressed or fully derepressed.

A limitation of the theory is that there is no obvious reason why the oscillations should have the same frequency as the cell cycle. To overcome this objection, it has been suggested that the oscillations are entrained by an event which is dependent on the cycle, for instance, a pulse of messenger RNA produced at the time of gene replication.

This theory, which has been primarily developed for prokaryotes, fits most of the data in synchronous bacterial cultures. In particular, it explains why sucrase in *Bacillus subtilis* is synthesised continuously when repressed and periodically when derepressed (Masters & Donachie, 1966). There is no inherent reason, however, why the theory should not also apply to eukaryotes, and a case rather similar to sucrase has been found with ribulose 1,5-diphosphate carboxylase in *Chlorella* (Molloy & Schmidt, 1970). This enzyme is synthesised continuously at high growth rates when it may be fully derepressed, and periodically at low growth rates where some degree of repression is likely to be operating. On the other hand, glucosidases in budding yeast are step enzymes both with and without induction (Halvorson *et al.* 1966). A possible explanation here is that these enzymes may be controlled by oscillations in the level of general catabolite repression which would be less specific than end-product repression. It is also, of course, difficult to tell when an enzyme is *fully* derepressed. Nevertheless, it is hard to believe that oscillatory repression is the control mechanism for all step enzymes. Stebbing (1972), for instance, finds that ornithine transcarbamylase appears to be fully repressed or constitutive in fission yeast, yet is synthesised in steps. He also finds no significant fluctuations during the cycle in the pool amino acids. Arginase and ornithine transaminase in budding yeast show steps at the same stage of cycle both with induction by arginine and without, but there are no major cyclical changes in the arginine pool (Carter, Sebastian & Halvorson, 1971*a*).

The second theory of the control of periodic synthesis has been developed by Halvorson and his colleagues and stems from their work on budding yeast. It can be called 'linear reading' or 'sequential transcription' since it postulates that genes are transcribed during the cycle in the same order as their linear sequence on the chromosomes. An RNA polymerase would move along the chromosome once per cycle transcribing the genes in order. As a result a gene could only be transcribed for a short time each cycle.

Supporting evidence for this theory comes from several sources. Tauro & Halvorson (1966) studied the multiple M genes for α-glucosidase in budding yeast. The homozygote M_1M_1 and the heterozygote M_1m_1 both show a single enzyme step at the same point in the cycle, indicating that an increase in gene dosage at one locus does not affect the timing of the step. But the introduction of the other non-allelic genes M_2 and M_3 produces extra steps. There are two steps per cycle for two non-allelic genes and three steps for three genes. In a later paper, Tauro, Halvorson & Epstein (1968) compared step timings with gene positions for nine enzymes in budding yeast. The timings were consistent with linear reading from one end of the chromosome to the other – the best evidence coming from four enzyme genes on the fifth chromosome. Another persuasive result is that of Cox & Gilbert (1970) with two strains of budding yeast. The distance between two enzyme genes on the second chromosome is much greater in one strain than in the other and so also is the distance between the two steps in the cell cycle. Apart from yeast, there is good evidence for linear reading in *B. subtilis*. Both in the normal cell cycle with three enzymes (Masters & Pardee, 1965) and after synchronous spore germination with five enzymes (Kennett & Sueoka, 1971), the enzyme steps follow the same sequence as the appropriate structural genes on the chromosome.

Despite this evidence for linear reading, it is most unlikely to be the only mechanism for the control of enzyme synthesis. To start with, there are the many examples of continuous enzyme synthesis which have been mentioned earlier. There is the unlikely possibility that each enzyme is controlled by so many non-allelic genes that the multiple steps produced by these genes cannot be resolved. But if this suggestion is rejected, then there must be another mechanism which allows for continuous transcription (or, at the very least, for continuous translation). Then there is the question of inducibility of those enzymes which can be induced or de-repressed. Linear reading predicts that enzymes should only be inducible for a restricted period of the cycle. But there is good evidence that enzymes *can* be induced at all stages of the cycle in bacteria, fission yeast and *Chlorella*. Recently this has been shown in budding yeasts for two enzymes that normally show a step pattern of synthesis in the cycle (Carter, Sebastian & Halvorson, 1971*b*). So again another mechanism must be postulated which allows a gene to be transcribed at all stages of the cycle under the appropriate stimulus, even if it is only transcribed under normal conditions during a restricted period.

It would be satisfying if we could conclude that one or other of these theories is correct. But we clearly cannot do this now since there is not enough clear-cut evidence. My own feeling is that it may turn out, in the end, that both theories are partially correct with linear reading applying to some genes and oscillatory repression to others. It may even be that both mechanisms can apply to the same gene with linear reading producing the initial entrainment pulse and with repression ending the period of synthesis.

In addition, I think it a safe guess that mechanisms other than these two will be discovered in due course.

Another problem about enzymes should be considered here. It has rightly been stressed that what is being measured in these experiments with synchronous cultures is enzyme activity and not enzyme protein. We may therefore be misled in interpreting changes in enzyme activity as changes in the actual amount of enzyme. They might instead be due to changes in activators or inhibitors. We can accept the criticism and hope that the few cases where enzyme quantities have been measured with immunological techniques will be extended. But is it in fact likely that we are being misled? In some of the methods of enzyme assay, low molecular weight activators or inhibitors would be diluted to such an extent that they would be unlikely to be effective. Even if they were effective, their presence should be detected by distortion of activity/concentration plots. We should also consider the biological situation of the growing cell. During a cycle, the amount of a stable enzyme must double. Let us suppose that this doubling occurs stepwise but after a step in activity. We then have to postulate a sophisticated system of activation which would come into action at the time of the activity step, exactly double the activity, and then later compensate for the doubling of the enzyme protein by keeping the activity constant. If, on the other hand, we postulate continuous synthesis of the enzyme with a step in activity, the activation system has to be even more sophisticated and unlikely. Before the step, the system will have to produce increasing inhibition to keep the activity constant. It will then have to allow a doubling in activity, followed by further increasing inhibition to keep the constant activity. For these reasons, I prefer to make the assumption that in most cases enzyme activity does correspond to the amount of enzyme protein. If it is shown that there are sudden changes in activation, then we shall have an interesting time trying to explain how they in turn are controlled. Another point which should be remembered is that enzyme activity in a cell extract may not be the same as that in the living cell because of the different environment.

A final and deeper question about the patterns of enzyme synthesis is why so many of them are periodic. It is surprising that such a pattern is so common since, at first sight, it would seem that continuous synthesis is a much more logical way of increasing the components of a growing cell without upsetting its delicate balance. We can give no firm answer to this question, only some guesses. It may be that many enzymes are present in excess so that periodic synthesis does not upset a delicate balance. A straightforward on–off control operating for a limited time may be an easier system to evolve than a graded response throughout the cycle.

Periodic synthesis may be linked with morphogenetic changes in the cell or its organelles which we do not yet appreciate. It may also be the expression of a cellular clock whose timing is essential for progress through the cycle. Whatever the outcome of this question, it is clearly going to become important as we go deeper into the temporal complexity of the cell cycle.

CONCLUSION

The cell cycle mimics in miniature many of the processes that determine the development of higher organisms. It shows morphogenesis in the construction of new cellular organelles. It shows differentiation at the structural level during mitosis, and there are also some indications of this during interphase. At the chemical level, there is increasing evidence of periodic synthesis which means that the chemical composition of the cell is changing in an ordered way through the cycle. This can be called chemical differentiation and implies ordered gene expression. Periodic synthesis is established for DNA, for some small classes of cellular proteins such as histones and immunoglobulins, and for the majority of enzymes. Two theories have been put forward to explain the periodic synthesis of enzymes and there is evidence for and against them. Whatever their final status, their immediate interest is that they are the *only* theories at present which attempt to explain the temporal control of gene expression in any differentiating system.

REFERENCES

BALHORN, R., BORDWELL, J., SELLERS, L., GRANNER, D. & CHALKLEY, R. (1972). Histone phosphorylation and DNA synthesis are linked in synchronous cultures of HTC cells. *Biochemical and Biophysical Research Communication*, **46**, 1326–33.

BUELL, D. N. & FAHEY, J. L. (1969). Limited periods of gene expression in immunoglobulin-synthesizing cells. *Science, New York*, **164**, 1524–5.

BYARS, N. & KIDSON, C. (1970). Programmed synthesis and export of immunoglobulin by synchronized myeloma cells. *Nature, London*, **226**, 648–50.

CABIB, E. & FARKAS, V. (1971). The control of morphogenesis: an enzymatic mechanism for the initiation of septum formation in yeast. *Proceedings of the National Academy of Sciences, U.S.A.* **68**, 2052–6.

CAMERON, I. L. & PADILLA, G. M. (eds.) (1966). *Cell Synchrony*. New York & London: Academic Press.

CARTER, B. L. A., SEBASTIAN, J. & HALVORSON, H. O. (1971*a*). The regulation of synthesis of arginine catabolizing enzymes during the cell cycle in *Saccharomyces cerevisiae*. *Advances in Enzyme Regulation*, **9**, 253–63.

CARTER, B. L. A., SEBASTIAN, J. & HALVORSON, H. O. (1971*b*). Regulation of enzyme synthesis in synchronous cultures of yeast. *Bacteriological Proceedings*, **71**, 156.

COX, C. G. & GILBERT, J. B. (1970). Nonidentical times of gene expression in two strains of *Saccharomyces cerevisiae* with mapping differences. *Biochemical and Biophysical Research Communications*, **38**, 750–7.

10   J. M. MITCHISON

888888
DONACHIE, W. D. & MASTERS, M. (1969). Temporal control of gene expression in bacteria. In *The Cell Cycle, Gene–Enzyme Interactions*, ed. G. M. Papilla, G. L. Whitson & I. L. Cameron, pp. 37–76. New York & London: Academic Press.

ERLANDSON, R. A. & DE HARVEN, E. (1971). The ultrastructure of synchronized HeLa cells. *Journal of Cell Science*, **8**, 353–97.

FLICKINGER, C. J. (1967). The fine structure of the nuclei of *Tetrahymena pyriformis* throughout the cell cycle, *Journal of Cell Biology*, **27**, 519–29.

GONZÁLEZ, P. & NARDONE, R. M. (1968). Cyclic nucleolar changes during the cell cycle. I. Variation in number, size, morphology and position. *Experimental Cell Research*, **50**, 599–615.

HALVORSON, H. O., BOCK, R. M., TAURO, P., EPSTEIN, R. & LA BERGE, M. (1966). Periodic enzyme synthesis in synchronous cultures of yeast. In *Cell Synchrony*, ed. I. L. Cameron & G. M. Padilla, pp. 102–16. New York & London: Academic Press.

HALVORSON, H. O., CARTER, B. L. A. & TAURO, P. (1971). Synthesis of enzymes during the cell cycle. *Advances in Microbial Physiology*, **6**, 47–106.

KENNETT, R. H. & SUEOKA, N. (1971). Gene expression during outgrowth of *Bacillus subtilis* spores. The relationship between gene order on the chromosome and temporal sequence of enzyme synthesis. *Journal of Molecular Biology*, **60**, 31–44.

LEGROS CLARK, W. E. & MEDAWAR, P. B. (eds.) (1945). *Essays on Growth and Form*. Oxford: Clarendon Press.

MASTERS, M. & DONACHIE, W. D. (1966). Repression and the control of cyclic enzyme synthesis in *Bacillus subtilis*. *Nature, London*, **209**, 476–9.

MASTERS, M. & PARDEE, A. B. (1965). Sequence of enzyme synthesis and gene replication during the cell cycle of *Bacillus subtilis*. *Proceedings of the National Academy of Sciences, U.S.A.* **54**, 64–70.

MITCHISON, J. M. (1971). *The Biology of the Cell Cycle*. London: Cambridge University Press.

MOLLOY, G. R. & SCHMIDT, R. R. (1970). Studies on the regulation of ribulose-1,5-diphosphate carboxylase synthesis during the cell cycle of the eucaryote *Chlorella*. *Biochemical and Biophysical Research Communication*, **40**, 1125–33.

PADILLA, G. M., WHITSON, G. L. & CAMERON, I. L. (eds.) (1969). *The Cell Cycle: Gene–Enzyme Interactions*. New York & London: Academic Press.

PITELKA, D. R. (1969). Centriole replication. In *Handbook of Molecular Cytology*, ed. A. Lima-de-Faria, pp. 1199–218. Amsterdam & London: North Holland Publishing Co.

ROBBINS, E., JENTZSCH, G. & MICALI, A. (1968). The centriole cycle in synchronized cells. *Journal of Cell Biology*, **36**, 329–39.

ROBBINS, E. & SCHARFF, M. (1966). Some macromolecular characteristics of synchronized HeLa cells. In *Cell Synchrony*, ed. I. L. Cameron & G. M. Padilla, pp. 353–74. New York & London: Academic Press.

SALAS, J. & GREEN, H. (1971). Proteins binding to DNA and their relation to growth in cultured mammalian cells. *Nature, New Biology, London*, **229**, 165–9.

SCOTT, R. E., CARTER, R. L. & KIDWELL, W. R. (1971). Structural changes in membranes of synchronized cells demonstrated by freeze-cleavage. *Nature, New Biology, London*, **233**, 219–20.

SHEPHERD, G. R., NOLAND, B. J. & HARDIN, J. M. (1971). Histone acetylation in synchronized mammalian cell cultures. *Biochimica et Biophysica Acta*, **228**, 544–9.

STEBBING, N. (1972). Amino acid pool components as regulators of protein synthesis in the fission yeast, *Schizosaccharomyces pombe*. *Experimental Cell Research*, **70**, 381–9.

STEIN, G. S. & BORUN, T. W. (1972). The synthesis of acidic chromosomal protein during the cell cycle of HeLa S-3 cells. I. The accelerated accumulation of acidic residual nuclear protein before the initiation of DNA replication. *Journal of Cell Biology*, **52**, 292–307.

TAURO, P. & HALVORSON, H. O. (1966). Effect of gene position on the timing of enzyme synthesis in synchronous cultures of yeast. *Journal of Bacteriology*, **92**, 652–61.

TAURO, P., HALVORSON, H. O. & EPSTEIN, R. L. (1968). Time of gene expression in relation to centromere distance during the cell cycle of *Saccharomyces cerevisiae*. *Proceedings of the National Academy of Sciences, U.S.A.* **59**, 277–84.

THOMPSON, D'A. W. (1917). *On Growth and Form*. London: Cambridge University Press.

ZELLWEGER, A., RYSER, U. & BRAUN, R. (1972). Ribosomal genes of *Physarum*: their isolation and replication in the mitotic cycle. *Journal of Molecular Biology*, **64**, 681–91.

ZEUTHEN, E. (ed.) (1964). *Synchrony in Division and Growth*. New York: Interscience Publishers Inc.

THE MEASUREMENT OF THE
INTERMITOTIC PERIOD

By G. G. STEEL

Biophysics Division, Institute of Cancer Research,
Belmont, Sutton, Surrey

What we seek to measure in a growing cell population is determined by our conception of the processes that are taking place within it, of the mechanisms that we seek to elucidate. And measurements by themselves (for instance of the duration of the DNA-synthetic period) are only of general interest until they can be combined with measurements of other parameters into a theory that helps us 'understand' how a biological system works. Some parameters, for instance growth rate, are unambiguous and often directly measurable; others, of which 'cell cycle time' is an example, are only appropriate when the real cell population corresponds to a particular theoretical model.

Many growing cell populations are asynchronous, in that the proportion of cells seen in division (the mitotic index) does not show significant cyclic variation. Over a certain period of time the cell birth rate within a population may therefore be constant and if there is no influx of cells the growth rate will be determined by the difference between the birth rate and the rate at which cells are lost by death, differentiation or emigration. When, as is the case in most developing systems, a proportion of cells are continually changing their characteristics, we can either consider the population as a whole or a morphologically defined part of it, for instance a part to which mitotic activity is confined. The problem of understanding the kinetic state of such a population becomes more intricate when we ask how the constant mitotic index is maintained. What proportion of the cells ever divide and what period (intermitotic time) do they spend between divisions? At the present time the usual approach to this problem is to assume that the distinction between proliferating and non-proliferating is clear-cut and that the population may be seen as one with a constant 'growth fraction' of cells that have a measurable 'cell cycle time'. In many cases this conception may be appropriate, but not always.

The analogy with a human population is a useful one. In England we have an asynchronous, slowly-growing population consisting of children, the middle aged and the elderly. The growth rate is determined by the difference between the birth rate (plus the net immigration rate) and the death rate, and death occurs at all ages but predominantly in the elderly.

Reproduction does not occur until after childhood and some individuals never reproduce. The fact that we seldom try to describe this situation in terms of a 'reproductive cycle time' and a 'reproductive fraction' is partly because we know that the age at which mothers bear children varies so widely that the average is not practically useful.

The techniques of cell population kinetics and demography differ considerably; it is easier to attach labels to individual members of a human population than to a cell population. Thus, except for cell populations whose individuals can be followed in culture, cell kinetic methods must be based on statistical studies of subgroups of the whole population.

AGE DISTRIBUTIONS

Most cell kinetic parameters are either numbers of cells (mitotic or labelling indices) or time intervals (residence times in some defined category) and the link between them usually involves consideration of an age distribution. Age is defined as time since a cell was born at the end of telophase and the age distribution indicates the probability that a randomly-selected cell will have any given age (or the number of cells with ages within a given interval). Ideal cell populations can be conceived in which the age distribution is rectangular (age independent); this will be the case if the intermitotic time is constant and the only loss of cells consists in the loss of one daughter cell immediately after each mitosis. But in expanding populations the age distribution cannot be rectangular: there must be an excess of young cells over old. This is because for growth to take place there must be more cells leaving mitosis than entering it and the cells leaving and entering are respectively the youngest and oldest proliferating cells in the population. In the case of exponentially growing populations the shape of the age distribution is constant in time and can easily be calculated (Johnson, 1961; Steel, 1968).

An example of an experimentally-determined age distribution in a growing cell population is given in Fig. 1. The number of cells having any particular age falls continuously (apart from scatter) from the youngest to the oldest age groups and the shape of the distribution can be understood in terms of two components. Firstly, there is dispersion in the intermitotic times of individual cells, by which in this example some cells have an intermitotic time of up to 30 h even though the mean is 19.5 h. Secondly, a group of cells born at any particular time will, even before they divide, form a progressively smaller proportion of the whole cell population owing to its subsequent growth. The solid theoretical line in Fig. 1 indicates the effect of this on the age distribution of cells that have

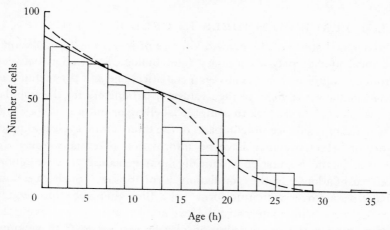

Fig. 1. Age distribution for an exponentially growing population of human amnion cells in culture (from Sisken & Morasca, 1965). The histogram shows the distribution measured by time-lapse cinemicrography. The solid line shows the expected age distribution if all cells had the average intermitotic time of 19.5 h. The broken line shows a theoretical age distribution which takes into account the spread in intermitotic times.

the average intermitotic time. The broken line shows the age distribution of a model which combines both growth and a spread of intermitotic times.

In populations of cultured cells which can be timed individually it is possible to measure directly the numbers of cells in any category (e.g. mitotic index) and their intermitotic time. When this cannot be done, the theoretical age distribution is an essential link in the calculations. The mitotic index of proliferating cells is given by the fraction of the area of the age distribution occupied by cells in mitosis. Similarly, the thymidine labelling index is given by the fractional area occupied by cells in the S phase. If non-proliferating cells are also present then the indices can be deduced similarly from an age distribution that includes both proliferating and non-proliferating cells (see Steel, 1968). In thymidine labelling work, the procedure may also be applied to calculate the increase in the labelling index of the population during continuous or repeated thymidine administration (Steel, Adams & Barrett, 1966).

THE PLACE OF MODELS IN CELL KINETICS

In the detailed study of the growth kinetics of in-vivo cell populations, mathematical models play a necessary role. Biological variation between cells produces a problem that is always a statistical one and the techniques that are available often yield average values of parameters for the population as a whole. When we seek to compare labelling or mitotic indices with measurements of residence time in parts of the intermitotic cycle a number of factors on which we may have no information potentially affect the calculations. Is the population kinetically homogeneous? Can we ignore inter-cell correlations? Is the population asynchronous? Such questions are usually passed over with a uniform 'yes'. But other questions require a choice between definite alternatives. If we know that cells are leaving the population, what is the age distribution of cells that are lost? If we know that cells are being sequestered in a non-proliferating state, at what age do they make the transition? Calculations that depend upon such factors can usually only be made on the basis of assumed answers to these questions.

When a number of assumptions is required there is great advantage in setting out in a formal way the type of model population that is being considered. The procedure is then as follows:

(i) Description of an assumed model.

(ii) Calculations on the basis of the model, using some experimentally determined parameters, leading to a prediction of other experimentally verifiable results.

(iii) Comparison of the theoretical predictions with the further experimental data,

leading either to

(iv) Confirmation of the model as an adequate description of the cell system as studied

or to

(v) Rejection or modification of the model and back to (i).

The basis of this procedure is the belief that if a model (incorporating uncheckable assumptions) can adequately simulate a body of experimental data then it can be accepted as a plausible hypothesis. A by-product is often the discovery of parts of the experimental data that cannot be understood in terms of a generally accepted model, leading to detailed study of the discrepancy and perhaps the discovery of unsuspected phenomena.

THE TECHNIQUE OF LABELLED MITOSES

A good example of the application of mathematical models to cell kinetic studies is in the analysis of results gained by the technique of labelled mitoses. This is the principal technique that has been used to study the timing of the intermitotic period and, as this Symposium is particularly concerned with this aspect of the kinetics of cell populations, it will be considered here in some detail.

The basis of the method was outlined by Howard & Pelc (1953) in their work on the ^{32}P labelling of bean roots, from which they introduced the notation G_1, S, G_2 for the three main subdivisions of interphase. The more elegant use of [^{3}H]thymidine was first described by Quastler & Sherman (1959) who were clearly aware that they were dealing with a method of studying the statistics of time parameters that showed considerable variation from cell to cell. They indicated that detailed mathematical techniques for analysing the results might be developed, but decided that their own experimental results did not warrant such treatment. In subsequent years the analysis of labelled mitoses data became caught up in the concept of the cell cycle and a common debate was whether one should read off the 'cell cycle time' from the intersections of successive rising limbs of the labelled mitoses curve with the 50% level or the 37% level or otherwise, and what to do when the second peak failed to reach the chosen level. Quastler's untimely death no doubt prevented him from spreading a more satisfactory attitude to the analysis of labelled mitoses data.

In a labelled mitoses experiment one labels the cohort of cells that at one time were making new DNA and then follows their movement through the mitotic cycle, using mitosis as a window. The cohort is initially synchronous and in many tissues a clear peak of labelled cells is observed in mitosis during the first day after labelling. This indicates that labelling was restricted to a cohort (i.e. that DNA synthesis occupied a discrete part of interphase) and that the cohort did not lose its synchrony within a few hours. The measurement of intermitotic time then becomes a matter of comparing the second peak in the curve with the first. If the second peak is high and not very different in shape from the first then the time between the peaks gives a good estimate of average intermitotic time. This situation is seldom observed. More commonly, the second peak is spread out in comparison with the first, indicating that within the cell population there is a spread of intermitotic times.

In such a situation, what dimension of the curve indicates the average intermitotic time? No simple answer can be given, because the form of the second peak is determined in a relatively complex way by the means and

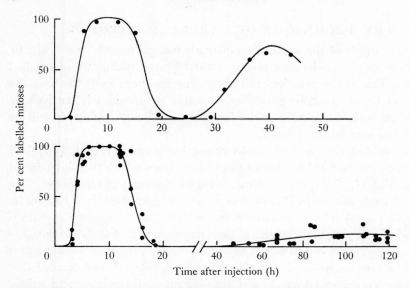

Fig. 2. Two examples of sets of labelled mitoses data that can be adequately
simulated by a theoretical model. The upper part shows the data of Muckenthaler
(1964) for grasshopper spermatogonia; the lower part shows the data of Brockwell
et al. (1972) for mouse corneal epithelium. The data are well fitted by the method
of Steel and Hanes (1971) whether the second peak is high or low and the parameters of the theoretical model can therefore be taken to be consistent with the
data (see Table 1).

variances of G_1, S and G_2 and also in principle by the form of each residence time distribution (which is invariably unknown). The analysis must
therefore be based on a mathematical model, as indicated in the previous
section. The procedure is to set up a model in which residence times in
G_1, S and G_2 are given defined distributions (these usually being assumed
to be independent) and an attempt is made to simulate the experimental
data by choosing suitable values for the means and variances of the
distributions. This approach has now been used by a number of authors
(Barrett, 1966; Brockwell & Trucco, 1970; Bronk, Dienes & Paskin, 1968;
Gilbert, 1972; Hartmann & Pedersen, 1970; Macdonald, 1970; Steel &
Hanes, 1971; Takahashi, Hogg & Mendelsohn, 1971). In general it is
found that the choice of mathematical form for the residence time distributions (normal, lognormal, gamma etc.) has an insignificant effect on the
shape of the theoretical labelled mitoses curve, as also has the assumption
of a constant or distributed mitosis phase. If the experimental data can be
satisfactorily simulated (Fig. 2), then the model is adequate and it contains
information on the mean and variance not only of the separate phases
but also of the whole intermitotic period.

Table 1. *Parameters of the intermitotic period for the theoretical labelled mitoses curves shown in Figs. 2, 3 and 4*

	G_1*	S*	G_2*	Median intermitotic time	Comment
Grasshopper spermatogonia (Fig. 2)	13 (14; 6)	12 (12; 1.4)	4.6 (4.8; 1.3)	30	Well-fitted
Corneal epithelium (Fig. 2)	89 (99; 47)	10.7 (10.8; 1.4)	3.7 (3.8; 0.7)	108	Well-fitted
Plucked mouse epidermis (Fig. 3)	32 (32; 1.9)	11.3 (11.5; 2.2)	2.2 (2.4; 1.1)	46	Slight discrepancy
Epiphyseal cartilage (Fig. 3)	6.4 (6.8; 2.5)	11.4 (11.6; 2.2)	4.2 (4.6; 2.0)	23	Not fitted
Intestinal epithelium (Fig. 4)	3.7 (3.9; 1.4)	8.4 (8.5; 1.0)	0.6 (0.9; 0.8)	13.2	Slight discrepancy
Chick embryo lens (Fig. 4)	0.3 (0.5; 0.5)	4.9 (5.0; 1.0)	3.1 (3.2; 1.0)	8.6	Not fitted

* Values given are the median phase duration in hours, together with the mean and standard deviation given thus (mean; S.D.)

If the model cannot simulate the experimental data satisfactorily, then in principle one must move on to stage (v) of the scheme outlined in the previous section. At least it is necessary to seek an explanation of the reason for an observed discrepancy. Lack of a good fit between theory and experiment should not necessarily be taken as an unfortunate state of affairs because it may provide an opportunity for detecting properties of the experimental system that were unsuspected. What is necessary, however, is to distinguish discrepancies that result from properties of the system from those that result from defects in experimental technique.

The simulation analysis of labelled mitoses data can be used whether or not the second peak is well-defined (Fig. 2) although the precision of the extracted parameters of course depends on the closeness to which the data define a particular curve. The analytical methods of Macdonald (1970) and Gilbert (1972) have the advantage of giving estimates of the precision of extracted values for the mean and variance of the phase durations.

DISCREPANCIES IN THE SIMULATION ANALYSIS OF LABELLED MITOSES DATA

In analysis by simulation it is usually wise to begin by assuming a simple cyclic model in which there are no cell-to-cell correlations and in which the phase durations are described by simple independent distributions. It may be assumed also that no cells are entering the population and that loss of

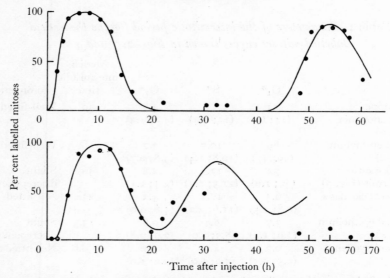

Fig. 3. Two examples of labelled mitoses data that cannot be simulated by a simple model. The upper part shows an example of 'limiting' in the data of Hegazy & Fowler (1972) for plucked mouse skin. The lower part shows an example of 'fade' in the data of Dixon (1971) for epiphyseal cartilage in new-born rats. The full lines are theoretical curves calculated by the method of Barrett (1966) for which the parameters are given in Table 1.

proliferating cells is independent of whether or not they are labelled. Most of the simulation models so far described (cited above) are of this type. When they are used to simulate real data, four types of discrepancy have been encountered (Steel, 1972):

(i) 'Fade'

This is the situation in which, when a theoretical curve has been found which fits the first peak, it is impossible to fit the data beyond the second rise, the data falling predominantly below the best theoretical curve. Theoretical curves based on conservative models are subject to the constraint that the areas associated with successive peaks must be equal. Fade implies a preferential loss of labelled cells, either for biological reasons (e.g. certain types of cell loss from the proliferative state or radiation effect from the tritium label) or by an autoradiographic artefact (cells reducing their radioactivity by division and thus falling below the grain count threshold). An example of fade seen in data on vertebral cartilage of new-born rats is given in Fig. 3, where it may well be due to the preferential loss of labelled cells in bone formation. An extreme example of fade is seen in the data of Toto & Dharwan (1966) for oral epithelium in the mouse.

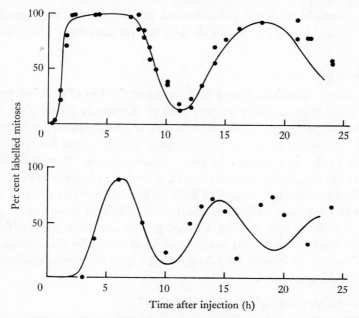

Fig. 4. Two further examples of labelled mitoses data that cannot be simulated by a simple model. The upper part shows an example of slight 'enhancement' in the data of Schultze *et al.* (1972) for mouse intestinal epithelium. The lower part shows an example of 'non-independence of phase durations' seen in the data of Zwaan & Pearce (1971) for presumptive lens cells of the chick embryo; the data appear to define a curve that has three peaks but these cannot be fitted even by assuming an essentially zero duration for G_1 (see Table 1). The theoretical curves are calculated by the method of Barrett (1966).

After the first peak, no labelled mitoses were seen for about four days, at which time a tiny second peak may have been observed. If this second peak was real, it could only indicate the intermitotic time of a small proportion of the cell population.

(ii) *'Enhancement'*

This is the opposite situation to fade, with data points beyond the first peak falling predominantly *above* the best theoretical curve. In general it is due to an increase during the experiment in the proportion of labelled proliferating cells. Possible mechanisms include certain forms of cell loss (in favour of unlabelled cells), or tritium reutilisation. Heterogeneity of the cell population may also cause enhancement; in populations that have a rapidly proliferating stem cell pool the labelled mitoses curve will tend to damp out to approximately the stem cell labelling index rather than the lower labelling index of the whole proliferating population (Tannock,

1968). An example of slight enhancement is seen in the very precise data of Schulze, Haack, Schmeer & Maurer for intestinal epithelium (Fig. 4).

(iii) 'Limiting'

This term covers situations where data points at the top of the first peak cannot be fitted (because they are too low) or conversely where they are too high in the first trough. Such discrepancies are only observed when the data closely define a smooth curve which rises steeply to the first peak or falls steeply to the first trough. In other situations a low first peak or a high trough can usually be attributed to large variation in the durations of G_2, S or G_1. Possible explanations of limiting include heterogeneity of the population (for instance such that, when most cells have completed G_2, a slow tail of cells is still coming through), also autoradiographic false negatives or false positives. An example of limiting at the first trough is the data of Hegazy & Fowler (1972) shown in Fig. 3 for mouse epidermis following plucking of hairs.

(iv) 'Non-independence of the phase durations'

This term covers three main situations. Firstly, where the second rise in the data is steeper than the first fall; this cannot be simulated without assuming negative correlation between individual phases of the cycle. Secondly, where the first trough is narrower than the mean duration of G_2 as is indicated by the first rise in the curve. This implies negative correlation between G_2 and S. Thirdly, the data seem to show clear peaks out to a time where the best theoretical curve must damp out completely. Only very reliable data can show that such peaks are real rather than the chance result of scatter in the data. An example is given in the data of Zwaan & Pierce (Fig. 4) for presumptive lens cells of the chick embryo.

When discrepancies occur it must be concluded that inferences about intermitotic time are to some extent unreliable. It may be judged that the discrepancies are not serious, in which case one has to work on the basis of a simulated curve that is a reasonable compromise (see, for example, the analysis of the data of Schultze et al. in Fig. 4).

MODIFICATIONS OF THE LABELLED MITOSES TECHNIQUE

In recent years there have been a number of interesting developments in the use of this technique. In rapidly proliferating cell systems where the S period takes up most of the intermitotic period a more precise determination of intermitotic time may be gained by using double labelling to identify a portion of the labelled cohort. This has been done by Schultze *et al.* (1972) who used [³H] and [¹⁴C]thymidine to pick out a three-hour cohort of singly ³H-labelled cells which they then followed through mitosis as usual. The resulting curve of singly ³H-labelled mitoses had a peak-width which corresponded well with the expected value of three hours and a median intermitotic time that was close to the value obtained from a conventional labelled mitoses curve.

In very slowly proliferating tissues the mitotic index is often too low to allow the labelled mitoses technique to be used. Burns & Tannock (1970) overcame this by a double-labelling method which used the S period as the window. They labelled specimens with [¹⁴C]thymidine and then at subsequent intervals gave [³H]thymidine just before sampling. Double-labelled cells were scored as an indication of those that had progressed through the whole intermitotic period. Another facet of the paper by Burns & Tannock is its use of a 'non-cycle' model in the analysis of results. They were able to simulate results on hamster cheek pouch epithelium using a model in which cells progressed uniformly through S, G_2 and mitosis but then returned to a 'G_0' pool from which they were randomly selected for a subsequent division.

Levi, Cowan & Cooper (1969) performed a labelled mitoses experiment on mouse bladder epithelium using chromosome preparations throughout. This enabled them to produce curves for cells of defined karyotype in a polyploid tissue and to make the remarkable observation that for diploid and tetraploid cells the timing of the mitotic cycle was the same.

In most in-vivo labelled mitoses work on animals, thymidine has been administered systemically. However, on larger animals (including man) a number of investigators have found it possible to use local infusion of thymidine to cutaneous sites that are subsequently excised (Downes *et al.* 1966; Young & De Vita, 1970).

ALTERNATIVE TECHNIQUES FOR THE STUDY OF THE INTERMITOTIC PERIOD

It should be clear from the foregoing discussion that the technique of labelled mitoses does not always give a straightforward and unequivocal indication of the duration of the intermitotic period. It nevertheless still remains the most direct and reliable method for in-vivo cell populations.

A number of alternative approaches have been used, some predating the labelled mitoses technique. In plant material, labelling of a proportion of a cell population has been accomplished by introducing specific poisons that give rise to morphological abnormalities. Thus Kihlman (1955) used 8-ethoxycaffeine to induce the formation of binucleate cells which could then be timed through a full intermitotic period. Similarly, Van't Hof (1965) used colchicine to induce polyploidy in root meristems and assumed that the time from colchicine administration to the peak incidence of polyploidy indicated the duration of one mitotic cycle. The result was a value that was 30 % smaller than that given by a labelled mitoses experiment. The obvious problem with such methods is that it is difficult to be sure that the damaging agent has not distorted the movement of cells through the proliferative cycle. Although the same criticism can be made of thymidine studies, it is a relatively simple matter to check for a 'thymidine effect' by performing part of the experiment with graded doses of thymidine and to study the dependence of the results on the dose of tracer.

The use of synchronised cell populations is an important technique, particularly applicable to cells and tissues in culture. Two main classes of synchronising method are available; those in which a synchronous sub-population of mitotic cells is selected and those in which the proliferation of the whole population is interrupted and subsequently released (see review by Nias & Fox, 1971).

Radioactive double-labelling techniques have attracted considerable attention, partly because of their theoretical possibilities and partly because of their potential saving in time. The earliest reports (Hilscher & Maurer, 1962; Wimber & Quastler, 1963) described how an estimate of the duration of the S period could be made using a single sample of tissue. This method has subsequently been widely used. Unfortunately, the reliability of the results is often open to question. The method depends on the observer being able to distinguish between ^{3}H-labelling and ^{14}C-labelling (or high and low ^{3}H-labelling intensities) in an autoradiograph. Special autoradiographic techniques have been introduced to assist the observer in this (Baserga & Nemeroff, 1962; Dawson & Field, 1962) but seldom has the discrimination been shown to be satisfactory. One important check is to show that as the time interval between administration of [^{3}H] and [^{14}C]-thymidine is increased, the initial increase in the proportion of singly ^{3}H-labelled cells is consistent with a straight line through the origin. Harriss & Hoelzer (1971) have shown that this may not be the case and that the use of one particular interval may be misleading.

THE STUDY OF THE KINETICS OF COMPLETE
CELL POPULATIONS

The methods so far discussed have been capable of producing information specifically about the proliferating cells in a population, i.e. those cells that are observed to enter DNA synthesis or mitosis. The problem of analysing a cell population as a whole involves the relationship of proliferating cells to non-proliferating cells, to differentiating cells and to cell immigration and emigration. The use of mathematical models can help considerably in clarifying the problem.

The simplest situation is that of an expanding cell population that is homogeneous except in regard to proliferative state. If non-proliferating cells are produced at division with constant probability, the population will tend to a condition of constant proliferative fraction and predictions can be made about the way in which the population should label under continuous thymidine infusion (Steel, Adams & Barrett, 1966). The mathematical model is then established by an attempt to simulate labelled mitoses and overall labelling data. A more complex problem is presented by populations in which cellular development is taking place. Experimentally it is necessary to select some criterion for judging the developmental stage that a cell has reached. In the relatively simple anatomical structure of the rat intestinal epithelium, Cairnie, Lamerton & Steel (1965a, b) were able to subdivide the population on spatial grounds assuming that cells proceeded in an orderly fashion up the crypt of Lieberkühn. The labelled mitoses technique was applied to subpopulations of defined developmental age and the results used to establish a model which contained predictions about the way in which cells move from the proliferating to non-proliferating state. A more complex model was later developed to simulate the reaction of the population to continuous irradiation (Cairnie, 1967). In differentiating cell populations, cell morphology can be used as an indicator of developmental stage. A good example of this is mammalian bone marrow. Tarbutt & Blackett (1968) have shown how the kinetics of the erythroid system as a whole can be understood on the basis of a sequential series of morphological stages.

The average proliferation rate of the whole of a cell population can usefully be studied by the so-called stathmokinetic method (Stevens Hooper, 1961). An agent is administered which specifically blocks cells that enter metaphase and the rate of accumulation of metaphases should indicate the mitotic rate (proportion of cells entering mitosis in unit time). Colchicine, or its purified form colcemid, has widely been used as the blocking agent, but other antimitotic agents have also been employed

(Gelfant, 1963; Cardinali, Cardinali & Blair, 1961). As with the double-labelling method, it is essential to perform proper checks to ensure that the method is reliable. The main hazards are that at low doses of the agent not all the cells will be arrested, while at high doses the movement of cells into mitosis may be inhibited, and arrested cells may degenerate to the point of being unrecognisable. The optimum dosage level may be narrow (Tannock, 1967). A number of comparisons have been made between thymidine methods and the stathmokinetic approach. The review by Bertalanffy (1964) showed broad agreement in data by various authors; the more detailed comparison by Clarke (1971) in the intestinal epithelium of the rat showed that the methods compare well.

It is important to recognise, however, that what is measured by the stathmokinetic method is not intermitotic time. What actually is measured is mitotic rate, from which can be calculated a 'turnover time' (the time taken to produce a number of cells equal to that already present). The turnover time will always be longer than the average intermitotic time of proliferating cells to an extent which increases with the proportion of non-proliferating cells.

Useful information on the structure of developing cell systems may be gained by the novel approach of producing complete thymidine labelling of the in-vivo cell population and subsequently examining the appearance of unlabelled cells (Haas, Werner & Fliedner, 1970). The advantage of this approach is that it gives information on the most slowly proliferating members of a cell population.

THE CONCEPT OF THE CELL CYCLE

The term 'mitotic cycle' was introduced to describe the sequence of morphological stages that a cell goes through from its birth to subsequent division: interphase, prophase, metaphase and telophase. In this sense it is a cycle that is common to all cells that undergo mitosis. Unfortunately, the term 'cycle' sometimes carries the implication of repetitive regularity and in precise quantitative considerations of proliferating cell systems it should be used cautiously.

Data so far gained by the technique of labelled mitoses can be divided into two categories. There are those sets of data that cannot be well simulated by a simple model of the type considered above. In general, it is not possible in such cases to draw precise conclusions about the timing of the mitotic cycle without a fuller understanding of factors that influence the data. The other category of data, those that can be well simulated by a simple model, lead to the conclusion that in proliferating cell systems

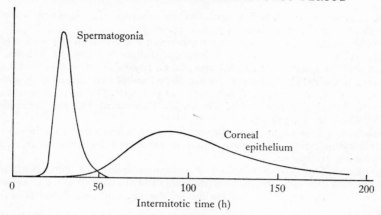

Fig. 5. Distributions of intermitotic times that correspond to the two theoretical labelled mitoses curves shown in Fig. 2, illustrating the breadth of the distributions that are implied by data that show high and low second peaks.

there is often considerable variation in the timing of the mitotic cycle. It is possible, using the mathematical model, to draw a distribution of inter-mitotic times that is consistent with any well-fitted set of data. Some examples are shown in Fig. 5. In some tissues the distributions are fairly narrow and the term 'cell cycle time' may be appropriate. In others, the distributions are broad and it may be misleading to think in terms of a cell cycle time (Steel, 1972). The time which cells spend in defined parts of the mitotic cycle may also show great variation within a population, as has been found in attempts to understand the yield of radiation-induced chromosomal aberrations in root meristems (Savage & Papworth, 1972). In some cases it may be appropriate to consider that the cells in one part of interphase form a resting 'pool' (Burns & Tannock, 1970).

REFERENCES

BARRETT, J. C. (1966). A mathematical model of the mitotic cycle and its application to the interpretation of percentage labelled mitoses data. *Journal of the National Cancer Institute*, **37**, 443–50.

BASERGA, R. & NEMEROFF, R. (1962). Two-emulsion autoradiography. *Journal of Histochemistry and Cytochemistry*, **10**, 628–35.

BERTALANFFY, F. D. (1964). Tritiated thymidine versus colchicine technique in the study of cell population cytodynamics. *Laboratory Investigations*, **13**, 871–86.

BROCKWELL, P. J. & TRUCCO, E. (1970). On the decomposition by generations of the PLM-function. *Journal of Theoretical Biology*, **26**, 149–79.

BROCKWELL, P. J., TRUCCO, E. & FRY, R. J. M. (1972). The determination of cell-cycle parameters from measurements of the fraction of labelled mitoses. *Bulletin of Mathematical Biophysics*, **34**, 1–12.

BRONK, B. V., DIENES, G. J. & PASKIN, A. (1968). The stochastic theory of cell proliferation. *Biophysical Journal*, **8**, 1353–98.

BURNS, F. J. & TANNOCK, I. F. (1970). On the existence of a G_0 phase in the cell cycle. *Cell and Tissue Kinetics*, **3**, 321–34.

CAIRNIE, A. B. (1967). Cell proliferation studies in the intestinal epithelium of the rat. Response to continuous irradiation. *Radiation Research*, **32**, 240–64.

CAIRNIE, A. B., LAMERTON, L. F. & STEEL, G. G. (1965a). Cell proliferation studies in the intestinal epithelium of the rat. *Experimental Cell Research*, **39**, 528–38.

CAIRNIE, A. B., LAMERTON, L. F. & STEEL, G. G. (1965b). Cell proliferation studies in the intestinal epithelium of the rat. II. Theoretical Aspects. *Experimental Cell Research*, **39**, 539–53.

CARDINALI, G., CARDINALI, G. & BLAIR, J. (1961). The stathmokinetic effect of vincaleukoblastine on normal bone marrow and leukemic cells. *Cancer Research*, **21**, 1542–4.

CLARKE, R. M. (1971). A comparison of metaphase arresting agents and tritiated thymidine autoradiography in measurement of the rate of entry of cells into mitosis in the crypts of Lieberkühn of the rat. *Cell and Tissue Kinetics* **4**, 263–72.

DAWSON, K. B. & FIELD, E. O. (1962). Differential autoradiography of tritium and another β-emitter by a double stripping film technique. *Nature, London*, **195**, 510–11.

DIXON, B. (1971). Cartilage cell proliferation in the tail-vertebrae of new-born rats. *Cell and Tissue Kinetics*, **4**, 21–30.

DOWNES, A. M., CHAPMAN, R. E., TILL, A. R. & WILSON, P. A. (1966). Proliferative cycle and fate of cell nuclei in wool follicles. *Nature, London*, **212**, 477–9.

GELFANT, S. (1963). Inhibition of cell division: a critical and experimental analysis. *International Review of Cytology*, **14**, 1–39.

GILBERT, C. W. (1972). The labelled mitoses curve and the estimation of the parameters of the cell cycle. *Cell and Tissue Kinetics* **5**, 53–63.

HAAS, R. J., WERNER, J. & FLIEDNER, T. M. (1970). Cytokinetics of neonatal brain cell development in rats as studied by the complete ^{3}H-thymidine labelling method. *Journal of Anatomy*, **107**, 421–37.

HARRISS, E. B. & HOELZER, D. (1971). DNA synthesis times in leukaemic cells as measured by the double labelling and percentage labelled mitoses methods. *Cell and Tissue Kinetics*, **4**, 433–41.

HARTMANN, N. R. & PEDERSEN, T. (1970). Analysis of the kinetics of granulosa cell populations in the mouse ovary. *Cell and Tissue Kinetics* **3**, 1–12.

HEGAZY, M. A. H. & FOWLER, J. F. (1972). Cell population kinetics of plucked and unplucked mouse skin. *Cell and Tissue Kinetics*, in press.

HILSCHER, W. & MAURER, W. (1962). Autoradiographische Bestimmung der Dauer der DNS-Verdopplung und ihres zeitlichen Verlaufs bei Spermatogonien der Ratte durch Doppelmarkierung mit C^{14-} und H^{3-}Thymidin. *Naturwissenschaften*, **49**, 352–54.

HOWARD, A. & PELC, S. R. (1953). Synthesis of deoxyribonucleic acid in normal and irradiated cells and its relation to chromosome breakage. *Heredity*, **6** suppl. 'Symposium on Chromosome Breakage'. London: Oliver and Boyd.

JOHNSON, H. A. (1961). Some problems associated with the histological study of cell proliferation kinetics. *Cytologia*, **26**, 32–41.

KIHLMAN, B. (1955). Chromosome breakage in *Allium* by 8-ethoxy-caffeine and X-rays. *Experimental Cell Research*, **8**, 345–68.

LEVI, P. E., COWAN, D. M. & COOPER, E. H. (1969). Induction of cell proliferation in the mouse bladder by 4-ethyl sulphonyl-naphthalene-1-sulphamide. *Cell and Tissue Kinetics*, **2**, 249–62.

MACDONALD, P. D. M. (1970). Statistical inferences from the fraction labelled mitoses curve. *Biometrika*, **57**, 489–503.

MUCKENTHALER, F. A. (1964). Autoradiographic study of nucleic acid synthesis during spermatogenesis in the grasshopper *Melanoplus differentialis*. *Experimental Cell Research*, **35**, 531–47.

NIAS, A. H. W. & FOX, M. (1971). Synchronisation of mammalian cells with respect to the mitotic cycle. *Cell and Tissue Kinetics*, **4**, 375–98.

QUASTLER, H. & SHERMAN, F. G. (1959). Cell population kinetics in the intestinal epithelium of the mouse. *Experimental Cell Research*, **17**, 420–38.

SAVAGE, J. R. K. & MILLER, M. W. (1971). Some problems of chromosomal aberration studies in meristems. In *The Dynamics of Meristem Cell Populations*, ed. M. W. Miller & C. C. Kuehnert. New York: Plenum Publishing Co.

SAVAGE, J. R. K. & PAPWORTH, D. G. (1972). The effect of variable G_2 duration upon the interpretation of yield–time curves of radiation-induced chromatid aberrations. *Journal of Theoretical Biology*, in press.

SCHULTZE, B., HAACK, V., SCHMEER, A. C. & MAURER, W. (1972). Autoradiographic investigation on the cell kinetics of crypt epithelium of the jejunum of the mouse: confirmation of steady state growth and constant frequency distribution of cells throughout the cycle. *Cell and Tissue Kinetics*, **5**, 131–46.

SISKEN, J. E. & MORASCA, L. (1965). Intrapopulation kinetics of the mitotic cycle. *Journal of Cell Biology*, **25**, 179–89.

STEEL, G. G. (1968). Cell loss from experimental tumours. *Cell and Tissue Kinetics*, **1**, 193–207.

STEEL, G. G. (1972). The cell cycle in tumours: an examination of data gained by the technique of labelled mitoses. *Cell and Tissue Kinetics*, **5**, 87–100.

STEEL, G. G., ADAMS, K. & BARRETT, J. C. (1966). Analysis of the cell population kinetics of transplanted tumours of widely differing growth rate. *British Journal of Cancer*, **20**, 784–800.

STEEL, G. G. & HANES, S. (1971). The technique of labelled mitoses: analysis by automatic curve-fitting. *Cell and Tissue Kinetics*, **4**, 93–105.

STEVENS HOOPER, C. E. (1961). Use of colchicine for the measurement of mitotic rate in the intestinal epithelium. *American Journal of Anatomy*, **108**, 231–44.

TAKAHASHI, M., HOGG, G. D. & MENDELSOHN, M. L. (1971). The automatic analysis of PLM curves. *Cell and Tissue Kinetics*, **4**, 505–18.

TANNOCK, I. F. (1967). A comparison of the relative efficiencies of various metaphase arrest agents. *Experimental Cell Research*, **47**, 345–56.

TANNOCK, I. F. (1968). The relation between cell proliferation and the vascular system in a transplanted mouse mammary tumour. *British Journal of Cancer*, **22**, 258–72.

TARBUTT, R. G. & BLACKETT, N. M. (1968). Cell population kinetics of the recognisable erythroid cells in the rat. *Cell and Tissue Kinetics*, **1**, 65–80.

TOTO, P. D. & DHAWAN, A. S. (1966). Generation cycle of oral epithelium in 400-day old mice. *Journal of Dental Research*, **45**, 948–50.

VAN'T HOF, J. (1965). Discrepancies in mitotic cycle time when measured with tritiated thymidine and colchicine. *Experimental Cell Research*, **37**, 292–9.

WIMBER, D. E. & QUASTLER, H. (1963). A C^{14} and H^3 thymidine double labelling technique in the study of cell proliferation in *Tradescantia* root tips. *Experimental Cell Research*, **30**, 8–22.

YOUNG, R. C. & DEVITA, V. T. (1970). Cell cycle characteristics of human solid tumours *in vivo*. *Cell and Tissue Kinetics*, **3**, 285–90.

ZWAAN, J. & PEARCE, T. C. (1971). Cell population kinetics in the chicken lens primordium during and shortly after its contact with the optic cup. *Developmental Biology*, **25**, 96–118.

CHANGES IN NUCLEAR AND CYTOPLASMIC ACTIVITY DURING THE CELL CYCLE WITH SPECIAL REFERENCE TO RNA

BY M. J. ORD*

Department of Biology, University of Southampton,
Southampton SO9 5NH

INTRODUCTION

Early biologists studying the cell focused their attention on its division, attributing the greatest importance to this dynamic process while passing over the period between divisions as a less important resting or interphase. In the 1950s, however, the emphasis placed on the physical separation of the cell into two daughter cells was eclipsed by a study of the biochemical events occurring during interphase. Illustration by autoradiography (Howard & Pelc, 1953) of the limits of the DNA synthetic period resulted in the subdivision of interphase into three phases, and concentrated interest on the experimental analysis of replication. It is only during the last few years, with the realization that other cell constituents could have active periods of defined limits, that an appreciable portion of experimental work has been diverted to filling the gap periods. One of the main instigators in the widening of our spectrum in this respect has been J. M. Mitchison, with work on continuous and discontinuous syntheses of cell proteins (1969). The extent to which RNA may also be discontinuously or sequentially transcribed has as yet to be adequately demonstrated.

Although the presentation of a four-phase cell cycle by Howard & Pelc was followed by a spate of work indicating the universal occurrence of such a cycle throughout the plant and animal kingdoms, it gradually became apparent that not all cells fitted into this well-tailored pattern. Notable exceptions were the extreme shortening of both G_1 and G_2 in the rapidly dividing cells of many early embryos, the indefinite extension of G_1 in many differentiated tissues and the complete lack of G_1 in some continuously dividing cells. The distortion of the four-phase cell cycle of early embryonic cells may be explained in terms of both prestored materials in the egg cytoplasm and on cleavage without growth. The presence of a long G_1 in differentiated tissues on the other hand can be explained by the inhibition of the signals initiating DNA replication, while the existence of a

* Member of the Toxicology Research Unit, Medical Research Laboratories, Carshalton, Surrey, England.

three-phase cell cycle in yeast, *Physarum*, *Amoeba* and some fast dividing lines of tissue culture cells may well result from the possible unimportance of G_1 activities to continuously dividing cells, or a spatial dissociation of the activities of division and DNA replication.

DNA SYNTHESIS IN THE CELL CYCLE OF *AMOEBA PROTEUS*

In *Amoeba proteus* the three-phase cell cycle is clearly defined, though there may be minor variations from one cell line to another, or from one laboratory to another where different culture media, techniques and temperatures are used in growing cells. A typical cycle consists of:

M   mitosis, approximately 1 % of the cell cycle
S   DNA replication, approximately 15–20 % of the cell cycle
G_2 occupying the remaining 80–84 % of the cell cycle

The pattern for DNA synthesis obtained from autoradiographic studies on groups of cells removed from large cultures at division and pulse labelled with [^{3}H]thymidine (Fig. 1) indicates that:

(*a*) DNA synthesis begins soon after division.

(*b*) The rate of synthesis is initially high, tailing off from 5 h and reaching G_2 level (Ord, 1968) generally by 9–10 h depending on the strain and on the temperature used for growing cells. A dip occurs at 2–3 h which may indicate a change in the rate of synthesis, but could also be due to fluctuations in precursor pools.

(*c*) A low basic incorporation of external thymidine, approximately 1 % that found at peak S, continues through G_2. This shows two small peaks at approximately 12 and 16 h.

DNA synthesis may for convenience be considered in three steps: (1) initiation, (2) main replication, (3) termination. By using homokaryons, made by transplanting the nucleus from a donor amoeba directly into a second host amoeba, it has been possible to learn considerably more about the individual roles of nucleus and cytoplasm in each of these steps.

The initiation of DNA synthesis

The beginning of S

Ten minute pulses of [^{3}H]thymidine were used to determine whether any gap intervenes between cytokinesis and the beginning of S (Fig. 2). Complete lack of a G_1 phase is indicated by the already maximal incorporation of external thymidine shown by cells exposed to radioisotope within

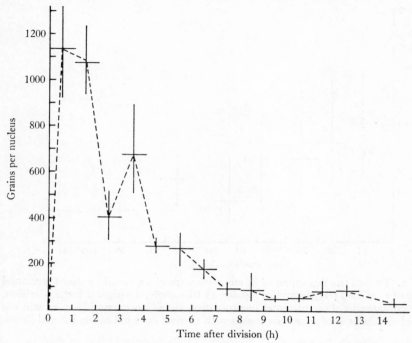

Fig. 1. The S phase in *Amoeba proteus* obtained using 1 h pulses (horizontal bars) of 0.25 mCi/ml [³H]thymidine (sp.act. 5 Ci/mmole). Grain counts are plotted as geometric means with fiducial limits given by vertical lines (Ord, 1968). Incorporation of [³H]thymidine is already maximal in early S and decreases as the cell passes through S, reaching its lowest incorporation level at 9–10 h. It is not known whether the 11–12 h small hump represents late replication of chromatin or replication of nucleolar DNA.

5 min of the completion of cytokinesis, or 15 min from the beginning of anaphase (Plate 1*a*). These cells would still be undergoing nuclear reconstruction. The high rate of [³H]thymidine incorporation found 5–15 min after cytokinesis may represent a high rate of DNA synthesis at the beginning of S, but more probably reflects the low size of the thymidine precursor pools as DNA replication begins. Small endogenous pools would result in maximal thymidine kinase activity (Fig. 3). Work with tissue culture cells has shown that nucleotide pools are not built up before S, but concomitant with it. Although there is an initial rapid build up of nucleotide pools at the beginning of S, the rate of DNA synthesis is no higher than during the subsequent S period (Adams, 1969; Adams, Berryman & Thomson, 1971).

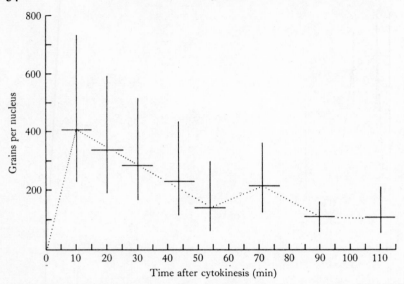

Fig. 2. To find the beginning of S, amoebae were given 10 min pulses (horizontal bars) of 1 mCi/ml [³H]thymidine (sp.act. 25 Ci/mmole) during 0–2 h after division. Autoradiographs show maximal incorporation already occurring in 5–15 min old amoebae. Curve is of geometric means with vertical lines indicating fiducial limits.

Inhibiting the initiation of S

Attempts to prevent the initiation of DNA synthesis once division had taken place proved impossible. No effect was apparent on the subsequent DNA replication period when cells were kept from division stage in 0.5×10^{-4} M (250 μg/ml) puromycin or 1.7×10^{-4} M (200 μg/ml) actino-mycin-D (ACT-D). The same was true when cells were exposed to 10^{-3} M *N*-methyl-*N*-nitroso urethane, a carcinogen which inhibits both RNA and protein synthesis in amoebae more rapidly than either puromycin or actinomycin, for 10 min immediately after division (Ord, Chatterjee & Bell, unpublished). Exposure of cells to RNA or protein inhibitors affected [³H]thymidine incorporation only when extended to a point 3 h or more prior to division. Such cells not only began to show decreased thymidine incorporation, but also poor nuclear reconstruction.

Mechanical manipulation of cells at or following cytokinesis also failed to prevent the inhibition of DNA synthesis. Thus incorporation of [³H] thymidine was not diminished during the S phase of cells which had one-third to one-half the volume of cytoplasm removed immediately after cytokinesis. Preventing cytokinesis either mechanically, or by using low pH Chalkley's solution, gave binucleate cells with both nuclei entering S at the expected time. A similar inability to prevent the induction of DNA synthesis has been observed in the *Xenopus* egg cytoplasm once it has been activated by the rupture of the germinal vesicle (Gurdon & Woodland, 1968).

Investigation of initiation signals

Heterophasic homokaryons were used to investigate the possible presence and duration of signals in the cytoplasm responsible for induction of DNA synthesis. However, when early, mid or late G_2 nuclei were put into S cells (Plate 1*b*) or into division spheres, DNA synthesis was not induced (Ord, 1969). Cell fusion experiments have confirmed the inability of S cytoplasm to reinitiate DNA synthesis in G_2 nuclei (Rao & Johnson, 1970). As no G_1 nuclei are available to allow recognition of cytoplasmic signals by premature induction of their DNA synthesis the presence of a 'switch on' signal responsible for initiating replication has not yet been established in *Amoeba*. However, work on tissue culture cells and *Xenopus* eggs has demonstrated that such signals exist (Rao & Johnson, 1970; Graham, Arms & Gurdon, 1966).

Cytoplasmic control over nuclear activity was recognized in *Amoeba* by investigating events in heterophasic homokaryons entering division. When a young nucleus from S or early G_2 was transplanted into an older G_2 cell, division of the host was delayed until the younger nucleus reached maturity. In all but a few 'early G_2 + late G_2' heterophasic homokaryons division of the two nuclei was synchronous and resulted in four nuclei all entering S simultaneously (Ord, 1971). The exceptions were homokaryons where the host nucleus was already committed to division before addition of the donor nucleus. In these cells the donor nucleus remained undivided although the host cell completed its division. Nuclear changes resulted if the donor nucleus was exposed sufficiently long to this pre-division stage, with some incorporation of [³H]thymidine by the donor nucleus as the host cell passed through S. Not infrequently such donor nuclei 'disappeared'. This could have been caused by premature chromosome condensation in the donor nucleus, as found in cell fusion experiments by Johnson and Rao (1970), or to extrusion of the donor nucleus by the host cell.

Where heterophasic homokaryons underwent division with two nuclei of widely different ages and size all daughter nuclei entered S in synchrony. The younger and smaller nuclei, however, frequently had a longer S period as shown by the longer period of [³H]thymidine incorporation. Since this occurred even when irregular cytokinesis resulted in one small and one large nucleus sharing the same cytoplasm, differences in size of endogenous cytoplasmic precursor pools could not be implicated. The shorter S of the older nucleus, and the extended S of the younger nucleus, would suggest that nuclei may normally be able to undergo some preparation for S before division, e.g. building up of nuclear DNA precursor pools. Differences in the length of the preceding cell cycle could account for some of the variations found during thymidine incorporation of equal-aged amoebae, since no attempt has been made in these experiments to synchronize the cell cycle preceding the cycle under study.

The maintenance of DNA synthesis

Lack of maintenance signals

A requirement for continual cytoplasmic signals to maintain DNA synthesis has not yet been satisfactorily identified in any cell, though there is some evidence that it may be necessary for macronuclear DNA synthesis in *Stentor* (De Terra, 1967). In *Amoeba*, once DNA synthesis has been initiated, cytoplasmic signals for its maintenance are not essential (Ord, 1971). Studies using two types of heterophasic homokaryons (i) early S nuclei transplanted into late S cells and (ii) S nuclei transplanted into G_2 cells (Plate 3a) indicate that once DNA synthesis has been initiated in a nucleus it continues regardless of its cytoplasmic environment.

Cytoplasmic regulation of [³H]thymidine incorporation

Peak S nuclei sharing the same cytoplasmic environment incorporate similar amounts of external thymidine (Plate 1c). Since nuclei in their own cytoplasm show widely different [³H]thymidine incorporation, even when they are of equal age, this suggests a regulating role of cytoplasm in the use of external thymidine.

Cytoplasmic regulation of thymidine incorporation could occur either through control of thymidine kinase activity and/or by the dilution of external thymidine by endogenous precursor pools. Studies using daughter cell pairs with one daughter deprived of food for 2–3 h before thymidine exposure to reduce the endogenous pools (Plate 1d, e), indicate the importance of the 'cytoplasmic state' (Ord, 1968). The smaller the pools the greater the incorporation of [³H]thymidine (Fig. 3). Furthermore, the addition of an S nucleus to '[³H]thymidine-exposed-cytoplasm' or of a '[³H]thymidine-labelled-nucleus' to an unlabelled S cell, resulting in each case in a small amount of label being incorporated by the unlabelled S nucleus, suggests that at least part of the thymidine nucleotide precursor pool is cytoplasmic. The similarity in incorporation of nuclei sharing cytoplasm suggests that thymidine kinase acts through the cytoplasm.

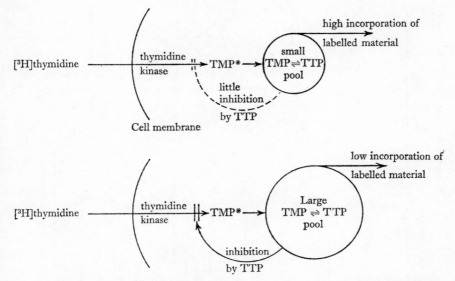

Fig. 3. Effect of pool size on the incorporation of exogenous [³H]thymidine by the cell. The larger the endogenous precursor pool of thymidine monophosphate (TMP), thymidine diphosphate (TDP) and thymidine triphosphate (TTP), the greater the dilution of the labelled nucleotide, and the greater the inhibition of thymidine kinase activity by the TTP of the pool (Quastler, 1963).

Termination of S

Transfer of S nuclei to G_2 cytoplasm showed the S nuclei able to continue to replicate their DNA for some hours. However, this did not continue indefinitely and, at the expected time for the donor nucleus, incorporation of [³H]thymidine ceased. Furthermore, a late S nucleus put into an early S amoeba stopped [³H]thymidine incorporation while the younger nucleus continued. These situations indicate that the S nucleus is able to terminate DNA replication without signals from the cytoplasm.

RNA SYNTHESIS IN THE CELL CYCLE OF *AMOEBA PROTEUS*

As yet information on change in synthesis of RNA through the cell cycle is relatively scarce. Studies using either pulse labelling with [³H]-uridine, or spectrophotometric measurements of total RNA, indicate in general that RNA synthesis is continuous through G_1–S–G_2, though a break in transcription usually occurs at mitosis (Mitchison, 1971). Minor exceptions to this pattern have been reported, e.g. a doubling of the intensity of RNA synthesis with the doubling of the genome in S in some

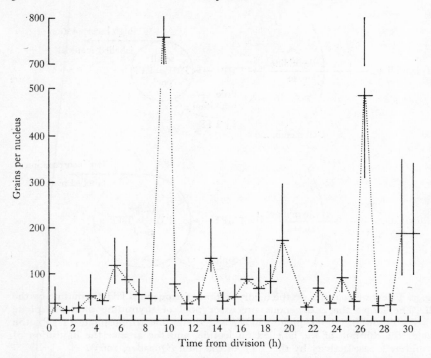

Fig. 4. Synthesis of RNA during the cell cycle as shown by autoradiographs of amoebae after 1 h pulses (horizontal bars) of 0.2 mCi/ml [³H]uridine (sp.act. 29.8 Ci/mmole). Curve is of geometric means with vertical bars indicating fiducial limits.

tissue culture lines (Zetterberg & Killander, 1965; Pfeiffer & Tolmach, 1968), a sharp reduction in RNA synthesis following DNA synthesis in *Physarum* (Braun, Mittermayer & Rusch, 1966).

Bacteria, with a haploid genome giving a genetically uniform culture of cells from a single individual, a single chromosome without associated protein, and direct translation of transcribed DNA, would seem ideal material for studies on the synthesis and subsequent fate of RNA. The short cell cycle, however, while allowing the cultivation of vast quantities of cells, makes almost impossible a recognition and separation of any sequential events occurring within it. In eukaryotes the time through which events of the cell cycle must pass is greater than for bacteria. Yeast has a cycle approximately four times, *Physarum* approximately sixteen times, *Amoeba* approximately one hundred times that of *E. coli*. Although *Amoeba* nuclei lack the perfect synchrony found in *Physarum*, and yields of synchronized cells are low compared with yeast or bacteria, the length of the cell cycle and synchrony of cells without chemical inhibition or shock treatment allows a separation of sequential events not easily obtained with other single cells.

If RNA synthesis in *Amoeba* is investigated using either continuous exposure to [³H]uridine or 3 h pulses through the cell cycle – i.e. pulses equivalent to 1/16th of the cell cycle – the grain count curve obtained shows a continuous incorporation of external uridine with only minor variations. If, on the other hand, [³H]uridine pulses of 1 h or ½ h duration are used – i.e. pulses representing 1/50th or 1/100th of the cell cycle – nuclear RNA synthesis is represented by a series of peaks and dips (Fig. 4).

Using short pulses the following observations have been made:

(1) The rate of incorporation of [³H]uridine changes continually as the cell passes through interphase, with uridine incorporation low during early S and late G_2.

(2) Periods of peak [³H]uridine incorporation occur chiefly during late S, early G_2 and mid G_2. Peaks are of short duration generally occupying only one or two 1 h pulse periods (Plate 2).

(3) Certain peaks are clearly defined and turn up regularly (8–9 h, 19–20 h, 26–27 h and 34–36 h); other peaks may be ill defined (1, 3, 5 h) or appear as a collection of several minor peaks rather than individual single peaks (13–17 h, 29–32 h).

(4) While some peaks are due to all amoebae having increased incorporation of [³H]uridine, others are due to increased incorporation in only some amoebae. If a peak were due to a very short burst of RNA synthesis, this could be explained by the degree of asynchrony which results when large numbers of division spheres are selected from cultures.

Interpretation of peaks and dips in RNA synthesis

A number of interpretations can be given to explain changes in the amount of labelled uridine incorporated during short pulse experiments by nuclei of different ages. The most obvious of these – and that now believed chiefly responsible for the peak-and-dip pattern in uridine incorporation in *Amoeba* – is that different quantities of RNA are synthesized at different times in the cell cycle. However, other explanations for a varying uridine incorporation pattern cannot be lightly dismissed, e.g. pool fluctuations (Mitchison, 1971), membrane permeability changes, changes in the class of RNA synthesized. These are discussed below in relation to work with *Amoeba*.

Pool fluctuations

So far, studies with amoebae have shown that changes in incorporation of uridine with large and small endogenous pools (as indicated by the quantity of *Tetrahymena* engulfed) are insignificant compared with those found when using thymidine in DNA studies. Well fed amoebae have somewhat lower peaks due to greater dilution of precursor pools; but, when food is withdrawn to give significant decrease in endogenous precursor pools, peaks tend to disappear, i.e. cell activity slows and only a low level of RNA synthesis is maintained.

Pool fluctuations due to use of nucleotides in DNA synthesis must also be taken into account. For example, the 8–9 h peak in late S could indicate a burst of RNA synthesis as the cell completes the replication of its DNA and other cell activities take on greater importance. Such an inverse DNA/RNA relationship has been found in *Xenopus* (Gurdon & Woodland, 1969). On the other hand this peak could be due to precursor pools diminished by DNA synthesis allowing greater use of external uridine. This explanation, however, is not supported by measurements of tissue culture cells which show nucleotide precursor pools increasing through S with large pools present at the end of it (Adams, Berryman & Thomson, 1971).

Membrane changes

If periodic fluctuations occur in the cell membrane permeability so varying the influx of substances from the external medium, then peaks and dips in uridine incorporation would result. It is not known whether changes occur in cell membrane permeability during the cell cycle, but changes do occur in the amoeba membrane as cultures deteriorate with age or over-feeding. Such changes may account for some of the variation found among individual amoebae, since in this work the large numbers of division spheres needed to cover the whole cell cycle meant that cells came from a number of culture dishes. A comparative measurement of the 9 h peak in two lines of *A. proteus*, strain P_{Da}, gave average incorporation of 1497 grains per nucleus for a healthy culture, but only 866 grains per nucleus (with proportional incorporation at all other peaks) for a deteriorating culture with fragile membrane of poor elasticity.

Class of RNA synthesized

Peaks and dips in uridine incorporation could be due to sequential changes in the class of RNA synthesized, or in the proportion of different RNAs transcribed at any specific time. A great deal of work has appeared on the characterization of RNAs. Two main classes are recognized (i) RNA which later becomes associated with ribosomes, pre-rRNA; (ii) DNA-like RNA, part of which (D-RNA$_1$) passes to the cytoplasm as messenger RNA and part of which (D-RNA$_2$) remains within the nucleus where it is degraded (Georgiev, 1966). Quantitative measurements suggest that rRNA forms approximately 80% of the total cell RNA (Mitchison, 1971). Since rRNA is more stable than D-RNA, however, such measurements may not reflect the quantities of each class of RNA synthesized at any specific time in the cell cycle.

Much evidence exists for a sequential pattern of synthesis of different classes of RNAs in the dividing egg (Gurdon & Woodland, 1969; Løvtrup-Rein, 1972). If such a pattern should also exist in the cell as it passes through interphase, peaks and dips in [^{3}H]uridine incorporation could be due to the higher AU nucleotide ratio of some D-RNA (Georgiev & Samarina, 1971; Aronson, Witt & Wartiovaara, 1972). Such a possibility will be checked using incorporation of either uridine + cytidine, or all four nucleotides.

Rate of RNA synthesis

Electron microscope studies on amoebae from different stages of the cell cycle showing a number of visible changes within the nucleus at different ages would support the supposition that changes in [^{3}H]uridine incorporation are due to changes in the rate of RNA synthesis at specific times in the cell cycle (Minassian & Bell, personal communication). These studies further suggest a difference in the time of synthesis of different RNAs, e.g. changes in nucleolar morphology correlate well with certain peak uridine incorporation periods suggesting that these peaks represent intense rRNA synthesis.

Further investigations involving both RNA and protein syntheses

Three experimental approaches have been used in further attempting to clarify peak periods of RNA synthesis. (1) Following the change in distribution of label over nucleus and cytoplasm with time. (2) Determining whether interaction between nuclei from different periods of the cell cycle or between nuclei in heterophasic cytoplasm can affect RNA synthesis. (3) Following changes in protein synthesis through the cell cycle, since synthesis of both messenger and ribosomal RNA may be accompanied or followed by peaks in protein activity.

The change in distribution of label between nucleus and cytoplasm with time

The change in distribution of labelled RNA between nucleus and cytoplasm was studied using different time intervals between [^{3}H]uridine

Table 1. *Comparison of nuclear and cytoplasmic label distribution at three peak-dip periods for cells fixed 1 h and 24 h after exposure to [³H]uridine*

		Chase 65 ± 20 min			Chase 24 h		
Age		Average nuclear count	Average cytoplasmic count	Ratio	Average nuclear count	Average cytoplasmic count	Ratio
9 h peak	8 h	84	10	8.4	124	46	2.7
	9 h	1131	213	5.3	635	290	2.2
	10 h	252	31	8	121	44	2.5
19 h peak	17 h	179	28	6.4	232	115	2
	18 h	182	38	5	545	217	2.5
	19 h	548	89	6.1	557	185	3
26 h peak	25 h	101	20	5	215	64	3.3
	26 h	821	106	7.7	1343	568	2.4
	27 h	437	75	5.8	370	174	2.1
	28 h	135	27	5	106	36	2.9

pulses and cell fixation. Table 1 shows the distribution of labelled RNA in cells 1 and 24 h after pulses. Where longer time intervals were used (48, 72 and 96 h) nuclear activity decreased to the extent where cytoplasmic RNA activity prevented any accurate estimations from autoradiographs.

In the present experiments behaviour at three peak–dip periods was determined and the following conclusions drawn:

(i) No significant measurable difference exists in the distribution of RNA between nucleus and cytoplasm with time in the three peak periods studied (9, 19 and 26 h). Nor are there significant differences in the dip periods which precede or follow them. While at 1 h after [³H]uridine treatment RNA is present in the nucleus in concentrations more than five times that of an equivalent volume of cytoplasm, by 24 and 48 h this nuclear concentration averages only 2.6 and 1.4 times that of the cytoplasm. Craig & Goldstein (1969), using transfer of nuclei labelled for several generations with RNA precursors, have shown that rRNA is nuclear in origin, but 50% moves out into the cytoplasm during the first 4 h and a further 30% during the following 20 h. Since in the present experiment the density of grains over the nucleus decreased as cytoplasmic label increased it seems likely that the change here represents chiefly a movement of nuclear RNA to cytoplasm not synthesis of cytoplasmic RNA. Since Craig & Goldstein's measurements show RNA moving rapidly from the nucleus during the first few hours, no significance can be attached to the variable ratio at 1 h until experiments have been repeated using more accurate timing of the chase period.

(ii) The marked increase in the nuclear count for 8, 17–18 and 25–26 h comparisons suggests that UMP–UTP pools are built up during the exposure period and continue to be used during the subsequent chase period. Increase in nuclear RNA during chase periods has been found by other workers following the distribution of labelled RNA with time, e.g. Neyfakh, Kostomarova & Burakova (1972).

(iii) Since the nuclear label incorporated during peak periods does not disappear rapidly in any of the peaks studied, it is unlikely that these represent synthesis of largely rapid-turnover D-RNA such as that found in the periods of intense RNA metabolism of sea urchins (Aronson, Witt & Wartiovaara, 1972).

Control of cytoplasm over nuclear RNA synthesis

When studying DNA synthesis one of the major facts which emerged was the independence of each nucleus once replication had begun. That is, DNA synthesis continued to completion regardless of the cytoplasm by which it was surrounded or the phase of other nuclei with which it shared that cytoplasm. With RNA synthesis, homokaryons show cytoplasm to have considerable control over the incorporation of [³H]uridine. This could indicate direct cytoplasmic control over transcription, or signals passing through cytoplasm, i.e. interaction between nuclei sharing the same cytoplasm.

The results obtained in three experiments with homokaryons are given in Table 2. Little difference exists in incorporation of [³H]uridine by homophasic nuclei sharing the same cytoplasm, whether nuclei are identical or sibling nuclei of equal age. The average ratio between heterophasic nuclei sharing the same cytoplasm is 1.34. Since nuclei in their own cytoplasm show differences in [³H]uridine incorporation of some 10–20 fold, sharing of cytoplasm reduces the variation in incorporation dramatically (Plate 3b). If results obtained for heterophasic homokaryons are separated into those with $S + G_2$ nuclei and those with $G_2 + G_2$ nuclei, it can be seen that there is a closer agreement if both nuclei are from G_2 (i.e. 1 : 1.27 compared with 1 : 1.37). Since in homokaryons the smaller nucleus (generally the S nucleus) has the lower count, this would suggest that lower incorporation is due either to a gene dosage effect (Mitchison, 1971), or an inability of non-replicated DNA to transcribe. Since S nuclei in these experiments were always part way through S, with a large proportion of their DNA already replicated, a 1 : 2 ratio would not be likely if gene dosage were implicated.

Multinucleate heterophasic homokaryons give the same agreement in nuclear labelling as found in binucleate cells. In many cases donor nuclei were from three or more different aged amoebae (Plate 4a). Similar incor-

Table 2. *A comparison of grain counts after [³H]uridine exposure between the two nuclei of binucleate homokaryons, with nuclei of similar (homophasic homokaryons) and dissimilar (heterophasic homokaryons) age*

Type of binucleate	Number	Nuclear/nuclear ratio
Homophasic homokaryons		
Identical nuclei (by inhibition of cytokinesis)	30	1:1.18
Sibling nuclei (by nuclear transfer)	22	1:1.2
Heterophasic homokaryons		
$G_2 + G_2$ (nuclei of different ages but both in G_2)	32	1:1.27
$S + G_2$ (nuclei from different cell cycle phases)	68	1:1.37

poration in such cases would support a possible cytoplasmic control, though dominance by one nucleus cannot be excluded.

Protein synthesis during the cell cycle

Initial protein cell cycle experiments were similar to those used in studies of RNA. Labelling was by exposure to the amino acid [³H]leucine which, as it is common to most protein, should reflect overall changes in protein synthesis. Grain counts over cytoplasm showed that with long pulses – one-sixteenth of the cell cycle per pulse – incorporation of leucine was continuous through the cell cycle, while with short pulses – i.e. one-fiftieth or one-hundredth of the cell cycle per pulse – incorporation of [³H]leucine was resolved into a peak-and-dip pattern, though always with some basic incorporation.

With protein, results using the autoradiographic squash technique (Ord, 1968) could be suspect owing to small variations in the volume of cytoplasm under the fixed area counted, and to the possibility that loss of acid-soluble protein during fixation could give dips in grain counts if these proteins were synthesized during limited periods. The peak-and-dip pattern for [³H]leucine incorporation in amoebae was confirmed in scintillation-counting experiments. This technique gives an average count per amoeba of the total [³H]leucine incorporation, but can give no indication of individual variation among amoebae or of the distribution of label between nucleus and cytoplasm.

In many cases peaks in protein synthesis coincided with peaks in RNA synthesis. This was particularly noticeable for the 8–9, 19–20 and 26 h peaks. Bell & Chatterjee (personal communication) using [³H]leucine incorporation in a detailed study of two peaks occurring in the 15–20 h age groups established that while base-line protein synthesis continued in the anucleate cytoplasm, the peaks in protein synthesis depended upon the presence of nuclear activity. Removal of the nucleus, or inhibition of RNA synthesis by ACT-D, eliminated both peaks. Though this suggests some peaks may be of mRNA, followed by immediate translation, at least some coincident peaks could be due to synthesis of ribonucleoprotein accompanying peaks in rRNA synthesis.

Little information has been obtained concerning the presence of radioactive protein in nuclei. Whole cell autoradiography is unsatisfactory for studies of nuclear protein activity owing to the intrusion of highly labelled cytoplasm between nucleus and emulsion. Initial experiments using transfer of labelled nuclei to unlabelled host cytoplasm followed by autoradiography suggest that nuclei of early S cells contain far more radioactive protein than nuclei from late S or mid G_2. It may be possible to design future experiments to identify such protein and to establish where it is synthesized.

Goldstein & Prescott (1967) in studies using transfer of protein-labelled nuclei to unlabelled host cells have been able to separate nuclear protein of amoebae into two physiologically different fractions: (i) protein which migrates rapidly between nucleus and cytoplasm, or between nuclei sharing the same cytoplasm – rapid migrating protein (RMP) – and (ii) protein which does not migrate, or moves only very slowly from the nucleus – slow turnover protein (STP). By exposing amoebae to labelled amino acids for three generations they were able to label all protein and from the behaviour of labelled nuclear protein in homokaryon nuclei estimated that STP represents approximately 60% of the protein present in the cell nucleus (Goldstein & Prescott, 1967). Though no proof is yet available, they believe that STP may represent proteins forming the nuclear envelope, structural proteins of the nucleolus, protein of nascent ribosomes and various chromosome proteins, while RMP may possibly be involved in regulatory processes (Goldstein & Prescott, 1966).

In the present work short pulse experiments followed by transfer of nuclei from labelled donor cells to unlabelled host cells should show when in the cell cycle the STP and RMP proteins are synthesized. Grain counts over autoradiographs of homokaryons with donor nucleus labelled before transfer indicate, by movement of protein between nuclei, that both STP and RMP are synthesized at each period investigated (Plate 4*b*). As yet only part of the cell cycle has been studied, i.e. seven pulse periods, 2 peak S, 2 mid S, 1 late S and 2 mid G_2 periods. A difference was observed in the ratio of STP and RMP at different periods suggesting that at particular times in the cell cycle a higher proportion of the protein synthesized is STP. Since at present, however, figures are based on only 15–20 transfers per pulse, more work is needed before this observation is confirmed. Repetition of experiments is particularly necessary in view of the discrepancy found by Goldstein and Prescott in the ratios of RMP/STP when counts for nuclei made using an autoradiography method (Byers, Platt & Goldstein, 1963) were compared with counts obtained for isolated nuclei in a gas flow counter (Goldstein & Prescott, 1967). Since with short pulse experiments radioactivity is too low for individual nuclear counts to be made on a gas flow or scintillation counter, autoradiographic techniques had to be used. Attempts are currently being made to reduce loss of acid soluble protein.

CONCLUSION

Exposure of *Amoeba* to short pulses of radioactive precursors throughout the cell cycle (each pulse one-fiftieth or one-hundredth of the cell cycle) has shown that the incorporation of [^{3}H]thymidine occurs chiefly during the first quarter of the cell cycle with a single major incorporation period, but that the incorporation of [^{3}H]uridine and [^{3}H]leucine occurs in a series of peak-and-dip periods. [^{3}H]uridine peaks-and-dips could represent precursor pool fluctuation or membrane permeability changes, but experiments tend to support the supposition that they represent changes in the rate of RNA synthesis and possibly changes in the class of RNA synthesized. Since use of longer pulses in amoebae reduced peaks into a few small humps, lack of peaks and dips in other cell cycle studies may be due to inadequate spatial separation with time.

In short pulse HeLa cell experiments RNA synthesis, though continuous, shows a doubling of precursor incorporation as the cell passes through S (Pfeiffer & Tolmach, 1968). Lower capacity of G_1 and S DNA to transcribe RNA could account for the lack of peaks in such detailed experiments, since only towards the end of S and in G_2 can peaks be clearly resolved in *Amoeba*. In tissue culture cells G_2 represents only a small portion of the cell cycle. The smaller capacity of G_1 DNA to transcribe, and its altering capacity through S, may dominate RNA synthesis measurements.

If a doubling of the number of genes transcribing RNA can cause the clearly measured effect found by Pfeiffer & Tolmach (1968) in HeLa cells, it should be borne in mind that *Amoeba* is believed to be highly polyploid. If all copies of a gene transcribe together this could magnify enormously changes in incorporation which are too small to be recognized in diploid cells. It may be that *Amoeba* facilitates the resolution of a sequential pattern of RNA synthesis which occurs universally but is not generally revealed using conventional crude experimental measuring techniques.

The author would like to thank Mr R. F. Legg for technical assistance with autoradiography and photography.

REFERENCES

ADAMS, R. L. P. (1969). The effect of endogenous pools of thymidylate on the apparent rate of DNA synthesis. *Experimental Cell Research*, **56**, 55–8.

ADAMS, R. L. P., BERRYMAN, S. & THOMSON, A. (1971). Deoxyribonucleoside triphosphate pools in synchronized and drug inhibited L$_{929}$ cells. *Biochimica et Biophysica Acta*, **240**, 455–62.

ARONSON, A. I., WITT, F. H. & WARTIOVAARA, J. (1972). Characterization of pulse-labeled nuclear RNA in sea urchin embryos. *Experimental Cell Research*, **72**, 309–24.

BRAUN, R., MITTERMAYER, C. & RUSCH, H. P. (1966). Ribonucleic acid synthesis *in vivo* in the synchronously dividing *Physarum polycephalum* studied by cell fractionation. *Biochimica et Biophysica Acta*, **114**, 527–35.

BYERS, T. J., PLATT, D. B. & GOLDSTEIN, L. (1963). The cytonucleoproteins of *Amoebae*. I. Some chemical properties and intracellular distribution. *Journal of Cell Biology*, **19**, 453–66.

CRAIG, N. & GOLDSTEIN, L. (1969). Studies on the origin of ribosomes in *Amoeba proteus*. *Journal of Cell Biology*, **40**, 622–32.

DE TERRA, N. (1967). Macronuclear DNA synthesis in *Stentor* : regulation by a cytoplasmic initiator. *Proceedings of the National Academy of Sciences, U.S.A.* **57**, 607–14.

GEORGIEV, G. P. (1966). Metabolism of nuclear RNA fractions. In *The Cell Nucleus Metabolism and Radiosensitivity*, pp. 79–85. London: Taylor & Francis.

GEORGIEV, G. P. & SAMARINA, O. P. (1971). D-RNA containing ribonucleoprotein particles. In *Advances in Cell Biology*, ed. D. M. Prescott, L. Goldstein, & E. McConkey, vol. II, pp. 47–110. New York: Appleton-Century-Crofts.

GOLDSTEIN, L. & PRESCOTT, D. M. (1967). Proteins in nucleocytoplasmic inter-actions. I. The fundamental characteristics of the rapidly migrating proteins and the slow turnover proteins of the *Amoeba proteus* nucleus. *Journal of Cell Biology*, **33**, 637–44.

GOLDSTEIN, L. & PRESCOTT, D. M. (1966). Protein interactions between nucleus and cytoplasm. In *The Control of Nuclear Activity*, ed. L. Goldstein, pp. 273–98. Englewood Cliffs, New Jersey: Prentice-Hall.

GRAHAM, C. F., ARMS, K. & GURDON, J. B. (1966). The induction of DNA synthesis by frog egg cytoplasm. *Developmental Biology*, **14**, 349–81.

GURDON, J. B. & WOODLAND, H. R. (1968). The cytoplasmic control of nuclear activity in animal development. *Biological Reviews*, **43**, 233–67.

GURDON, J. B. & WOODLAND, H. R. (1969). The influence of the cytoplasm on the nucleus during cell differentiation, with special reference to RNA synthesis during amphibian cleavage. *Proceedings of the Royal Society*, B, **173**, 99–111.

HOWARD, A. & PELC, S. R. (1953). Synthesis of desoxyribonucleic acid in normal and irradiated cells and its relation to chromosome breakage. *Heredity, London*, **6** (Suppl), 261–73.

JOHNSON, R. T. & RAO, P. N. (1970). Mammalian cell fusion: Induction of premature chromosome condensation in interphase nuclei. *Nature, London*, **226**, 717–22.

LØVTRUP-REIN, H. (1972). Nuclear RNA and protein synthesis during the early sea urchin development. *Experimental Cell Research*, **72**, 188–94.

MITCHISON, J. M. (1969). Enzyme synthesis in synchronous cultures. *Science, New York*, **165**, 657–63.

MITCHISON, J. M. (1971). *The Biology of the Cell Cycle*. London: Cambridge University Press.

48 M. J. ORD

NEYFAKH, A. A., KOSTOMAROVA, A. A. & BURAKOVA, T. A. (1972). Transfer of RNA from nucleus to cytoplasm in early development of fish. *Experimental Cell Research*, **72**, 223–32.

ORD, M. J. (1968). The synthesis of DNA through the cell cycle of *Amoeba proteus*. *Journal of Cell Science*, **3**, 483–91.

ORD, M. J. (1969). Control of DNA synthesis in *Amoeba proteus*. *Nature, London*, **221**, 964–6.

ORD, M. J. (1971). The initiation, maintenance and termination of DNA synthesis: a study of nuclear DNA replication using *Amoeba proteus* as a cell model. *Journal of Cell Science*, **9**, 1–21.

PFEIFFER, S. E. & TOLMACH, L. J. (1968). RNA synthesis in synchronously growing populations of HeLa S_3 cells. 1. Rate of total RNA synthesis and its relationship to DNA synthesis. *Journal of Cell Physiology*, **71**, 77–94.

QUASTLER, H. (1963). Effects of irradiation on synthesis and loss of DNA. In *Actions chimiques et biologiques des radiations*, ed. M. Haissinsky, pp. 149–85. Paris: Masson et Cie.

RAO, P. N. & JOHNSON, R. T. (1970). Mammalian cell fusion: 1. Studies on the regulation of DNA synthesis and mitosis. *Nature, London*, **225**, 159–64.

ZETTERBERG, A. & KILLANDER, D. (1965). Quantitative cytochemical studies on interphase growth. II. Derivation of synthesis curves from the distribution of DNA, RNA and mass value of individual mouse fibroblasts *in vitro*. *Experimental Cell Research*, **39**, 22–32.

EXPLANATION OF PLATES

PLATE 1

(a) Autoradiograph showing nuclear incorporation of [³H]thymidine at the beginning of S. The amoeba was exposed for 10 min (from 5–15 min after cytokinesis) to 1 mCi/ml [³H]thymidine (sp.act. 25 Ci/mmole).

(b) Nuclei of a heterophasic homokaryon containing a G_2-24 h-old donor nucleus in an S-$3\frac{1}{2}$ h-old amoeba when exposed to 0.5 mCi/ml [³H]thymidine (sp.act. 5 Ci/mmole). The G_2 nucleus has not incorporated [³H]thymidine while the S nucleus, which was replicating its DNA, was heavily labelled.

(c) Homophasic homokaryon with seven S nuclei in binucleate S cytoplasm exposed for 1 h to 0.25 mCi/ml [³H]thymidine (sp.act. 5 Ci/mmole). This shows the similar incorporation of [³H]thymidine by nuclei sharing the same cytoplasm. Though all nuclei are from peak S, their age varied by up to 1 h.

(d, e) Nuclei from a daughter pair of amoebae separated after division for 4 h. One daughter was given *Tetrahymena* (the food organism), the second was kept without food to decrease its endogenous precursor pools. (The presence of food vacuoles shows such amoebae are not starved.) Both cells were exposed at $4\frac{1}{2}$ h age to [³H]thymidine 0.25 mCi/ml (sp.act. 5 Ci/mmole) for 1 h. The 'non-fed' daughter (d) has incorporated far more of the [³H]thymidine than the 'fed' daughter (e), showing that decreasing the endogenous pools increased the use made by the cell of the exogenous thymidine.

PLATE 2

Autoradiographs of amoebae after exposure for 1 h to 0.2 mCi/ml [³H]uridine (sp.act. 29.8 Ci/mmole) showing typical peak-and-dip labelling periods. (a) 8 h dip period; (b) 9 h peak period; (c) 25 h dip period; (d) 26 h peak period.

PLATE I

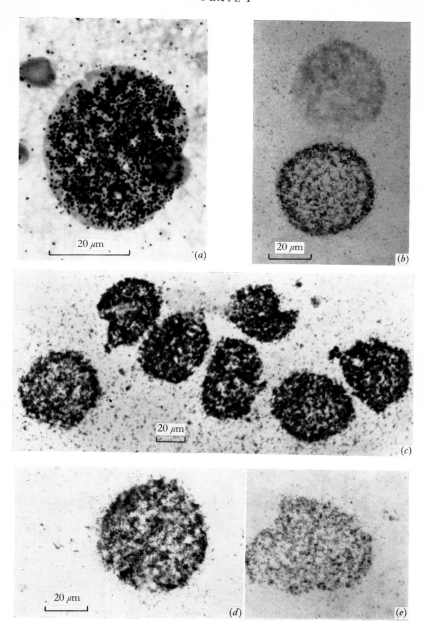

PLATE 2

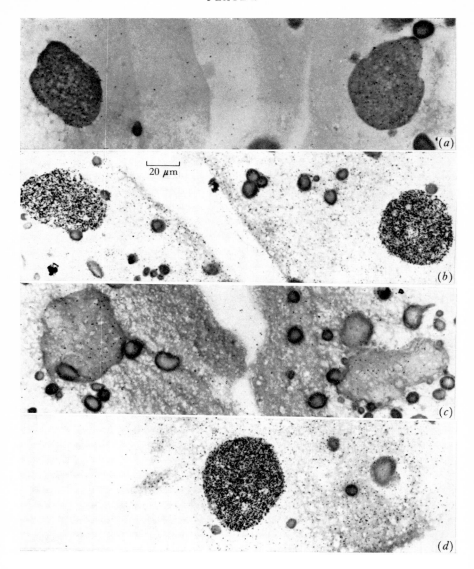

20 μm

(a)

(b)

(c)

(d)

PLATE 3

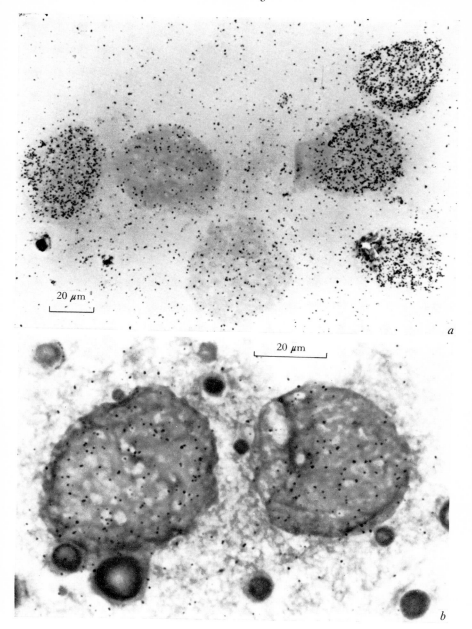

20 μm

20 μm

a

b

PLATE 4

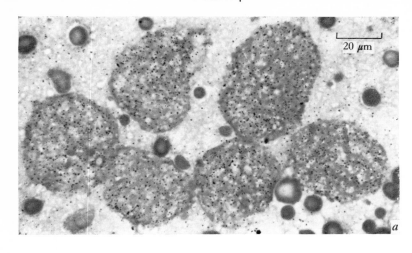

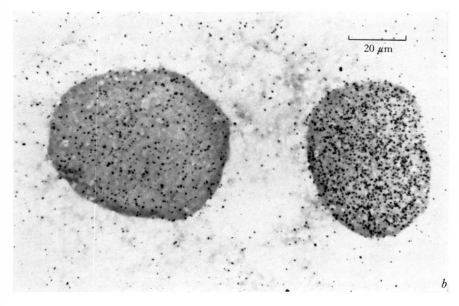

PLATE 3

(*a*) Four S nuclei transferred to a G_2 binucleate amoeba and exposed to 0.25 mCi/ml [^{3}H]thymidine (sp.act. 5 Ci/mmole) for 1 h when S nuclei were 4 h old and G_2 nuclei 18 h old. The autoradiograph shows the two central G_2 nuclei with no significant label, but each of the four S nuclei has incorporated [^{3}H]thymidine indicating a continuation of replication by the S nuclei in the G_2 cytoplasm.

(*b*) Heterophasic homokaryon exposed to 1 mCi/ml [^{3}H]uridine (sp.act. 30 Ci/mmole) for $1\frac{1}{4}$ h when nuclei were 5 h (mid S) and 9 h (late S) old. Both nuclei have incorporated similar amounts of labelled uridine.

PLATE 4

(*a*) Heterophasic homokaryon containing nuclei of three different ages (two 6 h, two 25 h and two 29 h) exposed to 1 mCi/ml [^{3}H]uridine (sp.act. 5 Ci/mmole) for 1 h. Though nuclei in their own cytoplasm would have incorporated widely different amounts of [^{3}H]uridine, sharing cytoplasm of a 29 h binucleate amoeba, all nuclei incorporated similar amounts of [^{3}H]uridine.

(*b*) A homokaryon in which the donor nucleus had been exposed for 1 h (from 29–30 h age) to [^{3}H]leucine before transfer to an unlabelled host amoeba of equal age. The autoradiograph, made 24 h after the transfer, shows the movement of labelled nuclear protein from donor to host nucleus.

CELL DEVELOPMENT AND DIFFERENTIATION IN THE CILIATE *STYLONYCHIA MYTILUS*

BY D. AMMERMANN

Zoologisches Institut der Universität, Abt. für Zellforschung,
D-7400 Tübingen, Germany

INTRODUCTION

Most ciliates possess two different types of nuclei: one or more micronuclei and one macronucleus. The micronuclei are diploid, condensed and do not contain nucleoli. The macronuclei of most ciliates contain much more DNA than the micronuclei and they possess a large number of nucleoli. The macronuclei control the main metabolic processes of the cell. In some species the micronuclei have no essential function during the vegetative cell cycle, but in certain others they have at least some functions (Ammermann, 1970). The main role of the micronuclei becomes apparent during the sexual reproduction or conjugation (Fig. 1). The major consequence of the conjugation is the degeneration of the existing macronucleus and the formation of a new one from the synkaryon of each partner cell.

Many species which can conjugate also have a limited life expectancy. Maupas (1888) was one of the first to show that some ciliates can reproduce asexually only for a limited time. When they do not conjugate within a certain period of time, senescence follows and the clones finally die. Preer (1968) points out that the macronucleus is responsible for this ageing process. The conclusion can be drawn that the main purpose of the conjugation is the replacement of the 'aged' macronucleus by a new one. But why does the macronucleus age and the micronucleus not? What differences between these two nuclei are responsible for this different behaviour?

Until now the macronucleus has largely been considered to be a polyploid, genetically identical edition of the micronucleus. Therefore only a simple endopolyploidisation could be expected during the development of the so-called macronuclear anlage (the name of the nucleus after the division of the synkaryon until the macronucleus is completed). However, when I observed the development of the macronuclear anlage of the hypotrich ciliate *Stylonychia mytilus*, I found a rather complicated picture with a lot of events which were difficult to interpret. Investigations of the

[51]

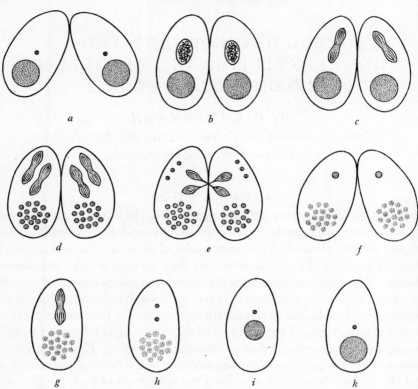

Fig. 1. Scheme of conjugation in the ciliates. *a*, two animals begin to fuse. *b*, the micronuclei start meiosis with the prophase I. *c*, anaphase of the first meiotic division. *d*, anaphase of the second meiotic division. The macronucleus falls into pieces. *e*, three of the nuclei degenerate, the fourth divides once into two gametic nuclei. One of them, the stationary nucleus, stays in the animal. The other, the migratory nucleus, migrates into the partner cell. *f*, the stationary and the migratory nuclei fuse to form the synkaryon. Then the partners separate from each other. *g*, the synkaryon in the exconjugants divides once into two nuclei. *h*, one of the nuclei is the new micronucleus. The other is the macronuclear anlage, which develops to a new macronucleus (*i, k*, see Fig. 3). The remainder of the old macronucleus is resorbed.

last few years have made some of them a bit clearer, and I think there are now, at least in *Stylonychia mytilus*, reasons to doubt whether the macronucleus is polyploid (i.e. whether it contains more than two complete genomes) and whether it is genetically like the micronucleus. This point is, of course, important for the question of why the macronucleus ages.

In the first section I will briefly discuss the cell cycle and the life cycle of the ciliates, especially of *Stylonychia mytilus*. The next section will demonstrate how the exconjugants of *S. mytilus* develop, and in the remaining section I will discuss some conclusions and open questions which arose from these investigations. The results and discussions are based mainly on *Stylonychia mytilus*. Some results found in the related species, *Euplotes aediculatus*, have been included for comparison. Both species belong to the Hypotricha. The results of investigations of ciliates from other groups are not included because their development is apparently completely different.

THE CELL CYCLE AND LIFE CYCLE OF CILIATES

The ciliates reproduce asexually by division. The time from one division to the next, i.e. the length of the cell cycle, is different in different species. Moreover, it depends, within the same species, upon external factors (e.g. temperature, quality of food, etc.) and internal factors. Because the duration of the different stages (G_1, S, G_2, and D) is often different for the macronuclei and the micronuclei, it is not possible to divide the cell cycle in the usual manner. The macronucleus may, for example, be in late S phase, while the micronucleus is still in G_1. Fig. 2 gives a survey of the cell cycle of four species. It is apparent that the DNA synthesis of the macronuclei and the micronuclei can occur independently of each other.

Stylonychia mytilus should be noted especially because its macronucleus – like those in all hypotrich ciliates – has a peculiar mode of DNA synthesis. A pair of so-called 'replication bands' appear at the outer tips of the macronuclei and migrate once across them. In these replication bands DNA synthesis takes place, and after the replication bands have traversed the macronuclei once, the DNA is doubled. Since homologous genes are distributed throughout the macronucleus and are not replicated simultaneously (Ammermann, 1971), and since the RNA synthesis is interrupted only in the replication bands, the metabolism of the cells is virtually unaffected during DNA synthesis.

Like other ciliates, *Stylonychia mytilus* has a life cycle. After conjugation the animals pass through a period of immaturity, during which they can multiply by division but are unable to conjugate. It lasts for about 200 divisions. This stage is followed by maturity, during which the animals continue to divide but are also able to conjugate. It lasts for about 400 divisions. If conjugation does not take place during this period, senescence follows. The senescent animals have a reduced ability to conjugate, high mortality of exconjugants, and a prolonged cell cycle, to mention only a few conspicuous events. Because the clones cannot conjugate intraclonally or start autogamy, they finally die. This ageing process is like that of other ciliates and has remarkable parallels to the senescence of higher organisms.

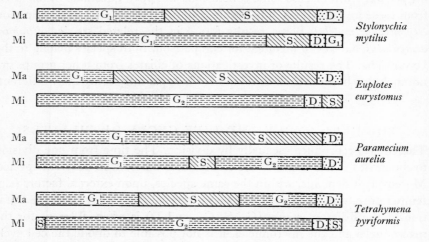

Fig. 2. The cell cycles of four ciliates: *Stylonychia mytilus* (Ammermann, 1970), *Euplotes eurystomus* (Kimball & Prescott, 1962), *Paramecium aurelia* (Kimball & Perdue, 1962), *Tetrahymena pyriformis* (Prescott & Stone, 1967). Ma = Macronucleus, Mi = Micronucleus.

I mentioned above that the ageing of the ciliates can be looked upon as ageing of their macronuclei. The hypotheses of ageing can be classified into two groups. One group starts from the assumption that the structure and the non-mitotic mode of division of the macronucleus causes the ageing. A random segregation, for example, of the chromosomes of the macronucleus would cause a limited life expectancy (for discussion of this point, see Ammermann, 1971). Neither the structure of the macronucleus nor its mode of division are yet understood. One should, however, keep in mind that there is a large number of ciliates which do not show ageing processes. In *Tetrahymena* there are certain strains which do not age whilst others do. (For details, see Preer, 1968.) It is improbable that there are basic differences between these strains in the structure of the macronucleus and the mode of its divisions.

A second group of hypotheses presumes that senescence is programmed in the life cycle, as is, for example, maturity. This assumption fits better with the above mentioned fact that some ciliates have a limited life cycle, others not.

These considerations show that it is necessary to obtain more information about the macronucleus. One way to learn something about the macronucleus of *Stylonychia mytilus* is to follow its development. This is the aim of the following section.

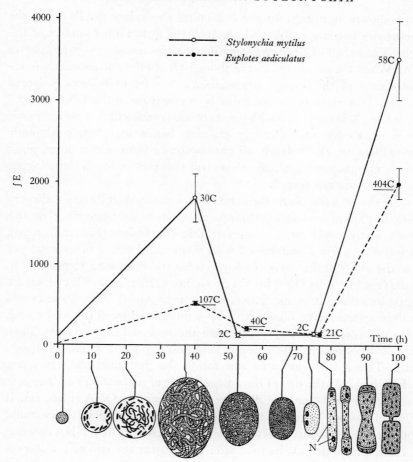

Fig. 3. Schematic diagram of the macronuclear anlagen development in *Stylonychia mytilus*. The curve shows the development of the DNA content in the macronuclear anlagen of this species and, for comparison, of *Euplotes aediculatus*. For the measurement of the DNA content with a microspectrophotometer Feulgen-stained nuclei were used. 1 C is the DNA content of a haploid micronucleus in G_1, $\int E$ the integral of the extinction in arbitrary units, and N the nucleoli. For further details see Ammermann (1971).

THE DEVELOPMENT OF THE NEW MACRONUCLEUS

When the two partners of a conjugating pair separate, all daughter nuclei of the synkaryon are similar. But, after a few hours, one of the nuclei enlarges and starts to develop into a macronucleus. The major events of development are similar in *Stylonychia mytilus* and *Euplotes*. Since *Stylonychia* is the more favourable species for cytological investigations, the description is based mainly on this species. The sequence of development is shown in Fig. 3. In the beginning the anlage swells while the chromosomes become visible and are arranged at the inner surface of the nucleus. The anlage of *Stylonychia mytilus* contains about 300 chromosomes ($=2$ n). When the exconjugants are about 15 h old the chromosomes move to the centre of the anlage, despiralise, and start to become polytene (Plate 1*a*). This event is accompanied by an increase in the DNA content of the anlage. It is now filled with giant chromosomes which show aperiodic patterns of weakly and strongly staining bands and heterochromatic regions (Plate 1*b*, *c*). Probably all chromosomes form one or more giant collective chromosomes, which are several millimetres long. Homologous chromosomes are not paired.

During the next few hours the giant chromosomes disintegrate and every band of the chromosomes seems to form an isolated compartment (Plate 1*d*). Electron microscopic studies support this conclusion (Kloetzel, 1970). This transformation is connected with a spectacular loss of over 90 per cent of the DNA of the anlage in *Stylonychia* and of around 80 per cent in *Euplotes* (Fig. 3). This DNA breaks down into smaller molecules which are excreted into the culture medium (Ammermann, 1969). After these events the anlage appears to be filled with a large number of small (around 1 μm), isolated granules. This stage lasts until the end of the third day. Then drastic changes occur. The exconjugants develop a new mouth and start feeding. The anlage elongates and forms the first nucleoli, which are especially large. At the tips of the anlage, a pair of replication bands appear and migrate over the anlage, doubling the amount of DNA (Plate 1*e*). It is followed by another pair of replication bands, and new bands appear and double the DNA content until the anlage has reached its final DNA content, and, connected with this, its final size. This marks the end of the macro-nuclear development.

There are two events during this development which are apparently important, but precisely what happens is not known. (1) We do not know whether during the polytenisation the whole genome with all the chromosomes replicates equally or whether some chromosomes or regions are over-replicated while others stay under-replicated. (2) We do not know whether, during the following DNA loss, whole genomes are lost or whether only parts of most or all genomes leave the anlage while others remain multiple (that would be a kind of diminution). Thus, it is not known whether the macronucleus anlage after the DNA loss is still genetically identical with the micronucleus or whether it differs from them. This is the reason why I avoid the term *polyploid* for the macronucleus of *Stylonychia*.

One way to find an answer to these questions is to compare the starting point of the developmental process, the micronucleus, with the end point, the macronucleus. If we find differences between them, this would give us some suggestions. These questions lead us to the next section.

DOES THE DEVELOPMENT LEAD TO DIFFERENT NUCLEI?

The questions which were raised at the end of the last section could be answered if one knew whether or not the macronuclei and micronuclei contain the same DNA. This question has been tackled in the last few years by two groups; by Prescott, Bostock, Murti, Lauth & Gamow in Boulder, Colorado, and by our group (von Berger, Hennig, Steinbrueck, & Ammermann). The results, although not in agreement on all points, lead to the conclusion that the macronuclei and the micronuclei of *Stylonychia mytilus* are indeed different:

(1) With equilibrium density centrifugation of DNA from macronuclei and micronuclei, Bostock & Prescott (1972) showed that macronuclear DNA consists of a single density component, while micronuclear DNA is resolved into at least four separate components of different density. These results were supported by melting curves which showed that macronuclear DNA melts as if it is a single component DNA while micronuclear DNA melts as if it is a mixture of several DNAs with different base composition. Our preliminary results (analytical centrifugation of DNA in $CsSO_4$ gradients after heavy metal binding to the DNA) showed that the macronuclear DNA contains at least two major components, while micronuclear DNA contains several (more than four). Macronuclear DNA is therefore much more homogeneous in its $C+G$ content.

(2) Prescott, Bostock, Murti, Lauth & Gamow (1971) described another difference between the DNA of macronuclei and micronuclei. They found an unusually low molecular weight of 1.15×10^6 daltons for macronuclear DNA. The average length of macronuclear DNA molecules, measured with an electron microscope after spreading, was $0.8 \ \mu m$ (range from 0.2–$2.2 \ \mu m$). The micronuclear DNA, however, was much longer, and too long to be measured. The shortest piece of micronuclear DNA they found was $15 \ \mu m$. We also found rather low molecular weights of macronuclear DNA (double-stranded in neutral NaCl: 2.24×10^6 daltons) and micronuclear DNA (same conditions: 5.37×10^6 daltons). When the DNA was spread with a new, especially protective method (von Berger, unpublished results), the DNA of macronuclei and micronuclei showed short (down to $1 \ \mu m$), medium-sized, and long (up to over $100 \ \mu m$) fibrils. The long fibrils were present even in DNA prepared from

macronuclei of emicronucleated strains, which excludes a contamination of the macronuclear DNA by micronuclear DNA. We conclude, therefore, that the low molecular weight may be an artifact due to preparation, and that there is no difference between the nuclei as far as DNA size is concerned. Perhaps the macronuclear DNA has preferential breaking points.

(3) Bostock and Prescott (1972) mention that a renaturation experiment with macronuclear DNA did not indicate large portions of repetitive DNA sequences in this nucleus. Our results from similar experiments confirmed that macronuclear DNA seems not to contain a large portion of repetitive DNA. Micronuclear DNA, however, contains a large amount of repeated DNA sequences.

These results give strong evidence that there are differences in the composition of macronuclear and micronuclear DNA. It can be assumed that these differences arise during the polytenisation by disproportional replication and/or during the DNA loss which occur during the development of a new macronucleus. To confirm this it would be interesting to investigate the DNA of the macronuclear anlage before and after the DNA loss, but up to now it has not been possible to get enough synchronised, clean DNA from macronuclear anlagen.

OPEN QUESTIONS

The development of the macronuclear anlage of *Stylonychia mytilus* raises some questions which cannot yet be answered. I want to discuss some of them here. Especially interesting are the giant chromosomes. To date it is not clear what significance they have for the exconjugants. In a previous communication (Ammermann, 1968) I mentioned that it was not possible to demonstrate RNA synthesis with [³H]uridine, although the exconjugants take up this precursor. Meanwhile this result has been confirmed several times with different dosages of precursors, different lengths of pulses, and different exposure times. On the other hand, Rao (1968) demonstrated RNA synthesis in the macronuclear anlage of *Euplotes woodruffi* in a stage in which they very likely contain giant chromosomes. I never found nucleoli, puffs or other functional structures, but Gil, Alonso & Perez-Silva (1972) described 'nucleolus-like bodies' in the anlage of *Stylonychia* (without a reference, however, to whether or not they contain RNA), which are visible in the electron microscope.

Recently, I performed some experiments with Actinomycin D (AMD). Fifty μg/ml block the DNA-dependent RNA synthesis of vegetative cells and conjugating pairs (Sapra, unpublished observations). Exconjugants were treated for twelve hours during different times of their development.

PLATE I

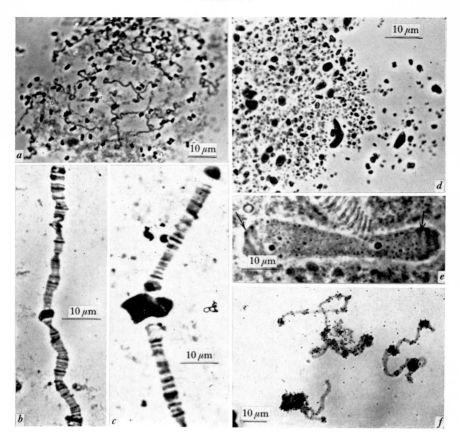

(*a*) Anlage, about 20 h old, with short spiralised and long despiralised chromosomes.
(*b*) Part of a giant chromosome of an anlage, about 40 h old. (*c*) As (*b*) with a large
heterochromatin block. (*d*) Content of an anlage, about 45 h old, with big and small
granules deriving from big (heterochromatic) and small bands. (*e*) Anlage, about
80 h old, with two replication bands (arrows) and two large and numerous small
nucleoli. (*f*) Giant chromosomes, denatured and then hybridised in-situ with RNA
made from macronuclear DNA in-vitro.

(*Facing p.* 58)

At the beginning the development of the macronuclear anlage is not affected until the chromosomes are despiralised. But then the development stops in the presence of AMD. After its removal the development proceeds normally. Application of AMD in later stages stops the development immediately. From these results it could be concluded that RNA synthesis, which is important for the further development, starts when the chromosomes are despiralised and the giant chromosomes develop. In any case, however, AMD causes a delay of the development even in dosages which do not affect the RNA synthesis and the development of the macronuclear anlage, and this effect could not be explained until now. The question of whether there is RNA synthesis during development needs further investigation.

The impressive heterochromatic regions in the giant chromosomes of the anlage (Plate 1c) are very interesting. One would suppose that they contain large amounts of over-replicated, repetitive DNA. From in-situ hybridisation experiments it was possible to obtain some information about it. The incubation of denatured giant chromosomes with an in-vitro synthesised [³H]RNA results in strong labelling due to hybrid formation in some 'normal' bands and in most of the heterochromatic blocks (Plate 1f). From the hybridisation conditions it must be assumed that these hybrids indicate repeated DNA sequences only. Since the labelled RNA was produced on a DNA template derived from macronuclei, this experiment shows that macronuclei must indeed contain such DNA sequences. These sequences are clearly highly repetitive in the heterochromatin of the anlage, but they are under-replicated during the second polyploidisation phase and therefore rare and not easily detectable by renaturation in the macronucleus.

At the end of this section I would like to return to one of the questions mentioned in the first one. Some information is now available which makes it probable that the macronucleus is differentiated and restricted in its potentialities, like somatic nuclei of certain Metazoa (e.g. *Ascaris*). We do not know, however, whether this is the reason why they age. It would be very interesting to compare macronuclei and micronuclei of a ciliate which does not have a limited life expectancy and to see if there are differences between the nuclei there, too.

I am indebted to Dr Prescott and his co-workers who sent me their manuscripts for reading.

This work was supported by the Deutsche Forschungsgemeinschaft.

REFERENCES

AMMERMANN, D. (1968). Synthese und Abbau der Nucleinsäuren während der Entwicklung des Makronukleus von *Stylonychia mytilus* (*Protozoa, Ciliata*). *Chromosoma, Berlin*, **25**, 107–20.

AMMERMANN, D. (1969). Release of DNA breakdown products into the culture medium of *Stylonychia mytilus* exconjugants (*Protozoa, Ciliata*) during the destruction of the polytene chromosomes. *Journal of Cell Biology*, **40**, 576–7.

AMMERMANN, D. (1970). The micronucleus of the ciliate *Stylonychia mytilus*; its nucleic acid synthesis and its function. *Experimental Cell Research*, **61**, 6–12.

AMMERMANN, D. (1971). Morphology and development of the macronuclei of the ciliate *Stylonychia mytilus* and *Euplotes aediculatus*. *Chromosoma, Berlin*, **33**, 209–38.

BOSTOCK, C. J. & PRESCOTT, D. M. (1972). Evidence of gene diminution during the formation of the macronucleus in the protozoan, *Stylonychia*. *Proceedings of the National Academy of Sciences, U.S.A.* **69**, 139–42.

GIL, R., ALONSO, P. & PEREZ-SILVA, J. (1972). Ultrastructure of the macronuclear anlage in *Stylonychia mytilus*. *Experimental Cell Research*, **72**, 509–18.

KIMBALL, R. F. & PERDUE, S. W. (1962). Quantitative cytochemical studies of nucleic acid synthesis. *Experimental Cell Research*, **27**, 405–15.

KIMBALL, R. F. & PRESCOTT, D. M. (1962). Deoxyribonucleic acid synthesis and distribution during growth and amitosis of the macronucleus of *Euplotes*. *Journal of Protozoology*, **9**, 88–92.

KLOETZEL, J. A. (1970). Compartmentalization of the developing macronucleus following conjugation in *Stylonychia* and *Euplotes*. *Journal of Cell Biology*, **47**, 395–407.

MAUPAS, E. (1888). Recherches expérimentales sur la multiplication des infusoires ciliés. *Archives de Zoologie Expérimentale et Générale* (2), **6**, 165–277.

PREER, J. R. (1968). Genetics of Protozoa. In *Research in Protozoology*, ed. Tse-Tuan Chen, vol. 3, pp. 234–7. Oxford & New York: Pergamon Press.

PRESCOTT, D. M., BOSTOCK, C. J., MURTI, K. G., LAUTH, M. R. & GAMOW, E. (1971). DNA of ciliated protozoa. I. Electron microscopic and sedimentation analyses of macronuclear and micronuclear DNA of *Stylonychia mytilus*. *Chromosoma, Berlin*, **34**, 355–66.

PRESCOTT, D. M. & STONE, G. E. (1967). Replication and function of the protozoan nucleus. In *Research in Protozoology*, ed. Tse-Tuan Chen, vol. 2, pp. 119–46. Oxford & New York: Pergamon Press.

RAO, M. V. N. (1968). Macronuclear development in *Euplotes woodruffi* following conjugation. *Experimental Cell Research*, **49**, 411–19.

THE CELL CYCLE IN *CHLORELLA*

BY P. C. L. JOHN, W. McCULLOUGH, A. W. ATKINSON, JR.,
B. G. FORDE AND B. E. S. GUNNING

Department of Botany, Queen's University,
Belfast, BT7 1NN, N. Ireland

Photosynthetic unicellular algae can be readily synchronised with respect to growth and division by employing regimes of alternating light and dark (Lorenzen, 1970). They therefore provide suitable subjects for the investigation of the patterns of biosynthesis during the cell cycle and their determining mechanisms. Considerable evidence has already accumulated that synthesis of enzymes during the cell cycle can be discontinuous but this information is drawn largely from studies of the yeasts (Mitchison, 1971) where, in *Saccharomyces cerevisiae* in particular, the amounts of individual enzymes usually increase stepwise during a period of synthesis which occurs once during each cell cycle (Halvorson *et al.* 1971).

We have employed the unicellular green alga *Chlorella fusca*-8p (*C. pyrenoidosa* 211-8p, Cambridge) to test whether temporal controls of enzyme synthesis are found. We have also performed concurrent ultra-structural analyses which permit the correlation of metabolic activities with quantitative stereological estimates of the volumes of cellular organelles and with the progress of structural change.

SUMMARY OF EVENTS IN THE CELL CYCLE OF *CHLORELLA*

The cell cycle of our strain is described in terms of the conventional markers in Fig. 1. Growth in mass occurs, after a lag of about 1 h, during the 15 h period of illumination. The period of DNA synthesis (S) occupies 6 h beginning at about 10 h in the cycle and the alga shows no G_2 period since mitosis to produce the first two daughter nuclei (M_1) precedes the second round of DNA replication which is then followed by the second mitosis (M_2) to produce the four daughter nuclei usual under our conditions of culture (McCullough & John, 1972*a*). Mitosis is usually complete by 17 h and again flows without pause into the process of cytokinesis and wall formation, yielding four discrete autospores capable of independent growth by 20 h. If the normal 9 h period of darkness is imposed the cells will be held in check for a further 4 h until they are reilluminated, when they are able to grow and so explode the mother cell envelope (Plate 2*c*).

[61]

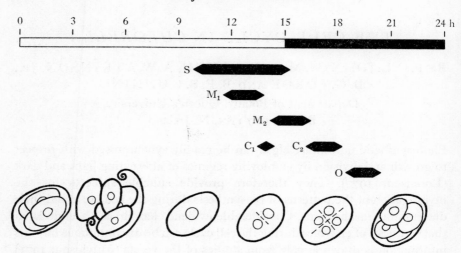

Fig. 1. Some key events in the cell cycle of *Chlorella* 211-8p grown on a 15 h light, 9 h dark regime (top). The period of DNA synthesis (S) commences before the first nuclear division (M_1), and continues until the second nuclear division (M_2). Microtubules appear in the cytoplasm just before the first round of centriole duplication (see Plate 1*a*) at the beginning of M_1, and later form the first spindle. After M_1 they reappear in the cytoplasm in the plane of the first septum at cytokinesis (C_1). With the onset of M_2 the microtubules disappear from the cytoplasm, septum formation is arrested, and they appear in the M_2 spindle. They reappear in the cytoplasm at the final cytokinesis (C_2) and disappear completely when cleavage is complete. Wall formation follows in two stages (Plates 1, 2*a*, *b*), only the first of which is included here – the assembly of the outer, trilaminar component (O). Autospore enlargement, commencing when the next light period starts, ruptures the mother cell wall at about 4 h (Plate 2*c*).

Other key events in the cycle are included in Fig. 1 and Plate 1. Cells of this strain of *Chlorella* possess a pair of centrioles (Atkinson, Gunning, John & McCullough, 1971), the duplication of which at 12 h initiates the first prophase (Plate 1*a*). Microtubules are present from this time onward, either in the intranuclear spindle (the nuclear envelope persists) or, in the short period between M_1 and M_2 and in the period between M_2 and the completion of cytokinesis at about 18 h, in the cytoplasm especially near the developing septa. They subsequently disappear, but we do not know whether this temporal control of microtubule assembly and disassembly reflects formation and destruction of the constituent protein.

TEMPORAL CONTROLS OF AUTOREGULATED
ENZYMES

There is no unequivocal evidence that a single stepwise increase is the normal pattern of enzyme accumulation during the cell cycle of eukaryotes other than yeast. Among the algae, Molloy & Schmidt (1970) have observed that a stepwise pattern of increase in ribulose 1,5-diphosphate carboxylase activity in a thermophilic strain of *Chlorella* could be obliterated by a change in growth rate. In *Chlamydomonas* Kates & Jones (1967) have indeed demonstrated the stepwise increase of five autoregulated enzymes, but this observation is of uncertain significance since the pattern may have been unnaturally imposed by the intermittent illumination employed (Mitchison, 1971). We have followed the pattern of activity for three autoregulated enzymes. To test whether temporal controls are naturally employed by the cell, we have assayed cells developing for one or two cell cycles in continuous light, following synchronisation in light and darkness.

Fig. 2 shows the pattern of accumulation of total soluble protein, DNA and the individual enzymes, succinic dehydrogenase, phosphoenolpyruvate carboxylase and cytochrome oxidase. The cultures were obtained in synchronous division by the regime of 15 h light and 9 h dark with dilution to a density of 5×10^6 cells per ml at the end of each dark period (McCullough & John, 1972a). In each case the solid symbols represent samples taken from a culture undergoing its first subsequent division in continuous light and the open symbols represent samples taken from a culture (derived from the same parent culture and sampled simultaneously) which had received no dark period in the previous cycle. This culture had been diluted to the standard density of 5×10^6 cells per ml at the beginning of autospore release, but because the period of carbon starvation which occurs during the dark had been omitted, this occurred earlier at 20 h in the previous cycle of continuous light.

Two measurements confirm that an earlier initiation of growth occurs in daughter cells formed in continuous light and not delayed by a dark period. The period of DNA synthesis is advanced by 4 h (Fig. 2b), and the accumulation of soluble protein (5 min, 130 000 g, supernatant) shows a plateau during cytokinesis which also occurs 4 h sooner, Fig. 2a, b. Total protein (2 min, 2000 g, supernatant) continues to increase during cytokinesis, presumably reflecting the extensive synthesis of plasma membrane, endoplasmic reticulum (seen in Plate 1b) and microtubules at this time (Atkinson et al. 1971, 1972).

It is striking that, in contrast to the smooth increase in soluble protein,

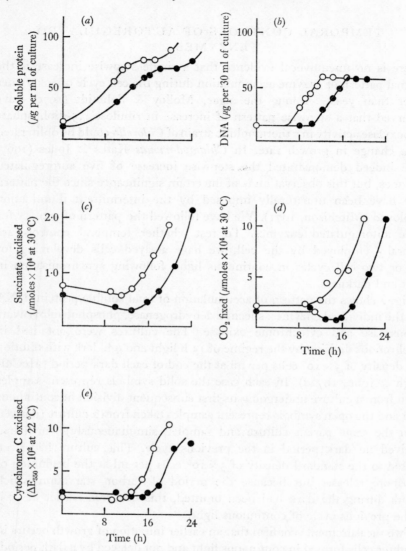

Fig. 2. The pattern of synthesis of (a) soluble protein not sedimentable by 5 min at 130000 g, (b) DNA, (c) succinate dehydrogenase activity, (d) phosphoenol pyruvate carboxylase activity. (e) cytochrome oxidase activity in two cultures derived from a parent culture synchronised with the standard light and dark regime. One culture (solid symbols) was undergoing its first cycle of division in continuous light, the other (open symbols) was undergoing its second cycle in continuous light. Enzyme activities were assayed according to conventional procedures (Forde & John, 1973) and all activities are expressed per ml of culture per min of assay. DNA was purified and assayed as described previously (McCullough & John, 1972a).

the individual enzymes increase discontinuously (Fig. *2c, d, e*). This stepwise increase in activity has the consequence that the contribution of each enzyme to total cell protein is subject to extreme fluctuation. For each enzyme the specific activity falls to less than a third of its maximum value, before the beginning of each stepwise accumulation.

There is no evidence, however, that the temporal control is imposed upon the cells by the method of synchronisation since these measurements were made in the absence of synchronising pressure and show clearly that the temporal restrictions were unchanged even into the second cycle of uninterrupted growth. It is also possible to discount the dilution of the culture entering its second cycle of continuous growth as providing an external stimulus, since when the experiment was repeated with dilution at 24 instead of 20 h in the first cycle of continuous growth, an identical pattern of control was observed.

The timing of the stepwise increases in the individual enzymes is of particular interest since they are held strictly in phase with other biochemical events. The stepwise increase in amount of all three enzymes is brought forward by approximately 4 h in conformity with the advancement of DNA synthesis and protein accumulation which occurs in the absence of the dark phase. This observation throws some light upon the mechanism of the temporal restriction of gene expression. In particular, it is difficult to reconcile with any simplified form of the hypothesis of oscillatory end product repression, which in prokaryotes is believed to produce self-maintaining oscillations in gene expression even in the absence of gene duplication (Masters & Donachie, 1966). Oscillatory repression, similarly, does not readily account for the observation that the same pattern of accumulation of these three enzymes of intermediary and energy metabolism occurs even during a period of darkness, when the cell is deprived of photosynthetically-derived energy and carbon (Forde & John, 1973).

These data are, instead, more readily reconciled with the hypothesis that gene expression is programmed within the cycle and regulated by either the sequential transcription of genes as is believed to occur in yeast (Halvorson, Carter & Tauro, 1971) or by the sequential release of stable mRNAs from controls at translation, as suggested by Sussman (1970). We shall consider these possibilities subsequently.

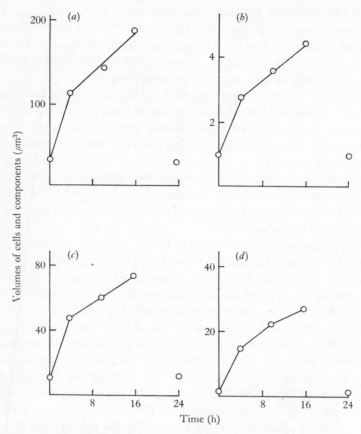

Fig. 3. The pattern of increase in volume during the cell cycle of (a) the whole cell, (b) the mitochondrion, (c) the chloroplast. (d) starch grains. Volume calculations were made from estimates of areas in fifty random sections according to a standard procedure based upon recommendations of Weibel (1969) and described in detail by Atkinson (1972), who shows that there is a single mitochondrial reticulum.

THE INTEGRATION OF TEMPORALLY REGULATED ENZYMES INTO RESPIRATORY METABOLISM

The observation that the synthesis of both succinate dehydrogenase and cytochrome oxidase is restricted by the cell to the period of the cycle following mitosis, prompts the question of whether the total synthesis of mitochondrial material follows the pattern of these two components of the mitochondrial inner membrane and whether, if any rate-limiting activities are subject to these temporal controls, the organelle will be capable of increasing the rate of its metabolic activity in proportion to the growth of the cell.

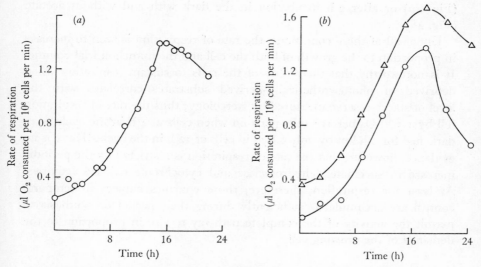

Fig. 4. (a) The rate of endogenous respiration during the cell cycle. Cells were taken from a standard growth cycle and were resuspended in fresh aerated growth medium to a density of 6.6×10^7 per ml for the measurement of oxygen consumption in the dark at 25 °C using an oxygen electrode (Rank Bros., Bottisham, Cambs., England). A linear rate of oxygen uptake was established within 2 min and this was monitored during a 10 min period. (b) The effect of 2 h incubation without carbon source (O–O) and with 25 mM sodium acetate (△–△), on the rate of respiration in cells taken from a standard growth cycle. The cells were resuspended before measurement of oxygen uptake, as in (a) above.

To measure the synthesis of mitochondrial material we have made use of the fact that the area of an organelle in random ultra-thin sections of cells is proportional to the volume of the organelle (Weibel, 1969). The mitochondrial volume (Fig. 3b) increases in proportion to the volume of the cell (Fig. 3a) and remains between 2.3 and 3.1 % of total cell volume throughout the cell cycle. This is the usual pattern among organelles which we have measured, including the chloroplast, seen in Fig. 3c. An exception is the quantity of starch reserve, depleted during the process of cytokinesis in the dark period and replenished by photosynthesis during the succeeding light period (Fig. 3d). In principle then, the growth of organelles resembles that of total cell protein in being essentially continuous although the individual component enzymes may be synthesised discontinuously.

To test for the possibility that discontinuity of enzyme synthesis might impose restrictions upon the metabolic activity of organelles we have measured the aerobic respiration of cells during the cell cycle. Respiration was monitored as oxygen uptake, immediately after darkening the cells

(Fig. 4*a*) or after 2 h incubation in the dark with and without acetate (Fig. 4*b*).

Under all of these conditions, the rate of respiration is seen to increase in proportion to the growth of both the cell and the mitochondrial volume. It is noteworthy that the ability of the cells to sustain respiration when deprived of photosynthetically-derived substrates correlates with the level of starch reserve measured by stereology; thus the rate of respiration fell by 1.5 μl O_2 per 10^9 cells per min when cells at 4 h in the cycle were darkened for 2 h but by only 0.8 μl in cells at 12 h in the cycle. There is no evidence, however, that the rate of respiration is restricted by the periodic increase in succinate dehydrogenase and cytochrome oxidase activities. At least for respiration, therefore, those enzymes subject to temporal control are accumulated sufficiently during their period of synthesis to permit the activity of the complete pathway to rise in proportion to the demands of the growing cell.

THE ROLE OF *DE NOVO* PROTEIN SYNTHESIS AND OF CONTROLS AT TRANSCRIPTION AND POST TRANSCRIPTION

We have employed inhibitors of protein and RNA synthesis in experiments to determine the role of such synthesis in governing the increase in activity of enzymes. Because inhibitors will have extreme structural repercussions if applied during cytokinesis, we have investigated their effect on enzymes synthesised earlier in the cycle.

The enzyme isocitrate lyase is particularly suitable for this study because it is synthesised in large amounts, it has been purified (John & Syrett, 1967) and the enzyme protein can be identified and estimated after disk electrophoresis (John & Syrett, 1968). Furthermore, the enzyme is stable unless made redundant by catabolite repression (John, Thurston & Syrett, 1970) and so changes in the rate of turnover can fortunately be discounted as a control mechanism.

Synthesis of the enzyme is adaptive, and follows darkening of the cells with the provision of acetate as the sole carbon source (Syrett, Merrett & Bocks, 1963). However, the synthesis of the enzyme is subject to strict temporal control with a sharp maximum of adaptive synthesis at 6 h in the cycle and with little or no synthesis for at least 6 h in cells between 10 and 18 h in the cycle (McCullough & John, 1972*a*).

When the soluble proteins of cells adapted to growth on acetate for 6 h are separated by disk electrophoresis (Plate 3), isocitrate lyase protein is detected only in the samples which show isocitrate lyase activity (Fig. 5)

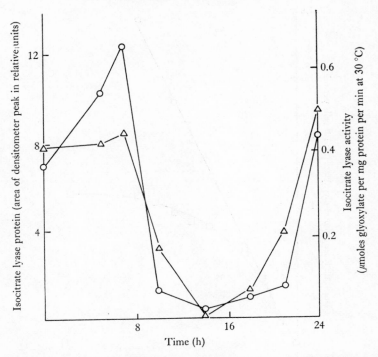

Fig. 5. The amount of isocitrate lyase protein (O–O) and of isocitrate lyase activity (△–△) produced during 6 h incubation with acetate as sole carbon source. Protein was prepared for electrophoresis as detailed in the legend to Plate 3. Enzyme activity and protein was measured as described previously (McCullough & John, 1972a).

and there is a good correlation between the yield of enzyme protein and the catalytic activity of the enzyme ($P = 0.001$). There is no evidence of quantitatively proportionate amounts of enzyme precursor protein remaining latent in the cells unable to produce isocitrate lyase activity. Therefore, for this enzyme, temporal control does not appear to reside at the level of precursor activation, although this mode of control is known in other organisms (Holzer, 1969; Cazzulo, Sundaram & Kornberg, 1970). This conclusion is strengthened by the observation that in cells maximally adaptive for the enzyme, the increase in enzyme activity can be rapidly halted by the presence of cycloheximide or the proline analogue azetidine-2-carboxylic acid (McCullough & John, 1972b).

Since the increase in enzyme activity does require de-novo synthesis of the enzyme protein it is possible to test whether this in turn depends upon the concurrent transcription of unstable mRNA, as is supposed by the hypotheses of oscillatory gene repression and sequential gene transcription,

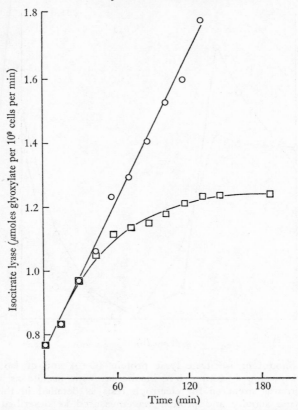

Fig. 6. The synthesis of isocitrate lyase in cells which were previously removed at 6 h in the cycle and incubated for 4 h with 5 mM sodium acetate in the dark. During the experiment the cells were incubated with no further addition (O–O), or with 1 mM 6-methyl purine from time o (□–□). Samples were removed and frozen for subsequent assay of isocitrate lyase as described previously (McCullough & John, 1972*b*).

or whether it depends instead upon the release of stable mRNA from control of its translation. The concurrent transcription of unstable mRNA might be detected by using the declining rate of enzyme synthesis following inhibition of further mRNA synthesis as a measure of the persistence of mRNA, as has been done with other microbial and higher plant cells (Nakada & Magasanik, 1962; Kepes, 1967; Leive & Kollin, 1967; Chrispeels & Varner, 1967; Gayler & Glasziou, 1968). Unfortunately, conventional inhibitors of RNA synthesis are ineffective against our alga (Syrett, 1966), presumably because they do not penetrate into the cell. However, 6-methyl purine, which is a useful inhibitor of RNA synthesis in higher plants (Chrispeels & Varner, 1967; Gayler & Glasziou, 1968), is an effective

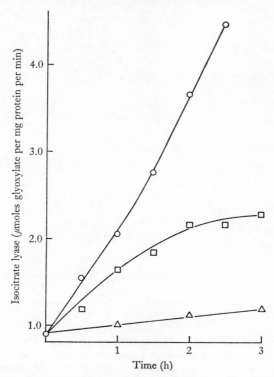

Fig. 7. The synthesis of isocitrate lyase in cells which were previously removed at 6 h in the cycle and incubated for 4 h with 5 mM sodium acetate in the dark. Incubation was continued during the experiment with no further addition O–O; or with the addition at time o of 1 mM 6-methyl purine □–□; or 30 mM glucose △–△. Protein was extracted as detailed in the legend to Plate 3 and isocitrate lyase assayed as described previously (McCullough & John, 1972*b*).

inhibitor of functional RNA synthesis in this alga, where it acts by competing with adenine for incorporation into nucleic acid without interfering with respiration (McCullough & John, 1972*c*). Following the application of 6-methyl purine to cells maximally adaptive for isocitrate lyase, the synthesis of total protein declines with a half life of 66 min reflecting the mean stability of mRNA in these cells. If the synthesis of isocitrate lyase is followed in the same cells, however, its synthesis is seen to decline with a half-life of 33 min (Fig. 6), presumably revealing that the messenger for this enzyme is slightly less stable than the average. Certainly there is no evidence that synthesis of the enzyme is supported by stable mRNA, as seems to be the case for synthesis of several enzymes in *Dictyostelium* and of tyrosine transaminase in rat hepatoma cells (Sussman & Sussman, 1969; Tomkins *et al.* 1970).

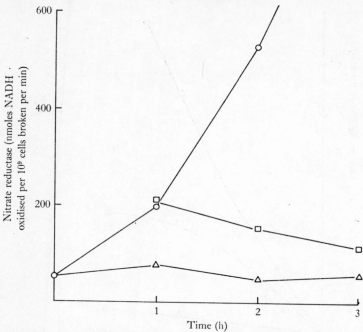

Fig. 8. The synthesis of nitrate reductase in cells grown until 12 h in the cell cycle
with ammonium sulphate as nitrogen source and then transferred at the start of the
experiment to medium containing potassium nitrate as sole nitrogen source.
Subcultures contained either no further addition (O–O), or the addition at time o,
of 1 mM 6-methyl purine (□–□); or 3 μg per ml of cycloheximide (△–△).
Nitrate reductase was extracted and assayed as described by Morris & Syrett
(1963). There is no synthesis of isocitrate lyase in comparable experiments where
6-methyl purine is present from the beginning of inducing conditions.

We therefore reject the hypothesis that the acquisition of the ability
to synthesise this enzyme is due to the release of a control operating at the
post-transcriptional level upon a stable mRNA, and conclude that synthesis
depends upon the release of genetic information in order to sustain the
synthesis of an unstable mRNA. It may be that this transcription is
controlled by changes in the inherent availability of the gene, as is believed
to occur in the sequential transcription of the yeast genome, but we cannot
strictly eliminate the possibility that the genetic information is controlled
by oscillatory repression. Even though it is clear that acetate enters the
metabolism of the cell throughout the cycle (since it is able to support
respiration above the starved rate (Fig. 4b)) it may be that its ability to
alter the concentration of the gene repressor varies during the cycle.

It is remarkable that although controls at the post-transcriptional level
do not govern the adaptive state of the cell, this control can be employed

PLATE I

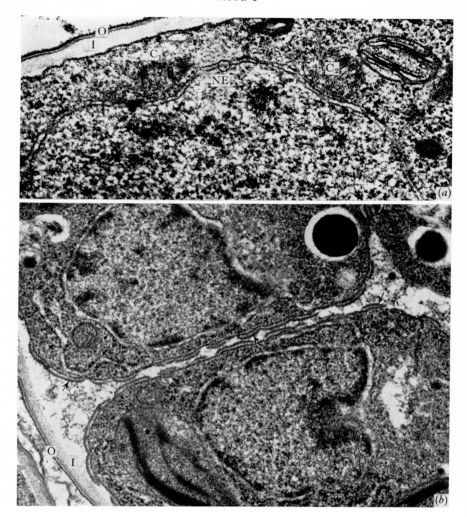

(a) *Chlorella* 211-8p: portion of a cell in prophase. The centriole pair has duplicated (C_1, C_2) and migration is in progress around the nuclear envelope (NE) to place one pair at each pole of the future intranuclear spindle. The inner fibrillar (I) and the outer trilaminar (O) components of the wall are also seen. $\times 44\,500$; 12 h cell.

(b) A stage of wall formation following cytokinesis. Trilaminar plaques (arrows) appear outside the plasma membranes of the naked daughter autospores, all within the parental wall (I and O). Other characteristic features of this part of the cycle are the peripheral chromatin in the nuclei and the abundant and somewhat distended endoplasmic reticulum and nuclear envelope. $\times 33\,400$; 20 h cell.

PLATE 2

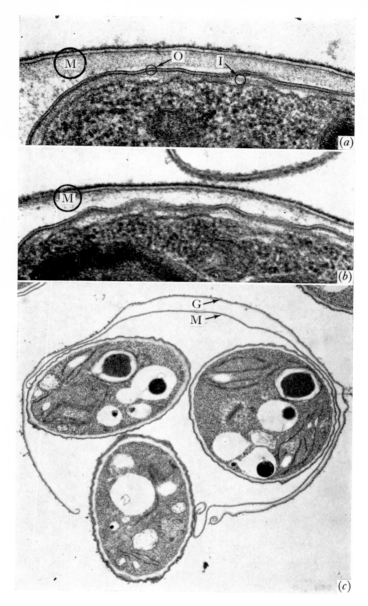

(a), (b) Later stages of wall formation. In (a) the plaques have fused to give a complete trilaminar sheath (O) around the autospore, and the first portions of the microfibrillar inner layer of the wall (I) are developing. The mother cell wall (M) is intact. × 54 000; 21 h cell. In (b) the main difference is in the mother cell wall (M) where the inner layer is being broken down concomitant with synthesis of the corresponding layers in the autospores. × 54 000; 22 h cell.

(c) Autospore release. This example shows both a mother cell wall (M) and a grandmother cell wall (G) from the previous cell cycle. × 3 500; 4 h cells.

PLATE 3

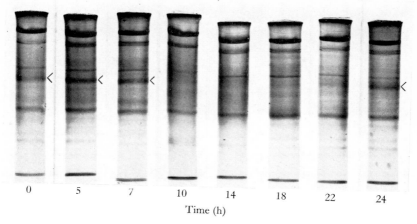

Time (h)

Soluble proteins resolved by disk electrophoresis from cells removed at intervals from a synchronous culture and exposed for 6 h to 25 mM sodium acetate in the dark. For incubation with acetate the cells were pelleted by centrifugation at 3000 *g* for 3 min before being resuspended at the original density of 5×10^6 cells per ml in fresh medium. After 6 h aeration at 20 °C with 0.5 % (v/v) CO_2 in air the cells were washed in 0.025 M imidazole HCl buffer (pH 6.8) at 4 °C and resuspended in this buffer at a density of 8×10^7 cells per ml, then broken by passage twice through a French pressure cell at 4 °C. Debris was discarded by centrifugation at 100000 *g* for 10 min at 4 °C and the supernatants diluted to contain 0.6 mg protein per ml. 0.2 ml volumes of the diluted protein, made 10 % (w/v) glycerol, were subject to electrophoresis. Isocitrate lyase protein is arrowed.

by the alga when isocitrate lyase is made redundant by catabolite repression. If glucose is provided for cells making the enzyme at the maximum rate the synthesis ceases immediately, although mRNA is clearly present (Fig. 7). We have therefore investigated other enzymes to test whether their synthesis can be initiated without the synthesis of new mRNA. Intriguingly, the adaptive enzyme nitrate reductase, which is induced by nitrate and repressed by ammonium in this alga (Morris & Syrett, 1963), is synthesised maximally by cells at 12 h in the cycle, but can be formed by protein synthesis in these cells without RNA synthesis (Fig. 8). Thus, for this enzyme, there does appear to be latent mRNA, although when nitrate is substituted entirely for ammonium the lower yield of enzyme in the cells treated with 6-methyl purine implies that synthesis of new mRNA also occurs.

THE FLOW OF METABOLIC INTERMEDIATES INTO CELL WALL COMPONENTS

Ultrastructural investigation has revealed that a new cell wall appears round the naked autospores formed by cytokinesis before the rupture of the protective mother cell wall. The first stage of this process is the appearance (Plate 1b), growth, and fusion (Plate 2a) of small trilaminar plaques, to form a complete sheath, within which a microfibrillar layer, digestible by cellulase, is later deposited (Plate 2b). Concurrently, the corresponding layer of the mother cell wall is digested (Atkinson *et al.* 1972).

The outer trilaminar layer confers upon the cells an immunity to enzymic attack. It is even resistant to acetolysis (digestion in 10% (v/v) sulphuric acid in acetic anhydride at 100 °C). The resistant material is probably the poly-terpenoid compound sporopollenin, as judged by comparison under infra-red spectral analysis with sporopollenin from *Lycopodium clavatum*.

We were able to turn the extreme resistance of the sporopollenin to advantage by carrying out chemical purification of this material from cells pulse-labelled or pulsed and chased with [U-^{14}C]acetate. The highest efficiency of ^{14}C incorporation into sporopollenin fractions, prepared at the end of the cycle, was obtained from 1 h periods of pulse labelling applied not during the actual process of trilaminar wall assembly, but 6–8 h earlier, after the cells had become committed to division by entering the S phase (Fig. 9a).

The material which becomes labelled at 14 h may be sporopollenin but in a non-polymerised form since it does not appear in the fraction prepared immediately after the 1 h pulse (Fig. 9b). It certainly appears to be

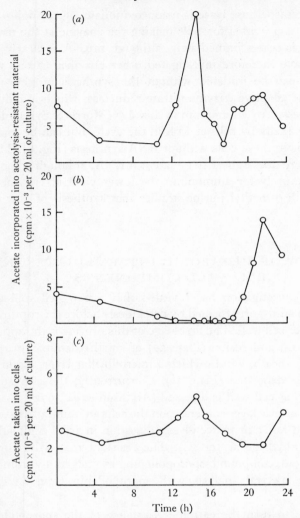

Fig. 9. The incorporation of [U-14C]acetate into autolysis-resistant material during the 24 h cell cycle, measured in (a) after exposure for 1 h to [14C]acetate and then continued incubation until the end of the cycle and in (b) at the end of 1 h exposure to [14C]acetate. The amount of [14C]acetate taken into the cells during the 1 h exposure is shown in (c). Details of extraction are given by Atkinson, Gunning & John (1972).

at least a specific precursor of sporopollenin since 5.1 % of the acetate assimilated at this time (Fig. 9c) is committed to the synthesis of the sporopollenin fraction, which is itself only 0.6 % of the dry weight of the cell. On a carbon for carbon basis the data imply a more than eight-fold preference for the flow of incorporated 14C towards sporopollenin synthesis.

These data, therefore, provide a striking example of programmed temporal controls in metabolic flow, leading to an integrated sequence of biosynthetic events. It will be interesting to determine whether the temporal controls of enzyme synthesis, which seem incidental to the regulation of respiration, do provide a mechanism for the sequential changes in metabolic flow so clearly necessary to carry out the ultrastructural events of the cell cycle.

REFERENCES

ATKINSON, A. W. JR. (1972). *Ultrastructural Studies on Chlorella.* Ph.D. Thesis, The Queen's University, Belfast.

ATKINSON, A. W. JR., GUNNING, B. E. S. & JOHN, P. C. L. (1972). Sporopollenin in the cell wall of *Chlorella* and other algae: ultrastructure, chemistry and incorporation of ^{14}C acetate, studied in synchronous cultures. *Planta, Berlin,* 107, 1–32.

ATKINSON, A. W. JR., GUNNING, B. E. S., JOHN, P. C. L. & McCULLOUGH, W. (1971). Centrioles and microtubules in *Chlorella. Nature, London,* 234, 24–5.

CAZZULO, J. J., SUNDARAM, T. K. & KORNBERG, H. L. (1970). Mechanism of pyruvate carboxylase formation from the apo-enzyme and biotin in a thermophilic *Bacillus. Nature, London,* 233, 1103–5.

CHRISPEELS, M. J. & VARNER, J. E. (1967). Hormonal control of enzyme synthesis: on the mode of action of gibberellic acid and abscisin in aleurone layers of barley. *Plant Physiology, Lancaster,* 42, 1008–16.

FORDE, B. G. & JOHN, P. C. L. (1973). The stepwise accumulation of autoregulated enzyme activities during the cell cycle of the eucaryote *Chlorella. Experimental Cell Research,* in press.

GAYLER, K. R. & GLASZIOU, K. T. (1968). Plant enzyme synthesis: decay of messenger RNA for peroxidase in sugar-cane stem tissue. *Phytochemistry,* 7, 1247–51.

HALVORSON, H. O., CARTER, B. L. A. & TAURO, P. (1971). Synthesis of enzymes during the cell cycle. *Advances in Microbial Physiology* 6, 47–106.

HOLZER, H. (1969). Regulation of enzymes by enzyme-catalysed chemical modification. *Advances in Enzymology,* 32, 297–326.

JOHN, P. C. L. & SYRETT, P. J. (1967). The purification and properties of isocitrate lyase from *Chlorella. Biochemical Journal,* 105, 409–16.

JOHN, P. C. L. & SYRETT, P. J. (1968). The estimation of the quantity of isocitrate lyase protein in acetate adapted cells of *Chlorella pyrenoidosa. Journal of Experimental Botany,* 19, 733–41.

JOHN, P. C. L., THURSTON, C. F. & SYRETT, P. J. (1970). The disappearance of isocitrate lyase enzyme from cells of *Chlorella pyrenoidosa. Biochemical Journal,* 119, 913–19.

KATES, J. R. & JONES, R. F. (1967). Periodic increases in enzyme activity in synchronised cultures of *Chlamydomonas reinhardtii. Biochimica et Biophysica Acta,* 145, 153.

KEPES, A. (1967). Sequential transcription and translation in the lactose operon of *Escherichia coli. Biochimica et Biophysica Acta,* 138, 107–23.

LEIVE, L. & KOLLIN, V. (1967). Synthesis, utilization and degradation of lactose operon mRNA in *E. coli. Journal of Molecular Biology,* 24, 247–59.

LORENZEN, H. (1970). Synchronous cultures. In *Photobiology of Microorganisms,* ed. P. Halldal, pp. 187–212. New York: Wiley Interscience.

McCULLOUGH, W. & JOHN, P. C. L. (1972a). A temporal control of the *de novo* synthesis of isocitrate lyase during the cell cycle of the eukaryote *Chlorella pyrenoidosa*. *Biochimica et Biophysica Acta*, **269**, 287–96.

McCULLOUGH, W. & JOHN, P. C. L. (1972b). Control of *de novo* isocitrate lyase synthesis in *Chlorella*. *Nature, London*, **239**, 402–5.

McCULLOUGH, W. & JOHN, P. C. L. (1972c). The inhibition of functional RNA synthesis in *Chlorella pyrenoidosa* by 6-methyl purine. *New Phytologist*, **71**, 829–37.

MASTERS, M. & DONACHIE, W. D. (1966). Repression and the control of cyclic enzyme synthesis in *Bacillus subtilis*. *Nature, London*, **209**, 476–79.

MITCHISON, J. M. (1971). *The Biology of the Cell Cycle*. London: Cambridge University Press.

MOLLOY, G. R. & SCHMIDT, R. R. (1970). Studies on the regulation of ribulose-1,5-diphosphate carboxylase synthesis during the cell cycle of the eukaryote *Chlorella*. *Biochemical and Biophysical Research Communications*, **40**, 1125–33.

MORRIS, I. & SYRETT, P. J. (1963). The development of nitrate reductase in *Chlorella* and its repression by ammonium. *Archiv für Mikrobiologie*, **47**, 32–41.

NAKADA, D. & MAGASANIK, B. (1962). Catabolite repression and the induction of β-galactosidase. *Biochemica et Biophysica Acta*, **61**, 835–7.

SUSSMAN, M. (1970). Model for quantitative and qualitative control of mRNA translation in eucaryotes. *Nature, London*, **225**, 1245–6.

SUSSMAN, M. & SUSSMAN, R. (1969). Patterns of RNA synthesis and of enzyme accumulation and disappearance during cellular slime mould cytodifferentiation. *Symposium of the Society for General Microbiology*, **19**, 403–35.

SYRETT, P. J. (1966). The kinetics of isocitrate lyase formation in *Chlorella*: evidence for the promotion of synthesis by photophosphorylation. *Journal of Experimental Botany*, **17**, 641–54.

SYRETT, P. J., MERRETT, M. J. & BOCKS, S. M. (1963). Enzymes of the glyoxylate cycle in *Chlorella vulgaris*. *Journal of Experimental Botany*, **14**, 249–64.

TOMKINS, G. M., MARTIN, D. W. JR., STELLWAGEN, R. H., BAXTER, J. D., MAMONT, P. & LEVINE, B. B. (1970). Regulation of specific protein synthesis in eucaryotic cells. *Cold Spring Harbor Symposia in Quantitative Biology*, **35**, 635–40.

WEIBEL, E. R. (1969). Stereological principles for morphometry in electron microscopic cytology. *International Review of Cytology*, **25**, 235–302.

RNA SYNTHESIS DURING THE CELL CYCLE IN *PHYSARUM POLYCEPHALUM*

BY W. D. GRANT

Department of Genetics, University of Leicester,
Leicester LEI 7RH

INTRODUCTION

Physarum polycephalum is a plasmodial slime mould which can be cultured in a defined axenic medium. Certain characteristics make it a useful organism for studies on the biochemical events associated with growth and differentiation (reviewed recently by Rusch, 1970). In particular, the diploid, macroscopic, plasmodial stage of the life cycle exists as a syncytium within which many millions of nuclei undergo mitosis with a high degree of natural synchrony. Provided the culture conditions are correct, plasmodia exhibiting good synchrony may be grown up to 14 cm in diameter (Mohberg & Rusch, 1969a), enabling milligram quantities of material from any stage in the cell cycle to be readily obtained. Under defined conditions in the laboratory, the plasmodium will undergo mitosis with a precise and predictable intermitotic time of some 8–12 h.

The plasmodial or 'growth' stage can be switched by altering the conditions of culture (reviewed by Daniel, 1966), to a sporulation, or 'differentiation' state, which also occurs in a highly synchronous fashion and is quite distinct from the 'growth' phase. Sauer (Sauer, Babcock & Rusch, 1969a, b; 1970) has recently extensively reported on the biochemical events leading up to, and during sporulation. In addition to the complex series of events which may take place leading to the production of haploid spores, the plasmodial stage also has the potential for going through what is probably a less complicated series of changes under adverse conditions, resulting in the formation of a resistant sclerotium or spherule stage. The biochemical events leading up to this stage, which may or may not be a true differentiation, have been extensively examined by several workers (Hüttermann & Chet, 1971; Hüttermann, Porter & Rusch, 1970a; Hüttermann, Elsevier & Eschrich, 1971; Chet & Rusch, 1969, 1970a, b, c; Goodman, Sauer, Sauer & Rusch, 1969; Lestourgeon & Rusch, 1971; McCormick, Blomquist & Rusch, 1970; Sauer, Babcock & Rusch, 1970), but any discussion of the processes of spherulation and sporulation is outside the scope of this article.

As well as lending itself readily to examination by conventional biochemical techniques, the plasmodium is a suitable system for a more theoretical

[77]

approach to the aspects of the cell cycle, such as the control of nuclear division. A series of experiments, started some ten years ago, show that plasmodia at different stages in the cell cycle can be fused to give a fusion product which undergoes mitosis at an intermediate time (see Rusch, 1970). Basic observations of this type are similar to those made in mammalian and other cell fusion studies (reviewed recently by Johnston & Rao, 1971), and mathematical models of control systems may be readily constructed with resulting predictions which are probably more readily testable in the plasmodial fusion system, than in other systems available (Grant, in preparation).

Probably the most important recent development in studies with *Physarum polycephalum* has been the establishment of the basis of a sound genetic approach to the cell cycle. Previously (Dee, 1960, 1962, 1966a), the genetics of the mating-type system involved in the haploid–diploid gameto-plasmodial transition in the life cycle had been extensively worked out, as had the compatibility system in plasmodial fusions (Poulter & Dee, 1968), and the genetics of an actidione-resistant mutant (Dee, 1966b; Dee & Poulter, 1970), but until recently no useful markers have been available for cell cycle studies. Two basic problems existed; first, no mutagenic technique had been described; second, since plasmodia are diploid, in the usual course of events formed from the fusion of two amoebal gametes of opposite mating types, and therefore heterothallic, any mutations obtained using haploid amoebae require to be dominant to be expressed in the plasmodium. However, in this laboratory Wheals (1970) has isolated and characterised a homothallic plasmodium derived from the fusion of two genetically identical amoebae, and a mutagenic technique is now available (Haugli & Dove, 1972). Utilising these important advances, Wheals (1971) has characterised several mutant strains unable to complete the amoebal–plasmodial transition, and amino-acid-requiring mutants of homothallic plasmodia have now been isolated (Dee, Holt & Wheals, in preparation) These mutants, and others now being isolated, should provide powerful tools in the elucidation of processes in the cell cycle.

Precisely where *P. polycephalum* should be placed in the eukaryotic hierarchy is also unclear. It could be argued that it has more affinity with the animal kingdom (see Rusch, 1970), but such categorisations are basically unhelpful. The high chromosome counts reported (Ross, 1966; Koevenig & Jackson, 1966), and the similarity of histone patterns to those of avowedly higher eukaryotes (Mohberg & Rusch, 1969b), suggest merely that the organism is a highly specialised eukaryote.

It should always be borne in mind that whereas technically, *P. polycephalum* plasmodia are extremely suitable for cell cycle studies, caution should be applied in any comparison with other synchronous systems. In particular, since the growth mode is syncytial, no cell division takes place. and thus presumably nuclear division is the point of realignment for all processes in the cell cycle. The method of preparing large synchronous plasmodia is by fusion of many microplasmodia which are at all stages of the cell cycle (Guttes & Guttes, 1964), giving good synchrony of DNA synthesis and mitosis in the fusion product after one post-fusion cell cycle. Mitchison (1971) has argued that in the absence of any information about how tightly coupled to DNA synthesis or mitosis other cell cycle processes are, then there is no reason to suppose that other processes achieve such good synchrony as rapidly as DNA synthesis. However, since most work on the cell cycle in *P. polycephalum* is carried out on the third or fourth cell cycle after fusion of microplasmodia, the author, at least, feels that substantial uncoupling of other cell cycle processes from DNA synthesis and mitosis is unlikely after this time (some 24–32 h).

It should also be noted that the strain of *P. polycephalum* most used for biochemical experiments has been grown for many years vegetatively, without completing the life cycle. At this time it is not clear if the ploidy level in such vegetatively propagated strains is the same as that in plasmodia freshly prepared from the fusion of gametes.

It is possible that there may be substantial biochemical differences between 'old' and 'new' plasmodia – for example are 'new' plasmodia, recently produced from fusion of sexual gametes, comparable with the most commonly used strain in lacking a G_1 period? With the advent of a suitable genetic system, and thus the inevitable proliferation and use of genetically useful strains, problems like this will have to be re-examined. However, until the present, all the biochemical information gathered about *P. polycephalum* has been obtained from the M_3C_4 strain developed at the McArdle Laboratory in Madison, and this article is concerned with this strain, in which the basic parameters of the cell cycle in such plasmodia are well established – there is no G_1 period, DNA synthesis starting immediately after telophase, reaching full rate after 10 minutes, the total DNA synthetic period occupying about one-third of the intermitotic period (see Rusch, 1970). The biochemical events reported in this communication take place during this plasmodial, or 'growth' phase, of the life cycle.

MAJOR SPECIES OF RNA
Cytoplasmic RNA

The principal ribosomal species of RNA were first detected on sucrose gradients by Mittermayer, Braun & Rusch (1964), and characterised on sucrose gradients shortly afterwards as having S values of 27 and 17 (Braun, Mittermayer & Rusch, 1966a). However, it is clear from these studies, which utilised a modified phenol extraction procedure, that extensive problems with degradation during extraction existed, and it is probably only recently that techniques have been developed which enable undegraded RNA to be reproducibly extracted from plasmodia (Melera, Chet & Rusch, 1970; Zellweger & Braun, 1971a; Jacobson, 1971). Recent electrophoretic studies have indicated that the S values of the principal ribosomal species are probably 26 and 19 (Melera et al. 1970; Zellweger & Braun, 1971a), although Jacobson (1971) feels that the S value of the heavier component is likely to be closer to 25.

Fig. 1a shows the profile obtained by the co-electrophoresis of P. polycephalum cytoplasmic ribosomal RNA with several different marker species, and examinations of this type have established that the molecular weights of the principal species are from $1.29-1.43 \times 10^6$ for the 26 S, and $0.65-0.76 \times 10^6$ daltons for the 19 S, determinations from different laboratories varying slightly (Melera et al. 1970; Zellweger & Braun, 1971a; Jacobson, 1971). Zellweger and Braun (1971a), in particular, have made an extensive study of the technical aspects of RNA extraction, and shown that careful electrophoresis of RNA extracts indicates that in addition to the 26 S, 19 S, and 4–5 S peaks readily observed, there are probably three additional stable species of RNA in total RNA extracts. Figs. 1b and 1c illustrate these species, namely a 30 S and a 15 S detectable in 2.4 % acrylamide gels, and a 6 S species detectable in 10 % gels. Base analysis of the 30 S and 15 S peaks indicated that they were remarkably similar in composition to the 26 S and 19 S peaks, suggesting that the 30 S may be a precursor of ribosomal RNA, and the 15 S a species of degradation product. The 6 S peak, however, was quite different in composition from any of the other species, and since it could be found in isolated ribosomes, it may be similar to the small species found in mammalian cells (Pène, Knight & Darnell, 1968), and peas (Sy & McCarty, 1970), hydrogen-bonded to the larger ribosomal species.

Until the recent studies of Jacobson (1971), despite a substantial number of publications concerned with pulse-labelling of RNA in P. polycephalum (Mittermayer et al. 1964; Braun, Mittermayer & Rusch, 1966a, b; Chet & Rusch, 1970b; Zellweger & Braun, 1971a) no evidence

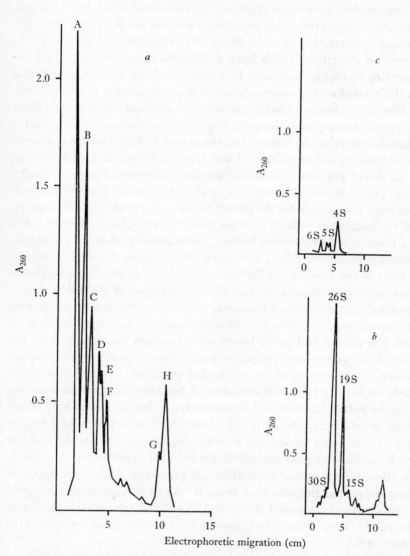

Fig. 1 (*a*) Electrophoretic separation in 2.4% polyacrylamide gel of total RNA from *Physarum*, *E. coli* and HeLa cells. A, 28S HeLa; B, 26S *Physarum*; C, 23S *E. coli*; D, 19S *Physarum* E, 18S HeLa; F, 16S *E. coli*; G, 5S three species; H, 4S three species.

(*b*) Electrophoresis in 2.4% polyacrylamide gel of total RNA from *Physarum*.

(*c*) Electrophoresis in 10% polyacrylamide gel of RNA from *Physarum* (all figures taken from Zellweger & Braun, 1971*a*, © Academic Press).

for a large molecular weight precursor or ribosomal RNA has emerged, with the exception of the stable 30 S peak of Zellweger & Braun (1971a).

In other systems, it has been found that ribosomal precursor RNAs are bound in structures which have a different resistance to phenol deproteination, and which also may be partially soluble in both phenol and water. This results in a concentration of such structures at the phenol–water interface when a classical phenol extraction is carried out (Perry *et al.* 1970), subsequent recovery requiring careful separation, and a hot phenol extraction, to release the ribosomal RNAs. However, despite careful extraction and separation of this type, Zellweger & Braun (1971a), failed to detect any high molecular weight precursors. Recent work by Jacobson (1971) has established that in *P. polycephalum* further technical problems are responsible for the failure to find ribosomal RNA precursors. One of the basic points which emerges from this work is that despite rapid lysis of plasmodia in extraction buffers, extensive degradation, specifically of ribosomal precursor RNA, still takes place, unless the extractions are carried out in the presence of Zn^{2+} which acts as a ribonuclease inhibitor. Another important observation was that short pulses of [^{3}H]uridine were incorporated into particulate material, which survived the usual phenol-based extraction procedures, and which sedimented through the interface between the phenol and water layers, and was thus lost totally from the interface layer unless centrifugation conditions were extremely precise. Microscopic examination of the particulate material indicated that it was probably made up of nucleolar remnants. A hot phenol extraction was then used on the particulate material to release the pulse-labelled RNA, which migrated on gels and gradients as a number of high molecular weight species making up two basic peaks, one a heterodisperse 45–35 S peak, and the other a homodisperse peak slightly larger than the cytoplasmic 26 S RNA. Fig. 2 illustrates a typical separation on gradients. Kinetic patterns of labelling supported the idea that these RNAs in the particulate material were precursors of ribosomal RNA, and in addition, these high molecular weight RNAs were methylated, and had a base composition similar to ribosomal RNA.

Jacobson (1971) has suggested that one of the explanations of the heterodisperse nature of the 45–35 S RNAs, when compared with HeLa 45 S RNA, may be due to very rapid processing of intermediates derived from the slowest moving of the species on gels. Attempts to slow down the rate of processing by cycloheximide treatment indicated that a major 40 S species was present, and in addition, a small amount of 42 S, and 35 S RNA. The order of precursors was much less well-defined from kinetic experiments than that observed with HeLa RNA precursors, perhaps as a

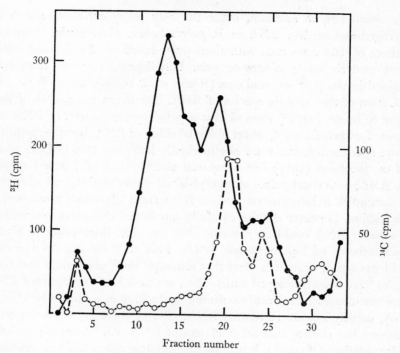

Fig. 2. Sedimentation on linear 5–20% sucrose gradients of ³H-pulse-labelled RNA extracted from particulate material. The marker is [¹⁴C]cytoplasmic ribosomal RNA from *Physarum*. Closed circles denote ³H; open circles denote ¹⁴C. (from Jacobson, 1971).

consequence of the increased ratio between synthesis time and the time required for processing precursors, as compared with HeLa precursors. Jacobson argues (1971) that the 40 S precursor is probably analogous to the HeLa 45 S precursor, and the 35 S to the HeLa 41 S precursor, and it is clear from a consideration of the molecular weights that a considerable proportion, perhaps 30% of the 40 S precursor for example, is not conserved during processing.

In general terms, the observations of Jacobson support the idea that ribosomal RNAs in *P. polycephalum* are synthesised via a sequential processing of a large precursor, probably around 45 S, through 42 S, 40 S, and 35 S steps, followed by a cleavage which gives rise to a 19 S ribosomal RNA which is transported out of the nucleolus, and a precursor slightly larger than the 26 S RNA, which is still retained in the nucleolus, presumably preparatory to final processing and transportation. However, a more detailed and more rigorous investigation is necessary to determine the primary transcription product and its subsequent processing.

Jacobson has, in addition, made the only really extensive search for heterogeneous nuclear RNA in *P. polycephalum*. Here again, there is a plethora of data concerned with short pulse-labelling of RNA and subsequent analysis on gradients or gels. Polydisperse patterns have been obtained in the past, on gradients (Braun *et al.* 1966*a*), but with no clear indication of the possible species of RNA. The extensive studies of Zellweger & Braun (1971*b*) showed that a polydisperse pattern of RNA was obtained on gels following short pulse-labelling of RNA, but the mobilities of the species detected were not markedly less than those of ribosomal RNAs. Jacobson (1971) has compared short pulses of labelled uridine into RNA, with short pulses of methyl-labelled methionine, and attempted to distinguish heterogeneous nuclear RNA from ribosomal precursors in this fashion. However despite carefully monitored extraction procedures, no class of RNA could be detected with the size distribution or kinetic characteristics of heterogeneous RNA. This may be due to technical problems again, since it is clear that although there are claims that techniques have been developed which allow extraction of undergraded RNA, these techniques have been monitored by looking at a particular class of RNA, usually mature ribosomal RNA. Clearly there is no guarantee, as Jacobson has shown, that other classes of RNA will be equally resistant to degradation. Of course, it may be that heterogeneous RNA as described from other systems does not exist in *P. polycephalum*, but it is more likely that technical difficulties, perhaps combined with a remarkably high turnover rate, are responsible for the failure to detect this class of RNA.

Mitochondrial RNA

Despite a substantial number of publications concerning various aspects of RNA synthesis in *P. polycephalum*, there are no reports of mitochondrial RNA being detected in whole RNA extracts, even using the sensitive separation technique of gel electrophoresis. Recently, however, competition hybridisation experiments with 4 S RNA have suggested that some of the 4 S RNA extractable from plasmodia is mitochondrial in origin (Sonnenshein, Newlon & Holt, in preparation). In particular, since the organism contains a relatively high proportion of mitochondrial DNA (Braun & Evans, 1969; Holt & Gurney, 1969), it might have been anticipated that an equivalently high proportion of mitochondrial RNA would be detected in electrophoretic studies of whole RNA extracts.

In this laboratory, an extensive investigation of mitochondrial RNA has been carried out recently, and mitochondrial ribosomal RNA has been characterised for the first time, following the development of a high yield

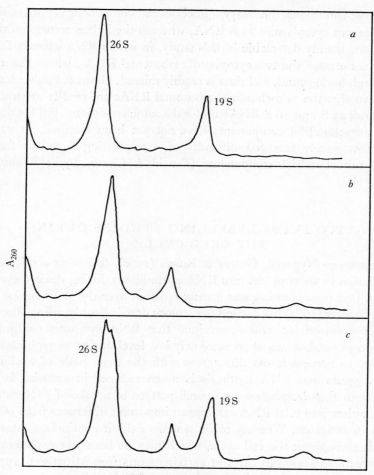

Electrophoretic migration

Fig. 3 (*a*) Electrophoresis in 3% polyacrylamide gels of total RNA from *Physarum*.

(*b*) Electrophoresis in 3% polyacrylamide gels of RNA extracted from isolated mitochondria of *Physarum*.

(*c*) Co-electrophoresis in 3% polyacrylamide gels of total RNA and mitochondrial RNA of *Physarum*. A more complete account of this appears in Grant & Poulter (1972, *J. Mol. Biol.*, in press). Mitochondria were isolated by the method described in Grant & Poulter, RNA extracted by the method of Jacobson (1971) and electrophoresis carried out on 100 × 6 mm gels as described by Melera, Chet & Rusch (1970) for $1\frac{3}{4}$ h at $7\frac{1}{2}$ mA/gel.

isolation procedure for mitochondria (Grant & Poulter, 1972, *J. Mol. Biol.*, in press). Fig. 3 shows the co-electrophoresis of mitochondrial and cytoplasmic RNAs on 3% polyacrylamide gels. It can be seen that under these conditions, the heavy mitochondrial component is not readily separable from cytoplasmic 26 S RNA, whereas the lighter mitochondrial component, readily detectable in this study, in whole RNA extracts falls in an area between the two cytoplasmic ribosomal RNAs, where there is often a high background, and thus is readily missed. Grant & Poulter have also shown that the mitochondrial ribosomal RNAs are readily separable from *E. coli* 23 S and 16 S RNAs under the same conditions, but S values for the mitochondrial components have not yet been ascribed, in view of the now widely accepted difficulty of extrapolating S values from gels for mitochondrial components (Dawid & Chase, 1972; Edelman *et al.* 1971).

IN-VIVO PULSE-LABELLING STUDIES DURING THE CELL CYCLE

Some years ago Nygaard, Guttes & Rusch (1960), following orotic acid incorporation in short pulses into RNA pyrimidines during the cell cycle, observed that incorporation was greatly reduced over the short period of mitosis. Kessler (1967) has carried out a more detailed study, using shorter pulses of radioactivity, and determined that RNA synthesis continued during prophase, but was at an extremely low level during metaphase and anaphase. In general terms this agrees with the large body of evidence which suggests that RNA synthesis is much reduced in mitosing cells. However, in *P. polycephalum* the overall pattern of uptake of [³H]uridine in short pulses into total RNA exhibits an important difference from other systems investigated. Whereas other systems exhibit continuous rates of synthesis throughout the cell cycle, which may be linear, or exponential, perhaps with increases in rates at certain points (see Mitchison, 1971), *P. polycephalum* is unique, in that the rate of [³H]uridine incorporation rises from a low value at mitosis, to a first maximum roughly corresponding with the end of S phase, falls again markedly some 4–5 hours after mitosis, and rises to a second, G_2 peak of activity some 2–3 hours before the subsequent mitosis, falling again at mitosis. These results which were first reported by Mittermayer *et al.* (1964), illustrated in Fig. 4a, have been subsequently repeated several times (Braun *et al.* 1966b; Sauer, Goodman, Babcock & Rusch, 1969). This biphasic pattern of rates of [³H]uridine incorporation is likely to represent true differences in the rate of RNA synthesis during the cell cycle, since the evidence there is suggests that RNA precursor pools do not markedly alter during the cell cycle, with the possible exception of a short period round about mitosis (Chin & Bernstein, 1968; Sachsenmaier *et al.* 1969).

It is also clear that a proportion of the RNA synthesised at any time in the cell cycle may be due to synthesis in cytoplasmic organelles. Since the mitochondrial DNA of *P. polycephalum* has conclusively been shown to be replicated throughout the cell cycle (Holt & Gurney, 1969; Braun & Evans, 1969), it might be expected that mitochondrial RNA would be synthesised in a similar fashion, although there appears to be no general rule from other systems, since yeast RNA appears to be replicated continuously (Fan & Penman, 1970), whereas mammalian systems may exhibit periodic synthesis (Pica-Mattoccia & Attardi, 1971). Fig. 4*b* shows the results of an experiment carried out to determine rate of RNA synthesis in *P. polycephalum* mitochondria over the cell cycle. Plasmodia were pulsed with high specific activity [³H]uridine at various times during the mitotic cycle, mitochondria isolated, and the specific activity of mitochondria RNA measured. RNA was extracted by the method of Jacobson (1971) from several of the preparations, and run on 3% gels as previously described in Fig. 3, to determine cytoplasmic contamination. However, after a 15 min pulse of [³H]uridine, the patterns of radioactivity obtained on gels were polydisperse, with no real indication of the derivation of the RNA. Preparations of mitochondria isolated by the method of Grant & Poulter (1972, *J. Mol. Biol.*, in press) contain very little cytoplasmic ribosomal contamination, as judged by analysis of mitochondrial RNA on gels (Fig. 3), and by various biochemical analyses (Grant & Poulter). Nevertheless, it might be argued that the majority of radioactivity incorporated in short pulses, presumably nuclear, would be released during the mitochondrial isolation procedure which causes lysis of nuclei, thus perhaps contaminating mitochondria. Accordingly, at hourly intervals over the cell cycle, plasmodia were pulse-labelled with high specific activity [³H]-uridine for 15 min, then transferred to unlabelled medium for 45 min, after which time mitochondria were isolated, and RNA extracted. Under these conditions, on analysis of RNA on gels, radioactivity appeared in mitochondrial ribosomal RNA in all cases. A typical example is shown in Fig. 5, which indicates, that on the basis of the distribution of radioactivity on the gel, some cytoplasmic ribosomal RNA is present as a contaminant, but it proved to be less than 20% of the total radioactivity incorporated in all cases. It might be expected from the quantity of cytoplasmic ribosomal RNA contamination, as measured by radioactivity, that an ultraviolet (u.v.) trace would also indicate the cytoplasmic species. However, in this laboratory, cytoplasmic ribosomal RNA contamination as indicated by such a u.v. trace has never been seen in RNA extracted from preparations of mitochondria made as described by Grant & Poulter (1972, *J. Mol. Biol.*, in press), and the contamination shown in Fig. 5 is presumably a

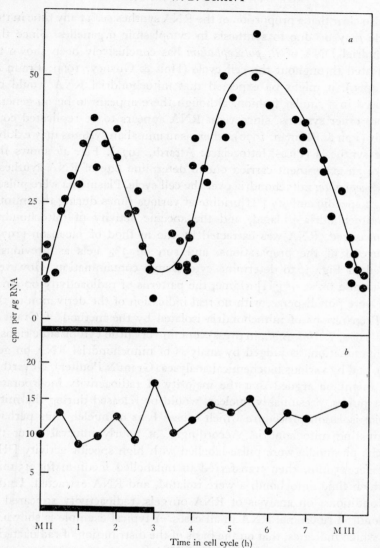

Fig. 4. (*a*) Uptake of 10 min pulses of [³H]uridine into RNA at different times of the cell cycle. Each point represents the specific activity of one whole stationary culture. MII = second mitosis after fusion of microplasmodia to form a macroscopic stationary plasmodium; MIII = third mitosis after fusion. MII–MIII lasts 8–10 h. The black bar along the abscissa is a schematic representation of the duration of S, assuming an intermitotic time of 8–10 h (redrawn from Mittermayer, Braun & Rusch, 1964).

(*b*) Uptake of 15 min pulses of [³H]uridine into mitochondrial RNA at different times of the cell cycle. Each point represents the specific activity of mitochondria isolated from one whole stationary culture. The black bar along the abscissa is a representation of the expected duration of S.

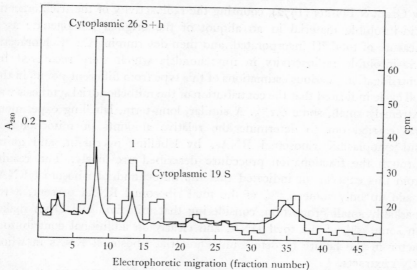

Fig. 5. Distribution of radioactivity in RNA extracted from isolated mitochondria, after a 15 min pulse of [³H]uridine, followed by a 45 min chase. Isolation of mitochondria, extraction of RNA, and gel electrophoresis as outlined in Fig. 3. The example shown is taken from a point in the cell cycle 6 h after mitosis. Polyacrylamide gels were frozen in dry ice, cut into 2 mm slices, and digested overnight in 0.5 ml 30 % (v/v) H_2O_2 containing one drop 880 ammonia. 10 ml of Beckman scintillator (toluene butyl-PBD-PBBO incorporating BB-3 solubiliser) was added to each slice. h indicates the heavier mitochondrial component, l the lighter mitochondrial component.

reflection of much more rapid synthesis of cytoplasmic ribosomal RNA, as compared with mitochondrial ribosomal RNA.

Thus it is likely that mitochondrial RNA synthesis is linear over the cell cycle. It is also indicated from Fig. 4*b* that there is no depression of mitochondrial synthesis during mitosis, but the process of mitosis is extremely rapid, and it is possible that a very short period exists where no mitochondrial synthesis takes place. The approximate contribution of mitochondrial synthesis to total RNA synthesis was also determined by pulse-labelling a plasmodium for thirty minutes with high specific activity [³H]uridine, and isolating mitochondria by the method described

---

Fig. 4 (*b*) cont.

Stationary cultures from MII–MIII were pulsed with 50 µCi/ml [³H]uridine (20000 mCi/mmole) for 15 min, mitochondria isolated by the method of Grant & Poulter (1972, *J. Mol. Biol.*, in press). Washed mitochondria were added to 2–3 vol ice-cold trichloracetic acid, and the radioactivity in trichloracetic acid-insoluble material determined by filtration and washing on Whatman GF/C glass-fibre disks, followed by subsequent scintillation counting in a toluene-based scintillation fluid.

by Grant & Poulter (1972), counting the radioactivity of the trichloracetic acid-insoluble material in an aliquot of the original homogenate as a measure of total ^3H incorporated, and then determining the trichloracetic acid-insoluble radioactivity in mitochondria which were recovered by centrifugation. Various estimations of this type from different points in the cell cycle indicated that the contribution of the mitochondrial synthesis was extremely small, some 0.1 %. A similar, long-term, labelling experiment was carried out to determine the relative amounts of mitochondrial and cytoplasmic ribosomal RNAs, by labelling overnight, and going through the fractionation procedure described previously. The results from this experiment indicated that the mitochondrial ribosomal RNAs made up only some 0.5 % of the total ribosomal RNAs present, a remarkably small proportion, considering that mitochondrial DNA makes up some 9 % of the total DNA, and clearly an additional contributory factor in the failure to detect mitochondrial ribosomal RNAs in whole RNA extracts.

The remarkable biphasic pattern of overall RNA synthesis has been investigated by several workers, with a view to establishing a relationship between this phenomenon and other processes in the cell cycle. Some years ago, Mittermayer, Braun, Chayka & Rusch (1966) reported that short pulses of labelled amino acids at various times during the cell cycle exhibited the same biphasic pattern of rates of uptake, but recently Brewer (1969) has shown that polysomes isolated from different points in the cell cycle show no difference in their ability to incorporate amino acids, and failed to detect the breakdown in polysome profiles shortly after mitosis reported by Mittermayer, Braun, Chayka & Rusch (1966). At the moment the relationship of these results to the biphasic pattern of RNA synthesis is not clear. Certainly there is no evidence which correlates the RNA pattern with enzyme levels in the cell cycle for example, the enzymes so far described exhibiting linear or stepwise increases (Hütterman, Porter & Rusch, 1970b; Braun & Behrens, 1969), with the exception of thymidine kinase (Sachsenmaier & Ives, 1965), and NAD pyrophosphorylase (Solao & Shall, 1971), which exhibit maximum activity in a periodic fashion, coinciding with the DNA synthetic period. The only clear correlation between the RNA biphasic pattern and other processes in the cell cycle was demonstrated by Sauer, Goodman, Babcock & Rusch (1969), who showed that polyphosphate in the nucleus was synthesised in an inverse pattern to RNA, but it is clear from the data described that polyphosphate is probably involved in RNA synthesis.

Another line of investigation has been to determine differences between the RNAs synthesised in the S and G_2 peaks of the biphasic pattern, and

Mittermayer *et al.* (1964) originally observed that the G_2 peak appeared to be more sensitive to actinomycin D inhibition. Braun *et al.* (1966*a*) have analysed pulse-labelled RNA from various points in the cell cycle, using sucrose gradients, showing that ribosomal and transfer RNAs were made throughout the cell cycle, with the possible exception of the period around mitosis. Polydisperse sucrose gradient patterns with high molecular weight material were seen in material isolated after short pulse-labelling throughout the cell cycle, and there was an indication that the pattern changed before mitosis, but it is clear that extensive degradation problems existed. These types of experiment have recently been repeated in much finer detail by Zellweger & Braun (1971*b*), using the more sensitive technique of gel electrophoresis and a much better RNA extraction technique, without revealing any reproducible differences during the mitotic cycle. Zellweger & Braun (1971*b*) also attempted to detect changes in pulse-labelled RNA from different points in the cell cycle using DNA/RNA competition hybridisation, again without detecting any changes. However, it has been pointed out that in eukaryotes it is likely that DNA/RNA hybridisation is only useful in the case of highly repetitious species, and the results of Zellweger and Braun do not rule out transcriptional control of non-repetitious genes. It might also be reiterated at this point that technical problems with RNA extraction may well cause the loss of heterogeneous nuclear RNA, which presumably includes messenger. Recently, some information has become available with regard to the repetitive nature of *P. polycephalum* DNA from Cot studies and hydroxyapatite binding (Britten & Smith, 1971). A partial reannealing curve of *P. polycephalum* DNA is shown in Fig. 6. It is clear from this, that at least 60% of the DNA is highly repetitious, although the nature of all the repetitive sequences is not known yet. However, from the hydroxyapatite binding studies of Britten & Smith, at least part of the repetitive sequences is due to the heavy nuclear satellite DNA which is detectable on CsCl gradients (Holt & Gurney, 1969; Braun & Evans, 1969), although clearly not all. The information concerning Cot values greater than 1.0 is not yet available. Theoretically, the region of single copy sequences could be predicted, and purification carried out by hydroxyapatite chromatography, provided the Cot value is not too high, as Fig. 6 suggests it is unlikely to be. A repeat of the hybridisation experiments of Zellweger & Braun (1971*b*) using essentially single-copy DNA might provide evidence for transcriptional control of non-repetitive genes.

Cummins, Weisfeld & Rusch (1966) have used a different approach for determining differences in RNA synthesised at different points during the cell cycle. Basically, the technique involved pulsing plasmodia with [32]P

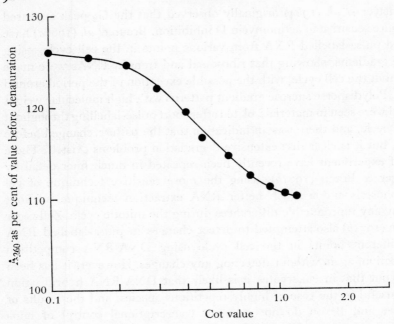

Fig. 6. Cot curve of *Physarum* nuclear DNA. Ordinate: A_{260} relative to value before denaturation (total hyperchromicity 133%); abscissa: Cot (mol.sec per litre). *Physarum* DNA was isolated by extraction of nuclei with sodium dodecyl sulphate, and purified by CsCl gradients. DNA in 2.0 M-NaCl was denatured by adding 2.0 M-NaOH until the pH was 13.0. After 10 min, 2.0 M-HCl was added until the pH was 7.0. The solution was incubated at 70° in a spectrophotometer, and the absorbance at 260 nm observed as a function of time. This graph, provided by Dr H. Matthews, although previously unpublished, has been discussed recently (*Abstracts of the Communications 7th Meeting European Biochemical Society*, Varna 1971, p. 182).

at various times in the cell cycle, extracting the RNA, and analysing hydrolysates of the RNA using several chromatographic techniques. The data obtained showed only small differences in the relative amounts of the four nucleotides in RNA extracted from different points in the cell cycle, but it is claimed that there was a significant change in the ratio of adenylic/guanylic acid residues as the cell cycle progressed. The widely quoted summary of these results is shown in Fig. 7a. However, if all the data from Cummins *et al.* (1966) are replotted in a similar fashion, indicating the standard errors for each of the points, the trend from high A/G just after mitosis to low A/G just before mitosis as shown in Figure 7a is much less pronounced. This is illustrated in Fig. 7b. Furthermore, if the cytidylic/uridylic acid ratio is plotted in a similar fashion, from the same data, indicated in Fig. 7c, the variation in individual experiments

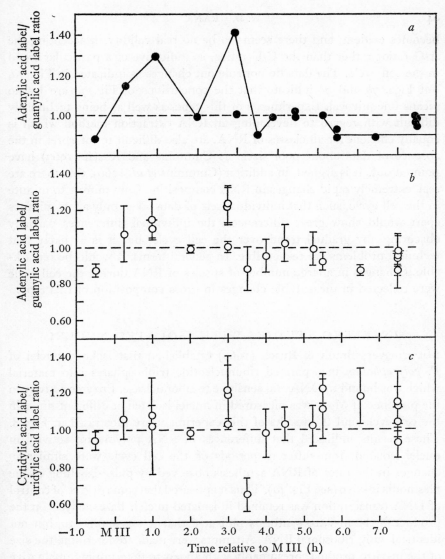

Fig. 7. (*a*) Fluctuations of ³²P label in adenylic acid and guanylic acid in RNA synthesised in-vivo at various times during the cell cycle (taken from Cummins, Weisfeld & Rusch, 1966).

(*b*) Fluctuations of ³²P label in adenylic acid and guanylic acid in RNA synthesised in-vivo at various times during the cell cycle (replotted from the data of Cummins, Weisfeld & Rusch, 1966). Approximate error bars for ratios were arrived at as follows. The proportional error in a ratio was calculated as the sum of the proportional errors in numerator and denominator. This was multiplied by the ratio to give an absolute value for the error in a ratio.

(*c*) Fluctuations of ³²P label in cytidylic acid and uridylic acid in RNA synthesised in-vivo at various times during the cell cycle. Error bars calculated as in (*b*).

MIII = third mitosis after fusion of microplasmodia to form macroscopic stationary culture.

becomes evident, and there seems to be no real validity in accepting the A/G ratios rather than the C/U ratios, as indicative of a particular trend in the cell cycle. The data do not rule out changes as indicated in Fig. 7a, but Figs. 7b and 7c indicate that the conclusions of Fig. 7a are by no means unequivocal. Experiments of this type, as well as being technically difficult with regard to developing an RNA extraction method which is equally efficient for all classes of RNA, are also difficult to interpret in the absence of data about pool sizes, as Cummins and Rusch (1967) have pointed out. It is argued, in addition (Cummins et al. 1966), that there are real, extremely rapid changes in RNA composition from minute to minute in the cell cycle, such that individual sets of data taken only a few minutes apart would show great differences, the individual differences partially obscuring any gradual trend over the cell cycle, but it is as likely that technical problems are responsible. In general terms, it would be remarkable if changes in a large number of species of RNA during the cell cycle were reflected in measurable changes in gross composition of this type.

IN-VITRO STUDIES WITH ISOLATED NUCLEI

Mittermayer, Braun & Rusch (1966) established that isolated nuclei of *P. polycephalum* incorporated ribonucleotide triphosphates into material which was bound to DNA and sensitive to ribonuclease. Enzyme activity in the presence of Mg^{2+} was measured in nuclei isolated at different times in the cell cycle and the results of this experiment are illustrated in Fig. 8. These results indicated that differences in RNA polymerase activity in nuclei isolated from different periods of the cell cycle were similar to changes in the rates of RNA synthesis observed by pulse-labelling of the plasmodia in-vivo (see Fig. 4a). Thus it appeared that some degree of control of DNA transcription was retained in isolated nuclei. Base analysis on the product of the in-vitro synthesis indicated that it was similar to, but not identical with, ribosomal RNA. Attempts were made to determine the size of the in-vitro product, presumably with a view to determining qualitative differences in the product synthesised by nuclei isolated from different stages in the cell cycle, but the size of the product proved to be extremely small, regardless of the period of incubation. This is attributed to the characteristics of the system (Mittermayer, Braun & Rusch, 1966), but it is equally likely to be degradation, since isolated nuclei extensively degrade purified ribosomal RNA after a short incubation (Grant, unpublished observation), and attempts to characterise the product of in-vitro synthesis after a very short incubation with high specific activity ribonucleotide triphosphates, using the carefully controlled RNA extraction conditions

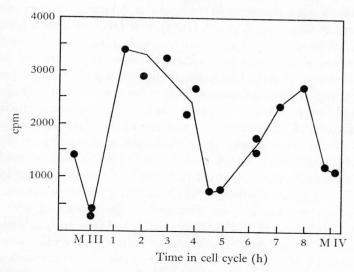

Fig. 8. Course of RNA synthesis in nuclei isolated at different times of the cell cycle. Each point represents the incorporation of [³H]UTP into RNA of nuclei from one stationary culture during a 10 min incubation period. MIII = third mitosis after fusion of microplasmodia to form a macroscopic stationary culture: MIV = fourth mitosis after fusion of microplasmodia (from Mittermayer, Braun & Rusch, 1966).

described by Jacobson (1971), yield essentially the same results (Grant, unpublished observation). The fact that activity in isolated nuclei followed a pattern similar to the changes in the rate of uridine incorporation into intact plasmodia suggested that changes in transcription in-vivo might be mimicked in-vitro in isolated nuclei from the same period of the cell cycle. Cummins & Rusch (1967) have attempted to determine changes, by incubating isolated nuclei with all four nucleotide precursors, and then determining nearest neighbour frequencies in the RNA product. On the basis of a large number of estimations made from nuclei isolated at different stages in the cell cycle, several facts emerged: (*a*) a variance analysis showed that despite substantial differences within individual sets of data, certain dinucleotides fluctuated in amount significantly during the cell cycle; (*b*) nucleotides were not randomly distributed along the newly synthesised RNA chain on the basis of the observed frequencies, and that only a limited portion of the total DNA was transcribed, since the average composition of the RNA synthesised was midway between that of DNA and ribosomal RNA. Cummins (1969) has repeated this type of experiment, comparing the transcription of purified *P. polycephalum* DNA and isolated nuclei, using purified *E. coli* RNA polymerase, and shown quite different dinucleotide frequencies in the products; (c) much less conclusive data suggested to Cummins & Rusch (1967), that DNA sequences beginning with guanylic or cytidylic acid were transcribed increasingly over the cell cycle, whereas sequences beginning with adenylic or uridylic acid decreased. However, it should be pointed out that out of twenty experiments purporting to show a linear association between nearest neighbour frequency and time in the cell cycle, only five have a correlation coefficient with a significance of better than 5%, and two of those are diametrically opposed in result. Nevertheless, the data do not rule out the principle formulated by Cummins & Rusch (1967) that sequences beginning with adenylic acid decrease throughout the cell cycle, whereas those beginning with guanylic acid increase, although the data are by no means unequivocal. The correlations claimed for the dinucleotide frequencies of uridylic acid and cytidylic acids are less convincing still, and it is unclear if there is any trend over the cell cycle. However, Cummins (1969) on the basis of these data, and the ^{32}P pulse-labelling data discussed earlier, has made the calculation that the RNA synthesised in plasmodia just after mitosis is two-thirds DNA-like and one-third ribosomal RNA-like, whereas just before mitosis the RNA is two-thirds ribosomal RNA-like and one-third DNA-like.

Recent work in this laboratory has provided some indirect evidence that the approximations made by Cummins (1969) may prove to be reasonably correct, indicating that the second, G_2 peak of the biphasic pattern of RNA synthesis may be due mainly to ribosomal RNA synthesis. The rationale behind the experiments derived from the observation that two types of RNA synthesis activity may be observed in isolated nuclei, depending on the conditions of the assay system. One type, localised in the nucleoplasm, has been shown in several different systems to be markedly stimulated in high salt concentration, giving rise to a product which resembles DNA in base composition (Maul & Hamilton, 1967; Pogo, Littau, Alfrey & Mirsky, 1967; Younger & Gelboin, 1970). The location of the activity, and the composition of the product, strongly suggest that it may be involved in messenger RNA synthesis, or at least heterogeneous RNA synthesis. The other type of RNA synthesis activity is confined to the nucleolus, unstimulated by high salt concentration, and gives rise to a product which is largely ribosomal (Younger & Gelboin, 1970). An additional important difference between the two types of activity is that nucleolar synthesis is insensitive to the inhibitor α-amanitin, whereas nucleoplasmic synthesis is completely inhibited (Roeder & Rutter, 1967; Jacob, Sajdel & Munro, 1970).

In their original observations, Mittermayer, Braun & Rusch (1966)

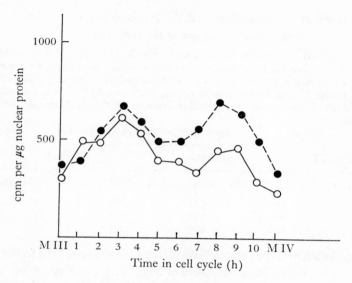

Fig. 9. Incorporation of [³H]UTP into trichloracetic acid – insoluble material by nuclei isolated at different times in the cell cycle. Open circles indicate incorporation in the presence of 1 mM-Mg^{2+}; closed circles indicate incorporation in the presence of 0.4 mM-Mn^{2+} + 60 mM-$(NH_4)_2SO_4$. MIII = third mitosis after fusion of microplasmodia, MIV = fourth mitosis after fusion of microplasmodia.

RNA synthesis was assayed by a modification of the method of Mittermayer, Braun & Rusch (1966). The standard assay system consisted of 0.25 M sucrose; 0.05 M Tris-maleate pH 7.8; 2 mM mercaptoethanol; 0.5 mM GTP, CTP; 1.0 mM ATP; 0.1 μM [³H]UTP (10 Ci/mmole). For assay in the presence of Mg^{2+}, 1 mM-$MgCl_2$ was added, and for assay in high salt concentration, 0.4 mM-$MnCl_2$ + 60 mM-$(NH_4)_2SO_4$ was added. Incubation was for 60 min at 37 °C, and the reaction was stopped with 2–3 vol 5% trichloracetic acid. Trichloracetic acid-insoluble material was washed on membrane filters, dried and radioactivity determined by scintillation counting. A full account of this will appear shortly (Grant, 1972).

assayed nuclear RNA synthesis in low Mg^{2+} concentrations, and did report stimulation of synthesis in the presence of relatively high Mn^{2+}, but confined their observations to G_2 nuclei. In this laboratory a more extensive examination of this type has been carried out (Grant, 1972). The results are summarised in Fig. 9 which shows the effect of high salt, in this case 0.06 M-$(NH_4)_2SO_4$ + 0.4 mM-$MnCl_2$, on RNA synthesis in nuclei isolated at different points during the cell cycle.

The basic points which emerge from these observations are that the biphasic S and G_2 peak pattern of synthesis is clearly seen, both in the presence of Mg^{2+} and Mn^{2+} + $(NH_4)_2SO_4$, and that very little stimulation of S peak nuclei occurs in the presence of Mn^{2+} + $(NH_4)_2SO_4$, whereas

4

Table 1. *Effect of α-amanitin on RNA synthesis in nuclei isolated from S and G_2 regions of the cell cycle*

Nuclei were isolated at mitosis $II + 3$ h, and mitosis $III - 2\frac{1}{2}$ h, for S and G_2 points respectively. Nuclei were assayed for RNA synthesis activity for 60 min at 37 °C in the presence of either 1 mM-Mg^{2+} or 0.4 mM-Mn^{2+} + 60 mM-$(NH_4)_2SO_4$ as described under Fig. 9. α-amanitin at 1.5 mg ml^{-1} in 0.85% saline was added to give the desired concentrations. A fuller account of this will appear shortly (Grant, 1972).

	cpm incorporated per μg protein			
Concentration of α-amanitin (μg ml^{-1})	S nuclei		G_2 nuclei	
	Mg^{2+}	$Mn^{2+} + (NH_4)_2SO_4$	Mg^{2+}	$Mn^{2+} + (NH_4)_2SO_4$
0	688	723	406	796
1	294	326	421	437
10	196	209	391	406

marked stimulation of G_2 peak nuclei is evident. This suggested the following points: (*a*) S peak nuclei in the presence of Mg^{2+} may be carrying out both types of RNA synthesis, and at least the nucleoplasmic portion of the total synthesis observed is likely to be going on at maximum rate, in view of its failure to be stimulated further in the presence of high salt; (*b*) it should be possible to determine the relative amount of nucleoplasmic synthesis in S peak nuclei by using the inhibitor α-amanitin; (*c*) a substantial part of the capability for nucleoplasmic synthesis must be repressed in G_2 peak nuclei in the presence of Mg^{2+}, since a large stimulation is observed in the presence of high salt; (*d*) the level of nucleoplasmic synthesis in G_2 peak nuclei in the presence of Mg^{2+} should be determinable in the presence of α-amanitin, and it would be predicted that the increased synthesis in the presence of high salt should be totally sensitive to α-amanitin.

Nuclei from approximately the midpoints of the S and G_2 peaks were isolated, and RNA synthesis assayed under the conditions described in Table 1. It can be seen that some 50% of the activity observed in S nuclei is inhibited by α-amanitin, whereas RNA synthesis in G_2 nuclei in the presence of Mg^{2+} is practically uninhibited in the presence of α-amanitin, the increase in activity observed in the presence of high salt being totally sensitive to α-amanitin. This strongly suggests that assayed under identical conditions, i.e. in the presence of Mg^{2+}, G_2 peak nuclei exhibit almost entirely nucleolar-type synthesis activity, whereas S peak nuclei exhibit a substantial proportion of both nucleolar and nucleoplasmic RNA synthesis activity. A further indication of the almost entirely nucleolar RNA synthesis activity of G_2 peak nuclei was obtained by isolating nucleoli from this period in the cell cycle, and comparing the characteristics of the

Table 2. *A comparison of RNA synthesis in nuclei and nucleoli isolated from the G_2 region of the cell cycle*

Nuclei and nucleoli were isolated as described in the text at MIII–$2\frac{1}{2}$ h, and assayed for RNA synthesis activity at 37 °C as previously described. The basic assay medium pH 7.8 described under Fig. 9, containing 1 mM-Mg^{2+}, was used throughout, except where additions to, or subtractions from the standard assay mixture are indicated in the table. Preincubation with deoxyribonuclease indicates preincubation of nuclei or nucleoli with 100 μg ml^{-1} pancreatic deoxyribonuclease at 37 °C for 30 min prior to addition to RNA synthesis assay medium. The amount of incorporation of [^{3}H]UTP by nucleoli or nuclei in basic assay medium was normalised to 100%, and incorporation under the other conditions described expressed as a percentage of this. However, the specific activities of nucleolar preparation were some five to six times higher than those of nuclear preparation.

	% incorporation of [^{3}H]UTP	
Changes in basic assay medium	G_2 nuclei	nucleoli
pH 7.2	46.2	52.9
pH 7.4	53.1	60.0
pH 7.6	81.7	85.7
pH 7.8	100.0	100.0
pH 8.0	95.1	93.1
pH 8.2	77.3	74.2
+0.1 μg ml^{-1} actinomycin D	85.2	82.9
+1.0 μg ml^{-1} actinomycin D	48.0	42.7
+10.0 μg ml^{-1} actinomycin D	8.1	5.0
+10.0 μg ml^{-1} α-amanitin	97.3	99.1
+0.4 mM-Mn^{2+}+60 mM-$(NH_4)_2SO_4$	219.8	112.1
Preincubation with deoxyribonuclease	5.2	6.1
−ATP, CTP, GTP	12.7	4.9
Incubation at 25 °C	63.1	59.7

RNA-synthesising systems in nucleoli and G_2 nuclei. Initially, the procedure of Mohberg & Rusch (1971) was used to isolate nucleoli, but the procedure was somewhat lengthy, subjecting nucleoli for extended periods to high sucrose concentration. However, in this laboratory, several strains of *P. polycephalum* have been isolated which have extremely large nuclei and nucleoli. One of these strains, 501, has nucleoli which are equivalent in size to 'normal' *P. polycephalum* nuclei, and these can be isolated essentially by the nuclear isolation procedure of Mohberg & Rusch (1971), which is much more rapid. Nuclei have also been isolated by the same procedure, using shorter homogenisation times, and this strain has been used for the determinations shown in Table 2. It can be seen that isolated nucleoli readily incorporated nucleotide triphosphates into RNA, and like other systems described, the synthesis is not stimulated by high salt, and insensitive to α-amanitin.

It is clear from Table 2 that the characteristics of RNA synthesis in isolated G_2 nuclei in the presence of Mg^{2+} are extremely similar to those of synthesis in isolated nucleoli, with regard to pH optimum, actinomycin D sensitivity, α-amanitin insensitivity, etc.

It is tempting to suggest that these results obtained from in-vitro experiments using nuclei from different points in the cell cycle, assayed under the same conditions, may reflect the in-vivo situation, suggesting substantial changes in different polymerase activities during the cell cycle. Thus there may be a substantial peak of nucleoplasmic, and therefore probably messenger RNA synthesis, shortly after mitosis, thereafter dropping to a much lower level, whereas nucleolar synthesis carries on at a high rate throughout the cell cycle, with the exception of mitosis, rising to a peak late in G_2. Table 1 suggests that very little nucleoplasmic synthesis goes on in G_2, although the techniques will only detect relatively large changes in polymerase activity. It is clear, however, that some messenger RNA synthesis must go on in G_2, since the results of Sachsenmaier, Fournier & Gürtler (1967) have demonstrated that the last messenger for mitosis is transcribed $1\frac{1}{2}$ h before mitosis. At the moment attempts are being made in this laboratory to determine differences in the RNAs synthesised in G_2 and S peak nuclei, both in the presence and absence of high salt, although the high ribonuclease activity of isolated nuclei presents considerable problems.

REPLICATION OF RIBOSOMAL CISTRONS DURING THE CELL CYCLE

It has been shown in several laboratories that DNA preparations from *P. polycephalum* show three different density bands on CsCl gradients: the main bulk of the DNA sediments as a band of density 1.700 g cm^{-3}, a lighter band, identified as mitochondrial DNA (Evans, 1966; Guttes, Hanawalt & Guttes, 1967) of density 1.686 g cm^{-3}, and a heavy band of density 1.714 g cm^{-3} which is found in the nucleus (Gurney & Holt, 1969; Braun & Evans, 1969). Selective extraction procedures have been devised to enrich for the satellite DNAs, and Fig. 10 shows analytical centrifugation profiles of DNAs prepared in different ways.

The mitochondrial DNA has been conclusively shown to be replicated throughout the cell cycle (Gurney & Holt, 1969; Braun & Evans, 1969), and similar observations have been made for the heavy nuclear satellite (51% G + C compared with 41% G + C for main band DNA) (Holt & Gurney, 1969; Braun & Evans, 1969). Suggestions that the heavy satellite might code for ribosomal RNA have been made by Guttes &

Guttes (1969) who showed that it was likely to be localised in the nucleolus, and by Zellweger & Braun (1971b) who showed that the heavy side of main-band DNA hybridised better with ribosomal RNA. Recently, Zellweger, Ryser & Braun (1972) have shown that DNA extracted from isolated nucleoli is markedly enriched in heavy satellite DNA (Fig. 11), confirming the location of the satellite, and that the satellite was replicated throughout the cell cycle, with the exception of about one hour after mitosis. Experiments showed that 32 % of the heavy satellite hybridised with ribosomal RNA, and that the ribosomal cistrons increased in amount during the G_2 period of the cell cycle. Newlon, Shaw & Holt (in preparation) have independently confirmed this observation. Zellweger *et al.* (1972) have calculated that there are probably 1000 to 2000 ribosomal cistrons per diploid nucleus, although how the number changes during the cell cycle is not yet known. Sonnenshein, Shaw & Holt (in preparation) have performed a similar series of experiments, and calculated that the number of ribosomal cistrons per diploid nucleus is likely to be about 300, a much lower figure than that given by Zellweger *et al.* (1972), the difference being probably attributable to different DNA extraction procedures or hybridisation techniques. Sonnenshein *et al.* have also calculated that there are probably 1000 4 S cistrons which are coded for by the bulk DNA, and 1000 5 S cistrons.

Thus the evidence is clear that ribosomal cistrons increase in number

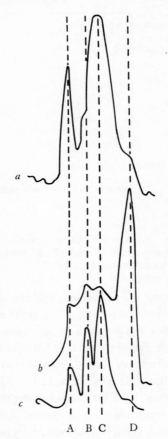

A B C D

Fig. 10. Analytical caesium chloride centrifugation profiles of *Physarum* DNA prepared by different extraction procedures. (*a*) bulk DNA; (*b*) selective extraction procedure of a whole plasmodium followed by Sephadex chromatography to enrich for mitochondrial DNA; (*c*) selective extraction of nuclear DNA to enrich for heavy nuclear satellite DNA. The dotted vertical lines represent A. *Pseudomonas aeruginosa* marker DNA (1.727 g cm^{-3}; B. *Physarum* nuclear satellite (1.714 g cm^{-3}); C. *Physarum* major nuclear DNA (1.700 g cm^{-3}); D. *Physarum* mitochondrial DNA (1.686 g cm^{-3}). (From Braun & Evans, 1969.)

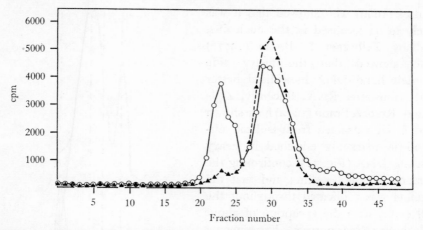

Fig. 11. Caesium chloride gradient of continuously labelled DNA extracted from isolated nuclei or nucleoli. ▲, nuclear DNA; ○, nucleolar DNA. (From Zellweger, Ryser & Braun, 1972.)

during the G_2 period, and the question arises as to whether the differential increase in rate of RNA synthesis observed in-vivo during G_2, if it really is due to the synthesis of predominantly ribosomal RNA, may be due to a gene dosage effect. This question is as yet unanswered, although the preliminary observations of Rao & Goncharoff (1969) indicate that, whereas 5′-fluorodeoxyuridine (FUDR) added during the S period markedly reduced RNA synthesis during this period, suggesting that freshly replicated molecules of DNA are transcribed immediately, the addition of FUDR at the same level in G_2 had essentially no effect on RNA synthesis during this period. However it is possible that heavy satellite replication is much less sensitive to inhibition by FUDR than bulk DNA, in view of its higher $G + C$ content, and experiments are in progress in this laboratory to determine this, and thus to investigate the effect of ribosomal cistron replication on G_2 RNA synthesis.

SUMMARY AND CONCLUSIONS

The myxomycete *Physarum polycephalum* clearly has many advantages as a system for studying events in the cell cycle, but there are certain peculiarities to the system outlined and argued in the Introduction to this article. To these peculiarities must be added another. If *P. polycephalum* really is a highly evolved eukaryote, and most workers would say that it was, then where is the heterogeneous nuclear RNA found in other eukaryotic systems? The avowed technical difficulties involved in working with RNA in this organism may account for the absence of this class of compound.

It is possible that the class exists, exhibiting rather unusual turnover properties for example, or perhaps synthesised only at one stage in the cell cycle, S for instance, or alternatively, and heretically, it may not exist at all. In any event, the failure to find it at the moment is a stimulus to much thought and experimentation.

There are a number of other points more specifically mentioned in this article which merit further discussion.

(1) In-vivo pulse-labelling experiments indicate that there are two maxima in rates of RNA synthesis during the cell cycle, one roughly corresponding with the end of S phase, the other in G_2. Analyses of the products of pulse-labelling indicate that ribosomal RNA is synthesised throughout the cell cycle, with the exception of a very short period during mitosis.

(2) In-vitro studies with isolated nuclei indicate that this biphasic pattern is mimicked by levels of polymerase activity in nuclei isolated from different points in the cell cycle, indicating that some degree of control of RNA synthesis is retained in isolated nuclei.

(3) Some idea of the contribution of ribosomal RNA synthesis to the biphasic pattern observed in-vivo, and in-vitro, is indicated by the fact that nuclei isolated from a region in the cell cycle roughly corresponding with the S peak exhibit substantial amounts of both nucleoplasmic and nucleolar RNA synthesis activity, whereas nuclei isolated from the G_2 peak region of the cell cycle exhibit mainly nucleolar synthesis.

(4) There is some equivocal evidence from analyses of the total composition of RNA synthesised in-vivo in short pulses, and from nearest neighbour analyses on RNA synthesised in isolated nuclei from different times in the cell cycle, that changes in transcription take place over the cell cycle, suggesting that ribosomal RNA may be synthesised increasingly as the cell cycle progresses.

(5) The conclusions outlined in (3) and (4) make it likely that the sharp differential increase in total RNA synthesis observed in in-vivo pulse-labelling experiments during G_2 is mainly due to increased ribosomal RNA synthesis. There is an increase in the number of ribosomal cistrons over this portion of the cell cycle, and the two may be connected.

(6) *P. polycephalum* DNA is highly repetitious, and this is probably the reason that attempts to show changes in transcription over the cell cycle using hybridisation have been unsuccessful.

(7) Ribosomal RNA precursors in *P. polycephalum* are comparable in size to HeLa ribosomal precursors, but they are probably processed much more rapidly. They are considerably larger than the ribosomal RNA precursors which have been found in non-warm blooded eukaryotes.

(8) Mitochondrial RNA synthesis is probably continuous throughout the cell cycle, but makes up only a tiny fraction of the total synthesis observed. Whereas mitochondrial DNA makes up 9% of the total DNA, only 0.5% of the total ribosomal RNA is mitochondrial in origin.

It is clear that differences in transcription over the cell cycle must occur, and some of the data presented here suggest that the control of different types of RNA polymerase may be important in bringing this about. It is likely that different classes of RNA will be detected if the hybridisation experiments reviewed here are repeated using single-copy DNA. Obviously, the ultimate in the analysis of transcription during the cell cycle is to relate a particular transcriptional event to the appearance of an identifiable protein, and this has yet to be done.

In conclusion, it is clear that many purely biochemical approaches can be designed, but the author would argue that more fundamental insight into the control of the ordered progression of basic processes during the cell cycle requires an additional genetic approach, and this applies to the cell cycle field as a whole, as well as to studies on RNA synthesis specifically. The development of a useable genetic system in *P. polycephalum* in this laboratory and elsewhere, and the demonstrated availability of mutants should prove of paramount importance, and strengthen the claims made over the years that *P. polycephalum* was an 'ideal' system for studying growth and differentiation. At this time, no other system exists which is of the eukaryotic complexity of *P. polycephalum*, has naturally occurring mitotic synchrony on a macroscopic scale, and has the possibility of obtaining mutants, with subsequent genetic analysis. To take some specific examples of how this line of approach may develop, the search for mutants will probably progress at two levels: (*a*) auxotrophic mutants have already been shown to be obtainable, and a valine-requiring mutant has been analysed in this laboratory (Dee, Wheals & Holt, in preparation). Such mutants which are not specific for the cell cycle will clearly be of use in analysing some of the basic biosynthetic processes and the relationship of growth to the nuclear division cycle, in a much more satisfactory way than previous investigations using inhibitors; (*b*) specific cell cycle mutants should be obtainable, although it is difficult to see at the moment any way round the massive screening operation which would be required, although it might be possible to enrich for DNA synthesis mutants using photo-sensitisation in the presence of bromodeoxyuridine, as the preliminary experiments of Haugli (1971) suggest. Theoretically it ought to be possible to obtain temperature-sensitive mutants, both in the 'growth' phase and the 'differentiation' phase, which have lesions in processes at particular points in the temporal order of events, using much the same reasoning and experimentation as Hartwell (Hartwell, 1971; Culotti & Hartwell, 1971) has elegantly instigated using *Saccharomyces cerevisiae*.

In general terms, such an initiation of what might be described as a second phase of cell cycle studies with *P. polycephalum*, utilising a combination of conventional biochemical techniques, and cell cycle mutants, should have great implications for the cell cycle field as a whole.

I am particularly in debt to Dr David Jacobson for allowing me to publish material from his Ph.D. Thesis, to Dr Ned Holt, Dr Jennifer Dee, Dr Gale Sonenshein, and Dr Carol Newlon, for access to unpublished information, and to Dr Harry Matthews for providing the Cot curve illustrated in Fig. 6. I am also greatly indebted to my colleagues, Dr Alan Wheals, Dr Robert Semeonoff and Dr Geoff Turnock, for assisting in developing some of the arguments in this article, and for advice and suggestions during its preparation.

REFERENCES

BRAUN, R., MITTERMAYER, C. & RUSCH, H. P. (1966*a*). Sedimentation patterns of pulse-labelled RNA in the mitotic cycle of *Physarum polycephalum*. *Biochimica et Biophysica Acta*, **114**, 27–35.

BRAUN, R., MITTERMAYER, C. & RUSCH, H. P. (1966*b*). Ribonucleic acid synthesis *in vivo* in the synchronously dividing *Physarum polycephalum* studied by cell fractionation. *Biochimica et Biophysica Acta*, **114**, 527–35.

BRAUN, R. & BEHRENS, K. (1969). A ribonuclease from *Physarum*. Biochemical properties and synthesis in the mitotic cycle. *Biochimica et Biophysica Acta*, **195**, 87–98.

BRAUN, R. & EVANS, T. E. (1969). Replication of nuclear satellite and mitochondrial DNA in the mitotic cycle of *Physarum*. *Biochimica et Biophysica Acta*, **182**, 511–22.

BREWER, E. N. (1969). Regulation of protein synthesis through the cell cycle in *Physarum polycephalum*. *Journal of Cell Biology*, **43**, 32A.

BRITTEN, R. J. & SMITH, J. F. (1971). The nuclear satellite in slime mould. *Carnegie Institute of Washington Year Book*, **69**, 518–21.

CHET, I. & RUSCH, H. P. (1969). Induction of spherule formation in *Physarum polycephalum* by polyols. *Journal of Bacteriology*, **100**, 673–8.

CHET, I. & RUSCH, H. P. (1970*a*). RNA and protein synthesis during germination of spherules of *Physarum polycephalum*. *Biochimica et Biophysica Acta*, **224**, 620–2.

CHET, I. & RUSCH, H. P. (1970*b*). RNA differences between spherulating and growing microplasmodia of *Physarum polycephalum* as revealed by sedimentation patterns and DNA–RNA hybridisation. *Biochimica et Biophysica Acta*, **209**, 559–68.

CHET, I. & RUSCH, H. P. (1970*c*). Differences between hybridisable RNA during growth and differentiation of *Physarum polycephalum*. *Biochimica et Biophysica Acta*, **213**, 478–83.

CHIN, B. & BERSTEIN, I. A. (1968). Adenosine triphosphate and synchronous mitosis in *Physarum polycephalum*. *Journal of Bacteriology*, **96**, 330–7.

CULOTTI, J. & HARTWELL, L. H. (1971). Genetic control of cell division cycle in yeast. III. Seven genes controlling nuclear division. *Experimental Cell Research*, **67**, 389–401.

CUMMINS, J. E., WEISFELD, G. E. & RUSCH H. P. (1966). Fluctuation of ^{32}P distribution in rapidly labelled RNA during the cell cycle of *Physarum polycephalum*. *Biochimica et Biophysica Acta*, **128**, 240–8.

CUMMINS, J. E. & RUSCH, H. P. (1967). Transcription of nuclear DNA in nuclei isolated from plasmodia at different stages of the cell cycle of *Physarum polycephalum*. *Biochimica et Biophysica Acta*, **138**, 124–32.

CUMMINS, J. E. (1969). Nuclear DNA replication and transcription during the cell-cycle of *Physarum*. In *The Cell Cycle*, ed. G. M. Padilla, G. L. Whitsun & I. L. Cameron, pp. 141–58. New York & London: Academic Press.

DANIEL, J. W. (1966). Light-induced synchronous sporulation of a myxomycete – the relation of initial metabolic changes to the establishment of a new cell state. In *Cell Synchrony*, ed. I. L. Cameron & G. M. Padilla, pp. 117–52. New York & London: Academic Press.

DAWID, I. B. & CHASE, J. (1972). Mitochondrial RNA in *Xenopus laevis*. II. Molecular weights and other physical properties of mitochondrial ribosomal and 4S RNA. *Journal of Molecular Biology*, **61**, 217–32.

DEE, J. (1960). A mating-type system in an acellular slime mould. *Nature, London*, **185**, 780–1.

DEE, J. (1962). Recombination in a myxomycete *Physarum polycephalum* Schw. *Genetical Research, Cambridge*, **3**, 11–23.

DEE, J. (1966a). Multiple alleles and other factors affecting plasmodial formation in the true slime mould *Physarum polycephalum* Schw. *Journal of Protozoology*, **13**, 610–16.

DEE, J. (1966b). Genetic analysis of actidione-resistant mutants in the myxomycete *Physarum polycephalum* Schw. *Genetical Research, Cambridge*, **8**, 101–10.

DEE, J. & POULTER, R. T. M. (1970). A gene conferring actidione resistance and abnormal morphology on *Physarum polycephalum* plasmodia. *Genetic Research, Cambridge*, **15**, 35–41.

EDELMAN, M., VERMA, I. M., HERZOG, R., GALUN, C. & LITTAUER, U. Z. (1971). Physico-chemical properties of mitochondrial ribosomal RNA from fungi. *European Journal of Biochemistry*, **19**, 372–8.

EVANS, T. E. (1966). Synthesis of a cytoplasmic DNA during the G_2 interphase. *Biochemical and Biophysical Research Communication*, **22**, 678–83.

FAN, H. & PENMAN, S. (1970). Mitochondrial RNA synthesis during mitosis. *Science, New York*, **168**, 135–8.

GOODMAN, E. M., SAUER, H. W., SAUER, L. & RUSCH, H. P. (1969). Polyphosphate and other phosphorus compounds during growth and differentiation of *Physarum polycephalum*. *Canadian Journal of Microbiology*, **15**, 1325–31.

GRANT, W. D. (1972). The effect of α-amanitin and $(NH_4)_2SO_4$ on RNA synthesis in nuclei and nucleoli isolated from *Physarum polycephalum* at different time during the cell cycle. *European Journal of Biochemistry*, **29**, 94–8.

GRANT, W. D. & POULTER, R. T. M. (1972). Rifampicin-sensitive RNA and protein synthesis by isolated mitochondria of *Physarum polycephalum*. *Journal of Molecular Biology*, in press.

GUTTES, E. W. & GUTTES, S. (1964). Mitotic synchrony in the plasmodia of *Physarum polycephalum*, and mitotic synchronisation by coalescence of microplasmodia. In *Methods in Cell Physiology*, vol. 1, ed. D. M. Prescott, pp. 43–54. New York & London: Academic Press.

GUTTES, E. W. & GUTTES, S. (1969). Replication of nucleolus-associated DNA during 'G$_2$ phase' in *Physarum polycephalum*. *Journal of Cell Biology*, **43**, 229–36.

GUTTES, E. W., HANAWALT, P. C. & GUTTES, S. (1967). Mitochondrial DNA synthesis and the mitotic cycle in *Physarum polycephalum*. *Biochimica et Biophysica Acta*, **142**, 181–94.

HARTWELL, L. H. (1971). Genetic control of the cell division cycle in yeast. IV. Genes controlling bud emergence and cytokinesis. *Experimental Cell Research*, **69**, 265–76.

HAUGLI, F. B. (1971). Mutagenesis, selection, and genetic analysis in *Physarum polycephalum*. Ph.D. Thesis, University of Wisconsin.

HAUGLI, F. B. & DOVE, W. F. (1972). Mutagenesis and mutant selection in *Physarum polycephalum*. *Molecular and General Genetics*, **118**, 109–24.

HOLT, C. E. & GURNEY, E. G. (1969). Minor components of the DNA of *Physarum polycephalum*. *Journal of Cell Biology*, **40**, 484–96.

HÜTTERMANN, A. & CHET, I. (1971). Activity of some enzymes in *Physarum polycephalum*. III. During spherulation (differentiation) induced by mannitol. *Archiv für Microbiologie*, **78**, 189–92.

HÜTTERMANN, A., PORTER, M. T. & RUSCH, H. P. (1970*a*). Activity of some enzymes in *Physarum polycephalum*. II. During spherulation (differentiation). *Archiv für Microbiologie*, **74**, 283–91.

HÜTTERMANN, A., PORTER, M. T. & RUSCH, H. P. (1970*b*). Activity of some enzymes in *Physarum polycephalum*. I. In the growing plasmodia. *Archiv für Microbiologie*, **74**, 90–100

HÜTTERMANN, A., ELSEVIER, S. M. & ESCHRICH, W. (1971). Evidence for the *de novo* synthesis of glutamate dehydrogenase during the spherulation of *Physarum polycephalum*. *Archiv für Microbiologie*, **77**, 74–85.

JACOB, S. T., SAJDEL, G. M. & MUNRO, H. M. (1970). Different responses of soluble whole nuclear RNA polymerase and soluble nucleolar RNA polymerase to divalent cations and to inhibition by α-amanitin. *Biochemical and Biophysical Research Communications*, **38**, 765–70.

JACOBSON, D. N. (1971). Isolation and characterisation of ribosomal precursor RNA of *Physarum polycephalum*. Ph.D. Thesis, Massachusetts Institute of Technology.

JOHNSTON, D. N. & RAO, P. V. (1971). Nucleo-cytoplasmic interaction in the achievement of nuclear synchrony in DNA synthesis and mitosis in multi-nucleate cells. *Biological Reviews*, **46**, 97–155.

KESSLER, D. (1967). Nucleic acid synthesis during and after mitosis in the slime mould *Physarum polycephalum*. *Experimental Cell Research*, **45**, 676–80.

KOEVENIG, J. L. & JACKSON, R. C. (1966). Plasmodial mitosis and polyploidy in the myxomycete *Physarum polycephalum*. *Mycologia*, **58**, 662–7.

LESTOURGEON, W. M. & RUSCH, H. P. (1971). Nuclear acidic protein changes during differentiation in *Physarum polycephalum*. *Science, New York*, **174**, 1233–7.

McCORMICK, J. J., BLOMQUIST, J. C. & RUSCH, H. P. (1970). Isolation and characterisation of a galactosamine wall from spores and spherules of *Physarum polycephalum*. *Journal of Bacteriology*, **104**, 1119–25.

MAUL, G. G. & HAMILTON, I. H. (1967). The intranuclear localisation of two DNA-dependent RNA polymerase activities. *Proceedings of the National Academy of Sciences, U.S.A.* **57**, 1371–8.

MELERA, P. W., CHET, I. & RUSCH, H. P. (1970). Electrophoretic characterisation of ribosomal RNA from *Physarum polycephalum*. *Biochimica et Biophysica Acta*, **209**, 569–72.

MITCHISON, J. M. (1971). *The Biology of the Cell Cycle*. London: Cambridge University Press.

MITTERMAYER, C., BRAUN, R. & RUSCH, H. P. (1964). RNA synthesis in the mitotic cycle of *Physarum polycephalum*. *Biochimica et Biophysica Acta*, **91**, 399–405.

MITTERMAYER, C., BRAUN, R. & RUSCH, H. P. (1966). Ribonucleic acid synthesis *in vitro* in nuclei isolated from the synchronously dividing *Physarum polycephalum*. *Biochimica et Biophysica Acta*, **114**, 536–46.

MITTERMAYER, C., BRAUN, R., CHAYKA, T. G. & RUSCH, H. P. (1966). Polysome patterns and protein synthesis during the mitotic cycle of *Physarum polycephalum*. *Nature, London*, **210**, 1133–7.

MOHBERG, J. & RUSCH, H. P. (1969a). Growth of large plasmodia of the myxomycete *Physarum polycephalum*. *Journal of Bacteriology*, **97**, 1411–18.

MOHBERG, J. & RUSCH, H. P. (1969b). Isolation of the nuclear histones of the myxomycete *Physarum polycephalum*. *Archives of Biochemistry and Biophysics*, **134**, 577–89.

MOHBERG, J. & RUSCH, H. P. (1971). Isolation and DNA content of nuclei of *Physarum polycephalum*. *Experimental Cell Research*, **66**, 305–16.

NYGAARD, O., GUTTES, S. & RUSCH, H. P. (1960). Nucleic acid metabolism in a slime mould with synchronous mitosis. *Biochimica et Biophysica Acta*, **38**, 298–306.

PÈNE, J. J., KNIGHT, E. & DARNELL, J. E. (1968). Characterisation of a new low molecular weight RNA in HeLa cell ribosomes. *Journal of Molecular Biology*, **33**, 609–23.

PERRY, R. P., CHENG, T.-Y., FREED, J. J., GREENBERG, J. R., KELLEY, E. & TARTOF, K. B. (1970). Evolution of the transcription unit of ribosomal RNA. *Proceedings of the National Academy of Sciences, U.S.A.* **65**, 609–16.

PICA-MATTOCCIA, L. & ATTARDI, G. (1971). Expression of the mitochondrial genome in HeLa cells. V. Transcription of mitochondrial DNA in relationship to the cell cycle. *Journal of Molecular Biology*, **57**, 615–21.

POGO, A. O., LITTAU, V. C., ALFREY, V. G. & MIRSKY, A. E. (1967). Modification of ribonucleic acid synthesis in nuclei isolated from normal and regenerating liver: some effects of salt and specific divalent cations. *Proceedings of the National Academy of Sciences, U.S.A.* **57**, 743–50.

POULTER, R. T. M. & DEE, J. (1968). Segregation of factors controlling fusion between plasmodia of the true slime mould *Physarum polycephalum*. *Genetical Research, Cambridge*, **12**, 71–9.

RAO, B. & GONCHAROFF, M. (1969). Functionality of newly synthesised DNA as related to RNA synthesis during the mitotic cycle in *Physarum polycephalum*. *Experimental Cell Research*, **56**, 269–74.

ROEDER, R. G. & RUTTER, W. J. (1969). Multiple forms of DNA-dependent RNA polymerase in eucaryotic organisms. *Nature, London*, **224**, 234–7.

ROSS, I. K. (1966). Chromosome numbers in pure and gross cultures of myxomycetes. *American Journal of Botany*, **53**, 712–18.

RUSCH, H. P. (1970). Some biochemical events in the life cycle of *Physarum polycephalum*. In *Advances in Cell Biology*, vol. 1, ed. D. M. Prescott, L. Goldstein & E. McConkey, pp. 297–328. New York: Appleton-Century-Crofts.

SACHSENMAIER, W. & IVES, D. H. (1965). Periodische Anderungen der Thymidine-kinase-Activität im synchronen Mitosecylus von *Physarum polycephalum*. *Biochemische Zeitschrift*, **343**, 399–406.

SACHSENMAIER, W., FOURNIER, D. V. & GÜRTLER, K. F. (1967). Periodic thymidine kinase production in synchronous plasmodia of *Physarum polycephalum*. Inhibition by actinomycin and actidione. *Biochemical and Biophysical Research Communications*, **27**, 655–60.

SACHSENMAIER, W., IMMICH, H., GRUNST, J., SCHOLZ, R. & BÜCHER, T. (1969). Free ribonucleotides of *Physarum polycephalum*. *European Journal of Biochemistry*, **8**, 557–61.

SAUER, H. W., BABCOCK, K. L. & RUSCH, H. A. (1969a). Sporulation in *Physarum polycephalum*. A model system for studies on differentiation. *Experimental Cell Research*, **57**, 319–27.

SAUER, H. W., BABCOCK, K. L. & RUSCH, H. P. (1969b). Changes in RNA synthesis associated with differentiation (sporulation) in *Physarum polycephalum*. *Biochimica et Biophysica Acta*, **195**, 410–21.

SAUER, H. W., BABCOCK, K. L. & RUSCH, H. P. (1970). Changes in nucleic acid and protein synthesis during sporulation and spherule formation in *Physarum polycephalum*. *Wilhelm Roux Archiv für Entwicklungmechanik der Organismen*, **165**, 110–24.

SAUER, H. W., GOODMAN, E. M., BABCOCK, K. L. & RUSCH, H. P. (1969). Polyphosphate in the life cycle of *Physarum polycephalum* and its relation to RNA synthesis. *Biochimica et Biophysica Acta*, **195**, 401–9.

SOLAO, P. B. & SHALL, S. (1971). Control of DNA metabolism in *P. polycephalum*. I. Specific activity of NAD pyrophosphorylase in isolated nuclei. *Experimental Cell Research*, **69**, 295–300.

SY, J. & McCARTY, K. S. (1970). Characterisation of 5.8S RNA from a complex with 26S ribosomal RNA from *Arbacia punctulata*. *Biochimica et Biophysica Acta*, **199**, 86–94.

WHEALS, A. (1970). A homothallic strain of the myxomycete *Physarum polycephalum*. *Genetics*, **66**, 623–33.

WHEALS, A. (1971). Mutants affecting plasmodial formation in a homothallic strain of *Physarum polycephalum*. Ph.D. Thesis, University of Leicester.

YOUNGER, L. R. & GELBOIN, H. V. (1970). The electrophoretic distribution of RNA synthesised *in vitro* by isolated rat liver nuclei. *Biochimica et Biophysica Acta*, **204**, 168–74.

ZELLWEGER, A. & BRAUN, R. (1971a). RNA of *Physarum*. I. Preparation and properties. *Experimental Cell Research*, **65**, 413–23.

ZELLWEGER, A. & BRAUN, R. (1971b). RNA of *Physarum*. II. Template replication and transcription in the mitotic cycle. *Experimental Cell Research*, **65**, 424–32.

ZELLWEGER, A., RYSER, U. & BRAUN, R. (1972). Ribosomal genes of *Physarum*. The isolation and replication in the mitotic cycle. *Journal of Molecular Biology*, **64**, 681–93.

THE DURATION OF MEIOSIS

By M. D. BENNETT

Plant Breeding Institute, Maris Lane, Trumpington,
Cambridge CB2 2LQ

INTRODUCTION

The duration of meiosis has been accurately determined in only a very few plant species. Most of the experiments have measured male meiosis; however, estimates of the duration of female meiosis in three higher plant species are given in this review.

A review of the methods and techniques used in meiotic timing experiments has already been made (Bennett, 1971). The methods fall into two general types, first, autoradiographic methods, and second, non-autoradiographic methods involving the sampling at known intervals of synchronously developing meiocytes. Autoradiographic techniques involve labelling meiocytes during premeiotic DNA synthesis (S) phase and then finding the minimum time before labelled cells reach the various stages of meiosis. Label has been supplied as [2-^{14}C]thymidine (Heslop-Harrison & Mac-Kenzie, 1967), and ^{32}P in phosphoric acid (Taylor & McMaster, 1954). However, [^{3}H]thymidine has been most frequently used (Marithamu & Threlkeld, 1966; Bennett, Chapman & Riley, 1971).

In plants it has often proved difficult to label meiotic chromosomes at a controlled time because of the obvious difficulties in uptake and translocation of isotopes associated with cellulose cell walls and the absence of a rapid transport system. Some novel techniques designed to overcome these difficulties have been used. For instance, Lima-de-Faria (1965) used the bract surrounding the florets in *Agapanthus umbellatus* as a cup to hold tritiated thymidine solution in direct contact with the anthers. Nevertheless, because of the difficulties associated with controlled uptake and incorporation of labelled solution in plants, non-autoradiographic sampling methods have also been frequently used (Ernst, 1938; Steinitz, 1944; Bennett *et al.* 1971).

Non-autoradiographic sampling methods can be conveniently subdivided into those in which meiosis was timed in anthers still attached to the plant (Bennett *et al.* 1971), and those where it was timed in cultured material. In the latter, whole florets or anthers have usually been cultured, for instance, in *Allium cepa* (Vasil, 1959) and *Trillium erectum* (Hotta & Stern, 1963). Ito & Stern (1967) successfully cultured columns of extruded meiocytes of *Lilium longiflorum* and of *Trillium erectum* from early leptotene

[111]

Table 1. *The duration of meiosis, nuclear DNA content, ploidy level and chromosome number in 31 species of higher plants*

Taxon	Duration of meiosis (h)	Nuclear DNA content (pg)	Reference to meiotic timing	Chromosome number
Diploids				
1. *Beta vulgaris*	24	4.1	This paper	18
2. *Antirrhinum majus*	24	5.5	Ernst, 1938	16
3. *Haplopappus gracilis*	24–36	5.5	Marithamu & Threlkeld, 1966	4
4. *Lycopersicum esculentum*	24–30	8.5	This paper	24
5. *Ornithogalum virens*	72	19.3	Church & Wimber, 1969	6
6. *Hordeum vulgare* var. Sultan	39	20.3	Bennett & Finch, 1971	14
H. vulgare var. Ymer	39	20.3	Finch & Bennett, 1972	14
7. *Triticum monococcum*	42	21.0	Bennett & Smith, 1972	14
8. *Rhoeo discolor*	48	23.8	Vasil, 1959	12
9. *Secale cereale*	51	28.4	Bennett *et al.* 1971	14
10. *Vicia faba*	72	44.0	Maquardt, 1951	12
11. *Allium cepa*	96	50.3	Vasil, 1959	16
12. *Tradescantia paludosa*	126	54.0	Stenitz, 1944; Taylor, 1949, 1950; Beatty & Beatty, 1953	12
13. *Tulbaghia violacea*	130	58.5	Taylor, 1953	12
14. *Endymion nonscriptus*	48	69.9	Wilson, 1959	16
15. *Convallaria majalis*	72	81.3	This paper	38
16. *Lilium henryi*	170	100.0	Pereira & Linskins, 1963	24
17. *Lilium longiflorum*	192	106.0	Taylor & McMaster, 1954; Ito & Stern, 1967	24
18. *Lilium candidum*	168	–	Sauerland, 1956	24
19. *Trillium erectum*	274	120.0	Ito & Stern, 1967	10
20. *Fritillaria meleagris*	400	233.0	Barber, 1942	24

Polyploids

21. *Capsella bursa-pastoris*	(4×)	This paper	18	2.6	32
22. *Veronica chamaedrys*	(4×)	This paper	20	2.8	28
23. *Alliaria petiolata*	(4×)	This paper	24	7.1	36
24. *Triticum dicoccum*	(4×)	Bennett & Smith, 1972	30	38.5	28
25. *Hordeum vulgare* var. Ymer	(4×)	Finch & Bennett, 1972	31	40.6	28
26. *T. aestivum* × *S. cereale* (polyhaploid)	(4×)	This paper	35	41.8	28
27. *Secale cereale*	(4×)	Bennett *et al.* 1971	38	56.8	28
28. *Tradescantia reflexa*	(4×)	Sax & Edmonds, 1933	144	144.9	24
29. *Triticum aestivum* var. Chinese Spring	(6×)	Bennett *et al.* 1971	24	54.3	42
T. aestivum var. Holdfast	(6×)	Bennett *et al.* 1972	24	54.3	42
30. *Triticale* var. Rosner	(6×)	Bennett & Smith, 1972	35	66.3	42
31. *Triticale* genotype A	(8×)	Bennett & Smith, 1972	21	82.7	56
genotype B	(8×)	Bennett & Smith, 1972	22	82.7	56

The duration of meiosis was measured at 18 °C in species 5 and at 20 °C in species 1, 4, 6, 7, 9, 13, 14, 15, 17, 21, 22, 23, 24, 25, 26, 27, 29, 30 and 31. No temperature was given for species 2, 3, 8, 10, 11, 12, 16, 18 and 28. Species 19 was measured at 15 °C and species 20 at 12–15 °C. However, the expected meiotic duration at 20 °C is given after adjustment of the recorded time assuming a Q_{10} for meiosis of about 2.3. The values for species 5, 10 and 20 are only approximate although of the right order.

to tetrad stage, an achievement which has not been repeated for meiocytes of other plant species.

MALE MEIOSIS

The duration of male meiosis in 31 species of higher plant are given in Table 1 together with the references from which the data are collated. Table 1 shows that plant species differ widely in the time they spend in meiosis even at constant temperature. Estimates for meiotic time at 20 °C range from about 18 h in *Capsella bursa-pastoris* to about 8 days in *Lilium longiflorum* (Ito & Stern, 1967). Since so few species have been examined for meiotic time, other species almost certainly exist with meiotic times both shorter than *Capsella* and longer than *Lilium*. Many species normally undergo meiosis at temperatures much lower than 20 °C. For instance, *Trillium erectum* often completes meiosis at temperatures just above freezing point and at 4 °C takes about 10 weeks to complete meiosis (Hotta & Stern, 1963). In *Larix decidua*, meiosis commences in the autumn and proceeds to diplotene at which stage development is halted for more than three months and does not recommence until the temperature reaches or exceeds 5 °C in the spring, whereupon meiosis proceeds to completion within about two days (Ekberg & Erickson, 1967).

Since plants differ so markedly in their meiotic times it is interesting to consider the factors which are known to affect this character.

Factors affecting the duration of meiosis

Nuclear DNA content

Table 1 gives the meiotic times and the nuclear DNA contents for 30 species. Meiotic times for 16 species were estimated by the author, the remainder were collated from the literature. The nuclear DNA contents of 22 species were measured by the author, using Feulgen photometry on squashes prepared by the method first described by McLeish & Sunderland (1961). Measurements were made using a Barr and Stroud type GN2 integrating microdensitometer. Nuclear DNA contents for the remaining species were collated from the literature. The DNA values for species given are either the 3 C amounts or the mean DNA content per cell, since in most species examined these do not differ greatly from each other (Bennett, 1971).

Fig. 1a illustrates the highly significant ($P < 0.001$) positive correlation between the duration of meiosis and nuclear DNA content in diploid species. All the points lie close to the regression line for nuclear DNA content on meiotic time with the exceptions of *Endymion nonscriptus* and

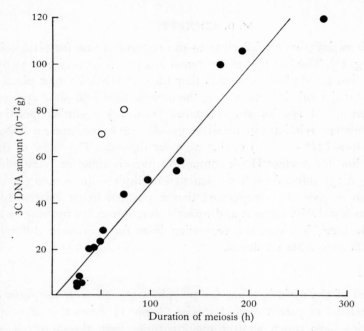

Fig. 1. (a) The relationship between the duration of meiosis and 3 C nuclear DNA amount in 15 diploid species ($r = 0.98$). ●, diploid species; ○, 'diploid' species which fall near the regression line for tetraploids (see (b)) not included in the diploid regression line. (These are *Endymion nonscriptus* and *Convallaria majalis*.)

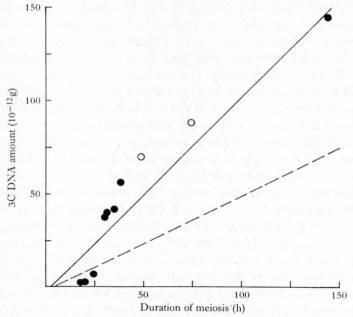

(b) The relationship between the duration of meiosis and 3 C nuclear DNA amount in 8 tetraploid species ($r = 0.96$). ——, regression line for tetraploid species; – – – regression line for diploid species; ●, tetraploid species included in the regression; ○ ,'diploid' species which fall near the regression line for tetraploids.

Convallaria majalis, which fall close to the regression line for tetraploid species (Fig. 1*b*). The fact that all the remaining points lie very near to the regression line for diploids indicates that nuclear DNA content plays a precise and major role in determining the meiotic time in diploid species.

When tetraploid species are compared (Fig. 1*b*) a similar positive linear correlation relationship between meiotic time and nuclear DNA content is found ($P < 0.001$) to that seen for diploids. The slope of the regression line for nuclear DNA content on meiotic time for tetraploids ($b = 1.07 \pm 0.13$) differs from that calculated for diploids ($b = 0.50 \pm 0.03$) being much steeper. It is suggested that a positive linear relationship between nuclear DNA content and meiotic duration exists for species at each ploidy level, but that the regression lines for species at different ploidy levels have different slopes.

Ploidy level

Comparison of the meiotic times for polyploid species with the meiotic times for related or parent diploid species (Table 1) shows that all polyploid species have much shorter meiotic times than do their related diploid species (Bennett & Smith, 1972; Finch & Bennett, 1972). Furthermore, in a series of related polyploids the departure from the meiotic duration for a diploid increased with increasing ploidy level. Thus, the tetraploid showed the least difference, the hexaploid an intermediate difference, and an octoploid the greatest difference from the related diploid species (Bennett & Smith, 1972). This effect of ploidy level (Fig. 2) has been found for both allo- and autopolyploids, and in polyhaploids (Dover & Bennett, unpublished). For instance, the duration of meiosis was 39 h in autotetraploid, and 51 h in diploid *Secale cereale,* while it was 31 h in the autotetraploid and 39 h in the diploid Ymer variety of *Hordeum vulgare* (Finch & Bennett, 1972). In the allopolyploid wheat series meiosis lasted 42 h in the diploid *Triticum monococcum,* 30 h in the tetraploid *T. dicoccum,* and only 24 h in two varieties of *T. aestivum* which is a hexaploid. In a polyploid containing 21 *T. aestivum* chromosomes and 7 *S. cereale* chromosomes the duration of meiosis, in the almost total absence of chromosome pairing, was about 35 h which is shorter than in either diploid wheat or rye species and intermediate between the estimates for tetraploid *T. dicoccum* (30 h) and tetraploid *S. cereale* (38 h). The effect of polyploidy giving faster meiosis is also seen in interspecific hybrids in which the chromosome number has been doubled using colchicine after hybridisation. Thus, in *Triticale* hybrids between wheat and rye species the meiotic time in two octoploids was much faster than in their related parent species (Table 1).

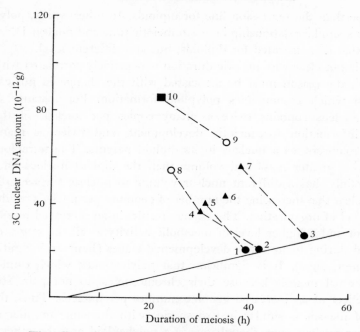

Fig. 2. The effect of polyploidy on meiotic duration in cereal species. 1, *Hordeum vulgare*; 2, *Triticum monococcum*; 3, *Secale cereale*; 4, *Triticum dicoccum*; 5, *Hordeum vulgare*; 6, *T. aestivum* × *S. cereale*; 7, *Secale cereale*; 8, *Triticum aestivum*; 9, *Triticale*; 10, *Triticale*. Ploidy level is indicated as follows: diploid, ●; tetraploid, ▲; hexaploid, ○; octoploid, ■. N.B.: the regression line for diploid species taken from Fig. 1 (*a*) is included for comparison.

In each polyploid species the duration of meiosis is much shorter than the expected time for a diploid species with a corresponding nuclear DNA content taken from the regression line for meiotic time on nuclear DNA content for diploids. For instance, both diploid *Allium cepa* and hexaploid *Triticum aestivum* have nuclear DNA contents of about 53 picograms (pg). The duration of meiosis is about 96 h in *A. cepa* but only 24 h in *T. aestivum*. Thus, with a constant nuclear DNA content, the duration of meiosis was about four times as long in the diploid as in the hexaploid.

Making a polyploid inevitably involves an increase in nuclear DNA content above that of the diploid parent species. Such an increase in nuclear DNA content is accompanied by a decrease in the duration of meiosis. At first sight this result appears to contradict the finding for diploid species that meiotic duration is positively correlated with nuclear DNA content. When tetraploid species alone are compared, however, a positive relationship between nuclear DNA content and meiotic time is found similar to that for diploids but the regression line for tetraploids has a steeper slope than the regression line for diploids. In other words, polyploids show a similar relationship between meiotic time and nuclear DNA content to that demonstrated for diploids, but at a different level.

While it is not clear why meiotic duration is negatively correlated with ploidy level, the reason must be associated with the change in nuclear organisation which accompanies polyploid formation. For instance, a tetraploid nucleus contains twice as many copies per nucleus of the genotypic information for meiotic development, and twice as many nucleolar organisers, as a nucleus in its diploid parents. The tetraploid nucleus has a greater mass and volume than the diploid nucleus and, almost certainly, has a different nuclear volume to nuclear surface area ratio. It is clear that increasing the number of genomes per nucleus results in a new level of organisation. The change results in an increased rate of development and a higher level of metabolic activity at all the stages of meiosis and during many other developmental stages (Bennett & Smith, 1972; Bennett, 1972). It is significant that polyhaploids, which cannot undergo normal meiosis because their chromosomes do not pair, still show the effect of an increased genome number per nucleus. Thus, the duration of meiosis is similar in two species with the same number of genomes per nucleus even though one is a polyhaploid and the other a polyploid with normal meiosis. Both types of polyploid have much shorter meiotic times than their related diploids.

Genotypic factors

The effect of polyploidy on meiotic time must in part be due to genotypic differences between diploids and related polyploids. While autopolyploids need not possess any qualitative genotypic differences from their parent diploids, they do have quantitative differences in the number of copies per nucleus of each gene. Allopolyploids have both quantitative and qualitative genotypic differences from their parent diploids. If the evidence available so far is typical, then it seems that qualitative genotypic differences are unimportant in determining the duration of meiosis compared with quantitative ones.

The duration of meiosis was compared in two varieties of *T. aestivum*, and in two varieties of *H. vulgare*, which differ in many genotypic characters (Bennett *et al.* 1972; Finch & Bennett, 1972). The duration of meiosis was the same in *T. aestivum* var. Chinese Spring and var. Holdfast at 15 °C and at 20 °C, yet these two genotypes differ markedly in their time to flowering. Similarly, both Sultan and Ymer, varieties of *H. vulgare*, had a meiotic time of about 39.2 h yet they differ markedly in floral morphology, one being two-rowed and the other six-rowed. If these results are typical,

Table 2. *The duration of meiosis in 6* Triticum aestivum *var. Chinese Spring genotypes grown at 20 °C*

Genotype	Meiotic duration (h)	2n
1. Euploid	24	42
2. Nullisomic 5B, tetrasomic 5D	30–32	42
3. Nullisomic 5D, tetrasomic 5B	24	42
4. Nullisomic 5B, 5C^u substitution	30–32	42
5. 5C^u disome addition	24–25	44
6. Nullisomic 7B	24	40

as seems likely, then meiotic duration is fairly constant within each species and does not vary from genotype to genotype.

The duration of meiosis has been measured for *Triticum aestivum* variety Chinese Spring in a euploid line, and in nullisomic, tetrasomic, addition and substitution lines for individual chromosomes. It is realised that variation involving the presence or absence of whole chromosomes is not what is normally understood by genotypic variation, however. Such differences must produce large genotypic changes involving many genes at once. Table 2 shows that individual chromosomes differ in their effect on meiotic duration (Bennett & Smith, in preparation). Whereas the duration of meiosis was 24 h in a euploid line, it was increased to about 30 h both in a line nullisomic for chromosome 5B but tetrasomic for chromosome 5D, and in a line with alien chromosomes from *Aegilops umbellulata* (C^u) substituted disomally for chromosome 5B. The absence of chromosome 5B apparently results in a decreased rate of meiotic development. Plants deficient for chromosome 5D but tetrasomic for 5B did not differ in their meiotic times from euploid plants, and neither did plants nullisomic for chromosome 7B. The absence of chromosome 7B does not apparently effect the duration of meiosis so presumably these chromosomes do not carry genes which determine this character. These results indicate that individual wheat chromosomes differ in their effects on meiotic behaviour. It is perhaps significant that 5B deficient plants have meiotic times equal to tetraploid wheat. It appears that the absence of one pair of chromosomes can have an equal effect to the absence of a whole wheat genome.

Temperature

Meiosis is very temperature sensitive, as is mitosis and the somatic cell cycle (Brown, 1951; Burhold & Van't Hof, 1971). Sax (1938) stated that meiosis in *Tradescantia* may be twice as long in winter as during summer.

Experiments to measure the duration of meiosis at different temperatures under carefully controlled conditions have recently been made. Wilson (1959) estimated the duration of meiosis in *Endymion nonscriptus* at each 5° interval over the range 0–30 °C. Over this range the duration of meiosis decreased with increasing temperature. The results for *Tradescantia paludosa* (Taylor 1949) for *S. cereale* and *T. aestivum* (Bennett, Smith & Kemble, 1972) and for *Trillium erectum* (Hotta & Stern, 1963; Kemp, 1964; Ito & Stern, 1967) reveal a similar relationship in each of these species between meiotic duration and temperature to that reported for *Endymion* by Wilson (*loc. cit.*). In *E. nonscriptus* meiosis was eighteen times as long at 0 °C as at 20 °C. In *T. aestivum* meiosis took 43, 24 and 18 h at 15, 20 and 25 °C respectively, while in *S. cereale* the corresponding times were 87.5, 51 and 39 h. It seems likely that in most plant species the duration of meiosis is negatively correlated with temperature over the range of temperatures normally encountered at the time when meiosis usually occurs.

Comparisons of data for meiotic duration at different temperatures in the four species for which data are available shows that the Q_{10} for meiosis varies between species. In *Trillium erectum* meiosis takes about 90 days at 1 °C (Hotta & Stern, 1963) and about 16 days at 15 °C (Ito & Stern, 1967) while in *E. nonscriptus* meiosis took 36 days at 0 °C and only 3.5 at 15 °C. Thus, in *T. erectum* meiosis is about six times as long at 0 °C as at 15 °C, while in *E. nonscriptus* it is about ten times as long. For the interval 15–25 °C the Q_{10} for meiosis was 2.8 in *E. nonscriptus*, 2.4 in *T. aestivum* and 2.25 in *Secale cereale* (Wilson, 1959; Bennett *et al.* 1972).

Over the temperature interval 15–25 °C the Q_{10} for meiosis (2.4) is much higher than for the cell cycle time in root-tip cells (1.7) (Bayliss, unpublished) in *Triticum aestivum*. The Q_{10} for meiosis in *T. aestivum* is between the Q_{10} values for somatic cell cycle time over the range 15–25 °C of 2.0 and 3.0 in *Pisum sativum* (Brown, 1951) and *Helianthus* (Burholt & Van't Hof, 1971) respectively.

The relative duration of the stages of meiosis

Individual meiotic stages occupy constant relative proportions of the total meiotic time in diploid species (Bennett, 1971). Thus, when the total meiotic duration is doubled the durations of all of its stages are also doubled (Fig. 3). This phenomenon is also demonstrated in polyploids (Bennett & Smith, 1972). Making a polyploid decreases the absolute duration of every stage but does not alter their relative durations. The phenomenon is also seen when the absolute duration of meiosis changes within a single species in response to temperature (Bennett *et al.* 1972).

Table 3. *The duration of meiosis and of individual stages (in hours) in several higher plant species*

Taxon		Total meiotic time	First meiotic prophase	Leptotene	Zygotene	Pachytene	Metaphase I–Telophase II inclusive
Diploids							
1. Antirrhinum majus		24	18.0	6–8.0	3.0	2.0	6.0
2. Lycopersicum esculentum		24–30	18.0	—	—	—	6.0
3. Hordeum vulgare		39	32.6	12.0	9.0	8.8	6.8
4. Triticum monococcum		42	34.0	—	—	—	8.0
5. Secale cereale		51	41.0	20.0	11.4	8.0	10.2
6. Allium cepa		96	72.0	—	—	—	24.0
7. Tradescantia paludosa		126	113.0	48.0	24.0	24.0	13.0
8. Tulbaghia violacea		130	102.0	—	—	—	28.0
9. Convallaria majalis		72	60.0	—	—	—	12.0
10. Lilium longiflorum		192	134.0	40.0	36.0	40.0	48.0
11. Trillium erectum		274	210.0	70.0	70.0	50.0	64.0
Polyploids							
12. Capsella bursa-pastoris	(4×)	18	13.0	—	—	—	5.0
13. Alliaria petiolata	(4×)	24	18.5	—	—	—	5.5
14. Triticum dicoccum	(4×)	30	22.5	—	—	—	7.5
15. Hordeum vulgare	(4×)	31	25.3	—	—	—	5.7
16. Secale cereale	(4×)	38	30.0	13.0	9.0	6.4	8.0
17. Triticum aestivum	(6×)	24	17.0	10.4	3.2	2.2	7.0
18. Triticale var. Rosner	(6×)	35	26.5	—	—	—	7.5
19. Triticale genotype A	(8×)	20.8	14.3	7.5	3.0	2.25	6.5
20. genotype B	(8×)	22	15.5	—	—	—	6.5

Values for species 5, 17 and 19 are taken from Bennett *et al.* (1971), for species 4, 14, 16, 18, 19 and 20 from Bennett & Smith (1972), for species 3 from Bennett & Finch (1971), for species 15 from Finch & Bennett (1972), for species 1 from Ernest (1938), for species 6 from Vasil (1959), for species 7 from Taylor (1950), for species 8 from Taylor (1953). Species 10 and 11 were estimated graphically from Ito & Stern (1967). The reference for species 1, 2, 9, 12 and 13 is this paper.

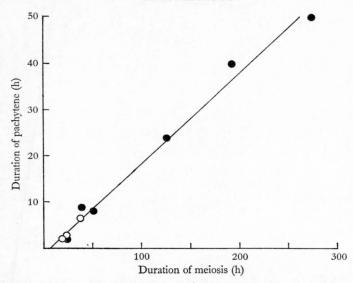

Fig. 3. The relationship between the duration of meiosis and the duration of pachytene in 9 species (r = 0.99). ●, diploid species; ○, tetraploid species.

Table 3 shows that first prophase of meiosis in higher plants occupies the major fraction of the total meiotic time; about between 66 and 75 per cent in most species. The inclusive duration of the meiotic stages after diakinesis together make up only about 25 to 33 per cent of the total meiotic time.

Data for the durations of individual stages are available for only a few higher plant species (Table 3), but they show that leptotene is invariably the longest single stage, closely followed by zygotene and pachytene. Zygotene is usually slightly longer than pachytene although sometimes the reverse is true. The stages after diplotene are all short but the shortest stages are invariably diakinesis, first and second anaphase, and first and second telophase.

The comparison of meiotic and somatic cell cycle times

The duration of the cell cycle which ends at first telophase of meiosis is always much longer than the minimum cell cycle time in root-tip meristem cells of the same species. In *Triticum aestivum* the minimum duration at 20 °C of the cell cycle which commences after premeiotic mitosis and ends after first telophase of meiosis is about 68 h compared with the cycle time in root-tip cells of about 12.5 h at the same temperature (M. W. Bayliss, unpublished). Thus, the meiotic cell cycle is about five times as long as the

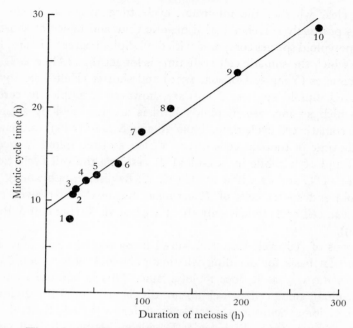

Fig. 4. The relationship between the duration of meiosis and cell cycle time in 10 diploid plant species (r = 0.99).

The references from which meiotic times are taken are given in Table 1. The references from which mitotic cycle times are taken are: sp. 1, Alfert & Das (1969); sp. 2, Van't Hof (1965); sp. 3, Sparvoli et al. (1966), Ames & Mitra (1966); sp. 4, this paper; sp. 5, Ayonoadu & Rees (1968); sp. 6, Webster & Davidson (1968); sp. 7, Van't Hof (1965); sp. 8, Van't Hof & Sparrow (1963); sp. 9, Stern & Hotta (1969); and sp. 10, Van't Hof & Sparrow (1963).

somatic cell cycle. This ratio is maintained for individual stages in the two cycles. In the root-tip cells the inclusive duration of metaphase to telophase was about 35 min (M. W. Bayliss) while the inclusive durations of first metaphase to first telophase of meiosis was about 160 min (Bennett et al. 1971), that is a ratio of about 1:4.7. Similarly, the duration of DNA synthesis phase (S) in root-tip cells was about 3 h while the duration of premeiotic S phase is between 12 and 15 h.

It has been shown several times now that the duration of the minimum cell cycle in root-tip cells is positively correlated with nuclear DNA content in comparisons of diploid species (Van't Hof & Sparrow, 1963; Evans & Rees, 1971). Since minimum mitotic cycle time and meiotic time both have a positive linear relationship with nuclear DNA content in diploid species they should show a similar relationship to each other. Fig. 4 shows that such is indeed the case.

It is not clear whether the minimum cycle time in somatic cells of polyploids is positively correlated with meiotic time and hence of shorter duration in polyploid species compared with their diploid parents. Examples are given in which the somatic cell cycle time is longer than (Alfert & Das, 1969), the same as (Yang & Dodson, 1970) and shorter (Kaltsikes, 1972) than in related diploid species. There are, however, examples in cereal species in which an increase in ploidy level is accompanied by a large reduction in somatic cell cycle time. Thus, Avanzi & Deri (1969) estimated the cell cycle time in root-tip cells of two varieties of tetraploid *Triticum durum* as 14 and 16 h, while in hexaploid *T. aestivum* the cell cycle time was 10.7 h at 24 °C and 12.5 h at 20 °C (M. W. Bayliss, unpublished). In the nonoploid endosperm cells of *T. aestivum* during the first seven cell cycles the mean cell cycle time is only about 4.5 h at 20 °C (Bennett & Rao, unpublished).

Comparisons of cell cycle times measured using root-tip cells may not provide the best basis for deciding whether polyploidy affects cell cycle time in the same way as it does meiotic time. This is because root-tip meristems contain some cells which do not divide at all, and the dividing cells do not divide synchronously and even differ in their individual cell cycle times (Clowes, 1965; Webster & Davidson, 1968). The cell cycle time obtained by autoradiography of root-tips is often only a crude approximation.

Measurements of cell cycle time in naturally occurring populations of somatic cells in which all the cells undergo mitosis synchronously and in which every cell divides would give a clearer picture of the effects of polyploidy on cell cycle time. Early endosperm development presents just such a situation, and experiments are under way to investigate the cell cycle times at 20 °C over the first six division cycles in related cereal species at different ploidy levels. Pollen development in cereals also presents a situation in which non-meiotic cells all undergo development almost synchronously through more than two cell cycles. In this case the duration of non-meiotic development exactly mirrors the duration of meiotic development (Bennett & Smith, 1972) and, therefore, the effect of polyploidy on non-meiotic development is identical with its effect on the rate and duration of meiotic development.

FEMALE MEIOSIS

As far as the author is aware, no estimates of the duration of female meiosis in plants have been published. It was suggested that the duration of female meiosis is longer than the duration of male meiosis in *Lilium* and *Fritillaria* (Fogwill, 1958) but this was not measured.

Experiments have recently been conducted (Bennett, unpublished) to estimate the duration of female meiosis in *Hordeum vulgare*, *Secale cereale*, and in *Triticum aestivum*. The method used depends upon knowledge of the duration of male meiosis and its stages, and upon the fact that in these species each floret contains a single ovary with a solitary ovule containing only a single egg mother cell (e.m.c.). Within each floret development is almost synchronous in pollen mother cells (p.m.c's) both within and between the three anthers throughout meiosis. Consequently, the stage of development in the e.m.c. and the p.m.c's from a single floret can be exactly determined from separate Feulgen stained squashes of the anthers and the ovule. This is not possible in most species which have several e.m.c's per floret showing asynchronous meiotic development.

Comparisons of the stages of male and female meiosis in individual florets quickly shows whether or not they are synchronous. If they are not synchronous then the developmental interval between the stages of female and male meiosis can be quantified and expressed as hours of corresponding normal male development, since the exact durations of the developmental stages before, during and after male meiosis have been carefully determined (Bennett *et al.* 1971; Bennett & Finch, 1972). The comparison can be made graphically by plotting the stages of male and female development on axes with individual stages given their absolute duration as measured for normal male meiosis. If meiosis is synchronous in male and female meiocytes from individual florets then all the points should fall on or very close to a line passing through the origin having a slope of one. If meiosis is not synchronous and is initiated earlier in one sex than the other but has the same duration in both, the calculated regression line would still have a slope of one, but would cut one axis at a point equivalent to the time by which female meiosis either precedes or follows male meiosis. If meiosis is asynchronous and proceeds at a different rate in the two sexes then the slope of the regression line for female meiosis on male meiosis will have a slope other than unity.

Plots of the meiotic stages in p.m.c's and e.m.c's from individual florets of *Triticum aestivum*, *Hordeum vulgare* and *Secale cereale* grown at 20 °C show that individual stages of meiosis occur in the e.m.c. within not more than 24 h of their occurrence in p.m.c's. In *T. aestivum* male and female meiosis occur almost synchronously within each individual floret at all stages of meiosis. Female meiosis must, therefore, be of similar duration to male meiosis in this species. The duration of female meiosis is very similar to the duration of male meiosis in both *H. vulgare* and *T. aestivum* and meiosis is initiated in the e.m.c. almost synchronously with its initiation in p.m.c's in the same floret. In *S. cereale*, however, each meiotic

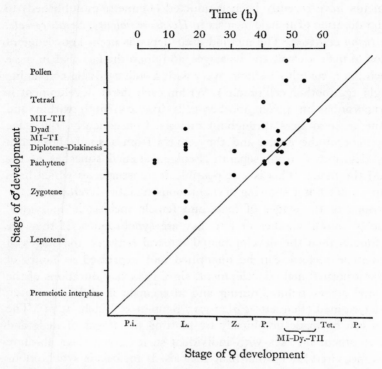

Fig. 5. The relationship between the stages of meiotic development in pollen mother cells (o meiosis) and egg mother cells (o meiosis) in individual florets of *Secale cereale*.

stage is initiated in the e.m.c. about 15 h before its initiation in p.m.c's in the same floret (Fig. 5).

It would be premature to suggest that female meiosis normally has a duration very similar to that of male meiosis in all higher plant species. Nevertheless, the results obtained for cereal species do suggest that this may often be the case. It is perhaps to be expected that male and female meiosis would have very similar durations in a single species, since both are subject to constant effects by many of the characters which determine meiotic time including nuclear DNA content, ploidy level and genotype.

DISCUSSION

Detailed studies of meiosis have been made in many organisms and, as a result, general conclusions have been made regarding the nature of the division, which apply to all higher organisms.

(1) 'There is uniformity of form and action in meiosis' (Darlington, 1940). By definition, meiosis is unique in always combining a reduction in nuclear DNA content to the 1 C amount with recombination of DNA from different homologous chromosomes. An essential feature, however, is that the means by which these ends are achieved do not vary from species to species, so that the same division stages occur in the same order in all species. Meiosis is a determinate process and, therefore, the comparison of meiosis in widely different species serves only to emphasise its essential invariability for most characters. A comparison of meiotic duration, however, shows that this is a most variable meiotic character and reveals important interspecific differences in meiotic behaviour.

(2) The stages of meiosis have constant relative durations (Bennett, 1971; Bennett & Smith, 1972) in higher plants. Changes in the duration of meiosis accompanying variation in nuclear DNA content, ploidy level and temperature do not alter the relative durations of the individual stages of meiosis. Whatever determines the duration of meiosis has, therefore, an equal effect in determining the duration of each individual stage.

(3) The rate of development during meiosis is slow. The approach of cells towards meiosis is marked by a gradual increase in the durations of successive cell cycles both in animal and plant species. This is seen in both male and female development and affects all stages of the somatic cell cycle. The rate of development remains very slow throughout meiosis and throughout gametogenesis (Bennett & Smith, 1972). Since slow development appears to be a constant feature of meiosis it prompts two questions. First, is slow development, both premeiotic and during meiosis, an essential prerequisite for meiosis to occur? Second, what is the direct cause of slow development during meiosis? Unfortunately it is not possible to answer these questions. In seeking an explanation for slow development during meiosis, however, it should be remembered that all stages of development have protracted durations during and before meiosis. Consequently, whatever theory is propounded to explain slow development at one stage of development, for instance premeiotic S phase, must also explain the slow rate of development at all the other stages of development before and during meiosis.

(4) Meiosis is subject to atypical genotypic control. 'Meiosis is a process which usually progresses to its termination irrespective of abnormalities of chromosome or cell behaviour that may arise' (Riley & Bennett, 1971). Faults in meiotic behaviour, for instance, failure to pair at zygotene or

failure of wall formation isolating the products of meiotic division, are not corrected before the process proceeds to later stages. This behaviour indicates that meiosis is subject to highly canalised control which lacks feedback mechanisms to ensure the normal completion of each stage before a later stage is initiated. It is important to note that in *Triticum aestivum* × *Secale cereale* polyhaploid plants with 28 chromosomes, in which meiotic chromosome pairing is almost completely absent so that meiosis is very abnormal, the time between the start of leptotene and the formation of tetrad walls did not differ from that expected for a tetraploid species with the hybrid nuclear DNA content (Dover & Bennett, unpublished). This result shows that the total duration of meiosis is not affected by failure of normal behaviour at some individual meiotic stages. It must be concluded, therefore, that the factors which determine the total duration of meiosis are independent of whether meiosis is normal or not.

There is an increasing body of evidence showing that events during meiosis are determined at stages during premeiotic interphase (Buss & Henderson, 1971; Bayliss & Riley, 1972). It seems possible that meiosis is controlled by long-lived messenger RNA molecules transcribed before meiosis is initiated during premeiotic interphase. If this were so it might explain why mistakes in meiosis are not corrected before the initiation of later stages, and why abnormal meiosis has the same duration as normal meiosis.

(5) Meiosis is subject to non-genotypic control. In higher plants with a single ploidy level the duration of meiosis is positively correlated with nuclear DNA content. In plant species which differ in their 4 C values by more than 1:100 (Bennett, 1972) the stages of meiosis and their order are essentially the same. There is, therefore, no need for any new genotypic information for the control of meiosis in species with different nuclear DNA contents. In higher plants at one ploidy level, the major factor determining the rate of development during meiosis, and hence the duration of meiosis, is the mass of nuclear DNA. It appears, therefore, that nuclear DNA can affect the rate of meiotic development indirectly and independently of its information content, apparently by the physical and mechanical consequences of its mass. If so then meiosis is subject to a control of its duration and rate which may operate independently of transcription of genotypic information.

Conditions of the nucleus affecting the phenotype independently of the informational content of its DNA have been defined as nucleotypic (Bennett, 1971, 1972). The nucleotype is determined by the physical properties of the nucleus and its chromosomes including their mass and volume. Many nuclear characters including chromosome and nuclear volume, and chromosome and nuclear dry mass have been shown to have a positive linear relationship with nuclear DNA content in wide inter-specific comparisons (Rees *et al.* 1966; Jones & Rees, 1968; Sparrow & Evans, 1961; Pegington & Rees, 1970; Paroda & Rees, 1971). Furthermore, the chromosomal histone content is directly proportional to nuclear content. While the nucleotype is determined by all of these characters, many of which probably affect meiotic time, it is convenient to consider the nucleotypic effects of DNA contents alone, since nuclear DNA content is probably the causal character in its correlation with many other nuclear characters (Bennett, 1972).

REFERENCES

ALFERT, M. & DAS, N. K. (1969). Evidence for control of the rate of nuclear DNA synthesis by the nuclear membrane in eukaryotic cells. *Proceedings of the National Academy of Sciences, U.S.A.* **63**, 123–8.

AMES, I. H. & MITRA, J. (1966). The mitotic cycle time of *Haplopappus gracilis* root tip cells as measured with tritiated thymidine. *Nucleus*, **9**, 61–6.

AVANZI, S. & DERI, P. L. (1969). Duration of the mitotic cycle in two cultivars of *Triticum durum*, as measured by ³H-thymidine. *Caryologia*, **22**, 187–94.

AYONOADU, U. W. & REES, H. (1968). The regulation of mitosis by B-chromosomes in rye. *Experimental Cell Research*, **52**, 284–90.

BARBER, H. N. (1942). The experimental control of chromosome pairing in *Fritillaria*. *Journal of Genetics*, **43**, 359–74.

BAYLISS, M. W. & RILEY, R. (1972). Evidence of pre-meiotic control of chromosome pairing in *Triticum aestivum*. *Genetical Research, Cambridge*, in press.

BEATTY, J. W. & BEATTY, A. V. (1953). Duration of the stages in microspore development and in the first microspore division of *Tradescantia paludosa*. *American Journal of Botany*, **40**, 593–6.

BENNETT, M. D. (1971). The duration of meiosis. *Proceedings of the Royal Society, London*, B, **178**, 259–75.

BENNETT, M. D. (1972). Nuclear DNA content and minimum generation time in herbaceous plants. *Proceedings of the Royal Society, London*, B, **181**, 109–35.

BENNETT, M. D., CHAPMAN, V. C. & RILEY, R. (1971). The duration of meiosis in pollen mother cells of wheat, rye and *Triticale*. *Proceedings of the Royal Society, London*, B, **178**, 259–75.

BENNETT, M. D. & FINCH, R. A. (1971). Duration of meiosis in barley. *Genetical Research, Cambridge*, **17**, 209–14.

BENNETT, M. D. & SMITH, J. B. (1972). The effects of polyploidy on meiotic duration and pollen development in cereal anthers. *Proceedings of the Royal Society, London*, B, **181**, 81–107.

BENNETT, M. D., SMITH, J. B. & KEMBLE, R. (1972). The effect of temperature on meiosis and pollen development in wheat and rye. *Canadian Journal of Genetics and Cytology*, **14**, in press.

BROWN, R. (1951). The effects of temperature on the duration of the different stages of cell division in the root-tip. *Journal of Experimental Botany*, **2**, 96–110.

BURHOLT, D. R. & VAN'T HOF, J. (1971). Quantitative thermal-induced changes in growth and cell population kinetics in *Helianthus* roots. *American Journal of Botany*, **58**, 386–93.

BUSS, M. E. & HENDERSON, S. A. (1971). Induced bivalent interlocking and the course of meiotic chromosome synapsis. *Nature, New Biology*, **234**, 243–6.

CHURCH, K. & WIMBER, D. E. (1969). Meiosis in *Ornithogalum virens* (Lileaceae): Meiotic timing and segregation of ³H-thymidine labelled chromosomes. *Canadian Journal of Genetics and Cytology*, **11**, 573–81.

CLOWES, F. A. L. (1965). The duration of G₁ phase of the mitotic cycle and its relation to radiosensitivity. *New Phytologist*, **64**, 355–9.

DARLINGTON, C. D. (1940). The prime variables of meiosis. *Biological Reviews*, **15**, 307–21.

EKBERG, I. & ERIKSSON, G. (1967). Development and fertility of pollen in three species of *Larix*. *Hereditas*, **57**, 303–11.

ERNST, H. (1938). Meiosis und crossing over. Zytologische und genetische Untersuchungen am *Antirrhinum majus*. L. *Zeitschrift für Botanik*, **33**, 241–94.

EVANS, G. M. & REES, H. (1971). Mitotic cycles in dicotyledons and monocotyledons. *Nature, London*, **233**, 350–1.

FINCH, R. A. & BENNETT, M. D. (1972). The duration of meiosis in diploid and autotetraploid barley. *Canadian Journal of Genetics and Cytology*, **14**, in press.

FOGWILL, M. (1958). Differences in crossing-over and chromosome size in the sex cells of *Lilium* and *Fritillaria*. *Chromosoma, Berlin*, **9**, 493–504.

HESLOP-HARRISON, J. & MACKENZIE, A. (1967). Autoradiography of soluble [2-¹⁴C] thymidine derivatives during meiosis and microsporogenesis in *Lilium* anthers. *Journal of Cell Science*, **2**, 387–400.

HOTTA, Y. & STERN, H. (1963). Inhibition of protein synthesis during meiotic development and its bearing on intra-cellular regulation. *Journal of Cell Biology*, **16**, 259–79.

ITO, M. & STERN, H. (1967). Studies of meiosis *in vitro*. I. *In vitro* culture of meiotic cells. *Developmental Biology*, **16**, 36–53.

JONES, R. N. & REES, H. (1968). Nuclear DNA variation in *Allium*. *Heredity, London*, **23**, 591–605.

KALTSIKES, P. J. (1971). The mitotic cycle in an amphiploid *Triticale* cultivar and its parental species. *Canadian Journal of Genetics and Cytology*, **13**, 656–62.

KEMP, C. L. (1964). The effects of inhibitors of RNA and protein synthesis on cytological development during meiosis. *Chromosoma, Berlin*, **15**, 652–65.

LIMA-DE-FARIA, A. (1965). Labelling of the cytoplasm and the meiotic chromosomes of *Agapanthus* with ³H-thymidine. *Hereditas*, **53**, 1–18.

McLEISH, J. & SUNDERLAND, N. (1961). Measurement of deoxyribonucleic acid (DNA) in higher plants by Feulgen photometry and chemical methods. *Experimental Cell Research*, **24**, 527–40.

MARITHAMU, K. M. & THRELKELD, S. F. H. (1966). The distribution of tritiated thymidine in tetrad nuclei of *Haplopappus gracilis*. *Canadian Journal of Genetics and Cytology*, **8**, 603–12.

MAQUARDT, H. (1951). Die Wirkung der Röntgenstrahlen auf die Chiasmafrequenz in der Meiosis von *Vicia faba*. *Chromosoma, Berlin*, **4**, 232–8.

PARODA, R. S. & REES, H. (1971). Nuclear DNA variation in *Eu-sorghums*. *Chromosoma, Berlin*, **32**, 353–63.

PEGINGTON, C. & REES, H. (1970). Chromosome weights and measures in the Triticinae. *Heredity, London*, **25**, 195–205.

PEREIRA, A. S. R. & LINSKINS, H. F. (1963). The influence of glutathione and glutathione antagonists on meiosis in excised anthers of *Lilium henryi*. *Acta botanica neerlandica*, **12**, 302–14.

Rees, H., Cameron, F. M., Hazarika, M. H. & Jones, G. H. (1966). Nuclear variation between diploid Angiosperms. *Nature, London*, **211**, 828–30.

Riley, R. & Bennett, M. D. (1971). Meiotic DNA synthesis. *Nature, London*, **230**, 182–4.

Sauerland, H. (1956). Quantitative Untersuchungen von Röntgeneffekten nach Bestrahlung verschiedener Meiosisstadien bei *Lilium candidum* L. *Chromosoma, Berlin*, **7**, 627–54.

Sax, K. (1938). Chromosome aberrations induced by X-rays. *Genetics*, **23**, 494–515.

Sax, K. & Edmonds, H. W. (1933). Development of the male gametophyte in *Tradescantia reflexa*. *Botanical Gazette*, **95**, 156–63.

Sparrow, A. H. & Evans, H. J. (1961). Nuclear factors affecting radiosensitivity. 1. The influence of nuclear size and structure, chromosome complement, and DNA content. In *Fundamental aspects of radiosensitivity, Brookhaven Symposia in Biology*, **14**, 76–100.

Sparvoli, E., Gay, H. & Kaufman, B. P. (1966). Duration of the mitotic cycle in *Haplopappus gracilis*. *Caryologia*, **19**, 65–71.

Steinitz, L. (1944). The effect of lack of oxygen on meiosis in *Tradescantia*. *American Journal of Botany*, **31**, 428–43.

Stern, H. & Hotta, Y. (1969). Biochemistry of meiosis. In *Handbook of Molecular Cytology*, Ed. A. Lima-de-Faria, pp. 520–39. Amsterdam: North-Holland Publishing Company.

Taylor, H. (1949). Increase in bivalent interlocking and its bearing on the chiasma hypothesis of metaphase pairing. *Journal of Heredity*, **40**, 65–9.

Taylor, H. (1950). The duration of differentiation in excised anthers. *American Journal of Botany*, **37**, 137–43.

Taylor, H. (1953). Incorporation of phosphorus 32 into nucleic acids and proteins during microgametogenesis of *Tulbaghia*. *Experimental Cell Research*, **4**, 169–79.

Taylor, H. & McMaster, R. D. (1954). Autoradiographic and microphotometric studies of DNA during microgametogenesis in *Lilium longiflorum*. *Chromosoma, Berlin*, **6**, 489–521.

Van't Hof, J. (1965). Relationships between mitotic cycle time duration, S period duration and the average rate of DNA synthesis in the root tip meristem cells of several plants. *Experimental Cell Research*, **39**, 48–58.

Van't Hof, J. & Sparrow, A. H. (1963). A relationship between DNA content, nuclear volume, and minimum mitotic cycle time. *Proceedings of the National Academy of Sciences, U.S.A.* **49**, 897–902.

Vasil, I. K. (1959). Cultivation of excised anthers *in vitro* – effect of nucleic acids. *Journal of Experimental Botany*, **10**, 339–408.

Webster, P. L. & Davidson, D. (1968). Evidence from thymidine-H^3 labelled meristems of *Vicia faba* of two cell populations. *Journal of Cell Biology*, **39**, 332–8.

Wilson, J. Y. (1959). Duration of meiosis in relation to temperature. *Heredity, London*, **13**, 262–7.

Yang, D. P. & Dobson, E. O. (1970). The amounts of nuclear DNA and the duration of DNA synthetic period (S) in related diploid and autotetraploid species of oats. *Chromosoma, Berlin*, **31**, 309–20.

MITOTIC CYCLES IN ROOT MERISTEMS

By P. W. BARLOW

Agricultural Research Council Unit of Developmental Botany,
181A Huntingdon Road, Cambridge CB3 0DY

INTRODUCTION

The meristem of the primary root lies at the tip of the root and is the source of cells for both the main body of the root and for the root cap. I shall describe some of the techniques for studying the cell cycle in root meristems and the results so obtained. I shall also discuss possible ways in which the proliferative capacity of cells may be controlled in normal roots, and the knowledge we have gained about the properties of cells in the meristem from experimental interference with the cell cycle. The terms G_1, S, G_2 and M will be used throughout to identify phases of the cell cycle (Howard & Pelc, 1953), and need little further comment. However, I shall use the term G_0 to define the phase occupied by a cell that is not participating in the cell cycle, though such a cell may do so after any length of time in this phase.

THE ONSET OF MITOTIC ACTIVITY IN THE ROOT MERISTEM

The cell in the seed

The cells in the meristem of the ungerminated embryo within the seed may be thought of as being in the G_0 phase. It is only when the seed takes up water that germination can start and the cells resume mitotic activity. In some species, such as *Lactuca sativa*, *Helianthus annuus* and *Pinus pinea*, the nuclei in the dry embryo all contain the 2C DNA content (Brunori & D'Amato, 1967; Brunori, Georgieva & D'Amato, 1970), but in many other species there is a small proportion of nuclei with a 4C DNA content (Avanzi, Brunori & Giorgi, 1966; Brunori & Ancora, 1968; Brunori, Avanzi & D'Amato, 1966) and the proportion of 2C and 4C nuclei differs from tissue to tissue (Avanzi, Brunori & D'Amato, 1970). Table 1 summarises the nuclear conditions of cells in the radicles of ungerminated seeds of a number of species.

The two classes of nuclei, with 2C and 4C DNA contents, are a consequence of the cessation of mitotic activity in the developing embryo. As the embryo in the fruit or grain matures, so it loses water; DNA synthesis is the first phase of the mitotic cycle to cease during maturation and at lower moisture content mitosis stops (Brunori, 1967). The nuclei with a 2C DNA content are in cells that ceased cycling at the G_1 phase, those with a 4C content are cells that stopped in the G_2 phase or had started to become endopolyploid. The occasional occurrence of 8C nuclei in dry embryos of *Vicia faba* indicates that endopolyploidy arises in embryogeny. Sometimes a cell can be arrested before it has completed a round of DNA synthesis as nuclei with DNA contents between 2C and 4C have been found in *Triticum durum* (Avanzi *et al.* 1963) and *Pisum sativum* (Bogdanov, Liapunova & Sherudilo, 1967). It seems plausible that the rate at which the water content decreases in the developing seed governs the frequency of occurrence of nuclei with a 4C and intermediate (2C–4C) DNA contents.

Table 1. *Nuclear conditions in root tips of ungerminated seeds*

| | Per cent nuclei (±S.E.) with the following DNA contents: | | | No. of | |
Species	2C	4C	8C	seeds	Author
Allium cepa var. Bianci di Maggio	98.1 ± 0.5	1.9 ± 0.5	0	10	Brunori & Ancora (1968)
A. cepa var. Rossa Fiorentina	98.1 ± 0.5	1.9 ± 0.5	0	10	
Lactuca sativa	100	0	0	10	Brunori &
Pinus pinea	100	0	0	10	D'Amato (1967)
Triticum vulgare var. Cappelli	90	10	0	1	Avanzi *et al.* (1966)
T. vulgare var. Azizah	82.5	17.5	0	1	
T. vulgare					
Primary Root	75.2 ± 4.2	24.8 ± 4.2	0	5	Avanzi *et al.*
1st pair seminal roots	92.4 ± 0.9	7.6 ± 0.9	0	5	(1970)
2nd pair seminal roots	98.4 ± 1.6	1.6 ± 1.6	0	5	
Vicia faba major	81.0 ± 1.5	17.4 ± 1.5	1.6*	10	Brunori *et al.*
V. faba minor	70.3 ± 4.9	29.4 ± 5.0	0.3†	10	(1966)

* In 8 out of 10 radicles.

† In 1 out of 10 radicles.

When the seed is put to germinate the cells do not immediately enter a new phase of mitotic activity, indeed, DNA synthesis and mitosis are among the last activities to be resumed as growth recommences. The evidence suggests that the machinery for meristem activity has first to be reassembled or reactivated. In the dry seed of wheat (*Triticum vulgare*) there are RNA molecules and ribosomes that can be used for protein synthesis within 30 min from the start of water uptake (Marcus, Feeley & Volcani, 1966; Chen, Sarid & Katchalski, 1968; Weeks & Marcus, 1971; Schultz, Chen & Katchalski, 1972) and only after a further 12 h is new mRNA made. DNA synthesis is initiated after 8 h of germination and depends on proteins made in the germinating embryo prior to this time, one of which may be DNA polymerase (Mory, Chen & Sarid, 1972). Similarly, in *Vicia faba* it is likely that DNA synthesis is initiated by proteins specified for, and translated from, preformed RNA as no new RNA is made until the time that new DNA itself is made (Jakob & Bovey, 1969). These preformed RNAs and ribosomes are probably stable in the dry seed, occasionally remarkably so as the oldest known viable seed (of a *Canna* sp.) was 550 years before it germinated (Nakayama & Sívori, 1968).

Table 2. *Cytological changes in root tips during germination of* Zea mays

Zea mays heterozygous at Yg_2, Shz, B and Wx loci; 22 °C (from van de Walle & Bernier, 1969, and van de Walle, 1971)

Time (h)	RNA Onset of [^{3}H]uridine labelling in:	DNA Onset of [^{3}H]thymidine labelling in:
4–8	Chromatin	—
6–9	—	Cytoplasm
8–10	Cytoplasm and nucleolus	—
33–36	—	Some nuclei

Zea mays INRA 258; 16 °C (from Deltour, 1970, and Deltour & Bronchart, 1971)

Time (h)	RNA [^{3}H]uridine labelling in:	Nucleolus structure:
0	—	Compact and homogeneous
3	—	Granules appear
4	Chromatin	—
8	Nucleolus	Vacuoles appear with Granules inside
24	—	Vacuolar granules disappear
36	Cytoplasm	—

Certain physiological and cytological changes visible with the light and electron microscopes also take place in embryos to give them the appearance of mitotically active cells. The most obvious are in the uptake and location of radioactive RNA precursors in autoradiographs of germinating cells. Some of the changes observed by van de Walle & Bernier (1969) and Deltour (1970) for *Zea mays* are summarised in Table 2. Their results suggest that the first RNA is made from a template in the chromatin and the new molecules move to both cytoplasm and nucleolus; later, the nucleolus itself becomes capable of synthesising RNA and its activity may be a result of the structural changes that take place within it during imbibition (see Table 2). Similar results have been obtained by Delseny (1971) with *Vicia sativa*. The RNA made in the chromatin has a sedimentation coefficient characteristic of heterodisperse messenger RNA (van de Walle, 1971). The labelling of the cytoplasm by [³H]thymidine, seen by van de Walle & Bernier (1967, 1969) is so far unexplained though some of the label later appears in the nucleus when it becomes capable of DNA synthesis.

The first cell cycles during germination

The nuclei held with the 2C and the 4C DNA content in the dry embryo contribute to the first mitoses in the germinating seed. Those nuclei with the 4C DNA content proceed into mitosis before the 2C nuclei (Jakob & Bovey, 1969). Thus the 4C nuclei enter the G_2 phase and the 2C nuclei the G_1 phase of the reactivated cycle. They must have retained a 'memory', during their dry state, of which phase of the cell cycle they were in before becoming dormant as the 4C nuclei do not synthesise any new DNA prior to mitosis, while the 2C nuclei do.

The duration of the first mitotic cycle of germinating roots does not seem too different from the subsequent cycles – at least in *Allium cepa* (Bryant, 1969), although Bryant's data suggest that the duration of G_1 may be more variable in the first cycle than in later ones. This is to be expected if cells in different regions of the meristem enter the G_1 phase from G_0 at different rates. Unfortunately, there are no data on the sequence of reactivation of nuclei within the root meristem: for instance one may question, in view of the faster cycle and shorter G_1 of root cap initials when compared with other regions of the actively growing root of *Zea mays*, whether nuclei in this region of this species come into cycle quicker than elsewhere in the germinating seed. In the shoot meristem of germinating seeds of *Xanthium pennsylvanicum* the lateral portion of the meristem shows a greater proportion of cells entering S from G_0/G_1 during the first 60 h of imbibition than in the axial zone of the meristem (Rembur, 1970). Thus the two regions of the shoot come to possess the properties that they show in the fully developed vegetative apex (and probably possessed in the embryo before dormancy) during the first cell cycle of the germination period.

The early divisions in the root (and shoot) meristem during germination are partially synchronous. Synchrony of mitoses in roots of *Allium cepa*, developing either from dormant bulbs or seeds, have been investigated in this respect. Results of Bishop & Klein (1971), using seeds of *A. cepa* var. Southport Yellow Globe at 27 °C, suggest that the synchrony is a consequence of the synchronous entry of cells into mitosis after the start of imbibition, rather than a response to the time of day, or light–dark cycles. They find the rhythmic pattern of mitotic activity dies out after about the fourth wave of mitoses indicating that variability of cycle time within the meristem soon makes the population asynchronous. The average time between maxima and minima of mitotic index is about 6 h, but whether this time is equal to the duration of the mitotic cycle is not known; other work shows a cycle duration in *A. cepa* seedlings of similar developmental stage and the same temperature to be about 13 h (González-Fernández, Giménez-Martín & de la Torre, 1971). The waves of mitotic activity seen by Bishop & Klein may therefore represent the activity of more than one cell population: as, for example, in the 1–2 cm long primary root of *Sinapis alba* Clowes (1962) has shown that different regions of the meristem have maxima and minima of mitotic index at different times. In the germinating root of *Phaseolus vulgaris*, Gavaudan & Chazelas (1969) could identify rows of cells that had arisen, by a sequence of transverse divisions, from a common mother cell originally in the pre-germination embryo. The number of cells making up the lineage were multiples of two indicating synchronous divisions of the products of the original cell: indeed within the lineage synchronised mitotic figures sometimes could be seen. Gavaudan & Gastelier (1970) found a similar situation in the shoot apex, though here the planes of division give groups of new cells bounded by the wall of the common mother cell to which they give the term 'polycytes'. In the tunica of the apex one division occurred every 24 h, so that on the fifth day of germination a polycyte contained on average 16 cells.

THE CELL CYCLE IN THE ROOT MERISTEM OF *ZEA MAYS*

A pulse labelling experiment with young roots of Zea

The apex of maize (*Zea mays*) roots provides a good system for studying cell cycles. There is only one root per seed and the roots are relatively large and have a well defined boundary between cells of the root cap and the main body of the root; further, they are easily handled for experimental purposes, such as complete and clean removal of the root cap.

There are many methods for estimating the duration of the phases of the cell cycle that make use of radioactive precursors of DNA and the preparation of autoradiographs of the labelled tissue. These methods include continuous labelling, double labelling, silver grain number diminution and pulse labelling (all described by Wimber, 1963). Pulse labelling, first devised by Quastler & Sherman (1959), gives complete information about the duration and variability of each phase of the cycle which the other methods cannot give. However, a disadvantage of the method is that cells need to be sampled over a relatively long period of time, during which the cell cycle may be altering. Nevertheless, pulse labelling has been widely used even though the data gained, a curve showing the fraction of labelled mitoses at various times after the pulse, have often been submitted to a variety of *ad hoc* interpretations (see Mendelsohn & Takahashi (1971) for a critique). Recently renewed interest has been shown in this method of cycle analysis and many papers have been published dealing with the interpretation of the fraction labelled mitoses curve (Takahashi, 1968; Trucco & Brockwell, 1968; Bronk, 1969; Macdonald, 1970; Gilbert, 1972). I shall present data concerning the cell cycle in various regions of the meristem, obtained by pulse labelling, that have been analysed in collaboration with Dr Peter Macdonald.

The experimental details are to be published elsewhere (Barlow & Macdonald, in preparation) but, in brief, roots 1–2 cm long were pulse labelled for 30 min with [^{3}H]thymidine, washed and returned to water. Roots were then fixed at 2-hourly intervals and the fractions of labelled mitoses in different regions of the meristem at times after the pulse were scored from autoradiographs of median longitudinal sections. In Table 3 are the parameters of the cell cycle in a number of locations covering the whole region of mitotic activity in the apex. We have assumed a gamma distribution of phase duration, rather than a normal or any other type of distribution, but cycle parameters are not much different when a normal distribution is used (Macdonald, 1970; Barlow & Macdonald, in preparation) and I have already published (Barlow, 1971) some estimates using this distribution of phase durations.

A number of points from the data require comment. The initials of the cap columella have a negative value for the duration of G_1 and this confirms the conclusion of Clowes (1965) who also used the pulse labelling technique. The reality of a negative G_1 is confirmed by finding labelled telophases immediately at the end of a 30 min pulse (Clowes, 1967a and my own personal observation). The graphic fit of the cap columella initials' fraction labelled mitoses curve has been published previously (Macdonald, 1970, Fig. 1c), where it will be seen that the second and the start of the third wave of labelled mitoses are apparent over the 36 h fixation period. This is because there is less damping out of the curve here than in other regions owing to the smaller variance for cycle duration; however, this region of the cap columella initials is the easiest to locate precisely for the purposes of scoring labelled mitoses. I scored cells in the 3 or 4 tiers nearest the cap-quiescent centre junction. Phillips & Torrey (1971) have noted that, in the cap initials of cultured *Convolvulus* roots, cells in adjacent tiers differ in their cycle duration, those cells in the most proximal tier being the fastest. However, because the method they use for cycle determination (colchicine accumulation of metaphases) only measures average cycle duration it could

Table 3. *Phase durations (h) and standard error in regions throughout the meristem of roots 1–2 cm long (a gamma distribution has been assumed)*

	Phase	Cap columella initials	Cap periphery	Quiescent centre
	C	13.97 ± 0.68	22.47 ± 2.37	170
	G_1	-0.31 ± 0.58	6.25 ± 1.75	135
	S	7.58 ± 0.91	6.72 ± 0.79	16
	G_2	5.29 ± 0.46	7.28 ± 0.44	13
	M	1.42	2.21	6

		Stele	Cortex	Epidermis
Just above Q.C.	C	16.98 ± 1.26	42.84 ± 2.75	
	G_1	2.97 ± 1.05	21.84 ± 2.38	
	S	4.78 ± 0.42	10.24 ± 0.61	Not scored
	G_2	7.53 ± 0.30	6.44 ± 0.33	
	M	1.70	4.32	
200 μm from Cap – Q.C. boundary	C	20.42 ± 0.72	29.81 ± 5.87	25.00
	G_1	7.51 ± 0.74	17.55 ± 4.98	8.42 ± 0.58
	S	6.10 ± 0.28	4.87 ± 1.55	3.30 ± 0.46
	G_2	3.15 ± 0.18	4.13 ± 0.81	9.92 ± 0.55
	M	3.67	3.26	3.37
400 μm from Cap – Q.C. boundary	C	18.90 ± 0.97	24.86 ± 1.66	25.00
	G_1	6.34 ± 0.94	12.76 ± 1.68	14.18 ± 0.63
	S	5.56 ± 0.36	4.28 ± 0.45	3.11 ± 0.42
	G_2	3.51 ± 0.32	4.91 ± 0.70	5.22 ± 0.78
	M	3.49	2.90	2.49
700 μm from Cap – Q.C. boundary	C	20.57 ± 1.30	19.61 ± 1.00	22.97 ± 3.63
	G_1	10.33 ± 1.28	9.12 ± 1.12	13.39 ± 3.50
	S	3.97 ± 0.28	3.42 ± 0.32	2.77 ± 0.54
	G_2	3.42 ± 0.30	4.62 ± 0.66	4.05 ± 0.67
	M	2.85	2.47	2.76
1000 μm from Cap – Q.C. boundary	C	17.39 ± 0.88	18.55 ± 1.29	23.51 ± 2.25
	G_1	6.59 ± 1.02	8.66 ± 1.31	13.67 ± 2.46
	S	4.78 ± 0.75	3.33 ± 0.33	3.77 ± 0.70
	G_2	4.72 ± 0.76	5.07 ± 0.31	4.16 ± 1.09
	M	1.30	1.49	1.90

be that the proportion of proliferating cells in the total population of each tier changes, there being a higher proportion of proliferating cells in the proximal than in the distal tiers. In *Zea* the same situation may hold as there is a higher proportion of cells in S and mitosis in the tiers nearest the cap junction than in more distal cells of the initials (Table 4). By the 5–6th tier cells are no longer cycling and microdensitometry shows a 2C nuclear DNA value (Clowes, 1968; Barlow, 1969). The nuclei have been arrested at G_1/G_0 and will never divide again, though they continue to synthesise DNA (tiers 7–11, Table 4) and attain a 4C or even 8C DNA value (Clowes,

Table 4. *Labelling and mitotic indices in columella cells of the root cap*

Cells in tier number 1 are the most proximal of the cap (i.e. cap initial cells at the cap-quiescent centre boundary), the higher numbered tiers are more distal. Roots 1–2 cm long were labelled with [³H]thymidine (5 μC/ml) for 15 min and scoring was done in autoradiographs of longitudinal sections of 10 roots.

Cell tier	Labelling index (%)	Mitotic index (%)
1	60.8	12.5
2	43.5	4.3
3	35.1	2.7
4	17.8	0.9
5	9.0	1.0
6	9.6	0
7	23.7	0
8	17.4	0
9	19.5	0
10	14.3	0
11	14.7	0
12–13	5.3	0
14–15	5.0	0
16–17	1.8	0
18–21	0	0

1968). Mitotic cells at the periphery of the cap, encircling the columella initials, have a more variable and longer cycle.

The cells of the quiescent centre have an average cycle of about 160 h most of which is spent in G_1, though the durations of S and G_2 are also longer than in other regions. Clowes (1971) has shown that about 17% of the cells have a shorter than average cycle of about 40 h, while about 40% of the cells in the quiescent centre population are not cycling (G_0) and are held with a 2C DNA content. It is not easy to define the basiscopic surface of the quiescent centre so we do not know where the cells with different cycle properties are located. Quite probably cells of the stele and cortex become more 'quiescent' the closer they are to the pole of the root. This idea is supported by the finding that in cells of the cortex which I identified as being just above the quiescent centre have a longer S and G_1 duration than cells in more proximal cortical regions. Further, about 16% of the cells in the stele just above the quiescent centre are not cycling (Clowes, 1971). However, the computed estimate of the cycle parameters in this region, illustrated in Fig. 1, shows a shorter G_1 phase than anywhere else in the meristem except the cap columella initials. This apparent contradiction can be resolved by postulating that in the stele just above the quiescent centre there is a mixture of cells with different cycle durations, as Clowes (1971) suggested. The presence of cells with a long $G_2 + S$ duration (quies-

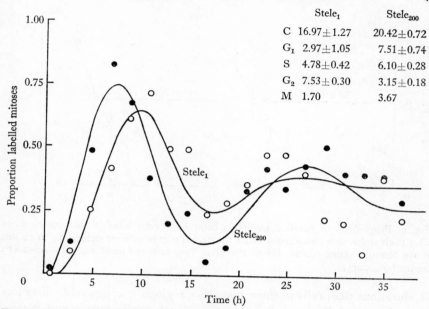

Fig. 1. Fraction labelled mitoses curves for the regions of stele just above the quiescent centre (stele 1) and 200 μm from the cap-quiescent boundary (stele 200).

cent centre-like cells) will tend to lengthen the duration of the first wave of labelled mitoses, while cells with a shorter cycle, about 70% of all cycling cells according to Clowes (1971), will contribute to the second ascending limb of the fraction labelled mitoses curve and so G_1 will be spuriously shortened.

To resolve this ambiguity which the pulse labelling method can cause, I estimated G_1 duration by another method. Roots were placed in an 0.1% caffeine solution for 1 h to induce a small proportion of binucleate cells (caffeine inhibits cell plate formation at telophase, López-Sáez, Risueño & Giménez-Martín, 1966); they were then returned to water and grown on for various times before labelling for 20 min with [³H]thymidine immediately prior to fixation. The first labelled nuclei in binucleate cells in the stele up to 50 μm from the quiescent centre appear 4.8 h after their induction and 50% are labelled after 7.8 h (Fig. 2a); these two values are the minimum and approximate mean duration of G_1. The latter value is about 1 h longer than in the stele 150 μm from the quiescent centre-cap boundary (Fig. 2b). The duration of G_1 in this latter region is in keeping with the value derived from analysis of the data from the pulse labelling experiment.

In the more basal regions of the meristem cell cycles are similar in both stele and cortex; epidermal cells, however, have a slightly longer cycle and

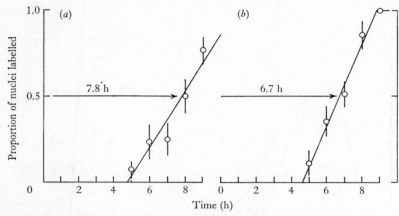

Fig. 2. Proportion of labelled nuclei in binucleate cells labelled at various times after their induction (between 0 and 1 hour). (a) region of stele extending to 50 μm above the quiescent centre. (b) in the stele 100–200 μm from the cap-quiescent centre boundary.

G_1 durations than cells in these two other regions at comparable distances from the tip. Adjacent cell populations, such as epidermis and cortex, having different rates of division present no problem to the integrity of the tissue as long as the cells in the two populations grow at the same rate.

Cells 1000 μm from the cap-quiescent centre boundary are nearly at the basal limit of the meristem. There is no sign that the cell cycle becomes slower before ceasing, so probably the decision whether to divide or not is an abrupt one: a cell either divides at its normal rate or it does not, there is no slow progress through a complete cycle.

In concluding this section dealing with pulse labelling data I should mention one of the assumptions made in computing the data. It was supposed that each cell that divides gives rise to two daughters that themselves will divide; thus A = 2, where A is the mean number of daughter cells that enter a new mitotic cycle per mitosis. If only one daughter divides then A = 1. Fig. 3 shows the effect of varying the assumed value of A on the estimated duration of different phases of the cell cycle. The data used were those of cortex 700 μm from the cap-quiescent centre boundary and the values for the duration of mitosis and its variance were kept constant. As the assumed value of A falls from 2 to 1.01, the estimated mean cycle duration (C) falls from 18.9 h to 18.15.

Ideally it would be useful to know the actual value for A in a root meristem, as presumably at some point towards the limit of meristematic activity the daughters of a divided cell either divide again or differentiate, so in this region A < 2. But it is not easy to estimate A experimentally.

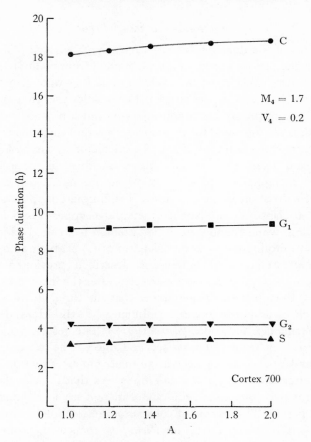

Fig. 3. The effect of varying the value of A on duration of phases of the mitotic cycle. The durations of mitosis (M_4) and its variance (V_4) are fixed at the values indicated.

One approach would be to induce binucleate cells, thus marking the products of a mitotic division, and looking to see what proportion of these binucleate cells subsequently divide. If synthesis of DNA in the two nuclei is taken as an indication of the preparation for mitosis, then the data of Benbadis, Levy & Pareyre (1971) provide an estimate of A for root apices of *Allium cepa*. Although their experiment was to estimate G_1 duration, their figure shows that the percentage of cells with two nuclei, induced by caffeine or quinazoline dione, that become labelled by continuous [³H]-thymidine feeding, plateaus at about 80%; this gives A = 1.6.

The mitotic cycle in aged roots

The estimates of the durations of cell cycles and their component phases described in the previous sections were obtained from young roots 1–2 cm long; I have also investigated the cycle in roots that were grown in water until they reached 100 cm in length. The results are of a preliminary nature as I was primarily interested in trying out a new method of cycle analysis with these roots and for this reason I present them here. The roots were transferred to a solution of 0.05 % colchicine and 0.7 μCi/ml [^{3}H]-thymidine and fixed after various times in this mixture. Colchicine accumulates metaphases and the rate of accumulation can be used to estimate cycle duration (Evans, Neary & Tonkinson, 1957; Clowes, 1961). The time that elapses before labelled metaphases appear gives an estimate of G_2 duration, while S and M can be obtained from the relations of the labelling and mitotic indices to the cycle time; G_1 is obtained by subtraction. (A pretreatment with caffeine, as described previously, could be used to determine G_1 duration more directly, after the method of Benbadis *et al.* (1971), though there is the danger that the three chemicals used to label the cycle could poison the cells (Giménez-Martín, Meza, de la Torre, González-Fernández & López-Sáez, 1971).) The results are shown in Fig. 4 and the conclusions summarised in Table 5. The rate at which both unlabelled and labelled metaphases accumulate indicates that they revert to interphase after about 4–5 h and so emphasises that for calculating rates of mitosis it is necessary to use the rate of metaphase accumulation over a brief initial period. This period of constant accumulation is similar to that found by Evans & Savage (1959) for *Vicia faba* roots at the same temperature, 21 °C; temperature very much influences the constant accumulation period.

Table 5. *Estimates of cycle and phase durations in regions of the meristem of roots* 100 *cm long (the value for* G_2 *is a minimum)*

Region...	Duration (h)				
	C	G_1	S	G_2	M
Cap initials	19.6	10.0	4.2	3.2	2.1
Quiescent centre	127.5	Not calculated		> 7.5	~ 5.5
Stele just above Q.C.	35.9	19.2	6.1	6.9	3.2
Stele 220 μm	20.8	8.2	6.7	2.7	3.2

Age has caused the mitotic cycle to become longer in all regions except the quiescent centre. It is the G_1 phase that is responsible for this increased duration. However the durations given are only approximate; to calculate

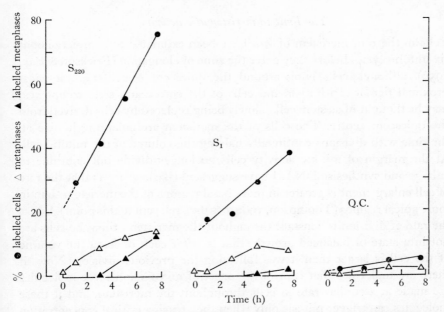

Fig. 4. The results of marking cells with a continuous treatment of colchicine and [³H]thymidine. Q.C. quiescent centre; S_1, stele just above Q.C. S_{220}, stele 220 μm from cap–Q.C. boundary.

them more accurately requires knowledge of the growth fraction. The longer cycle may be because the roots were grown in water and, because of their length, the meristem is far from its only (and diminishing) nutrient source in the endosperm; under these conditions the roots had shown no reduction in their growth rate. It is worth recalling that the *Zea* primary root meristem is ephemeral: that is, it would age and die naturally, the root system being continued by the lateral and adventitious roots.

THE CONTROL OF MITOTIC ACTIVITY IN MERISTEMS

Two points inevitably emerge from a study of the behaviour of cells within the meristem: first, what causes a meristematic cell to cease dividing and contribute to the zone of cell enlargement in the regions immediately proximal to the meristem of the main body of the root and distal to the cap initials; second, what is the reason for the slow cycle, and G_0 state, of cells in the quiescent centre? Much of what follows is, I hope, intelligent guesswork which in view of the lack of experimental results may at least serve to focus on possible and testable mechanisms of control.

The limit to meristematic activity

Cells in the root meristem of *Zea* have been estimated to complete about six mitotic cycles before they enter the zone of elongation (Erickson & Sax, 1956), although cells lying around the quiescent centre (i.e. the most proximal tier of cap initials and cells of the cortex and stele complexes) may be thought of as stem cells, slowly being replaced by cells derived from the quiescent centre. The cells of the meristem are enlarging in size all the time with divisions continually halving the volume of the mitotic cell. At the margin of mitotic activity cells no longer divide but continue to enlarge and synthesise DNA. I have suggested (Barlow, 1971) that the rate of cell enlargement is greater in more basal regions of the meristem than in more apical regions (Thompson, 1960, has data relevant to this point). But as the rate of division is constant throughout the meristem, then the cells are not in a state of balanced growth, that is, after each division the volume of the cell is larger than it was following the previous division. Now, if there are molecules that control the onset of mitosis and these molecules are made at a similar rate in cells throughout the meristem, and if these molecules can trigger mitosis only when they reach a critical concentration in the cell, then there will be a region in the population of cells in an unbalanced state of growth when the initiator molecules for mitosis no longer reach their critical concentration. When this happens the cell will not divide although it will continue to elongate. Differences in the rate of cell elongation along the length of the meristem may be controlled by an auxin gradient within the apex whose source is in the mature or ageing cells of the root and cap.

In many roots loss of mitotic activity is not accompanied by a loss of DNA synthetic ability and nuclei become polyploid as cells enter an endoreduplication cycle. The amount of DNA in nuclei of *Zea* roots can reach 8, 16 or 32C values (Swift, 1950). An exception is *Lactuca sativa* where the nuclei of cells leaving the meristem remain with the 2C amount of DNA (Brunori, 1971), suggesting that in this species the DNA synthetic and mitotic capacities are more closely coupled.

Changes in histone composition during cell differentiation have been reported by Kusanagi & Yanagi (1970) and Yanagi & Kusanagi (1970) in roots of *Hordeum vulgare* and *Allium sativum*. They used cytochemical and biochemical methods to show that meristematic nuclei have histones richer in lysine, while nuclei in elongated cells are richer in arginine. The nuclei of intermediate cells may stain for both types of histone. Whether these histone changes are correlated with any functional changes in nuclear metabolism, or whether they are due to differences in the rates of histone turnover, is not known. Corsi & Avanzi (1970) have also studied DNA-histone relations in maturing cells of *Allium cepa* root tips. They could not find a proportional increase in the amount of Fast Green-stained histone in pro-vascular cells as the nuclei increased in DNA content. Their evidence makes it unlikely that a rise in histone/DNA ratio represses mitosis. On the other hand, Kirk & Jones (1970) found that in nuclei of root meristems of *Secale* containing supernumerary chromosomes the histone/DNA ratio was increased compared to plants lacking the extra chromosomes. In plants with supernumerary chromosomes the duration of the mitotic cycle was lengthened (Ayonoadu & Rees, 1968), so at least in this species an increased amount of histone is associated with a slower mitotic cycle (see discussion on pp. 151 *et sqq.*). Whether a higher histone/DNA ratio is present in nuclei of the quiescent centre is not known, although in cells of dormant buds of *Tradescantia*, which are somewhat similar to quiescent centre cells, the nuclear histone/DNA ratio was lower than in the nuclei of actively dividing buds (Dwivedi & Naylor, 1969). Similarly, nuclei of the dividing meristem of *Triticum* had a higher histone/DNA ratio than the non-dividing meristematic nuclei in the dry seed (Innocenti, 1971).

The problem of the quiescent centre

The properties of cells within the quiescent centre have been reviewed by myself (Barlow, 1971) and Clowes (1967*a*, 1972*b*). We conclude that these cells are much less metabolically active than other cells of the meristem. I also reviewed (Barlow, 1971) some of the possible ideas that could explain why the cells of the quiescent centre divide so slowly and I came to the conclusion that in *Zea* and other species with a similar root anatomy the cap played a role in preventing quiescent centre cells from dividing. My argument was that cells of the cap columella initials grow only in a direction at right angles to the preferred direction of growth of cells at the pole of the epidermis, and this antagonism locks the cells of the epidermis and their neighbours in the cortex and stele (i.e. the cells that make up the quiescent centre) into a non-growing state. The inhibition of cell growth then prevents cell division, for which there is much evidence from the older cytological literature (see Hottes (1929) for a review). This theory is eminently testable in *Zea* because it is possible to remove the cap completely and by doing so remove the constraint to growth of quiescent centre cells at the exposed apex. I have been able to show that when the cap is removed the first thing that cells of the quiescent centre do is to grow, and later they divide. But they do not divide in a disorderly fashion. They have a controlled sequence of division that regenerates a new cap. Sometimes the decapping operation is unsuccessful, the cap remaining in place and one or two columella cells being killed; when this occurs only the cells in the quiescent centre adjacent to the dead cap grow and divide. It is also possible to bisect the root cap longitudinally and remove one half, so exposing only half the quiescent centre. In this case the group of quiescent centre cells at, and immediately behind, the exposed surface respond by growing and dividing; the group of cells in the quiescent centre that remains in contact with the intact cap appear to be unaffected by the operation. I take these observations to support my idea that the quiescent centre results from the suppression of cell growth imposed by the cap.

Clowes (1972b) has performed another type of surgical experiment that is of interest to the problem of the quiescent centre. He bisected the cap transversely and removed the mature cap cells distal to the cut. This operation led to an immediate stimulation of mitotic activity in the quiescent centre. How this comes about is not known. However, I would tentatively like to suggest that the root cap acts as a valve through which substances pass on their way to the exterior of the root; any interference with the integrity of the cap could lead to a surge of materials towards the apex and this may influence the mitotic activity of cells in the meristem. Some evidence that the cap can regulate the flow of substances in the root is suggested by findings of Pilet (1971) who removed the caps of *Lens* and *Zea* roots and after a four hour period found a 32 % and 23 % increase respectively in the amount of [^{14}C]indole acetic acid transported towards the root apex from a donor block placed 10 mm basal to the exposed tip.

I cannot leave a discussion of the quiescent centre without mentioning the so-called 'apical cell' that is found in the roots of some Pteridophytes (ferns) as these cells have many features in common with the quiescent centre. The apical cell is a prominent tetrahedral cell lying at the pole of the lineages of stele and cortex and, like the cells of the quiescent centre, appears only rarely to divide (D'Amato & Avanzi, 1965; Avanzi & D'Amato, 1967) though no rates of division are known for roots of any species of fern that have an apical cell. The apical cell also incorporates precursors of RNA and protein at a slower rate than the surrounding meristematic cells (Sossountzov, 1969; Avanzi & D'Amato, 1970). But unlike the cells of the quiescent centres of angiosperms apical cells become polyploid (with up to the 16C DNA content) during root development (Avanzi & D'Amato, 1967). When this cell does divide its descendents are also polyploid so creating a mixoploid meristem. The quiescent centre, on the other hand, remains a reservoir of diploid cells (the nuclei are mainly with the 2C DNA content) throughout root development. The factors that favour continued synthesis of DNA without intervening mitosis in the apical cell but suppress this in the quiescent centre cells are unknown.

THE CELL CYCLE IN MERISTEMS
AND GENETIC CHANGES

Gene mutations

In bacteria and yeast it has been possible to analyse events of the mitotic cycle through the use of mutations. Such an approach is useful for determining the position within the cell cycle at which the gene products required for a particular cycle event are synthesised, and also for revealing the extent to which events are either interrelated or independent. As an example of this use of mutations, Hartwell (1971) has found that in the yeast *Saccharomyces cerevisae* there are two genes that initiate and maintain DNA synthesis that act at different times during the cell cycle, but the completion of DNA synthesis is not required to initiate the formation of new buds (the daughter cells). Bud formation in these mutants is a periodic event and Hartwell suggests that a cellular 'clock' controls this part of the cell cycle. However, genes affecting nuclear division do disturb bud initiation (Culotti & Hartwell, 1971). In multicellular organisms new cell formation by cleavage or cell wall formation can be dissociated from nuclear division only under exceptional circumstances (Fankhauser, 1934; González-Fernández, Giménez-Martín & López-Sáez, 1970). Unlike the cases with micro-organisms, genetics has not yet been exploited to investigate the control of division in higher plants, although advances in cell and protoplast culture may make this feasible.

There are, however, mutations in some species that alter the behaviour of mitotic cells. In the seaweed *Ulva mutabilis* there are mutants that alter the shape of the whole plant by influencing the orientation and rate of cell division (Lövlie, 1964). In higher plants the influence of genes is confined to certain organs. Stebbins and his colleagues have made detailed analyses of the effect of certain genes on the reproductive spike of barley. They find that in plants expressing the genes *hooded* and *calcaroides* the planes of division are altered in the lemma awn so bringing about the mutant phenotype. It is unlikely that in barley these genes act directly on some facet of the cell cycle (although in *Ulva* this may be the case), it is more likely that they bring about a change in the *milieu* in which the cells divide; indeed Stebbins & Yagil (1966) and Stebbins & Price (1971) suggest that the production and interplay of growth regulators are disturbed and this is responsible for the change in the pattern of division. As yet there is no evidence that directly supports this hypothesis but if it is true then it means this change in growth regulator balance only manifests itself in this way at a particular time and place in development. But there are genes in *Zea mays* that influence the level of gibberellins in the whole plant. These genes are at the dwarf (d) locus, which when homozygous cause small plants that may be restored to normal size by spraying with gibberellic acid. Banerjee (1968) has made a study of the cell cycle in the root meristems of two of the dwarf mutants of *Zea*, d_1 and d_5, as well as in the respective F_1 heterozygotes, $d_1/+$ and $d_5/+$. From a pulse labelling experiment Banerjee concluded that in both mutants the duration of the cell cycle is lengthened by about 25% when compared with the respective phenotypically normal heterozygotes. Banerjee's data suggest that G_2 duration is longer than normal in both mutants; the S phase is lengthened also, particularly in the d_5 mutant. These results are an encouraging step towards analysing the role that natural growth regulators play in controlling cell division, particularly since gibberellins are thought to be synthesised in, and can govern the growth of, root tips (Jones & Phillips, 1966; Mertz, 1966; Sitton, Richmond & Vaadia, 1967). It would be interesting to use these dwarf mutants of *Zea* further: one question immediately suggested by Banerjee's findings is whether adding the missing gibberellins to the mutants restores the normal timing of the mitotic cycle as Mertz (1966) has shown that adding gibberellin to root tips of mutant d_1 restores its growth to a normal rate.

Changes in DNA content

Changes in the amount of nuclear DNA, and thus the number of genes present, may also influence the cell cycle in root meristems. A change in ploidy is one such change, aneuploidy is another. There is some doubt as to whether a change in ploidy alters the cell cycle, for, although the amount of DNA is altered, the sequence and rate of DNA chain replication would remain unaltered. Two types of observation have been made. The first is on cells induced to become tetraploid in a diploid meristem and the cycles of both populations are then compared. If there is any difference it is small (Table 6), but it could be argued that the tetraploid cells occur in small numbers and are surrounded by diploid cells which influence their behaviour so that they conform with that of the majority. Further, it is usually the first cell cycle in the induced tetraploid cells that is studied and this may not be the same as succeeding cycles. The second type of observation is where the cell cycle in a diploid plant is compared with the cycle in the autotetraploid derived from the same stock of seed. When the data of Skult (1969), who performed a pulse labelling experiment with root tips of diploid and colchicine-derived autotetraploid *Hordeum vulgare* seedlings, are inspected, it appears that the tetraploid has a slightly longer and more variable cycle than the diploid (Fig. 5).

There is a tendency for species with more DNA to have a longer

Table 6. *Durations of cell cycles in mixoploid meristems and diploid and colchicine-induced autotetraploid meristems. In the mixoploid meristems the percentage of each cell population is given where known*

Phase duration (h) in cells of different ploidy in mixoploid meristems

Species	Phase	2n (36%)	4n (63%)	4n (84%)	Author
Vicia faba	C	14.4	14.0	15.8	
	$G_1 + M/2$	2.5	3.8	3.8	Friedberg &
	S	7.4	7.2	7.5	Davidson (1970)
	$G_2 + M/2$	4.5	4.0	4.5	
		2n	2n+2n	4n	
			(binucleate) (40–60%)		
Allium cepa	C	13.5	14.0	14.5	Giménez-Martín *et al.* (1966)
		2n (85%)		4n (15%)	
Pisum sativum	C	16		13	van't Hof (1966),
	S	6		4–5	van't Hof & Ying
	G_2	6		4–6	(1964)
	G_1	7		3	

Diploid and autotetraploid meristems

		2n	4n	
Tradescantia paludosa	S	10.7	10.4	
Ornithogalum virens	S	7.9	7.9	Troy & Wimber (1968)
Cymbidium sp.	S	7.1	7.7	
Lycopersicum esculentum	S	7.2	7.6	

mitotic cycle than species with less DNA (van't Hof, 1965) and in this respect monocots behave differently to dicots (Evans & Rees, 1971). But there are cases where the linear relation between nuclear DNA content and cell cycle duration breaks down: this is when supernumerary, or B, chromosomes are present in the genome. These chromosomes show no homology with any of the regular, or A, chromosomes of the complement at meiosis. They are often heterochromatic and cause disturbances to the meiotic cycle in pollen and egg mother cells (Barlow & Vosa, 1970; Vosa & Barlow, 1972 and references cited therein). When they are present in mitotic cells they cause the cycle to be lengthened disproportionately to the extra amount of DNA they contribute to the nucleus. In root meristems of rye (*Secale cereale*) each additional B chromosome lengthens the duration of the cycle by 9% compared to plants lacking them (Ayonoadu & Rees,

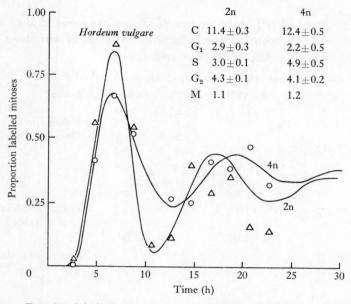

	2n	4n
C	11.4±0.3	12.4±0.5
G_1	2.9±0.3	2.2±0.5
S	3.0±0.1	4.9±0.5
G_2	4.3±0.1	4.1±0.2
M	1.1	1.2

Fig. 5. Fraction labelled mitoses curves for diploid and autotetraploid roots of barley. Data of Skult (1969).

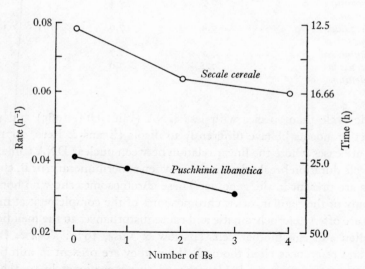

Fig. 6. The influence of B chromosomes on the rate of mitosis in root meristems of *Secale cereale* (data of Ayonoadu & Rees, 1968) and *Puschkinia libanotica* (data of Barlow & Vosa, previously unpublished).

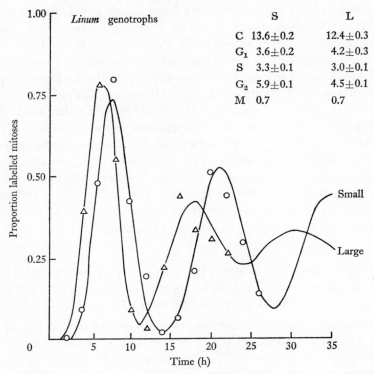

	S	L
C	13.6±0.2	12.4±0.3
G_1	3.6±0.2	4.2±0.3
S	3.3±0.1	3.0±0.1
G_2	5.9±0.1	4.5±0.1
M	0.7	0.7

Fig. 7. Fraction labelled mitoses curves for L and S genotrophs of flax. Data of Timmis (1971).

1968) and in root meristems of *Puschkinia libanotica* by 8% (Barlow & Vosa, unpublished) (Fig. 6).

It may be a general property of aneuploid cells to show a disturbed cell cycle, as in human cells in culture extra G group and X chromosomes also cause a disproportionate lengthening of cycle duration (Kaback & Bernstein, 1970; Barlow, 1972). In the case of rye the presence of B chromosomes causes an increase in the nuclear histone/DNA ratio (Kirk & Jones, 1970) and it may be that the extra histone brings about the slowing down of the cell cycle by reducing the rate of synthesis of RNA and protein essential for the maintenance of the cycle, or possibly by rendering the DNA less accessible to the molecules that bring about its replication.

There is one apparently anomalous case where an increased amount of DNA speeds up the mitotic cycle in root meristems. This occurs in flax (*Linum* var. Stormont Cirrus) where heritable changes in the amount of nuclear DNA can be induced in the original variety (the plastic genotroph) by growing the plants in soil of high nitrogen or high phosphate content;

the resulting plants are changed into large (L) or small (S) genotrophs respectively (Durrant, 1962). L plants have 16% more DNA than S plants (Evans, 1968). An analysis by Dr P. Macdonald of a pulse labelling experiment using roots of both genotrophs performed by Timmis (1971) shows that the average duration of the mitotic cycle in the L genotroph is shorter than in the S genotroph (Fig. 7). Although the duration of S phase is slightly longer in the S genotroph (which has *less* DNA than the L genotroph) the greatest difference is in the duration of the G_2 phase.

So far the behaviour and phenotypes of the two genotrophs are unexplained, as, indeed, is the nature of the expendable DNA, though Timmis believes that at least some of it is located in heterochromatin. If this expendable DNA codes for, or interacts with (by being located at the centromere), proteins needed for mitosis, such as those of the spindle apparatus, this could explain the shorter G_2 duration in L plants.

THE BEHAVIOUR OF THE CELL CYCLES IN ROOT MERISTEMS AT DIFFERENT TEMPERATURES AND AFTER IRRADIATION

The behaviour of cells after ionising radiation shows us that not all cells in the meristem respond in the same way and so leads us to uncover new properties of these cells. The way in which the mitotic cycle alters in response to changes of temperature allows us to study other aspects of the physiology of mitotic cells.

Temperature and the mitotic cycle

Roots of only a few species have been grown over a sufficient range of temperatures to afford a study of its effect on the cell cycle. Nevertheless, all the studies show that the mitotic cycle is fastest between 30 and 35 °C. At temperatures either above 35 °C or approaching 0 °C the cycle is very much slowed, although the roots are not killed. The techniques and cells used for estimating cycle duration have been different in each species whose roots have been used: López-Sáez, Giménez-Martín & González-Fernández (1966) measured the cycle in binucleate cells in *Allium cepa* and Murín (1966) measured that of colchicine-induced tetraploid cells in *Vicia faba*; only González-Fernández, Giménez-Martín & de la Torre (1971) and Burholt & van't Hof (1971) measure the cycle in diploid cells after [³H]thymidine marking. In Table 7 I have compiled data to show the effect of temperature on cycle duration.

The durations of the interphase periods at different temperatures have

Table 7. *The factor by which temperature increases the mitotic cycle duration in root tips compared to the minimum cycle duration* ($=1.00$)

Temp (°C)	Allium cepa binucleate cells	Allium cepa diploid cells	Helianthus annuus diploid cells	Vicia faba tetraploid cells
3	—	—	—	22.9
5	11.8	—	—	—
8	—	—	—	7.8
10	5.02	4.97	7.36	—
13	—	—	—	2.94
15	2.75	2.71	3.68	—
20	1.73	1.71	1.99	2.01
25	1.24	1.26	1.24	1.00
30	1.01	1.00	1.00	1.06
35	1.00	—	1.02	2.08
Min. cycle (h)	7.9	11.0	6.3	11.6
Author	López-Sáez et al. (1966)	González-Fernández et al. (1971)	Burholt & van't Hof (1971)	Murín (1966)

been examined in *Allium cepa* and *Helianthus annuus*. In *Allium* González-Fernández *et al.* (1971) found that the proportion of the cycle occupied by G_1, S, G_2 and M was constant at all temperatures used (10–30 °C). In contrast, Burholt & van't Hof (1971) found that in *Helianthus* the proportion of cells in S phase rose from about 0.5 at temperatures between 10 and 20 °C to 0.68 at 30 °C, the proportion of cells in G_1 showed a corresponding decrease while the proportions of cells in G_2 and M phases remained constant. But the data are not strictly comparable as the roots of the two species were grown in different ways: those of *Allium* were raised from bulbs at the experimental temperature while those of *Helianthus* were first raised at 21 °C and then transferred to the experimental temperature and it is possible the cells were not equilibrated to the new temperature before phase durations were determined. However, Wimber (1966) using *Tradescantia paludosa* roots raised at constant temperatures also showed a decrease in the proportion of cells in the G_1 phase at higher temperatures.

A short heat shock – the transfer of roots from 15 °C to 35 °C for 3 h and then returning them to 15 °C – has located a temperature-sensitive period in the middle of the S phase of *Allium cepa* (de la Torre *et al.* 1971). The 3 h shock applied at this time slows the mitotic cycle by 9 h and de la Torre *et al.* also find that prophase is lengthened by this treatment. The sensitive period in S corresponds to a portion of that phase when synthesis of DNA is slower (Fernández-Gómez, 1968) and is at a time that is also more sensitive to 2′-deoxyadenosine (Fernández-Gómez,

González-Fernández & Giménez-Martín, 1969). (A reduced rate of synthesis of DNA in mid-S has also been noted by Howard & Dewey (1962) in *Vicia faba* roots.) The disturbance to the mitotic cycle is accompanied by cytological changes in the nucleolus (de la Torre *et al.* 1971; Díez, Marin, Esponda & Stockert, 1970) and by the occasional failure of cytokinesis. How these changes come about is not clear since when plants are grown continuously at these temperatures the progress of cells through the mitotic cycle is not prevented. Burholt & van't Hof (1971) found that when [^{3}H]thymidine pulse labelled roots of *Helianthus* were grown at 38 °C cells could complete G_2 to reach M but could not reach M from G_1; labelled mitoses still appeared 18 h after the pulse indicating that in this species there may also be a period early in S phase sensitive to temperatures of 38 °C and above. These results mean that different phases of the cell cycle have different temperature sensitivities. Prolonged cold treatment (5 °C for 7 days) of meristems of *Zea* roots also led to nuclear damage and the loss of reproductive integrity of all cells except those of the quiescent centre (Clowes & Stewart, 1967) which, after returning the roots to 21 °C, were stimulated to a faster than normal cell cycle. This response is similar to that which occurs after acute X-irradiation.

Burholt & van't Hof show that the optimal temperatures for cell elongation and cell division are different. Mature cells in the *Helianthus* roots reach a greater length at 20 °C while the fastest cycle time occurs at 30–35 °C. Thus the maximal rate of root growth occurs at an intermediate temperature, 25 °C, and also at this temperature the number of meristematic cells is maximal.

Ionising radiation and the mitotic cycle

Ionising radiations have been used to advantage to probe events in the cell cycle of cultured animal cells. These studies have revealed that not all periods of interphase are equally sensitive to radiation, using mitotic delay as a criterion of radiation response (Sinclair, 1968). As far as plant cells have been used in radiation work the response to radiation has been studied mainly in tissues, such as root meristems, rather than cultured cells and so rather different aspects of radiosensitivity have been examined. Analyses of chromosome aberrations have been a favourite pursuit of many radiobiologists and conclusions have been drawn concerning the structure and arrangement of chromosomes within the interphase nucleus, as well as on the effectiveness of various types of radiation in inducing chromosome aberrations and mutations. But radiation work with meristems has also been important in uncovering differences in the behaviour of cells in different regions of the irradiated meristem and it is because of these differences that some cells within the root can reorganise a new meristem and so recover from the radiation event.

After giving root meristems of *Zea* an acute dose of X-radiation Clowes (1959) found that during the following eight days the number of cells engaged in DNA synthesis in the main body of the meristem became fewer than normal. By contrast, on the fourth day after irradiation the outer cells of the quiescent centre had started to synthesise DNA, and by the eighth day cells in the position of the quiescent centre were the only cells synthesising DNA. Clowes (1963a) then estimated changes in mitotic rates after irradiation and confirmed the inference from the early study that cell division was stimulated in the quiescent centre but was reduced elsewhere. Nine days after irradiation cells of the quiescent centre had returned to a low rate of division, while the cells in the cap initials and stele just above the quiescent centre had partly recovered their ability to divide. Fig. 8 illustrates the changes in rate of division for *Zea* and also for *Vicia faba* which shows a similar pattern of recovery. The recovery of mitotic cycle activity in regions around the quiescent centre is due to their repopulation by descendants of cells in the quiescent centre that had been stimulated to divide. At the anatomical level this is seen as pushing forward towards the apex of the thick wall at the cap-quiescent centre boundary and the apparent 'flow' of cells from the quiescent centre to give rise to a new root cap. Later a new cap-quiescent centre boundary is reformed showing, as in the regeneration experiments described earlier, the remarkable ability of cells within the root to recover their original pattern of growth and division.

The quiescent centre is able to provide cells to repopulate the meristem after irradiation because it is less sensitive to X-rays than the cells surrounding it. The reason given by Clowes (1965) for the relative insensitivity of cells of the quiescent centre is that it is the greater proportion of nuclei in the G_1 phase that confers this resistance. But this may not be the only reason, the nature of the cytoplasm may also confer resistance; for instance, mitochondria are smaller and fewer and this property has been considered to play a part in radiosensitivity in animal cells (Goldfeder, 1963). Further, the amount of nuclear damage in the quiescent centre, as judged by the fraction of cells with micronuclei, is not increased when the X-ray dose is increased (Clowes, 1963b, 1972b), this is in contrast to the behaviour of the cells of cap initials and stele. This might indicate that cells of the quiescent centre are more efficient at repairing radiation damage than cells of other regions and may be another contributory reason for their relative radioresistance. Unfortunately nothing is known of repair mechanisms of cells

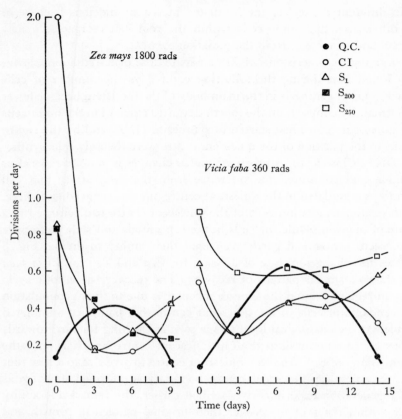

Fig. 8. The influence of an acute dose of X-radiation on the rates of mitosis in different regions of root meristems of *Zea mays* and *Vicia faba* on different days of the recovery period. Data of Clowes (1963*a*) for *Zea* and Clowes & Hall (1962) for *Vicia*.

within different regions of the meristem, but they probably exist (Berjak & Villiers, 1972) and may contribute to survival of inactive, but potentially mitotic, cells (Kovacs & van't Hof, 1972).

Neither is much known of the duration of the various phases of the cell cycle in different regions of the meristem after irradiation although Wimber (1966*b*) gives an account of the effect of prolonged gamma radiation on *Tradescantia* roots and finds G_1 and G_2 are particularly affected. Webster and van't Hof (1970) have found that an acute dose of gamma radiation delays the progress of cells through G_2 into mitosis (cf. heat radiation), but cells in G_1 are not prevented from entering the S phase, though their subsequent progress into mitosis is delayed (Kovacs & van't Hof, 1971). If these data can be related to the situation in *Zea* roots after acute X-irradiation then there seems to be no bar to the progress of irradiated G_1 cells into S: in the quiescent centre this progress must actually be stimulated. And this stimulation must hold for both the slowest and non-dividing G_0 cells. It is unlikely that this stimulation of division of quiescent centre cells is due to the abolition of division in the surrounding cells as when roots of Zea are exposed to continuous gamma radiation the rate of division of cells in the cap initials and stele is reduced but there is no corresponding increase in the rate of division of quiescent centre cells (Clowes & Hall, 1966). And in Vicia roots, during chronic irradiation, the cells of the quiescent centre are stimulated to divide while there is little, if any, effect on the duration of the cycle in the surrounding cells. In the Vicia root meristem there is considerable reorganisation of cell patterns with much disruption of cell lineages through cell death (Clowes & Hall, 1966). If, as I argued earlier, opposing directions of cell enlargement are responsible for maintaining the quiescent centre then I must argue that radiation changes the polarity of cell enlargement and division, especially in the cap initials, and this releases the constraint on the quiescent centre cells so allowing them to grow and divide. Recovery of the root from irradiation is then accompanied by a reorganisation of the original cell polarities and this imposes quiescence once again on cells at the boundary between the new cap initials and the main body of the root. In the chronically irradiated roots of Zea of Clowes & Hall (1966) the radiation dose may have been sufficient to disturb cell division but not the polarities of division or cell growth. In these roots, cells from the quiescent centre replace cells of the cap initials only when individual cells of this latter region die.

The effects of acute radiation I have described occur some days after the radiation event. But there are spectacular immediate events. Within 30 min after irradiating Zea roots with 1800 rads the mitotic index in the quiescent centre reaches four times the value of unirradiated roots (Clowes, 1970). Further, the number of quiescent centre cells capable of incorporating [^{3}H]thymidine increases two-fold and continues to rise for 2 h after irradiation. These two observations indicate that irradiation can immediately accelerate cell cycle events in the quiescent centre. The G_2 phase can be completed in 30 min and Clowes (1972a) claims that cells can come from the G_1 phase to mitosis in 46 minutes without synthesising DNA. These mitotic chromosomes have less than the 4C amount of DNA, though it must here be assumed that the radiation treatment has not caused any loss in the ability of the DNA to stain with Feulgen's reagent. The rise in labelling index must indicate the advancement of G_1 and G_0 cells into the S phase, though the possibility that some repair synthesis of DNA is occurring cannot be absolutely excluded. These events in cells of the quiescent centre after irradiation are quite different from what happens in other cells of the meristem and for this reason makes these cells of fascinating and challenging interest to all concerned with the properties of meristematic cell populations.

I would like to thank the following: Dr P. Macdonald for supplying data from which Figs. 1, 5 and 7 were drawn and for many valuable discussions, Mrs Nancy Amery for helping me compute the data presented in Fig. 3, Dr J. Timmis who supplied me with information on flax genotrophs for the computation of Fig. 7.

REFERENCES

AVANZI, S., BRUNORI, A. & D'AMATO, F. (1970). Nuclear conditions in the meristems of resting seeds of *Triticum durum*. *Wheat Information Service*, **30**, 5–6.

AVANZI, S., BRUNORI, A. & GIORGI, B. (1966). Radiation response of dry seeds in two varieties of *Triticum durum*. *Mutation Research*, **3**, 426–37.

AVANZI, S., BRUNORI, A., NUTI RONCHI, V. & SCARASCIA-MUGNOZZA (1963). Occurrence of 2C (G_1) and 4C (G_2) nuclei in the radicle meristems of dry seeds in *Triticum durum*. Its implications in studies on chromosome breakage and developmental processes. *Caryologia*, **16**, 553–8.

AVANZI, S. & D'AMATO, F. (1967). New evidence on the organization of the root apex in leptosporangiate ferns. *Caryologia*, **20**, 257–64.

AVANZI, S. & D'AMATO, F. (1970). Cytochemical and autoradiographic analysis on root primordia and root apices of *Marsilea strigosa*. *Caryologia*, **23**, 335–45.

AVANZI, S. & DERI, P. L. (1969). Duration of the mitotic cycle in two cultivars of *Triticum durum*, as measured by ^{3}H-thymidine labelling. *Caryologia*, **22**, 187–94.

AYONOADU, U. W. & REES, H. (1968). The regulation of mitosis by B-chromosomes in rye. *Experimental Cell Research*, **52**, 284–90.

BANERJEE, S. N. (1968). DNA synthesis in the root meristem cells of *Zea mays* dwarf mutants (d_1 and d_5). *Plant and Cell Physiology*, **9**, 557–81.

BARLOW, P. W. (1969). Cell growth in the absence of division in a root meristem. *Planta, Berlin*, **88**, 215–23.

BARLOW, P. W. (1971). Properties of cells in the root apex. *Revista de la Facultad de Agronomía, La Plata, Argentina*, **47**, 275–301.

BARLOW, P. W. (1972). Differential cell division in human X chromosome mosaics. *Humangenetik*, **14**, 122–7.

BARLOW, P. W. & VOSA, C. G. (1970). The effect of supernumerary chromosomes on meiosis in *Puschkinia libanotica* (Liliaceae). *Chromosoma, Berlin*, **30**, 344–55.

BENBADIS, M.-C., LÉVY, F. & PAREYRE, C. (1971). Méthode de détermination directe de la durée de la période G_1 du cycle cellulaire dans les méristèmes radiculaires. *Comptes rendus de l'Académie des sciences*, **273**, 352–5.

BERJAK, P. & VILLIERS, T. A. (1972). Ageing in plant embryos. II. Age-induced damage and its repair during early germination. *New Phytologist*, **71**, 135–44.

BISHOP, R. C. & KLEIN, R. M. (1971). Diurnal rhythms in mitosis of cells of the onion root tip meristem. *Canadian Journal of Genetics and Cytology*, **13**, 597–9.

BOGDANOV, Y. F., LIAPUNOVA, N. A. & SHERUDILO, A. I. (1967). Cell population in pea embryos and root tip meristem. Microphotometric and autoradiographic studies. *Tsitologiya*, **9**, 569–76. (In Russian.)

BRONK, B. V. (1969). On radioactive labeling of proliferating cells: the graph of labeled mitosis. *Journal of Theoretical Biology*, **22**, 468–92.

BRUNORI, A. (1967). Relationship between DNA synthesis and water content during ripening of *Vicia faba* seed. *Caryologia*, **20**, 333–8.

BRUNORI, A. (1971). Synthesis of DNA and mitosis in relation to cell differentiation in the roots of *Vicia faba* and *Lactuca sativa*. *Caryologia*, **24**, 209–15.

BRUNORI, A. & ANCORA, G. (1968). The DNA content of nuclei in the embryonic root apices of dry seeds of *Allium cepa* and their radiation response. *Caryologia*, **21**, 261–9.

BRUNORI, A., AVANZI, S. & D'AMATO, F. (1966). Chromatid and chromosome aberrations in irradiated dry seeds of *Vicia faba*. *Mutation Research*, **3**, 305–13.

BRUNORI, A. & D'AMATO, F. (1967). The DNA content of nuclei in the embryo of dry seeds of *Pinus pinea* and *Lactuca sativa*. *Caryologia*, **20**, 153–61.

BRUNORI, A., GEORGIEVA, J. & D'AMATO, F. (1970). Further observations on the X-irradiation response of G_1 cells in dry seeds. *Mutation Research*, **9**, 481–7.

BRYANT, T. R. (1969). DNA synthesis and cell division in germinating onion. II. Mitotic cycle and DNA content. *Caryologia*, **22**, 139–48.

BURHOLT, D. R. & VAN'T HOF, J. (1971). Quantitative thermal-induced changes in growth and cell population kinetics of *Helianthus* roots. *American Journal of Botany*, **58**, 386–93.

CHEN, D., SARID, S. & KATCHALSKI, E. (1968). Studies on the nature of messenger RNA in germinating wheat embryos. *Proceedings of the National Academy of Sciences, U.S.A.* **60**, 902–9.

CLOWES, F. A. L. (1959). Reorganization of root apices after irradiation. *Annals of Botany*, **23**, 205–10.

CLOWES, F. A. L. (1961). Duration of the mitotic cycle in a meristem. *Journal of Experimental Botany*, **12**, 283–93.

CLOWES, F. A. L. (1962). Rates of mitosis in a partially synchronous meristem. *New Phytologist*, **61**, 111–8.

CLOWES, F. A. L. (1963a). X-irradiation of root meristems. *Annals of Botany*, **27**, 343–52.

CLOWES, F. A. L. (1963b). Micronuclei in irradiated meristems. *Radiation Botany*, **3**, 223–9.

CLOWES, F. A. L. (1965). The duration of the G_1 phase of the mitotic cycle and its relation to radiosensitivity. *New Phytologist*, **64**, 355–9.

CLOWES, F. A. L. (1967a). Synthesis of DNA during mitosis. *Journal of Experimental Botany*, **18**, 740–5.

CLOWES, F. A. L. (1967b). The quiescent centre. *Phytomorphology*, **17**, 132–40.

CLOWES, F. A. L. (1968). The DNA content of the cells of the quiescent centre and root cap of *Zea mays*. *New Phytologist*, **67**, 631–9.

CLOWES, F. A. L. (1970). The immediate response of the quiescent centre to X-rays. *New Phytologist*, **69**, 1–18.

CLOWES, F. A. L. (1971). The proportion of cells that divide in root meristems of *Zea mays* L. *Annals of Botany*, **35**, 240–61.

CLOWES, F. A. L. (1972a). The control of cell proliferation within root meristems. In *The Dynamics of Meristem Cell Population*, ed. M. W. Miller & C. C. Kuehnert, pp. 133–45. New York: Plenum Publishing Company.

CLOWES, F. A. L. (1972b). Regulation of mitosis in roots by their caps. *Nature, New Biology, London,* **235,** 143–4.

CLOWES, F. A. L. & HALL, E. J. (1962). The quiescent centre in root meristems of *Vicia faba* and its behaviour after acute X-irradiation and chronic γ-irradiation. *Radiation Botany,* **3,** 45–53.

CLOWES, F. A. L. & HALL, E. J. (1966). Meristems under continuous irradiation. *Annals of Botany,* **30,** 243–51.

CLOWES, F. A. L. & STEWART, H. E. (1967). Recovery from dormancy in roots. *New Phytologist,* **66,** 115–23.

CORSI, G. & AVANSI, S. (1970). Cytochemical analyses on cellular differentiation in the root tip of *Allium cepa. Caryologia,* **23,** 381–94.

CULOTTI, J. & HARTWELL, L. H. (1971). Genetic control of the cell division cycle in yeast. III. Seven genes controlling nuclear division. *Experimental Cell Research,* **67,** 389–401.

D'AMATO, F. & AVANZI, S. (1965). DNA content, DNA synthesis and mitosis in the root apical cell of *Marsilea strigosa. Caryologia,* **18,** 383–94.

DELSENY, M. (1971). Aspects cytologique de métabolisme des ARN au cours de la germination de *Vicia sativa* L. *Revue générale de botanique,* **78,** 161–70.

DELTOUR, R. (1970). Synthèse et translocation de RNA dans les cellules radiculaires de *Zea mays* au début de la germination. *Planta,* **92,** 235–9.

DELTOUR, R. & BRONCHART, R. (1971). Changements de l'ultrastructure des cellules radiculaires de *Zea mays* au début de la germination. *Planta,* **97,** 197–207.

DÍEZ, J. L., MARIN, M. F., ESPONDA, P. & STOCKERT, J. C. (1970). Prenucleolar bodies in the cytoplasm of meristematic cells after thermal shock. *Experientia,* **27,** 266–7.

DURRANT, A. (1962). The environmental induction of heritable change in *Linum. Heredity,* **17,** 27–61.

DWIVEDI, R. A. & NAYLOR, J. M. (1968). Influence of apical dominance on the nuclear proteins in cells of the lateral bud meristem in *Tradescantia paludosa. Canadian Journal of Botany,* **46,** 289–98.

EVANS, G. M. (1968). Nuclear changes in flax. *Heredity,* **23,** 25–38.

EVANS, G. M. & REES, H. (1971). Mitotic cycles in dicotyledons and monocotyledons. *Nature, London,* **233,** 350–1.

ERICKSON, R. O. & SAX, K. B. (1956). Rates of cell division and cell elongation in the growth of the primary root of *Zea mays. Proceedings of the American Philosophical Society,* **100,** 499–514.

EVANS, H. J., NEARY, G. J. & TONKINSON, S. M. (1957). The use of colchicine as an indicator of mitotic rate in broad bean root meristems. *Journal of Genetics,* **55,** 487–502.

EVANS, H. J. & SAVAGE, J. R. K. (1959). The effect of temperature on mitosis and the action of colchicine in root meristem cells of *Vicia faba. Experimental Cell Research,* **18,** 51–61.

FANKHAUSER, G. (1934). Cytological studies on egg fragments of the salamander *Triton.* IV. The cleavage of egg fragments without the egg nucleus. *Journal of Experimental Zoology,* **67,** 349–93.

FERNÁNDEZ-GÓMEZ, M. E. (1968). Rate of DNA synthesis in binucleate cells. *Histochemie,* **12,** 302–6.

FERNÁNDEZ-GÓMEZ, M. E., GONZÁLES-FERNÁNDEZ, A. & GIMÉNEZ-MARTÍN, G. (1969). Interphase sensitivity to inhibition of DNA synthesis. *Mutation Research,* **8,** 419–21.

FRIEDBERG, S. H. & DAVIDSON, D. (1970). Duration of S phase and cell cycles in diploid and tetraploid cells of mixoploid meristems. *Experimental Cell Research,* **61,** 216–18.

GAVAUDAN, P. & CHAZELAS, S. (1969). Groupes et files cellulaires éléments de construction intermédiaires entre les cellules et les tissus dans l'organogénèse végétale. *Comptes rendus des séances de la Société de biologie*, **163**, 751–5.

GAVAUDAN, P. & GASTELIER, J. (1970). Organisation en groupes cellulaires de l'apex végétatif de *Phaseolus multiflorus* L. *Comptes rendus des séances de la Société de biologie*, **164**, 362–6.

GILBERT, C. W. (1972). The labelled mitoses curve and the estimation of the parameters of the cell cycle. *Cell and Tissue Kinetics*, **5**, 53–63.

GIMÉNEZ-MARTÍN, G., GONZÁLES-FERNÁNDEZ, A. & DE LA TORRE, C. (1971). Direct measurement of the G_1 phase in meristematic cells at two different temperatures. *Cell and Tissue Kinetics*, **4**, 563–7.

GIMÉNEZ-MARTÍN, G., GONZÁLEZ-FERNÁNDEZ, A. & LÓPEZ-SÁEZ, J. F. (1966). Duration of the division cycle in diploid, binucleate and tetraploid cells. *Experimental Cell Research*, **43**, 293–300.

GIMÉNEZ-MARTÍN, G., MEZA, I., DE LA TORRE, C., GONZÁLEZ-FERNÁNDEZ, A. & LÓPEZ-SÁEZ, J. F. (1971). Colchicine and caffeine in the estimation of cell flow and cycle time in *Allium* roots. *Cytologia*, **36**, 680–9.

GOLDFEDER, A. (1963). Radiosensitivity at the subcellular level. *Laval Médical*, **34**, 12–43.

GONZÁLEZ-FERNÁNDEZ, A., GIMÉNEZ-MARTÍN, G. & DE LA TORRE, C. (1971). The duration of the interphase periods at different temperatures in root tip cells. *Cytobiologie*, **3**, 367–71.

GONZÁLEZ-FERNÁNDEZ, A., GIMÉNEZ-MARTÍN, G. & LÓPEZ-SÁEZ, J. F. (1970). Cytokinesis at prophase in plants treated with ethidium bromide. *Experimental Cell Research*, **62**, 464–6.

HARTWELL, L. H. (1971). Genetic control of the cell division cycle in yeast. II. Genes controlling DNA replication and its initiation. *Journal of Molecular Biology*, **59**, 183–94.

HOTTES, C. F. (1929). Studies in experimental cytology. *Plant Physiology*, **4**, 1–30.

HOWARD, A. & DEWEY, D. L. (1961). Non-uniformity of labelling rate during DNA synthesis. *Experimental Cell Research*, **24**, 623–4.

HOWARD, A. & PELC, S. (1953). Synthesis of deoxyribonucleic acid in normal and irradiated cells and its relation to chromosome breakage. *Heredity*, **6**, suppl., 261–73.

INNOCENTI, A. M. (1971). Cytophotometric determination of histone content in cell nuclei of proliferating and non-proliferating root meristem cells of *Triticum durum*. *Caryologia*, **24**, 457–61.

JAKOB, K. M. & BOVEY, F. (1969). Early nucleic acid and protein syntheses and mitoses in the primary root tips of germinating *Vicia faba*. *Experimental Cell Research*, **54**, 118–26.

JONES, R. L. & PHILLIPS, I. D. J. (1966). Organs of gibberellin synthesis in light-grown sunflower plants. *Plant Physiology*, **41**, 1381–6.

KABACK, M. M. & BERNSTEIN, L. H. (1970). Biological studies of trisomic cells growing *in vitro*. *Annals of the New York Academy of Sciences*, **171**, 526–36.

KIRK, D. & JONES, R. N. (1970). Nuclear genetic activity of *B*-chromosome rye, in terms of the quantitative interrelationships between nuclear protein, nuclear RNA and histone. *Chromosoma, Berlin*, **31**, 241–54.

KOVACS, C. J. & VAN'T HOF, J. (1971). Mitotic delay and the regulating events of plant cell proliferation: DNA replication by a G_1/S population. *Radiation Research*, **48**, 95–106.

KOVACS, C. J. & VAN'T HOF, J. (1972). Changes in polysome activity related to radiation-induced mitotic delay in synchronized root meristem cells. *Radiation Research*, **49**, 530–42.

KUSANAGI, A. & YANAGI, T. (1970). Cytochemical determination of changes in nuclear histone content in differentiating root cells of barley and garlic. *Protoplasma*, **69**, 279–82.

LÓPEZ-SÁEZ, J. F., GIMÉNEZ-MARTÍN, G. & GONZÁLEZ-FERNÁNDEZ, A. (1966). Duration of the cell division cycle and its dependence on temperature. *Zeitschrift für Zellforschung und mikroskopische Anatomie*, **75**, 591–600.

LÓPEZ-SÁEZ, J. F., RISUEÑO, M. C. & GIMÉNEZ-MARTÍN, G. (1966). Inhibition of cytokinesis in plant cells. *Journal of Ultrastructure Research*, **14**, 85–94.

LÖVLIE, A. (1964). Genetic control of division rate and morphogenesis in *Ulva mutabilis* Föyn. *Comptes rendus des travaux du Laboratoire de Carlsberg*, **34**, 77–168.

MACDONALD, P. D. M. (1970). Statistical inference from the fraction labelled mitoses curve. *Biometrika*, **57**, 489–503.

MARCUS, A., FEELEY, J. & VOLCANI, T. (1966). Protein synthesis in imbibed seeds. III. Kinetics of amino acid incorporation, ribosome activation and polysome formation. *Plant Physiology*, **41**, 1167–72.

MENDELSOHN, M. & TAKAHASHI, M. (1970). A critical evaluation of the fraction labeled mitoses method as applied to the analysis of tumor and other cell cycles. In *The Cell Cycle and Cancer*, ed. R. Baserga, chapter 3. New York: Marcel Dekker.

MERTZ, D. (1966). Hormonal control of root growth. I. *Plant and Cell Physiology*, **7**, 125–35.

MORY, Y. Y., CHEN, D. & SARID, S. (1972). Onset of deoxyribonucleic acid synthesis in germinating wheat embryos. *Plant Physiology*, **49**, 20–3.

MURÍN, A. (1966). The effects of temperature on the mitotic cycle and its time parameters in root tips of *Vicia faba*. *Naturwissenschaften*, **53**, 312–13.

NAKAYAMA, F. & SÍVORI, E. M. (1968). Planta de 'Achira' ('*Canna*' sp.) obtenida de semilla de 550 años approximadamente. *Revista de la Facultad de Agronomía, La Plata, Argentina*, **44**, 73–82.

PHILLIPS, H. L. & TORREY, J. G. (1971). The quiescent center in cultured roots of *Convolvulus arvensis* L. *American Journal of Botany*, **58**, 665–71.

PILET, P.-E. (1971). Rôle de l'apex radiculaire dans la croissance, le géotropisme et le transport des auxines. *Bulletin de la Société botanique suisse*, **81**, 52–65.

QUASTLER, H. & SHERMAN, F. G. (1959). Cell population kinetics in the intestinal epithelium of the mouse. *Experimental Cell Research*, **17**, 420–38.

REMBUR, J. (1970). Etude autoradiographique et cytophotométrique de la synthèse du DNA au cours des premières heures de la germination chez le *Xanthium pennsylvanicum* Wallr. (Ambrosiacées). *Comptes rendus hebdomadaires des séances de l'Académie des sciences*, **271**, 908–11.

SCHULTZ, G. A., CHEN, D. & KATCHALSKI, E. (1972). Localization of a messenger RNA in a ribosomal fraction from ungerminated wheat embryos. *Journal of Molecular Biology*, **66**, 379–90.

SINCLAIR, W. K. (1968). Cyclic X-ray response in mammalian cells *in vitro*. *Radiation Research*, **33**, 620–43.

SITTON, D., RICHMOND, A. & VAADIA, Y. (1967). On the synthesis of gibberellins in roots. *Phytochemistry*, **6**, 1101–5.

SKULT, H. (1969). Growth and cell population kinetics of tritiated thymidine labelled roots of diploid and autotetraploid barley. *Acta Academiae åboensis*, B, **29**, 1–15.

SOSSOUNTZOV, L. (1969). Incorporation de précurseurs tritiés des acides nucléiques dans les méristèmes apicaux du sporophyte de la fougère aquatique, *Marsilea drummondii* A. Br. *Revue générale de botanique*, **76**, 109–56.

STEBBINS, G. L. & PRICE, H. J. (1971). The developmental genetics of the *cal-caroides* gene in barley. I. Divergent expression at the morphological and histological level. *Genetics*, **68**, 527–38.

STEBBINS, G. L. & YAGIL, E. (1966). The morphogenetic effects of the hooded gene in barley. I. The course of development in hooded and awned genotypes. *Genetics*, **54**, 727–53.

SWIFT, H. (1950). The constancy of deoxyribose nucleic acid in plant nuclei. *Proceedings of the National Academy of Sciences, U.S.A.* **36**, 643–54.

TAKAHASHI, M. (1968). Theoretical basis for cell cycle analysis II. Further studies on labelled mitosis wave method. *Journal of Theoretical Biology*, **18**, 195–209.

THOMPSON, S. H. (1960). Cellular development and morphogeny of the root tip of *Trillium grandiflorum*. *Botanical Gazette*, **121**, 215–20.

TIMMIS, J. (1971). Nuclear and genetic changes in *Linum*. Ph.D. Thesis, University of Wales.

TROY, M. R. & WIMBER, D. E. (1968). Evidence for a constancy of the DNA synthetic period between diploid–polyploid groups in plants. *Experimental Cell Research*, **53**, 145–54.

TRUCCO, E. & BROCKWELL, P. J. (1968). Percentage labeled mitoses curves in exponentially growing cell populations. *Journal of Theoretical Biology*, **20**, 321–37.

VAN DE WALLE, C. (1971). Caractérisation du premier RNA synthetisé au début de la germination de l'embryon de maïs. *Archives internationales de physiologie et de biochimie*, **79**, 852–3.

VAN DE WALLE, C. & BERNIER, G. (1967). L'incorporation de la thymidine tritiée dans le cytoplasme précéde l'incorporation dans le noyau cellulaire au cours de la germination de *Zea mays*. *Comptes rendus hebdomadaires des séances de l'Académie des sciences*, **265**, 1599–601.

VAN DE WALLE, C. & BERNIER, G. (1969). The onset of cellular activity in roots of germinating corn. *Experimental Cell Research*, **55**, 378–84.

VAN'T HOF, J. (1965). Relationship between mitotic cycle duration, S period duration and the average rate of DNA synthesis in the root meristem cells of several plants. *Experimental Cell Research*, **39**, 48–58.

VAN'T HOF, J. (1966). Comparative cell population kinetics of tritiated thymidine labeled diploid and colchicine-induced tetraploid cells in the same tissue of *Pisum*. *Experimental Cell Research*, **41**, 274–88.

VAN'T HOF, J. & YING, H.-K. (1964). Simultaneous marking of cells in two different segments of the mitotic cycle. *Nature, London*, **202**, 981–3.

VOSA, C. G. & BARLOW, P. W. (1972). Meiosis and B-chromosomes in *Listera ovata* (Orchidaceae). *Caryologia*, **25**, 1–8.

WEBSTER, P. L. & VAN'T HOF, J. (1970). Recovery of G_2 cells in pea root meristems: survival and mitotic delay following irradiation. *Radiation Botany*, **10**, 145–54.

WEEKS, D. P. & MARCUS, A. (1971). Preformed messenger of quiescent wheat embryos. *Biochimica et Biophysica Acta*, **232**, 671–84.

WIMBER, D. E. (1963). Methods for studying cell proliferation with emphasis on DNA labels. In *Cell Proliferation*, ed. L. F. Lamerton & R. J. M. Fry, pp. 1–17. Oxford: Blackwell.

WIMBER, D. E. (1966a). Duration of the nuclear cycle in *Tradescantia* root tips at three temperatures as measured with ^{3}H-thymidine. *American Journal of Botany*, **53**, 21–4.

WIMBER, D. E. (1966b). Prolongation of the cell cycle in *Tradescantia* root tips by continuous gamma irradiation. *Experimental Cell Research*, **42**, 296–301.

YANAGI, T. & KUSANAGI, A. (1970). Cytochemical and biological determination of changes in nuclear histone content during differentiation of barley root cells. *Development, Growth & Differentiation*, **12**, 1–12.

THE CELL CYCLE IN THE SHOOT APEX

By R. F. LYNDON

Department of Botany, University of Edinburgh,
Mayfield Road, Edinburgh, EH9 3JH

Almost all our knowledge of the cell cycle in the apical meristems of higher plants has been gained from the use of seedling roots. It is not hard to see why. The root apex is easily seen, actively growing roots can easily be obtained and manipulated in large numbers, and they readily absorb radioactive substances from solutions. In contrast, the shoot apex has been neglected. It is small, usually about 0.1 or 0.2 mm in diameter, it is enshrouded in the young leaves, and it is difficult to label with radio-isotopes. Nor does the shoot grow in a steady state, for it changes its shape as it initiates new leaves at regular intervals, each interval being a plastochron.

MEAN CELL GENERATION TIMES FOR THE WHOLE APEX

The production of new leaves at the shoot apex has the advantage that it provides a series of markers which represent a time scale against which to measure growth. The first estimates of the mean cell generation time (MCGT) appear to be those of Richards (1951) for the apices of *Dryopteris* and *Lupinus*. The cell doubling times were assumed to be the same as the volume doubling times which were calculated from the rates of radial displacement of leaf primordia at the shoot apex. The MCGTs estimated by this method range from 47 h for *Triticum* to 48 days for *Elaeis*, the oil palm (Table 1).

Direct determinations of the MCGT have been made by measuring the number of cells in the apical dome and the total number of cells produced by it in the course of a plastochron, i.e. the number of cells in the apical dome, the youngest primordium and its associated axial tissue. The MCGT in plastochrons for the cells of the apical dome can then be found. Since the rate of leaf initiation (the plastochron) is easily measured, the MCGT in hours can be calculated. The values obtained in this way are all one or more days (Table 1), and indicate that the cell cycle tends to be longer than in the root apex where it may be as short as 6 or 8 h (Barlow, this symposium). Only in *Vicia*, excised and grown in culture, has a MCGT as short as 8 h been recorded for the shoot (Ball & Soma, 1965). These

Table 1. *Mean cell generation time (MCGT) in vegetative shoot meristems*

Plant	MCGT (h)	Reference
Elaeis*	1150	Rees (1964)
Dryopteris*	480	Richards (1951)
Lupinus*	120	Richards (1951)
Chrysanthemum*	60	Schwabe (1971)
Chrysanthemum*	50	Berg & Cutter (1969)
Triticum*	47	Evans & Berg (1971)
Tradescantia	96	Denne (1966b)
Lupinus	72	Sunderland & Brown (1956)
Trifolium	64	Denne (1966a)
Secale	48	Sunderland (1961)
Lonicera	45	Edgar (1961)
Lupinus	32	Sunderland (1961)
Pisum	28	Lyndon (1968a)

An asterisk denotes those plants in which the MCGT was calculated from the rate of radial expansion at the apex. In the others, the MCGT was calculated from the rate of increase of cell number.

values are, of course, averages for the whole meristem and take no account of the variations in the length of the cell cycle which we may reasonably expect there to be in different parts of the apex.

CYCLE LENGTHS IN DIFFERENT REGIONS OF THE APEX

It has for long been thought that the cells at the extreme summit of the shoot apex are dividing slowly, except when the apex becomes floral, but the evidence for this has until recently been rather circumstantial. Much of it has depended on the assumption that the mitotic index is proportional to the rate of cell division (which is probably usually correct, see below), and that the metabolic activity of the cells could be inferred from their staining characteristics. The cells of the central zone, at the summit of the apical dome, characteristically stain more lightly and have a mitotic index about 50% lower than the cells further down on the flanks of the apex where the leaves are initiated (Nougarède, 1967).

The first direct measurement of the rates of cell division in the different regions of the shoot apex was made by Denne (1966a) who used the method of accumulation of colchicine-metaphases to measure rates of cell division in *Trifolium*. Although the accumulation of metaphases with colchicine gives a measure only of the number of cells dividing per unit time, the length of the cell cycle in the different regions of the apex can be calculated if we know what proportion of cells in each region is meristematic. What evidence we have suggests that all the cells in the shoot apex are meristematic. This is thought to be so because (at least in *Pisum*) there are no regions from which cell divisions are absent (Lyndon, 1970a), DNA is synthesised in all parts of the apex (Lyndon, 1972b), and all cells appear to be metabolically active in that they synthesise RNA (Lyndon, 1972a).

Measurements have also been made of the length of the cell cycle in the various regions of the apices of three other species by the method of metaphase accumulation, and these are summarised in Table 2. Cycle times for the subapical meristem and the region of the incipient pith have not been included because the cells in this region soon start to differentiate and the rate of division may slow down. Estimates of cycle times in this region are therefore liable to vary widely from species to species, according to the detailed structure of the apex (Denne, 1966a; Lyndon, 1970a).

Table 2. *Length of the cell cycle (hours) at the summit (central zone) and on the flanks (I_1 and I_2)* of the shoot meristem as measured from the accumulation of colchicine-metaphases*

Except where shown, all apices were vegetative

	Region of apex			
	Central zone	I_2	I_1	Reference
Trifolium	108	87	69	Denne (1966a)
Pisum	69	30	28	Lyndon (1970a)
Chrysanthemum	140		70	Berg (quoted in Gifford & Corson, 1971)
Datura	76		36	Corson (1969)
Datura (floral)	46		26	Corson (1969)

* Those parts of the apical dome where the next leaf primordium will be formed (I_1), and where the next one after that will be formed (I_2).

Although the cells of the central zone have a longer cycle than the cells on the flanks of these apices, it is only about twice as long (Table 2). This is not a very marked difference when we consider that in the root the difference in cycle times between the slowly dividing and faster dividing regions of the meristem is often sixfold or more (Barlow, this symposium). The cycle times for the pea apex were shorter than for the other plants, the flank cells (I_1 and I_2) having a cycle time of 28 to 30 h. It was noted that in the pea apex, in the presence of 0.5% colchicine, metaphases accumulated at a maximum rate of only about 1% per hour (Lyndon, 1970a). This was not sufficient to account for the known rate of cell division in the apex as a whole which was 2.5% per hour which had been calculated from the MCGT obtained from the rate of increase of cell number. The colchicine-metaphase data were therefore corrected to give values for the rates of division which correspond to the cycle times given in Table 2. It was concluded that the colchicine, at the concentration of 0.5% which was necessary to eliminate anaphases and telophases, was inhibiting entry of the cells into metaphase as well as exit from it. Corson (1969) also used 0.5% colchicine on the *Datura* apex but there was no independent check on the MCGT so it is possible that the values for *Datura* represent apices which have been inhibited to some extent by the colchicine. In the cases of *Trifolium* and *Chrysanthemum*, independent estimates gave MCGTs somewhat shorter than the cell cycles calculated from the rate of accumulation of colchicine-metaphases. The MCGT (from cell counts) was 64 h in *Trifolium*, whereas the shortest cycle time measured with colchicine was 69 h (Denne, 1966a). In *Chrysanthemum* the MCGT (volume doubling time) was 50 h but the shortest cycle time measured with colchicine was 70 h (Berg, quoted in Gifford & Corson, 1971).

Even if some of the values in Table 2 err on the side of too long a cell cycle, it seems probable that the shortest cycle in the shoot apex is usually not much less than a day. Whatever their accuracy these values are almost certainly correct in giving the relative lengths of the cell cycle in the slowly dividing and faster dividing cells of the apices, for it was shown for the pea that whatever their rates of accumulation, the relative numbers of colchicine-metaphases in different regions of the apex remained the same (Lyndon, 1970a).

In any measurements of the absolute length of the cell cycle or its phases by methods which involve treating the plant with chemicals or with radioisotopes, or in any way which could conceivably alter the rate of growth, it is obviously desirable, indeed essential, to have an independent check on the MCGT by measuring the rate of increase of cell number directly by counting cells.

The only phase of the cell cycle which could be measured in these experiments was mitosis, which proved to be remarkably constant for a given species. The length of time spent in mitotis was, in each plant, the same for all cells of an apex irrespective of the length of the cell cycle. Within an apex the mitotic index therefore appears to be proportional to the rate of cell division and inversely proportional to the length of the cell cycle when the plant is vegetative and growing under constant conditions. This means that most of the inferences that have been made about the relative rates of cell division in vegetative apices on the basis of mitotic index are fortuitously correct (Nougarède, 1967). However, such inferences are no substitute for direct measurements of rates of division or cycle times.

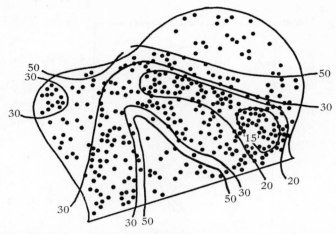

Fig. 1. Rates of cell division and length of the cell cycle in a median longitudinal section of the shoot apex of *Pisum*. Each point represents a colchicine-metaphase and the density of points is proportional to the rate of cell division (Lyndon, 1970*a*). The values are the approximate lengths of the cell cycle (h) for cells lying on the lines.

Values such as those in Table 2, which group together all the cells in a particular region of the apex, obscure the more detailed variation in cycle lengths which there is within an apex. These variations can be mapped by plotting the positions of colchicine-metaphases, since the density of the accumulated metaphases is proportional to the rates of cell division (Fig. 1). Average values for the density of colchicine-metaphases can be converted into the corresponding values for the lengths of the cell cycle. These have been superimposed as contours on the diagram of the pea apex (Fig. 1). We can see that the average values for the cell cycle (Table 2) do not do justice to the gradations in cycle length which there are within the apex, and that the fastest cell cycles may be as short as 15 h (this was at 23 °C).

In the shoot apex cells are continually being formed at the summit as growth occurs and so cells become displaced down the apical dome and eventually into the leaf primordia. Fig. 2 illustrates how a cell would have its cell cycle accelerated as it was displaced down the apical dome and into a leaf primordium in successive plastochrons. A cell would go through a period when each cycle was shorter than the last. In fact, apart from the cells at the extreme summit of the apex it seems unlikely that many cells are ever in a state in which the length of the cell cycle is constant for even a single cycle. Since it seems that all phases of the cell cycle (except mitosis) may be extended or shortened when the cycle length changes (see later) we may presume that the cells, in their transition from one region to another

Approx.
cycle length
(h)

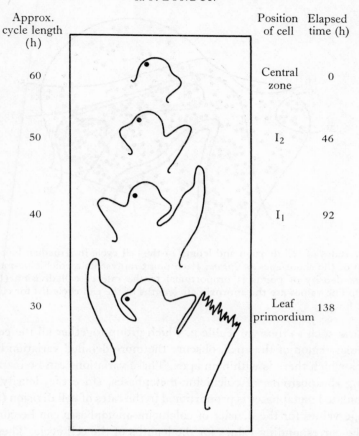

Position of cell	Elapsed time (h)
Central zone	0
I_2	46
I_1	92
Leaf primordium	138

60

50

40

30

Fig. 2. The dot represents a cell which is displaced down the apical dome, by the growth of the cells above it, and shows how it is found in different regions of the apex in successive plastochrons.

where the cycle time is shorter, will adjust by a general speeding up of their progress through all the phases of the cycle, except mitosis itself.

We ought to bear in mind that in long-term labelling experiments, which last for about a plastochron or more, a cell which is labelled in one part of the apex will be observed later in some quite different region.

SYNTHESIS DURING THE CELL CYCLE

The average composition of cells in all parts of the pea shoot apex is the same and is approximately 12 pg DNA, 9 pg RNA and 70 pg protein (Lyndon, 1970b). The composition of comparable cells in the pea root tip is 16 pg DNA, 79 pg RNA and 287 pg protein (Lyndon, 1968b). The cells of the shoot meristem have to accumulate much less RNA and protein and yet their cell cycle is almost certainly longer than that of the root meristem cells. The greater length of the cell cycle in the shoot apex is therefore not because more material has to be accumulated – on the contrary. The presumed slowness of RNA synthesis in the shoot apex is reflected in the slowness with which the nucleoli increase in size over the cell cycle and their much smaller size at preprophase than in the root (Lyndon, 1968b). The slower rate of nucleolar growth in the shoot apex cannot be ascribed to there being fewer nucleolar (ribosomal) genes per nucleus in the shoot than in the root, for such differences do not seem to exist (Ingle & Sinclair, 1972). The cells of the central zone, with a long cell cycle, increase their RNA and protein content at a slower rate than the cells on the flanks of the apex with a shorter cycle. But this does not necessarily imply a lower metabolic activity of the central zone cells. The uptake of precursors into RNA and protein can be as fast or faster than in the cells with a shorter cycle (Lyndon, 1972a) but it is not yet clear whether this implies a rapid rate of synthesis of RNA and protein (with a concomitant faster turnover). It may be merely that the central zone cells are more efficient at concentrating exogenous precursor molecules in the first place, or have smaller precursor pools which reach a higher specific activity than elsewhere in the apex.

LENGTHS OF THE PHASES OF THE CELL CYCLE

All the measurements considered so far have been of the whole cell cycle. Measurement of the lengths of the component phases of the cell cycle in the shoot apex, by the use of radioisotopes, has just been started. The reason for the lack of data is at least partly due to the difficulty of doing labelling experiments with the shoot apex. Firstly, it is difficult to get label in at all. Whereas the root is an absorbing organ which readily takes up solutes, this is not the case for the shoot. Often substances applied to the intact shoot apex hardly find their way into the plant. Clowes (1959) tried to overcome the problem by using aquatic plants which could be immersed in the radioactive solution, but, even so, the plants had to be immersed in the solution for one or more days in order to label the nuclei of the apex. Gifford, Kupila & Yamaguchi (1963) tried a number of ways of labelling the apex but, even with the most effective method, application of the radioactive solution directly onto the apical dome, the plants had to be exposed to the radioisotope for 24 h or longer to get labelled nuclei. Bernier & Bronchart (1963) also compared many methods of applying labelled compounds to shoot apices and they found the only effective method to be partial defoliation before applying the label. Nuclei then become labelled within 0.5 h. Autoradiographs show that the radioactive substances are probably taken up via the stumps of the excised leaves. Although this technique is successful it suffers from the disadvantage that the amount of mutilation to which the plant is necessarily subjected might well be producing a wound reaction in the form of an alteration of the length of the cell cycle and its various phases in the shoot meristem under investigation. The need to dissect every plant before application of label also seriously limits the number of plants that can be used in any experiment. This can be a serious limitation if it is necessary to score the percentage of mitotic figures which become labelled, since the mitotic index in the shoot apex is characteristically very low, of the order of one or two per cent.

A further difficulty is that the long cycle times in the shoot apex mean that experiments often have to go on for several days, which raises the possibility of radiation damage and consequent alterations of the cell cycle times. All these difficulties are compounded when we come to look at what may be some of the most interesting cells in the shoot apex, the slowly dividing cells of the central zone. Since the central zone may consist of only about 100 cells (e.g. in the pea) and since the mitotic index is least in this region, about 1 %, this means that there is, on average, only one mitotic figure per plant in this region. Therefore, in order to get any idea of the percentage of mitotic figures labelled, a large number of plants is needed, which as we have already seen is itself difficult.

In view of these difficulties it is perhaps hardly surprising that so far there are only two sets of measurements of the lengths of the component phases of the cell cycle in the shoot apex. The first of these illustrates some of the difficulties. Michaux (1969) measured the lengths of the component phases of the cell cycle in the water plant *Isoetes*, and her results are summarised in Table 3. The length of the whole cell cycle in the lateral zone (where leaves are initiated) was measured by supplying [³H]thymidine continuously by immersion of the plants in the solution. The length of the whole cell cycle could be measured only by the time taken to reach a maximum number (93 %) of labelled nuclei. The pulse-labelling method was tried but had to be abandoned because there were so few mitoses. The time spent in DNA synthesis (S) had to be estimated from the percentage of nuclei labelled 6 h after the label was introduced, for this was the earliest that labelled nuclei could be detected. The time spent in mitosis (M) was estimated from the mitotic index. $G_2 + M$ was taken as the time for the first labelled telophases to appear. G_1 was obtained by difference. Measurements of cycle phases in the central zone were much more restricted because only 12 % of the nuclei had become labelled after 36 h (so that the

Table 3. *Lengths of the phases of the cell cycle (hours) in the lateral regions of the apex of* Isoetes *(from Michaux, 1969)*

Phase of cell cycle...	Whole cycle	G_1	S	G_2	M
Length (h)	36	12.5	5.5	15	3

length of the whole cycle could not be found) and after 36 h necrosis of the apex set in and no more nuclei became labelled. Only $G_2 + M$ could be estimated, from the appearance of the first labelled anaphase, as 34 h, twice as long as in the lateral regions. Even assuming that the lengths of the other phases of the cell cycle in the central zone were no longer than in the lateral regions, the cycle time in the central zone would be 52 h, and was probably much more.

Table 4. *Duration (hours) of the cell cycle and its component phases in four regions of the apical bud of* Rudbeckia bicolor *(from Jacqmard, 1970)*

	Whole cycle	$G_1 + M/2$	S	$G_2 + M/2$
Peripheral zone	30.1	9.0	11.6	9.5
Central zone	–	–	19.2	14.8
Pith-rib meristem	30.5	6.8	15.9	7.8
Subapical pith	32.9	13.1	13.0	6.8

More complete measurements were made by Jacqmard (1970) on the apex of *Rudbeckia*. The apices were supplied for 30 min with [³H]thymidine which was then washed off with water. The percentage of labelled mitotic figures was scored at intervals during the subsequent 40 h and the values which were obtained for the lengths of the phases of the cell cycle are given in Table 4. Again, the length of the cell cycle and most of its phases in the central zone could not be obtained because of the long cycle time. Except for a lengthening of G_1 in the subapical pith, the length of the cell cycle and its phases was similar in all parts of the apex except the central zone where both S and G_2 were extended. The only other measurements of the length of a phase of the cycle is for the pea, in which S was estimated (from the proportion of nuclei labelled after 2 h) as 11 h in the central zone as opposed to 7 to 8 h elsewhere in the apex (Lyndon, 1972a). These meagre data suggest that S and G_2 in the central zone are both longer than in the peripheral regions of the apex, possibly twice as long. Since the whole cycle in the central zone is about double what it is elsewhere in the apex (Table 2) then G_1 would be expected to be about twice as long in the central zone as on the flanks of the apex.

Phases of the cell cycle in Pisum

In an attempt to get some idea of the lengths of the component phases of
the cell cycle in the central zone as well as in the rest of the meristem an
analysis of the pea shoot apex was made by the method of Mak (1965).
This consists essentially of labelling the apex with [³H]thymidine so that
the nuclei in S become labelled. The apices are then sectioned, stained with
Feulgen, and exposed to autoradiographic emulsion. The unlabelled
nuclei are those in G_1 and G_2 and can be distinguished by measurement of
their DNA content with a microdensitometer so that the proportions of G_1
and G_2 nuclei can be found. The proportion of cells in M is given by the
mitotic index, and in S by the proportion of labelled nuclei. The relative
lengths of the component phases of the cell cycle, which are related to the
number of cells in each phase, can then be calculated. When the length of
the cell cycle is known then the absolute lengths of the phases can also be
found.

For analysis each section was divided in the ways already described
(Lyndon, 1968a) into (1) the central zone, distinguished by its larger
nuclei which are therefore less intensely stained; (2) the regions on the
flanks of the apical dome, i.e. I_1 which will form the next leaf primordium
and I_2 which will form the next primordium after that; and (3) the primordium of the youngest leaf.

The DNA values of the unlabelled nuclei fell into two distinct groups
representing 2 C (up to 35 units of DNA) and 4 C nuclei (36 and more units
of DNA) (Fig. 3). The proportions of 2 C and 4 C nuclei in each of the
regions of the apex were similar to those of all the nuclei shown in Fig. 3.
It seemed possible that these might not be the true proportions of 2 C
and 4 C nuclei since not all nuclei could be measured and underestimation
of the numbers of 4 C nuclei could have occurred since these, being larger,
might be more likely to have been cut or to have overlapped other nuclei.
A check on the proportions of 2 C and 4 C nuclei was possible because the
nuclear diameters of 2 C and 4 C nuclei were sufficiently different to allow
the number of 2 C nuclei to be estimated from the frequency of nuclei
with diameters below the median for 2 C nuclei (Fig. 4). The diameters
of *all* nuclei in the sections were therefore measured (including cut and
overlapping nuclei) and their frequency distribution is shown in Fig. 5.
It is at once clear that the proportion of 2 C nuclei was not less, and may
perhaps be greater, than in the sample for which DNA was also measured
(Fig. 4). When the frequency distributions of nuclear diameters shown in
Figs. 4 and 5, and also for the separate regions of the apex, were plotted as
probability curves it was equally apparent that the proportions of 2 C

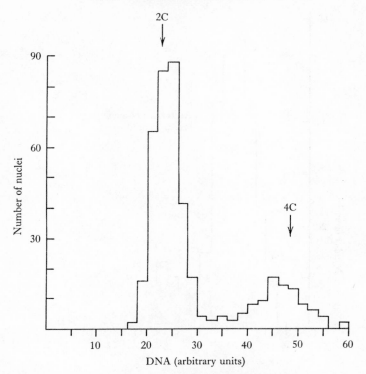

Fig. 3. The frequency of nuclei with different DNA contents in the shoot apex of *Pisum*. The arrows indicate the values for the DNA content of telophase nuclei (2 C; mean of 13) and prophase plus metaphase nuclei (4 C; mean of 41). The apices were fixed in 70% ethanol, which contracted the nuclei, making measurement of individual nuclei easier. The DNA contents of all unlabelled nuclei which were uncut, did not overlap other nuclei, and could be optically isolated, were measured (using a Barr and Stroud integrating microdensitometer) in a total of 8 median and adjacent longitudinal sections (12 for the central zone) taken from 4 plants. Sections were 10 μm thick.

and 4 C nuclei which were found in the sample in which DNA was also measured could be accepted and used in the calculations that follow.

The percentages of nuclei which were labelled in these sections are shown in Table 5, together with the proportions of nuclei in G_1, G_2 and M. The proportion of the whole cell cycle spent in each of these phases can be calculated using the formulae given by Nachtwey & Cameron (1968) which correct for the age gradient in a population of cells whose number is increasing exponentially. Using the values for the length of the cell cycle given by Lyndon (1970a) the lengths of the phases of the cycle can be calculated (columns A, Table 6). Before considering the significance of these values, it is worth comparing them with a further set of values (columns B, Table 6),

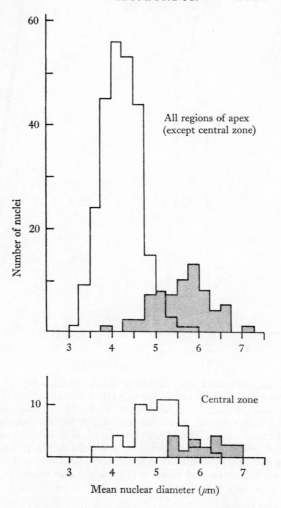

Fig. 4. The frequency of 2 C (clear) and 4 C (shaded) nuclei with different diameters in the sample of nuclei in which DNA was measured (shown in Fig. 3). Values for the central zone are shown separately because these nuclei appear to be larger than nuclei with the same amount of DNA in the rest of the apex.

which have been obtained by using published values for the mean DNA content per nucleus as a basis for calculating the proportions of G_1 and G_2 nuclei. Mean amounts of DNA per nucleus in the different regions of the pea apex were measured and expressed as multiples of the C amount (Lyndon, 1970b). Using the values for the percentages of nuclei in S and M given in Table 5, the proportions of G_1 and G_2 nuclei can be calculated as follows.

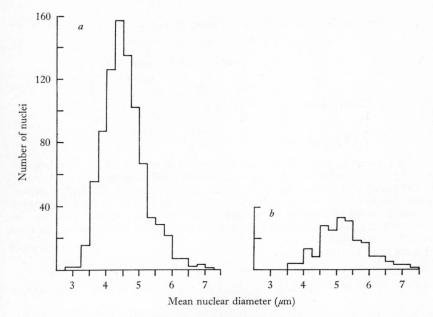

Fig. 5. The frequency of nuclei with different diameters when all unlabelled nuclei were measured in (a) all regions of the apex except the central zone, and (b) the central zone. The proportion of 2 C nuclei is at least as high as in the smaller sample in which DNA was measured (shown in Fig. 4).

Where c is the mean C value for DNA per cell, let s, m, and g be the fractions of nuclei that are in S, M and G_1 respectively. The DNA content of nuclei in M will be 4 C, in S is assumed to be on average 3 C, in G_1 will be 2 C and in G_2 will be 4 C. Then,

$$c = 3s + 4m + 2g + 4\,(1 - s - m - g),$$

and this reduces to
$$g = \frac{4 - s - c}{2},$$

and so gives the fraction of the total number of nuclei in G_1, from which the fraction in G_2 can also be calculated.

Applying this formula to the data of Lyndon (1970*b*) the percentages of cells in the phases of the cell cycle were obtained and the lengths of the phases calculated (columns B, Table 6). They turn out to be very similar to those in columns A, Table 6. If anything, G_2 is a little shorter by this second method, which is consistent with the possibility that the percentage of nuclei in G_2 may have been overestimated by the first method.

The values in Table 6 show that in the slowly dividing cells of the central zone of the pea all the phases of the cell cycle (except M) have been

Table 5. *Proportions of cells in the different phases of the cell cycle in four regions of the shoot apex of* Pisum

	Percentage of total number of cells in			
	G_1	S*	G_2	M†
Central zone	63	17	19	1
I_2	53	23	22	2
I_1	62	24	12	2
Leaf primordium	55	29	14	2

* The percentage of labelled nuclei in the sections. The nuclei were labelled for 1 h by the method of Lyndon (1972a). Although the sections were 10 μm thick, labelled nuclei and nuclear fragments at all levels in the section were recorded. Presumably this was because during the exposure the section and emulsion were desiccated so that all parts of the section were within 3 μm of the emulsion. On rehydration the sections regained their former thickness of 10 μm which they retained when mounted.

† Data from Lyndon (1970a).

Table 6. *Lengths of the phases of the cell cycle (hours) in four regions of the shoot apex of* Pisum

Columns A: data obtained by the method of Mak (1965). Columns B: values derived from published data (see text).

	Whole cycle	G_1		S		G_2		M	
		A	B	A	B	A	B	A	B
Central zone	69	38	37	13	13	17	18	1	1
I_2	30	13	15	7	8	9	6	1	1
I_1	28	15	15	8	8	4	4	1	1
Leaf primordium	29	14	16	9	10	5	2	1	1

extended, S by about 50%, G_1 by about 250%, and G_2 by 250% or more. This is the first direct evidence that in the slowly dividing cells at the summit of the shoot apex both G_1 and G_2 are extended. These values, together with those for *Rudbeckia* (Jacqmard, 1970), indicate that in the shoot apex the cells which are dividing more slowly than the rest tend to have the whole of interphase extended rather than a single phase as tends to be the case in the root (Barlow, this symposium). This may suggest that the control of the rate of cell division and of progress through the cell cycle is effected in different ways in the root and the shoot meristems. The situation may be different when the growth of the whole apex is inhibited, for in *Tradescantia* axillary buds which were inhibited by auxin nearly all the nuclei were held at the 2 C level (Naylor, 1958). Before we can extend our speculations we need information about the levels of growth substances in the root and the shoot, and how these might affect the different phases of the cell cycle.

To get information about the cell cycle in the shoot it should be a relatively straightforward matter to examine the meristems of different species, measure the average DNA content per nucleus by the methods already used (Lyndon, 1970b), find the percentage of nuclei labelled with a terminal label of [³H]thymidine, and measure the mitotic index. From these data, even if the length of the cell cycle is not known, it should be possible to estimate and compare the relative lengths of the phases of the cell cycle in different parts of the apex. Although measuring the phases of the cell cycle by labelling experiments alone has its difficulties, it may be easier if new methods were devised which relied on smaller numbers of plants and used terminal labelling where possible.

THE TRANSITION FROM VEGETATIVE TO FLORAL GROWTH

The cell cycle becomes speeded up during floral induction and evocation. Although this has been inferred mostly from the uncertain evidence of changes in the mitotic index (Nougarède, 1967) there is also some direct evidence that this happens. In lupin and in vernalised rye cell counts showed that the cell cycle became faster when the apices underwent transition from vegetative to floral growth, whereas in continued vegetative growth the cell cycle lengthened (Sunderland, 1961). In *Datura* the cell cycle in all parts of the apex was shortened to about 60 to 70% of that in the vegetative apex (Corson, 1969) but the differential between the cells at the summit of the apex and the cells on the flanks was maintained (Table 2). A very similar situation has been found in *Sinapis* (M. Bodson, personal communication). In addition to a speeding up of the cell cycle there may also be some synchronisation of the cells during floral induction, soon after the floral stimulus reaches the apex (Bernier, 1971). This is inferred from the slight increase in the mitotic index at this time and in *Sinapis* from the marked increase in the proportion of cells with the 2 C amount of DNA (Jacqmard & Miksche, 1971).

In *Rudbeckia* flowering is brought about by long days or applications of gibberellic acid (Jacqmard, 1965). In both cases the mitotic index and the percentage of nuclei which incorporated [³H]thymidine increased in the central zone and the subapical meristems, indicating that both M and S occupied a greater part of the cell cycle in the induced plants and hence G_1 or G_2 or both were reduced in proportion. In several other plants too, gibberellic acid treatment resulted in an increase in the proportions of cells in S and M even though the plants did not flower (Bernier, Bronchart, Jacqmard & Sylvestre, 1967).

The evidence we have so far suggests that changing levels of growth substances in the apex could bring about changes in the lengths of the cell cycle and in the rate of progress from one phase of the cycle to the next, but we do not yet know whether these changes in the cell cycle are essential or merely incidental to the morphogenetic processes which occur at the shoot apex.

REFERENCES

BALL, E. & SOMA, K. (1965). Effect of sugar concentration on growth of the shoot apex of *Vicia faba*. In *Proceedings of an International Conference on Plant Tissue Culture*, ed. P. R. White & A. R. Groves, pp. 269–85. Berkeley: McCutchan Publishing Corp.

BERG, A. R. & CUTTER, E. G. (1969). Leaf initiation rates and volume growth rates in the shoot apex of *Chrysanthemum*. *American Journal of Botany*, **56**, 153–9.

BERNIER, G. (1971). Structural and metabolic changes in the shoot apex in transition to flowering. *Canadian Journal of Botany*, **49**, 803–19.

BERNIER, G. & BRONCHART, R. (1963). Application de la technique d'histoautoradiographie à l'étude de l'incorporation de thymidine tritiée dans les méristèmes caulinaires. *Bulletin de la Société royale des Sciences Liège*, **32**, 269–83.

BERNIER, G., BRONCHART, R., JACQMARD, A. & SYLVESTRE, G. (1967). Acide gibbérellique et morphogénèse caulinaire. *Bulletin de la Société royale de Belgique*, **100**, 51–71.

CLOWES, F. A. L. (1959). Adenine incorporation and cell division in shoot apices. *New Phytologist*, **58**, 16–19.

CORSON, G. E. (1969). Cell division studies of the shoot apex of *Datura stramonium* during transition to flowering. *American Journal of Botany*, **56**, 1127–34.

DENNE, M. P. (1966a). Morphological changes in the shoot apex of *Trifolium repens* L. 1. Changes in the vegetative apex during the plastochron. *New Zealand Journal of Botany*, **4**, 300–14.

DENNE, M. P. (1966b). Diurnal and plastochronal changes in the shoot apex of *Tradescantia fluminensis* Vell. *New Zealand Journal of Botany*, **4**, 444–54.

EDGAR, E. (1961). *Fluctuations in Mitotic Index in the Shoot of Lonicera nitida*. Christchurch: University of Canterbury, New Zealand.

EVANS, L. S. & BERG, A. R. (1971). Leaf and apical growth characteristics in *Triticum*. *American Journal of Botany*, **58**, 540–3.

GIFFORD, E. M. & CORSON, G. E. (1971). The shoot apex in seed plants. *Botanical Review*, **37**, 143–229.

GIFFORD, E. M., KUPILA, S. & YAMAGUCHI, S. (1963). Experiments in the application of H³-thymidine and adenine-8-C¹⁴ to shoot tips. *Phytomorphology*, **13**, 14–22.

INGLE, J. & SINCLAIR, J. (1972). Ribosomal RNA genes and plant development. *Nature, London*, **235**, 30–2.

JACQMARD, A. (1965). Comparison des actions de la photopériode et de l'acide gibbérellique sur le méristème caulinaire de *Rudbeckia bicolor* Nutt. *Bulletin Société Française de Physiologie Végétale*, **11**, 165–70.

JACQMARD, A. (1970). Duration of the mitotic cycle in the apical bud of *Rudbeckia bicolor*. *New Phytologist*, **69**, 269–71.

JACQMARD, A. & MIKSCHE, J. (1971). Cell population and quantitative changes of DNA in the shoot apex of *Sinapis alba* during floral induction. *Botanical Gazette*, **132**, 364–7.

LYNDON, R. F. (1968a). Changes in volume and cell number in the different regions of the shoot apex of *Pisum* during a single plastochron. *Annals of Botany*, **32**, 371–90.

LYNDON, R. F. (1968b). The structure, function and development of the nucleus. In *Plant Cell Organelles*, ed. J. B. Pridham, pp. 16–39. London: Academic Press.

LYNDON, R. F. (1970a). Rates of cell division in the shoot apical meristem of *Pisum*. *Annals of Botany*, **34**, 1–17.

LYNDON, R. F. (1970b). DNA, RNA, and protein in the pea shoot apex in relation to leaf initiation. *Journal of Experimental Botany*, **21**, 286–91.

LYNDON, R. F. (1972a). Nucleic acid synthesis in the pea shoot apex. In *Proceedings of a Conference of the Hungarian Academy of Sciences*, in press.

LYNDON, R. F. (1972b). Leaf formation and growth at the shoot apical meristem. *Physiologie Végétale*, **10**, 209–22.

MAK, S. (1965). Mammalian cell cycle analysis using microspectrophotometry combined with autoradiography. *Experimental Cell Research*, **39**, 286–9.

MICHAUX, N. (1969). Durée des phases du cycle mitotique dans le méristème apical de l'*Isoetes setacea* Lam. *Comptes Rendues Académie Sciences, Paris*, Série D, **269**, 1396–9.

NACHTWEY, D. S. & CAMERON, I. L. (1968). Cell cycle analysis. In *Methods in Cell Physiology*, vol. 3, ed. D. M. Prescott, pp. 213–59. New York & London: Academic Press.

NAYLOR, J. M. (1958). Control of nuclear processes by auxin in axillary buds of *Tradescantia paludosa*. *Canadian Journal of Botany*, **36**, 221–32.

NOUGARÈDE, A. (1967). Experimental cytology of the shoot apical cells during vegetative growth and flowering. *International Review of Cytology*, **21**, 203–351.

REES, A. R. (1964). The apical organization and phyllotaxis of the oil palm. *Annals of Botany*, **28**, 57–69.

RICHARDS, F. J. (1951). Phyllotaxis: its quantitative expression and relation to growth in the apex. *Philosophical Transactions of the Royal Society*, B, **235**, 509–64.

SCHWABE, W. W. (1971). Chemical modification of phyllotaxis and its implications. In *Control Mechanisms of Growth and Differentiation, Symposia for the Society for Experimental Biology*, **25**, 301–22.

SUNDERLAND, N. (1961). Cell division and expansion in the growth of the shoot apex. *Journal of Experimental Botany*, **12**, 446–57.

SUNDERLAND, N. & BROWN, R. (1956). Distribution of growth in the apical region of the shoot of *Lupinus albus*. *Journal of Experimental Botany*, **7**, 127–45.

CHANGES IN ENZYME ACTIVITIES DURING THE DIVISION CYCLE OF CULTURED PLANT CELLS

By M. M. YEOMAN and P. A. AITCHISON

Botany Department, University of Edinburgh,
Mayfield Road, Edinburgh EH9 3JH

The concept of a cell cycle has been formed as a result of numerous studies with prokaryotic cells or undifferentiated cells of eukaryotic origin, which have shown that there is an ordered progression of events in the life of a cell from one division to the next. This progression has been well documented in bacteria (Donachie & Masters, 1969), yeasts (Mitchison & Creanor, 1969), animal (Petersen, Tobey & Anderson, 1969) and lower-plant cells (Schmidt, 1969; Rusch, 1969). The investigation of events occurring in a cell cycle necessitates an amplification stage, either in the measuring system, using very sensitive techniques to monitor changes in individual cells, or in that which is being measured, by using more material of identical nature. The former approach is limited in its usefulness, especially when applied to enzyme assays, by the paucity of techniques which allow quantitative assessment of enzymes in individual cells. The latter technique is that of using naturally synchronous or synchronised populations of cells, which magnifies the changes in individual cells to a level at which they can be easily measured. It is this approach that has been adopted in most cases.

The investigation of temporal relationships in the cell cycle of higher plants has been severely hindered by the lack of suitable synchronous systems. In part, this might be explained by the impossibility of employing techniques which collect cells at common points in the cell cycle when using multicellular aggregates. Secondly most plant tissues contain a diversity of specialised cells and might not be expected to yield cells with similar cycle durations or characteristics, even if they could be brought to a common starting point by some external agent.

A low level of synchrony occurs naturally in some plant tissues, but the extent of synchrony is too small to be useful (Erickson, 1964). Some success has been achieved in increasing the degree of synchrony by the use of inhibitors. Mattingly (1966) improved synchrony in *Vicia faba* root meristems by treatment with 5 amino-uracil, and Kovacs & van't Hof (1970) produced a high degree of mitotic synchrony in cultured pea roots with fluorodeoxyuridine (FUdR) accompanying carbohydrate starvation. Apart from the general criticisms that can be levelled at systems synchronised with inhibitors, these systems have the further disadvantage that a variety of cell types are present.

Hotta & Stern (1961, 1963a, b, 1965) have exploited a system which possesses a high degree of natural synchrony, the developing microspores of *Lilium longiflorum*. This system has the advantage that the very long cell cycle (*c.* 21 days between the first and second divisions of meiosis) allows a leisurely sampling programme. A marked periodicity in thymidine kinase (TdR kinase) was observed (Hotta & Stern, 1963a). The enzyme appeared briefly, before DNA synthesis, coincidentally with the appearance of an increased pool of deoxyribosides, and evidence was presented (Hotta & Stern, 1965) that the periodicity in appearance of the enzyme was due to induction by thymidine. The sudden increase in deoxyribosides was in turn explained by an increase in DNAase in the anther tissue surrounding the developing microspores. The increase in TdR kinase activity is not solely due to variation in the supply of an inducer for a permanently inducible system, as induction by supplied thymidine could only be demonstrated at the normal time of appearance of the enzyme in whole anthers, or shortly before (Hotta & Stern, 1965). The system, however, cannot be regarded as typical of higher plant cell cycles, being a meiotic division in a highly specialised cell type.

One recent line of attack on higher plant systems, successfully developed with animal cells, has been to overcome the complication of cellular interaction by separating the cells into single or 'near-single' cell suspensions. It might be possible to use selection techniques similar to those that have been applied to algal cultures in order to obtain homogeneous synchronous populations. Alternatively, it may be easier to induce synchrony in cell suspensions. Eriksson (1966) was able to partially synchronise suspension cultures of *Haplopappus gracilis*, but again using inhibitors (5 aminouracil and hydroxyurea) which caused chromosome damage at the concentrations used. Roberts & Northcote (1970) have partially synchronised a suspension culture of *Acer pseudoplatanus* with kinetin. A high degree of synchrony in tobacco cell suspensions has recently been achieved by the temporary withholding and later addition of kinetin, an essential growth requirement for tobacco cultures (Jouanneau, 1971). Sycamore cells may also be brought into a synchronous condition by a dilution 'shock' treatment (Wilson, King & Street, 1971). This synchrony is apparently maintained through several divisions but has the disadvantage that the low cell densities at which cultures are grown make extraction and estimation of enzymes very difficult.

An alternative system exists in higher plants in some largely homogeneous tissues in a quiescent condition that can be induced to embark on a course of rapid cell division by presenting them with growth substances. At least in one case, that of explants from the storage parenchyma of mature Jerusalem artichoke tubers cultured in a mineral medium containing 2,4-dichlorophenoxyacetic acid (2,4-D) there is an initial period of synchronous division in which a high percentage of the cells take part (Yeoman & Evans, 1967). This synchrony has completely disappeared by the fourth division. The system has the disadvantage that it may be difficult to distinguish between events reflecting the change from quiescence to growth, those due to wounding, and periodic changes specifically associated with the cell cycle. In this context it is important to establish which of the catalytic changes are related to division by an examination of the pattern of change in similar tissue not preparing for division.

Several increases in the activity of enzymes in cultured Jerusalem artichoke cells have been described and these may be categorised. The pattern of events depends on the conditions of culture. When incubated in water, the cells expand and some protein and RNA synthesis appear to be associated with this expansion (Masuda, 1966). Auxins may either promote expansion (e.g. Flood, Rutherford & Weston 1967) or initiate a different pattern of synthesis, leading to synchronous cell divisions and ultimately to the formation of a callus. Which direction is taken after auxin treatment depends on the method of preparation of the tissue and the composition of the medium, e.g. calcium concentration (Setterfield, 1963). The increases in levels of enzymes may be broadly divided into two groups, (1) Changes not related to cell division and (2) Changes related to cell division.

CHANGES NOT RELATED TO CELL DIVISION
Increases related solely to tissue damage

Fresh material from mature tubers contains detectable amounts of peroxidase activity. When disks are incubated for 24 h in buffer ($+$ 1 mM magnesium acetate) the measurable peroxidase activity increases markedly after a lag of about 8 h (Bastin, 1970). This increase seems to be dependent on protein synthesis as it is inhibited by cycloheximide (1.8×10^{-5} M). Phenol oxidase activity also increases over this time, and the increase is sensitive to 1.5×10^{-4} M chloramphenicol (generally a less effective inhibitor of protein synthesis in plants) as well as cycloheximide. However, Bastin (1968) showed that the rate of increase of both these enzymes, on a unit weight of tissue, increased with decreasing thickness of disks. This was interpreted as showing that the increase was directly dependent on the degree of injury imparted to the tissue, although it could equally well be explained by increased accessibility of the tissue to stimuli from the medium or increased gaseous exchange due to the increased surface area to volume ratio.

Acid phosphatase in the external layers of cells of artichoke explants also increases over 24 h independently of the presence of external growth factors. This enzyme, which was measured histochemically (Yeoman & Mitchell, 1970), is associated with the lysosomes of autolysing cells. In the same material a transient 5–8-fold increase in DNAase (Fig. 1) and a progressive 2-fold increase in thymidine monophosphatase also seem to be a property of excised material not associated with the cell cycle (Harland, 1971), as they were observed both in cultures which went on to divide and in those which did not. Acid phosphatase, thymidine monophosphatase (an alkaline phosphatase) and DNAase all increased in activity after excision, but in none of these cases was the response affected by the addition of 2,4-D and the phenomenon should probably be placed, along with increases in peroxidase and phenol oxidase, as responses to wounding. No information is available on whether protein synthesis is necessary for these increases.

Washed aerated disks also develop an invertase activity associated with cell walls (Edelman & Hall, 1965). This activity is absent from fresh tissue and developed to a lesser extent in the presence of chloramphenicol (33 % of control with 6.2×10^{-3} M chloramphenicol), fluorophenyl-alanine (44 %, 2 mM) or thiouracil (30 %, 1 mM). Similar increases in soluble ascorbic

---

Fig. 1. DNAase activity in explants during culture with and without 2,4-D. (Harland, 1971).

Assay mixtures contained, in a volume of 160 μl: 11 μmol tris-maleate buffer, pH 5.6; 40 μg bovine serum albumin (b.s.a.); 4.5 μmol β-mercaptoethanol (MCE); 1 μg native (Jerusalem artichoke) DNA (c. 6000 cpm); 0.75 μmol MgCl$_2$; 0.1 ml cell-free extract. After 10 min at 30 °C in a shaking water bath, the reaction was stopped by addition of 0.1 ml carrier DNA (0.2 mg) and 0.2 ml 20% (w/v) perchloric acid. After centrifugation (1000 g, 10 min at 0 °C), the supernatant was removed and an aliquot neutralised for scintillation counting. Total possible hydrolysis was measured by replacing the cell-free extract by a DNAase preparation (30 μg bovine pancrease DN-C DNAase 1).

For estimation of rates of DNA synthesis (○) batches of 16 explants were introduced into normal culture medium containing [6-^{3}H]thymidine (3 μCi/ml, 1.5–1.8×10^{-7} M) for 45 min. They were then washed and transferred to medium containing unlabelled thymidine (3×10^{-4} M) for 15 min. Explants were removed and fixed in 4 ml methanol. Total nucleic acid was extracted and hydrolysed in perchloric acid (Yeoman & Mitchell, 1970) and aliquots of the hydrolysate removed for scintillation counting.

Batches of 100 explants were incubated in mineral salts and 4% sucrose alone (1b) or with 10^{-6} M 2,4-D (1a). Hatching on the time axis, in this and subsequent figures, indicates the period of cell division.

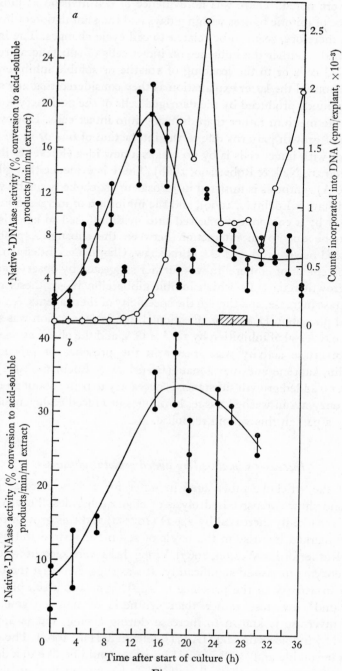

Fig. 1

oxidase were noted. There was no evidence of cell division as judged by appearance of mitotic figures within 3 days and the stimulation of invertase does not, therefore, seem to be related to cell cycle changes. The increase was ascribed to either the influence on intact cells of substances produced in wounded cells or to the leaching of volatile or soluble inhibitors. The authors favoured the latter explanation as they considered that any specific new compounds produced by the damaged cells of the periphery would be lost into the medium rather than diffusing into intact cells. However, the initiation of growth patterns due to the interaction of one or more wound substances with intact cells is by no means a new idea (Fosket & Roberts, 1965; Yeoman, Naik & Robertson, 1968). There is evidence that gibberellic acid (GA) synthesis is initiated in Jerusalem artichoke tissue by wounding (Bradshaw & Edelman, 1969), and the inhibition of increased invertase production by a component released into medium used to age disks was shown to be due to an interaction between the substance, probably a protein, and newly produced GA (Bradshaw, Chapman & Edelman, 1970). The involvement of gibberellin was further suggested by observations that several growth retardants which inhibit gibberellin biosynthesis reduced the invertase increase, and though the specificity of these agents, Amo-1618, CCC and phosphon D, may be questioned, the interpretation was supported by the reversal of inhibition by 10^{-4} M GA, and the observation that the rise in invertase activity was greater in the presence of precursors of gibberellin, kaurene and mevalonate (Bradshaw & Edelman, 1971). Thus although no added growth substance is necessary to bring about an increase in these enzymes in washed tissue, hormones produced endogenously may well play a part in this wound response.

Increases stimulated by added growth substances

Disks of the artichoke tuber aged in water for 3 days increase in fresh weight and show extensive hydrolysis of oligosaccharides. This expansion growth is markedly increased by 2,4-D (10^{-5} M) and this growth regulator caused a marked increase in the levels of soluble invertase and hydrolase (Flood, Rutherford & Weston, 1967). When disks were aged in water alone, neither enzyme increased significantly. It was suggested that the observed increase in activity in the presence of 2,4-D reflected a solubilisation of 'wall-bound' invertase rather than synthesis of new enzyme. 'Wall-bound' invertase is known to increase during ageing, but is apparently inhibited by indoleacetic acid (IAA) (Edelman & Hall, 1964). The opposite changes in soluble and 'wall-bound' invertase could be due to a decreased binding of invertase to the cell wall. The observation that no significant increase in total protein levels were noted over 3 days in 2,4-D medium was also held to suggest that possibly no new protein synthesis was necessary, although this contrasts with findings of other workers (e.g. Masuda, 1966). However, a simple transfer cannot account for the subsequent observation that under the conditions of Rutherford's cultures there was an increase in cell-wall bound invertase as well as soluble invertase, in the presence of 2,4-D. Furthermore, the increases in invertase and hydrolase were inhibited by actinomycin D (90% by 10 mg/l) added prior to or together with the auxin, which suggests that continued synthesis of mRNA and protein are necessary.

In contrast to the observation of Yeoman & Mitchell (1970), Palmer (1970) did not find any rise in activity of acid phosphatase with β-glycero-phosphate as substrate and only a small transient increase with ATP as substrate, in the absence of added growth substances. The enzyme was detectable in fresh material and showed a 3-fold increase in response to 2,4-D (3×10^{-5} M) or IAA (10^{-4} M). The following evidence was presented that the increase does not require synthesis of new enzyme protein: (a) cycloheximide (2×10^{-4} M) did not prevent the rise in phosphatase activity; (b) 2,4-D and 2,4,6-trichlorophenoxyacetic acid (2,4,6-T) were more or less equally effective in stimulating activity although 2,4,6-T is much less efficient an auxin than 2,4-D; (c) phosphatase could be activated by homogenising fresh disks without exposure to auxin. Palmer (1970) suggested that the effect of auxins was to cause a breakdown in compartmentalisation in the cells, perhaps releasing phosphatase from lysosomes. Similar autolysis was observed by Yeoman & Mitchell (1970) whether 2,4-D was present or not. The reason for this discrepancy is not clear, although possibly a difference in the preparation of the material is responsible. The excised material of Yeoman & Mitchell was introduced directly into the growth medium, whereas that of Palmer was first washed for 20 min in distilled water. This treatment might remove auxins or other substances released by the external layers of damaged cells, which would otherwise interact with the intact tissue and elicit the release of phosphatase in the absence of added auxin. If this interpretation is correct the stimulated phosphatase activity measured by Palmer is part of a wound response and the requirement for exogenous auxin an artefact.

CHANGES RELATED TO CELL DIVISION

When artichoke explants are incubated without ageing in the dark in a medium containing mineral salts, sucrose and an auxin (10^{-6} M, 2,4-D), a high proportion of the cells divide synchronously (Yeoman, 1970) and some

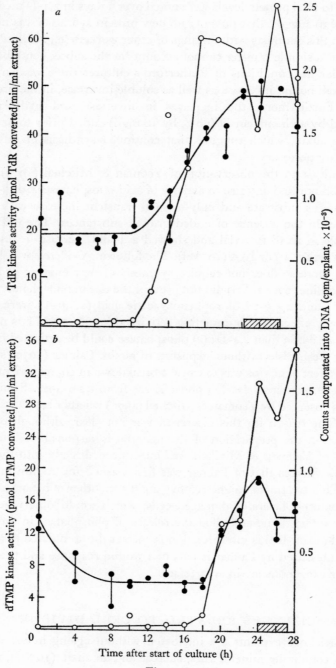

Fig. 2

parameters for example DNA (Mitchell, 1967), RNA (Yeoman & Davidson 1971) increase in a stepped fashion. Harland (1971) has shown that some enzymes increase in a more or less stepped manner before division. TdR kinase and dTMP kinase showed stepped increases which occurred coincidentally with or just after the onset of DNA synthesis (Fig. 2). DNA polymerase also increased after DNA replication had started, and continued to increase through the first division (Fig. 3) In the absence of 2,4-D, no increase in any of these activities could be detected.

The temporal precedence of initiation, at least, of DNA synthesis over these increased rates of enzyme synthesis (assuming for the present that increased activity does reflect an increased synthesis) could have several explanations:

(*a*) Transcription is restricted in these cells to the period of DNA replication, and transcription is followed soon afterwards by an increase in enzyme protein. Enzyme levels prior to 'S' must be due to stable mRNA or protein produced since the previous round of DNA synthesis.

(*b*) Transcription occurs at a constant rate depending solely on the number of appropriate gene copies and as these copies are replicated they immediately become active in transcription and the extra mRNA is immediately available from translation. This is similar to the gene dosage effect in bacteria where the expression of activity is limited by the number of gene copies specifying the appropriate mRNA. Observed levels of enzyme are due to the balance between synthesis and degradation of mRNA and protein, which changes only at 'S'.

Fig. 2. Thymidine kinase and thymidine monophosphate kinase in explants during culture in a medium with 2,4-D (Harland, 1971).

(*a*) TdR kinase. Assay mixtures contained, in a volume of 110 μl: 20 μmol phosphate buffer, pH 8.0; 20 μg b.s.a.; 2.25 μmol MCE; 3.87 nmol thymidine labelled with [6-^{3}H]thymidine (2.62 μCi); 0.25 μmol phospho-enol pyruvate (PEP); 0.1 mg PEP kinase; 50 μl cell-free extract. After 20 min at 37 °C in a shaking water bath, the reaction was stopped by placing reaction tubes in water at 100 °C for 2 min. Precipitated protein was removed by centrifugation (1000 g, 5 min) and dTMP in the supernatant was separated from excess substrate by subjecting a 10 μl aliquot to cellulose acetate paper electrophoresis at 4 °C in 0.05 M ammonium formate buffer, pH 3.5, 200 volts for 4 h. The air-dried papers were cut into 0.5 cm strips for scintillation counting.

(*b*) dTMP kinase. Assay mixtures contained in a volume of 100 μl; 20 μmol phosphate buffer, pH 8.0; 20 μg b.s.a.; 2.25 μmol MCE; 0.5 nmol [CH$_3$-^{3}H]thymidine monophosphate (0.5 μCi); 25 nmol MgCl$_2$; 0.41 μmol ATP; 0.25 μmol PEP; 100 μg PEP kinase; 50 μl cell-free extract. Incubation as for TdR kinase. Thymidine diphosphate in the supernatant was separated from excess substrate by high-voltage paper electrophoresis of a 30 μl aliquot in 0.5 M ammonium formate buffer, pH 3.5, 1000 volts for 1 h. Dried papers were cut into 1 cm strips for scintillation counting. The rate of DNA synthesis was estimated as described in Fig. 1.

7

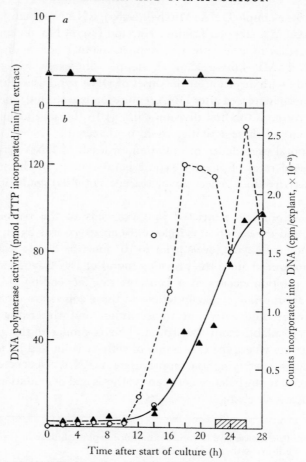

Fig. 3. DNA polymerase during culture without (*a*) and with 2,4-D (*b*) (Harland, 1971). Assay mixtures contained in a volume of 350 μl: 50.5 μmol tris-maleate buffer, pH 8.0; 40 μg b.s.a.; 8.46 μmol MCE; 3.6 nmol [2-^{14}C]thymidine triphosphate (164 nCi); 0.19 μmol each of dATP, dGTP, dCTP; 0.43 μmol ATP; 4.95 μmol $MgCl_2$; 1 μl denatured DNA; 100 μl cell-free extract. After 30 min at 30 °C, the reaction was stopped by addition of 25 μl b.s.a. (10 mg/ml) and 0.6 ml 7% perchloric acid. The precipitate after centrifugation (2000 g, 5 min) was successively dissolved in 0.3 ml 0.2 M-NaOH and reprecipitated with 0.6 ml 7% perchloric acid 4 times to remove excess dTTP. The pellet was finally dissolved in 1 ml 2 M-NH$_4$OH and counted on a planchet in a gas flow counter.

(*c*) Some other factor is limiting the level of the enzyme and, for this group, relaxation of the control shortly after the initiation of DNA synthesis is coincidental. For instance it might be due to a change in the structure of the nuclear envelope, increasing the flux of mRNA into the cytoplasm.

If either of the first two conditions apply generally in this system it might be predicted that an increase in the rate of synthesis of enzyme would only occur in the period following the onset of the 'S' period.

To test this the pattern of activity of some enzymes unrelated in function to the above group has been investigated. ATP-glucokinase (ATP-GK) was chosen as an enzyme of the glycolytic pathway and glucose 6-phosphate dehydrogenase (G6PDH) as an enzyme of the pentose–phosphate pathway. G6PDH was measured in crude extracts of artichoke explants by following the reduction of nicotine adenine dinucleotide phosphate (NADP). Activity was detected in cold-stored (2 °C) material and increased within a period of 30 h in culture. Closer examination of this increase revealed the pattern shown in Fig. 4b. Activity, as measured by the initial rate of reduction of pyridine nucleotide, is expressed as nmol NADP reduced/min/ml extract. Up to the time of cell division (c. 34 h) this is proportional to activity expressed on a per cell basis and is preferred to specific activity as total protein increases in a somewhat discontinuous manner in the tissue during this period. A typical value, however, for the initial specific activity is 60 nmol/min/mg protein. In a medium containing 2,4-D, G6PDH activity usually decreased initially. This decrease may reflect protein turnover, but apparent decreases in other parameters, e.g. cell number, RNA and total protein have been observed in the early stages of culture, and are believed to reflect the loss of cells from the periphery of explants. The fall was followed by a fairly sudden rise in activity, which in different experiments varied between a 60 and 150 % increase and occurred some 8–12 h after the start of culture. Thereafter the levels fluctuated to some extent but no progressive change appeared before the first wave of division. In the absence of 2,4-D, no significant increase in activity was observed (Fig. 4b). ATP-GK showed a very similar pattern (Fig. 4a). The specific activity in fresh material was very low, c. 0.1 nmol/min/mg protein and could not be measured by spectrophotometric means, in the presence of excess commercial G6PDH. Instead ATP-GK was assayed by a modification of the ion-exchange paper disk method of Newsholme, Robinson & Taylor (1967), in which the excess [^{14}C]glucose substrate was separated from G6P formed by washing it off DEAE-cellulose paper disks which were then counted in a scintillation counter. Details of the assays will be published elsewhere. This low activity probably accounts for the inability of Chong-Jin (1971) to measure hexokinase activity in extracts from artichoke tubers. A feature noted by this worker was a high endogenous rate of NADP reduction. We have not noted this in our extracts. The difference might be due to a physiological difference in the material, higher levels of NADP reductase or the presence of any substrate the oxidation of which may be linked to NADP. Alternatively such substrates may have been

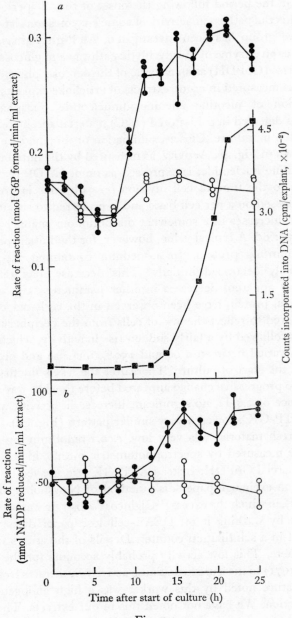

Fig. 4.

introduced by the use of coconut milk in the culture medium used by Chong-Jin.

Both G6PDH and ATP-GK increase in activity significantly before the onset of DNA replication as measured by the incorporation of tritiated thymidine (Fig. 4), i.e. in the G_1 period. In this respect they obviously differ from the enzymes reported by Harland (1971). Furthermore, the implication that the increased activity is in no way related to DNA synthesis was substantiated by the demonstration that FUdR present from the start of the culture period did not prevent the increase in activity of G6PDH, (Table 1).

Table 1. *The effect of FUdR on development of G6PDH activity*

Material	G6PDH (nmol NADP reduced/min/ml extract)	
	Expt. 1	Expt. 2
Fresh	51.2	51.9
Incubated without 2,4-D	51.6	44.2
Incubated with 2,4-D	69.2	90.2
Incubated with 2,4-D + FUdR (9×10^{-5} M)	64.0	92.6

If cycloheximide (2×10^{-5} M) was present, neither G6PDH nor ATP-GK increased significantly (Fig. 5). This suggests that the increases observed are probably due to protein synthesis rather than activation of pre-existing enzyme (though the possibility of a decreased rate of degradation cannot be eliminated (Trewavas, 1972).

Fig. 4. (a) ATP glucokinase in explants cultured with (●) and without (○) 2,4-D. Assay mixtures contained in a volume of 0.3 ml; 24 μmol tris-HCl buffer, pH 8.0; 40 μg b.s.a.; 7.2 μmol MCE; 30 nmol glucose labelled with [U-^{14}C]glucose; 0.32 μmol ATP; 2 μmol MgCl$_2$; 0.1 ml cell-free extract. After 30 min at 25 °C, 40 μl aliquots were spotted onto DEAE–cellulose paper disks and immediately dried under an infra-red lamp. Excess glucose was removed from the disks by washing for 10 min in distilled water and the dried disks were counted in scintillation fluid. Specific activity of the glucose was estimated by counting 40 μl of incubation mixture on unwashed disks. The rate of DNA synthesis (■) was estimated as described in Fig. 1, except that a pulse time of 20 min and a chase of 10 min were used. Cell division started after c. 28 h.

(b) Glucose 6-phosphate dehydrogenase in explants cultured with (●) and without (○) 2, 4-D. Assay mixtures contained, in a volume of 2 ml: 106 μmol tris-HCl buffer pH 8.0; 40 μg b.s.a.; 108 μmol MCE; 2 μmol G6P; 0.5 μmol NADP; 10 μmol MgCl$_2$; 10 μl cell-free extract. The reaction was followed by measuring the reduction of NADP at 340 nm in an Unicam SP800 spectrophotometer. Measurements were made before and after addition of G6P to the reagent cuvette, against a blank from which G6P was omitted.

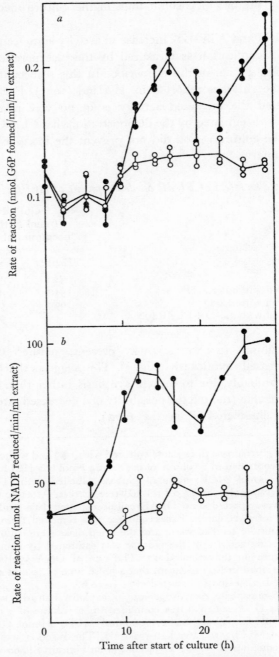

Fig. 5. The effect of cycloheximide on ATP-GK (5a) and G6PDH (5b) activities during culture of explants in 2,4-D. G6PDH and ATP-GK were assayed as described in Fig. 4. ●, Control; ○, with cycloheximide (2×10^{-5} M).

This putative demonstration of a stepped increase in enzyme synthesis in the pre-'S' phase means that no general obligate temporal link between DNA replication and the rate of enzyme synthesis exists in this population of dividing cells. In the cases of G6PDH and ATP-GK the controlling factor which determines the level of enzyme activity is not the amount of DNA present in the cell. Also this is clearly not the case for enzymes such as invertase in the artichoke under conditions in which large increases in activity are observed in expanding cells not synthesising DNA.

It is possible that DNA polymerase, TdR kinase and dTMP kinase are examples of a special condition, in that, during the 'S' period at least, the level of enzyme protein is determined directly by the amount of genetic information, and the observed levels are being restricted by no other control system. This would explain the correlation between levels of these enzymes and new DNA, which is particularly close for DNA polymerase (Jackson, private communication, 1972). As the increase in levels does not represent a strict doubling during 'S' it must be assumed that for the rest of the cell cycle some other factor controls the actual levels of enzyme (Varner, 1971). For other enzymes this applies throughout the cell cycle, and periodic increases would be due to relaxation of the limiting factor at that particular time.

REFERENCES

BASTIN, M. (1968). Effect of wounding on the synthesis of phenols, phenol oxidase and peroxidase in the tuber tissue of Jerusalem artichoke. *Canadian Journal of Biochemistry*, **46**, 1339–43.

BASTIN, M. (1970). Synthesis and degradation of peroxidase in slices of Jerusalem artichoke tuber cultivated in the presence of cycloheximide or anaerobically. *Comptes rendus de l'Académie des Sciences*, Série D, **271**, 1284–7.

BRADSHAW, M. J., CHAPMAN, J. M. & EDELMAN, J. (1970). Enzyme formation in higher plant tissue. A protein inhibitor of invertase synthesis secreted by tissue slices of plant storage organs. *Planta, Berlin*, **90**, 323–32.

BRADSHAW, M. J. & EDELMAN, J. (1969). Enzyme formation in higher plant tissue. The production of a gibberellin preceding invertase synthesis in aged tissue. *Journal of Experimental Botany*, **20**, 87–93.

BRADSHAW, M. J. & EDELMAN, J. (1971). The effects of growth substances and retardants on renewed growth processes of Jerusalem artichoke tuber tissues. *Journal of Experimental Botany*, **22**, 391–9.

CHONG-JIN, G. (1971). Respiratory enzyme systems in cultured callus tissues. *New Phytologist*, **70**, 389–95.

DONACHIE, W. D. & MASTERS, M. (1969). Temporal control of gene expression in bacteria. In *The Cell Cycle. Gene–Enzyme Interactions*, ed. G. M. Padilla, G. L. Whitson & I. L. Cameron, pp. 37–76. New York & London: Academic Press.

EDELMAN, J. & HALL, M. A. (1964). Effect of growth hormones on the development of invertase activity associated with cell walls. *Nature, London*, **201**, 296–7.

EDELMAN, J. & HALL, M. A. (1965). Enzyme formation in higher plant tissue. Development of invertase and ascorbic oxidase activities in mature storage tissue of *Helianthus tuberosus*. *Biochemical Journal*, **95**, 403–10.

ERICKSON, R. O. (1964). Synchronous cell and nuclear division in tissues of the higher plants. In *Synchrony in Cell Division and Growth*, ed. E. Zeuthen, pp. 11–37. New York: Wiley Interscience.

ERIKSSON, T. (1966). Partial synchronisation of cell division in suspension cultures of *Haplopappus gracilis*. *Physiologia plantarum*, **19**, 900–10.

FLOOD, A. E., RUTHERFORD, P. P. & WESTON, E. W. (1967). Effects of 2,4-dichloro-phenoxyacetic acid on enzyme systems in Jerusalem artichoke tubers and chicory roots. *Nature, London*, **214**, 1049–50.

FOSKET, D. E. & ROBERTS, L. W. (1965). A histochemical study of callus initiation from carrot tap root phloem cultivated in vitro. *American Journal of Botany*, **52**, 929–37.

HARLAND, J. (1971). Changes in the pattern of enzyme activities during the cell division cycle. Ph.D. Thesis, Edinburgh University.

HOTTA, Y. & STERN, H. S. (1961). Transient phosphorylation of deoxyribosides and regulation of deoxyribonucleic acid synthesis. *Journal of Biophysical and Biochemical Cytology*, **61**, 311–19.

HOTTA, Y. & STERN, H. S. (1963a). Molecular facets of mitotic regulation. I. Synthesis of thymidine kinase. *Proceedings of the National Academy of Sciences, U.S.A.* **49**, 648–54.

HOTTA, Y. & STERN, H. S. (1963b). Molecular facets of mitotic regulation. II. Factors underlying the removal of thymidine kinase. *Proceedings of the National Academy of Sciences, U.S.A.* **49**, 861–5.

HOTTA, Y. & STERN, H. S. (1965). Inducibility of thymidine kinase by thymidine as a function of interphase stage. *Journal of Cell Biology*, **25**, 99–108.

JOUANNEAU, J. P. (1971). Contrôle par les cytokinines de la synchronisation des mitoses dans les cellules de tabac. *Experimental Cell Research*, **67**, 329–37.

KOVACS, C. J. & VAN'T HOF, J. (1970). Synchronisation of a proliferative population in a cultured plant tissue. *Journal of Cell Biology*, **47**, 536–9.

MASUDA, Y. (1966). Auxin-induced growth of tuber tissue of Jerusalem artichoke. III. The relation to protein and nucleic acid metabolism. *Plant and Cell Physiology*, **7**, 75–91.

MATTINGLY, E. (1966). Synchrony of cell division in root meristems following treatment with 5-aminouracil. In *Cell Synchrony: Studies in Biosynthetic Regulation*, ed. I. L. Cameron & G. M. Padilla, pp. 256–68. New York & London: Academic Press.

MITCHELL, J. P. (1967). DNA synthesis during the early division cycles of Jerusalem artichoke callus cultures. *Annals of Botany, London*, **31**, 427–35.

MITCHISON, J. M. & CREANOR, J. (1969). Linear synthesis of sucrase and phosphatases during the cell cycle of *Schizosaccharomyces pombe*. *Journal of Cell Science*, **5**, 373–91.

NEWSHOLME, E. A., ROBINSON, J. & TAYLOR, K. (1967). A radiochemical enzymatic activity assay for glycerol kinase and hexokinase. *Biochimica et Biophysica Acta*, **132**, 338–46.

PALMER, J. M. (1970). The induction of phosphatase activity in thin slices of Jerusalem artichoke by treatment with indoleacetic acid. *Planta, Berlin*, **93**, 53–9.

PETERSEN, D. F., TOBEY, R. A. & ANDERSON, E. C. (1969). Essential biosynthetic activity in synchronous mammalian cells. In *The Cell Cycle. Gene–Enzyme Interactions*, ed. G. M. Padilla, G. L. Whitson & I. L. Cameron, pp. 341–60. New York & London: Academic Press.

ROBERTS, K. & NORTHCOTE, P. H. (1970). The structure of sycamore callus cells during division in a partially synchronised suspension culture. *Journal of Cell Science*, 6, 299–321.

RUSCH, H. P. (1969). Some biochemical events in the growth cycles of *Physarum polycephalum*. *Federation Proceedings*, 28, 1761–70.

SCHMIDT, R. R. (1969). Control of enzyme synthesis during the cell cycle of *Chlorella*. In *The Cell Cycle. Gene–Enzyme Interactions*, ed. G. M. Padilla, G. L. Whitson & I. L. Cameron, pp. 159–79. New York & London: Academic Press.

SETTERFIELD, G. (1963). Growth regulation in excised slices of Jerusalem artichoke tuber tissue. In *Cell Differentiation. Society of Experimental Biology Symposia*, 17, 98–126.

TREWAVAS, A. (1972). Control of the protein turnover rates in *Lemna minor*. *Plant Physiology*, 49, 47–51.

VARNER, J. E. (1971). The control of enzyme formation in plants. In *Control Mechanisms of Growth and Differentiation. Society for Experimental Biology Symposia*, 25, 197–206.

WILSON, S. B., KING, P. J. & STREET, H. E. (1971). Studies on the growth in culture of plant cells. XII. A versatile system for the large-scale batch or continuous culture of plant cell suspensions. *Journal of Experimental Botany*, 22, 177–207.

YEOMAN, M. M. (1970). Early development in callus cultures. *International Review of Cytology*, 29, 383–409.

YEOMAN, M. M. & DAVIDSON, A. W. (1971). Effect of light on cell division in developing callus cultures. *Annals of Botany, London*, 35, 1085–1100.

YEOMAN, M. M. & EVANS, P. K. (1967). Growth and differentiation of plant tissue cultures. II. Synchronous cell divisions in developing callus cultures. *Annals of Botany, London*, 31, 323–32.

YEOMAN, M. M. & MITCHELL, J. P. (1970). Changes accompanying the addition of 2,4-D to excised Jerusalem artichoke tuber tissue. *Annals of Botany, London*, 34, 799–810.

YEOMAN, M. M., NAIK, G. G. & ROBERTSON, A. I. (1968). Growth and differentiation of plant tissue cultures. III. The initiation and pattern of cell division in developing callus cultures. *Annals of Botany, London*, 32, 301–13.

REGULATION OF MACROMOLECULAR SYNTHESES IN SEA URCHIN OOGENESIS

By G. GIUDICE

Institute of Comparative Anatomy, University of Palermo, Italy

Oogenesis represents a very special moment of the cell cycle in which the cell is halted in the stage of diplotene of the meiotic prophase. This is also a very important moment for the future embryonic development, because during oogenesis several molecules are synthesised, and stored in the egg, so that the embryo can develop up to a certain stage with no need for further nuclear activity following fertilisation. An interesting case in this respect is that of the sea urchin egg which can develop up to the blastula stage even when the synthesis of RNA has been blocked by various means (Gross & Cousineau, 1964; Gross, Malkin & Moyer, 1964; Neyfakh, 1960; Giudice, Mutolo & Donatuti, 1968; Nemer, 1962; Bamberger, Martin, Stearns & Jolley, 1963; Crkvenjakov, Bajkovic & Glisin, 1970).

This implies that all the genetic information needed for the egg to develop up to blastula has been stored in the egg during oogenesis. It is therefore of interest to study the mechanisms of regulation of macromolecular syntheses during this stage. This has led Giudice and co-workers (1972) to develop a technique which allows the preparation in bulk of sea urchin oocytes (see Fig. 1; Plate 1). A cell fractionation method has also been devised that permits a satisfactory purification of germinal vesicles, nucleoli and cytoplasm.

Two main aspects of oocyte macromolecular synthesis, maturation of ribosomal RNA and the formation of histones, have been considered by Giudice and co-workers and will be briefly discussed in this article.

SYNTHESIS AND MATURATION OF RIBOSOMAL RNA

It has been demonstrated that there is little ribosomal RNA (rRNA) synthesis in the sea urchin egg up to the mesenchyme blastula stage, when it becomes several-fold accelerated (Nemer, 1963; Giudice & Mutolo, 1967; Giudice & Mutolo, 1969; Sconzo, Pirrone, Mutolo & Giudice, 1970; Sconzo & Giudice, 1971). Up to the blastula stage the metabolic needs of the embryo are met by a storage of rRNA present in the egg. Actually, according to a calculation presented by Sconzo *et al.* (1970), based on the measurement of the specific activity of the radioactive RNA precursor pool, the amount of rRNA synthesised by the sea urchin embryo in the

[203]

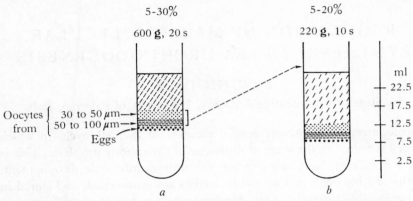

Fig. 1. Diagram showing the procedure for oocyte purification through two consecutive Ficoll gradients: (a) centrifuged 20 sec at 600 g on 5–30% gradient; (b) centrifuged 10 sec at 200 g on a 5–20% gradient. (From Giudice et al., Expl Cell Res. 72 (1972), 90–4.)

period between blastula and prism does not exceed 10% of that already present in the egg. It has therefore to be assumed that the oocytes are actively engaged in the synthesis of rRNA. This was first demonstrated by experiments of Gross, Malkin & Hubbard (1965) and Piatigorsky & Tyler (1967) in which radioactive RNA precursors, injected in the body cavity of sea urchin females, were recovered in the mature rRNA of the egg. These experiments did not, however, exclude the possibility that rRNA was synthesised in ovarian cells other than oocytes and then fed into the latter ones. This was disproved by experiments of Sconzo, Bono, Albanese & Giudice (1972) in which radioactive RNA precursors were administered to oocytes isolated by the method of Giudice et al. (1972). Under these conditions the radioactivity is almost entirely incorporated into the germinal vesicles and especially at the nucleolar level at least after 5 h of exposure to the isotope. This result is confirmed by autoradiography and is in agreement with earlier autoradiographic data obtained after injection of radioactive RNA precursor into the body cavity of the female (Ficq, 1964). The sedimentation and electrophoretic properties of the purified nuclear RNA labelled in isolated oocytes clearly show that it is entirely represented by rRNA precursor with the characteristics described in sea urchin gastrula by Sconzo et al. (1971) (see Fig. 2).

An interesting conclusion is, however, to be drawn when the kinetics of maturation of the rRNA precursor of gastrulae and oocytes are compared. It is actually apparent from the examination of Fig. 3 that at the gastrula stage the radioactivity from [³H]uridine is very quickly chased from the 33 S precursor into the product of its cleavage 28 S. A steady state labelling

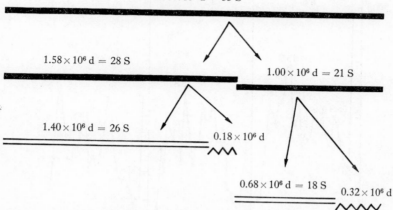

2.58×10^6 d = 33 S

1.58×10^6 d = 28 S

1.00×10^6 d = 21 S

1.40×10^6 d = 26 S

0.18×10^6 d

0.68×10^6 d = 18 S

0.32×10^6 d

Fig. 2. Tentative scheme for processing of the precursor of rRNA. The solid line indicates the immature products; the double line the mature products; the zig-zag line discarded fragments. (From Sconzo *et al.*, *Biochim. Biophys. Acta*, **232** (1971), 132, modified.)

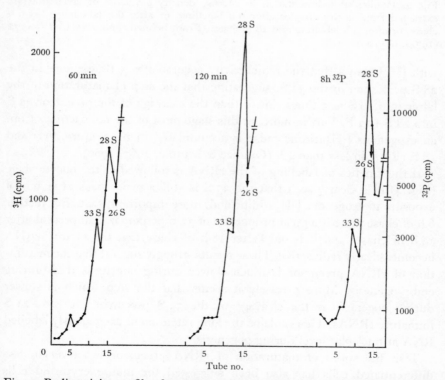

Fig. 3. Radioactivity profile of sucrose gradients of nucleolar RNA from gastrulae exposed to 0.5 μCi/ml of [5-³H]uridine (21 Ci/mmole) for 60 or 120 min, or to 4 μCi/ml of carrier-free ³²P for 8 h. The arrow points to the 26 S marker. (From Sconzo *et al.*, *Biochim. Biophys. Acta*, **232** (1971), 132.)

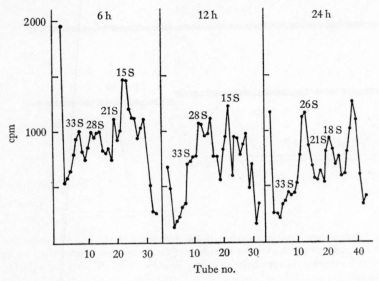

Fig. 4. Profiles of sedimentation in sucrose density gradients of labelled RNA extracted from entire oocytes after 6 h labelling, or after 6 h labelling plus 6 h chase, or after 6 h labelling and 18 h chase. (From Sconzo *et al.*, *Expl Cell Res.* **72** (1972), 95–100).

with [^{32}P] shows that the radioactivity accumulates 5 times more in the 28 S peak than in the 33 S, suggesting that the step of maturation of the latter into 26 S is 5 times slower than the cleavage of the precursor 33 S into 28 S + 21 S. Furthermore, at this stage most of the radioactivity from an exogenous [^{3}H]uridine pulse is accumulated in the mature 26 S and 18 S rRNA in less than 4 h (Giudice & Mutolo, 1967, 1969).

If the kinetics of labelling of the rRNA is followed at the oocyte stage (Fig. 4) it is clearly seen that the 33 S is still a major peak after 6 h of exposure to exogenous [^{3}H]uridine and, more important, that after further 6 h of chase there is a great proportion of 33 S percursor, still present after 18 h of chase; and it is only after 18 h of chase that the mature rRNA becomes clearly radioactive. These results strongly suggest that the maturation of rRNA precursor is much slower during oogenesis than during embryogenesis. More precisely it seems that the step which is slower during oogenesis is the cleavage of the 33 S precursor into 28 + 21 S immature rRNAs. This explains the long retention of exogenously labelled RNA at nucleolar level during oogenesis.

That the speed of maturation of rRNA precursor is slower in less differentiated cells has also been suggested for mouse erythroid cells (Fantoni & Ghiara, 1971). The maturation of rRNA precursor represents

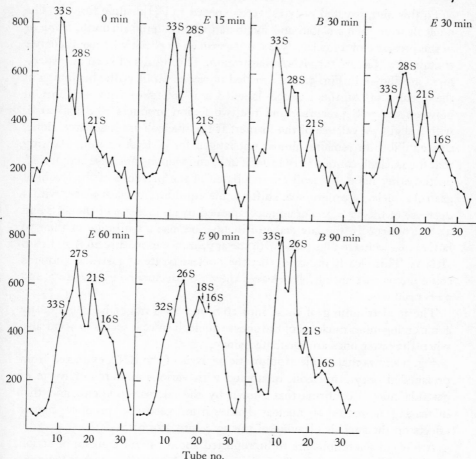

Fig. 5. Radioactivity profile of sucrose density gradients of RNA extracted from isolated nucleoli of oocytes that had been exposed to [5-^{3}H]uridine for 5 h. The isolated nucleoli have been incubated for the indicated times in a pH 7.4 Tris buffer containing 0.6 M sucrose, 15×10^{-4} M-Mg^{2+}, 8×10^{-2} M-NaCl, 18×10^{-5} M-CaCO$_3$ and 5×10^{-3} M mercaptoethanol. In the samples indicated with B the nucleoli were incubated alone; in those indicated with E, with the addition of a lysate of gastrula nuclei. (Giudice, C., Pirrone, A. M. & Roccheri M., manuscript in preparation.)

therefore a very important step at which level the rate of rRNA production can be regulated. Very little is known about the mechanisms which bring about the specific cleavage of the rRNA precursor. The availability of two homogeneous systems, i.e. oocytes and gastrulae, that show a markedly different rate of rRNA precursor cleavage, offers the possibility of an experimental approach to the problem of rRNA maturation.

To this aim, isolated oocytes were exposed to [³H]uridine for 5 h. The nucleoli were then isolated and incubated in a buffered medium, alone or in the presence of a nuclear lysate of gastrulae, to check whether the latter contained a 'factor' for rRNA maturation. The results of such an experiment are shown in Fig. 5. The labelled nucleolar RNA at the beginning of the incubation of the isolated labelled nucleoli shows an equilibrium between the 33 S precursor and the two fission products 28 S and 21 S, with a slight prevalence of the former. If the nucleoli are incubated alone, this equilibrium remains almost unchanged for at least 90 min, showing that the nucleoli contain very little, if any, endogenous nuclease activity. If, on the other hand, nucleoli are incubated in the presence of the lysate of gastrula nuclei, a progressive shift of this equilibrium takes place, with a decrease of the 33 S precursor accompanied by an increase of the 28 S and 21 S products. In the late stage of incubation also a trimming of the two latter ones seems to take place with the appearance of mature 26 S and 18 S rRNAs. This clearly indicates that the nuclear lysate of gastrulae contains some factor that specifically cleaves the 33 S precursor into the 28 S and 21 S products.

The final trimming of these into 26 S and 18 S might be operated by some endogenous nuclease of the oocyte nucleoli since it seems to occur also when the latter ones are incubated alone.

Fig. 6 shows that an electrophoretic analysis of the RNA extracted from prelabelled oocyte nucleoli, incubated with various amounts of lysate of gastrula nuclei, confirms that shown by the sucrose gradient, and that increasing amounts of nuclear lysate from gastrulae produce higher effects on the specific cleaving of the 33 S into 28 + 21 S.

A few considerations are to be reported about the mechanism of action of this 'maturation factor'. It is in principle possible that it cleaves specifically only the 33 S RNA contained in the nucleolar RNPs or that it is a non-specific nuclease able to cleave any kind of RNA and that the 33 S RNA is somehow more exposed to nuclease action in a specific point than the other nucleolar RNAs. These possibilities are susceptible to experimental test, and will be soon investigated. It has to be observed that the data reported in the figures have been normalised to the same number of counts per graph for a better comparison; it can, however, be excluded that the effect of the 'maturation factor' is simply due to a preferential non-specific nuclease effect on the 33 S RNA which leaves unaffected the 28 S and 21 S RNAs, because in this case one should observe an actual decrease of the 33 S RNA with production of fragments of unpredictable size, but not an increase of the 28 S and 21 S RNAs. On the contrary, we have always observed that, as the 33 S decreases upon incubation of the oocyte

PLATE I

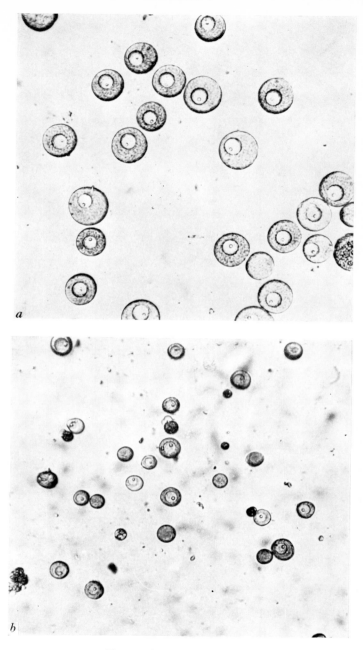

For explanation see p. 213

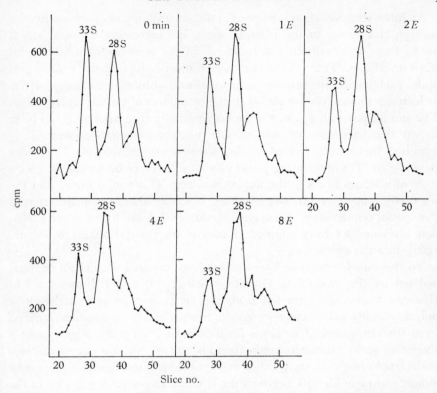

Fig. 6. Radioactivity profile of polyacrylamide gel electrophoreses of RNA extracted from isolated nucleoli of oocytes that had been exposed to [5-³H]uridine for 5 h. The isolated nucleoli have been incubated, under the conditions described in Fig. 5, for 30 min with various amounts of gastrula nuclei lysate. o min = nonincubated nucleoli; 1 E; 2 E; 4 E; 8 E = nucleoli incubated with 1, 2, 4, 8 parts of gastrula nuclei lysate. (Giudice, G., Pirrone, A. M. & Roccheri, M., manuscript in preparation.)

nucleoli with the nuclear lysate of gastrulae, the 28 S and 21 S RNAs increase.

SYNTHESIS OF HISTONES

The regulation of the histone synthesis is a problem of obvious interest, *per se*. The study of this subject in sea urchin embryos offers several advantages. It is in sea urchin embryos that a messenger RNA for histones has been first tentatively identified and partially purified from cytoplasmic polysomes (Nemer & Lindsay, 1969; Kedes, Gross, Cognetti & Hunter, 1969; Moav & Nemer, 1971). It has moreover been shown that this mRNA is synthesised on repetitious DNA sequences (Kedes & Birnstiel, 1971).

An interesting situation is present in the sea urchin embryo. Namely, the embryo cleaves up to the blastula stage, i.e. increasing the number of nuclei from one to about five hundred, in the presence of inhibitors of RNA synthesis. This means that either the blastula nuclei synthesised under such conditions contain no histones or that histone messenger RNA or histones themselves are stored in the cytoplasm of the unfertilised egg. The second possibility seems to be only partially true because it has been shown that synthesis of acid-soluble nuclear proteins is some 65% inhibited in blastulae raised in the presence of actinomycin D (Kedes *et al.* 1969). The interesting possibility can therefore be examined that a store of histones exists in the unfertilised egg. This would imply that the sea urchin oocytes synthesise histones. If so, we should conclude that in this special cellular stage, i.e. oogenesis, the synthesis of histones and DNA, that are usually tightly coupled (Kedes *et al.* 1969; Robbins & Borun, 1967), become decoupled.

To this aim Cognetti and co-workers have exposed sea urchin oocytes, isolated by the method of Giudice *et al.* (1972) to [^{3}H]lysine for 8 h. Histones were then extracted from the entire oocytes and analysed by polyacrylamide gel electrophoresis in comparison with histones purified from the chromatin of embryos labelled with [^{14}C]lysine. Figs. 7 and 8 show that some radioactive peaks found in the acid soluble oocyte proteins coelectrophorese with the purified embryonic histones. Actually there is an almost complete identity between the labelled acid-soluble proteins of the oocytes and the labelled histones extracted from the chromatin of the 16–32 blastomere stage; whereas there is only a partial but definite identity between the oocyte labelled acid-soluble proteins and the labelled histones from prism chromatin. The result is the same when the electrophoretic behaviour is dictated by the charge and the size of the analysed proteins as in the case of electrophoresis in acetate buffer–urea (Fig. 7) and when it is dictated only by the molecular weight, as in the case of SDS–acrylamide gel electrophoresis (Fig. 8). The same result was also obtained when the comparison between the labelled oocyte acid-soluble proteins and chromatinic histones of embryos was made by chromatography on Amberlite columns.

The results of these comparisons, obtained with three different procedures, allow one to safely conclude that the oocytes actively synthesize proteins entirely identical to the chromatinic basic proteins synthesized by the embryo between fertilisation and 32 blastomeres and identical to some of the prism histones. The differences in the electrophoretic and chromatographic pattern between histones of 32 blastomeres and prism confirm the observations of other authors (Benttinen & Comb, 1971; Johnson & Hnilica, 1971).

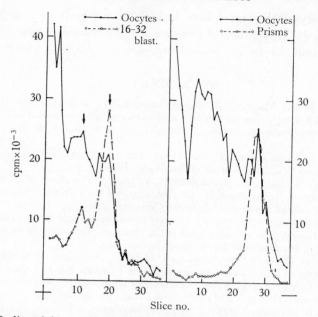

Fig. 7. Radioactivity profiles of the electrophoreses of ^{3}H-labelled oocyte basic proteins and ^{14}C-labelled basic proteins from the chromatin of embryos. Electrophoresis in 7.5 % acrylamide–ethylene diacrylate in 8 M urea and 0.1 M acetate pH 4.0. (Cognetti, G., Spinelli, G. & Vivoli, A., manuscript in preparation.)

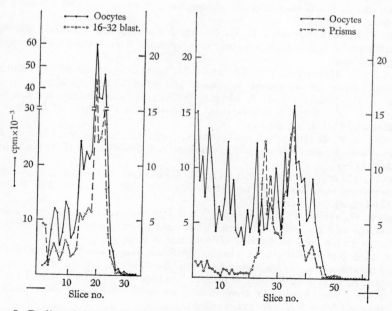

Fig. 8. Radioactivity profiles of the electrophoreses of ^{3}H-labelled oocyte basic proteins and ^{14}C-labelled basic proteins from the chromatin of embryos. Electrophoresis in 15 % acrylamide–bis acrylamide in the presence of 4 M urea and 0.1 % sodium dodecylsulphate, in Tris buffer pH 8.9. (Cognetti, G., Spinelli, G. & Vivoli, A. manuscript in preparation.)

In the oocytes therefore there is histone synthesis uncoupled from DNA synthesis. How much of histones are synthesised during oogenesis, a minute amount or an important store, is at present under investigation.

REFERENCES

BAMBERGER, J. W., MARTIN, W. E., STEARNS, L. W. & JOLLEY, W. B. (1963), Effect of 8-azaguanine on cleavage and nucleic acid metabolism in sea urchin, *Stronglyocentrotus purpuratus*, embryos. *Experimental Cell Research*, **31**, 266–74.

BENTTINEN, L. C. & COMB, D. G. (1971). Early and late histones during sea urchin development. *Journal of Molecular Biology*, **57**, 355–8.

CRKVENJAKOV, R., BAJKOVIC, N. & GLISIN, V. (1970). The effect of 5-azacytidine on development, nucleic acid and protein metabolism in sea urchin embryos. *Biochemical and Biophysical Research Communications*, **39**, 655–60.

FANTONI, A. & GHIARA, L. (1971). Control of pre-ribosomal RNA processing in differentiating yolk sac erythroid cells. *Abstracts Communications of the 7th Meeting of the European Biochemical Society*, 192.

FICQ, A. (1964). Effets de l'actinomycine D et de la puromycine sur le métabolisme de l'oocyte en croissance. *Experimental Cell Research*, **34**, 581–94.

GIUDICE, G. & MUTOLO, V. (1967). Synthesis of ribosomal RNA during sea urchin development. *Biochimica et Biophysica Acta*, **138**, 276–85.

GIUDICE, G. & MUTOLO, V. (1969). Synthesis of ribosomal RNA during sea urchin development. II. Electrophoretic analysis of nuclear and cytoplasmic RNAs. *Biochimica et Biophysica Acta*, **179**, 341–7.

GIUDICE, G., MUTOLO, V. & DONATUTI, G. (1968). Gene expression in sea urchin development. *Wilhelm Roux' Archiv*, **161**, 118–28.

GIUDICE, G., SCONZO, G., BONO, A. & ALBANESE, I. (1972). Studies on sea urchin oocytes. I. Purification and cell fractionation. *Experimental Cell Research*, **72**, 90–4.

GROSS, P. R. & COUSINEAU, G. H. (1964). Macromolecule synthesis and the influence of actinomycin on early development. *Experimental Cell Research*, **33**, 368–95.

GROSS, P. R., MALKIN, L. I. & HUBBARD, M. (1965). Synthesis of RNA during oogenesis in the sea urchin. *Journal of Molecular Biology*, **13**, 463–81.

GROSS, P. R., MALKIN, L. I. & MOYER, W. A. (1964). Templates for the first proteins of embryonic development. *Proceedings of the National Academy of Sciences, U.S.A.* **51**, 407–14.

JOHNSON, A. W. & HNILICA, L. S. (1971). Cytoplasmic and nuclear basic protein synthesis during early sea urchin development. *Biochimica et Biophysica Acta*, **246**, 141–54.

KEDES, L. H. & BIRNSTIEL, M. L. (1971). Reiteration and clustering of DNA sequences complementary to histone messenger RNA. *Nature, New Biology, London*, **230**, 165–9.

KEDES, L. H., GROSS, P. R., COGNETTI, G. & HUNTER, A. L. (1969). Synthesis of nuclear and chromosomal proteins on light polyribosomes during cleavage in sea urchin embryo. *Journal of Molecular Biology*, **45**, 337–51.

MOAV, B. & NEMER, M. (1971). Histone synthesis. Assignment to a special class of polyribosomes in sea urchin embryos. *Biochemistry*, **10**, 881–8.

NEYFAKH, A. A. (1960). A study of nuclear function in the development of the sea urchin *Strongylocentrotus droebachiensis* by radiational inactivation. *Doklady Akademii Nauk S.S.S.R.* **132**, 1458–61.

NEMER, M. (1962). Characteristics of the utilization of nucleosides by embryos of *Paracentrotus lividus*. *Journal of Biological Chemistry*, **237**, 143–9.

NEMER, M. (1963). Old and new RNA in the embryogenesis of the purple sea urchin. *Proceedings of the National Academy of Sciences, U.S.A.* **50**, 230–5.

NEMER, M. & LINDSAY, D. T. (1969). Evidence that the *s*-polysomes of early sea urchin embryos may be responsible for the synthesis of chromosomal histones. *Biochemical and Biophysical Research Communications*, **35**, 156–60.

PIATIGORSKY, J. & TYLER, A. (1967). Radioactive labeling of RNAs of sea urchin eggs during oogenesis. *Developmental Biology*, **21**, 13–28.

ROBBINS, E. & BORUN, T. W. (1967). The cytoplasmic synthesis of histones in HeLa cells and its temporal relationship to DNA replication. *Proceedings of the National Academy of Sciences, U.S.A.* **57**, 409–16.

SCONZO, G. & GIUDICE, G. (1971). Synthesis of ribosomal RNA in sea urchin embryos. V. Further evidence for an activation following the blastula stage. *Biochimica et Biophysica Acta*, **254**, 447–51.

SCONZO, G., PIRRONE, A. M., MUTOLO, V. & GIUDICE, G. (1970). Synthesis of ribosomal RNA during sea urchin development. III. Evidence for an activation of transcription. *Biochimica et Biophysica Acta*, **199**, 435–40.

SCONZO, G., VITRANO, E., BONO, A., DIGIOVANNI, L., MUTOLO, V. & GIUDICE, G. (1971). Synthesis of rRNA in sea urchin embryos. IV. Maturation of rRNA precursor. *Biochimica et Biophysica Acta*, **232**, 132–9.

SCONZO, G., BONO, A., ALBANESE, I. & GIUDICE, G. (1972). Studies on sea urchin oocytes. II. Synthesis of RNA during oogenesis. *Experimental Cell Research*, **72**, 95–100.

EXPLANATION OF PLATE

PLATE I

a, Oocytes from 50 to 100 μm; *b*, oocytes from 30 to 50 μm; both in the final stage of purification. (From Giudice *et al.*, *Expl Cell Res.* **72** (1972), 90–4.)

ORIGIN AND PROSPECTIVE SIGNIFICANCE OF DIVISION ASYNCHRONY DURING EARLY MOLLUSCAN DEVELOPMENT

By J. A. M. van den BIGGELAAR and
E. K. BOON-NIERMEIJER

Zoological Laboratory, University of Utrecht, Janskerkhof 3,
The Netherlands

CELL-LINEAGE AND MOSAIC DEVELOPMENT

The embryonic development of molluscs is characterized by a highly ordered division chronology leading to a well defined configuration of the blastomeres. By the first two cleavages the egg is divided into the quadrants A, B, C and D. After the third division the egg consists of the micromeres of the first quartet, $1a–1d$, and the larger vegetative macromeres, $1A–1D$. During further development a second, third and fourth quartet of micromeres is split off from the macromeres. The cell-lineage during the early cleavage period may be visualized by the division chronology in the egg of the pond snail, *Lymnaea stagnalis* (Fig. 1). After each division the macromeres are designated by a coefficient indicating the number of quartets formed from them. During further division of the micromeres the descendants are designated by a coefficient indicating the number of the quartet from which they arise and by a number of exponents each denoting a subsequent division; 1 is the cell nearer the animal pole and 2 the sister cell nearer the vegetative pole.

For a large number of molluscs the cell-lineage was studied around the turn of the century. It has been possible to trace the origin of different parts of the embryo to a certain stem cell or to a group of stem cells. Irrespective of minor differences, the cell-lineages of all species investigated are essentially comparable. The entoderm is derived from the macromeres $4A–4D$ and from the micromeres of the fourth quartet, $4a–4d$. Besides some endodermal structures the cell $4d$ produces the whole primary mesoderm. The ectoderm is formed by the first, second and third quartet. The head region of the adult animal arises from the first quartet of micromeres, mainly from the descendants of the cells $1a^1–1d^1$. After one or two divisions the remaining cells of the first quartet, $1a^2–1d^2$, stop dividing and develop into a band of ciliated cells, the larval prototroch. Eggs deprived of one or more blastomeres develop into partial embryos which lack the structures normally derived from the removed cells. Isolated blastomeres never

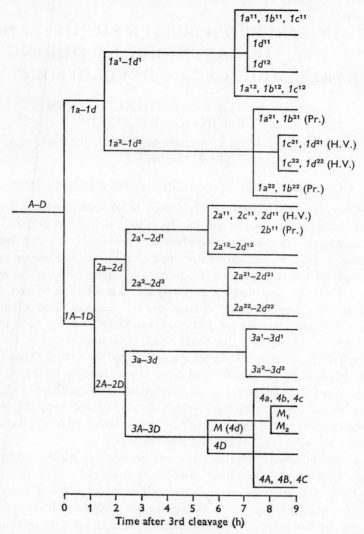

Fig. 1. Division chronology during normal development of *Lymnaea stagnalis*. Note the precocious division of *3D* into *4d* and *4D* and division asynchrony in the first quartet of micromeres. Cells of the head vesicle (H.V.) and the prototroch (Pr.) stop dividing.

develop into qualitative normal dwarf embryos. In many aspects isolated blastomeres behave as they would have done as a part of the whole embryo (Wilson, 1904*a*, *b*; Rattenbury & Berg, 1954; Clement, 1952). Therefore, the molluscan egg is considered as a mosaic of developmental capacities which is sub-divided along the diverse cell lines by a continued process of dichotomy starting at first cleavage. The programme of differentiation of the blastomeres is almost independent of correlative interactions between diverse types of cells. Especially the differentiation programme of blastomeres developing into larval structures has a strongly autonomous character, i.e. these blastomeres are capable to reach a high degree of self-differentiation. As it may be assumed that the nuclei of the different blastomeres have the same genomic content, the self-differentation of a blastomere lineage must be due to intrinsic properties determined and limited by the fragment of the egg cortex and ooplasm inherited by its stem cell early in embryogenesis.

DIVISION INEQUALITY AND DIVISION ASYNCHRONY

The formation of qualitative unequal blastomeres seems to be related to a differentiation in the cleavage rhythm, i.e. with the appearance of division asynchrony. In many molluscan species, especially in gastropods, the 4-cell stage is reached by an equal first and second division and in these cases the blastomeres A, B, C and D divide synchronously (e.g. in *Biomphalaria*, *Fiona*, *Limax*, *Littorina*, *Lymnaea*, *Neritina*, *Paludina*, *Patella*, *Physa*, *Planorbis*, *Trochus*). In other species the four quadrants are inequal; usually, D is the largest quadrant. This division inequality may be attended by the occurrence of a polar lobe formed at the vegetative pole and connected by a narrow stalk with the rest of the egg. At the 1- and the 2-cell stage the rest of the egg divides equally. Inequality arises only secondarily by the fusion of the polar lobe with the blastomere CD and D after the first and second cleavage, respectively. This type of delayed division inequality is not followed by division asynchrony (e.g. *Bithynia*, *Crepidula*, *Dentalium*, *Fulgur*, *Ilyanassa*). In *Umbrella* the blastomere AB is larger than CD, in *Pholas* and *Spisula* the AB blastomere is smaller than CD, nevertheless, the 8-cell stage is reached by synchronous divisions. In *Dreissensia*, *Aplysia* and *Unio* the unequal blastomeres AB and CD divide asynchronously already. In the cases in which the 8-cell stage is reached by synchronous divisions, the ooplasmic substances usually exhibit a radial distribution. Thus by the first two meridional cleavages four cytologically equal quadrants are formed. However, at the formation of the four quartets of micromeres the ooplasm is distributed differentially among the quartets of micromeres and the macromeres. In this way, the egg is built up of a number of groups of four cells arranged radially symmetrically along the egg axis. The formation of qualitatively different quartets of micromeres and macromeres is generally associated with the appearance of division asynchrony. Macromeres and micromeres have their own cleavage rhythm and thus division synchrony in the egg as a whole is lost. As a rule corresponding cells at the same level in the egg divide synchronously (e.g. the cells $1a–1d$; $2a^2–2d^2$). A great variety exists in the moment at which counterparts in the four quadrants start to divide asynchronously. However, there is one thing which almost all species have in common: the first deviation to the rule that corresponding cells divide synchronously is due to a differentiation in the cleavage rhythm of blastomeres in the D quadrant (more precisely in the future dorsal side of the embryo). Especially in gastropods division synchrony in corresponding cells is first lost at the formation of the fourth quartet of micromeres ($4a–4d$) from the macromeres $3A–3D$, thus at the vegetative pole of the egg. At the same time the egg loses its radial symmetry and according to Raven (1958, p. 107) 'bilaterality only becomes manifest with the formation of $4d^1$'.

Experimental data with respect to the relation between division inequality and division asynchrony are very scarce. For this reason special attention must be paid to the experiments of Clement on *Ilyanassa* (1952). In this embryo the micromere $1d$ is smaller than the other first quartet cells and the tempo of division of $1d$ and its descendants is slower than in the corresponding cells of the other quadrants. The micromere $2d^{11}$ is considerably larger than its counterparts $2a^{11}$, $2b^{11}$ and $2c^{11}$. The micromere $4d$ is filled with a clear cytoplasm and precedes the formation of the yolk-rich micromeres $4a$, $4b$ and $4c$ by about three hours. Thus, 'in the normal egg certain of the D quadrant micromeres differ from their counterparts in the other quadrants in either size, division tempo, visible composition or some combination of these features. If the polar lobe is removed at the first cleavage, these special features of the D quadrant micromeres fail to develop. In the lobeless egg the appearance, the division tempo and the cleavage pattern of the micromeres is the same in all quadrants' (Clement, 1952, p. 616).

It seems to be of interest to reconsider the observations on the formation of the micromere $1d$. According to Clement (1962) the material of the first polar lobe passes into the D cell. In a normal egg the larger D cell produces the smallest micromere, $1d$. In a lobeless egg the reduced D cell produces a larger micromere than it does normally. This implies that the polar lobe has an influence upon the position of the spindle in the D blastomere. Now the question arises whether the lower division tempo of $1d$ and its descendants is obtained by a passage of polar lobe material into $1d$, or whether this lower division tempo is due only to its smaller size as a result of a shifting of the spindle in the direction of the animal pole under the influence of the polar lobe. The latter possibility is in agreement with Raven's idea about the attainment of bilateral symmetry in a great variety of gastropods: 'the factors of bilateral symmetry do not directly affect the position of the cytoplasmic substances, but act primarily on the cleavage pattern by way of determining the spindle positions' (Raven, 1958, p. 107). In both cases, however, the special features of the $1d$ micromere would be due to intrinsic properties determined only by the inheritance of a special fragment of the egg cortex and ooplasm. Generalizing, it may be assumed that the division tempo is primarily determined by intrinsic properties of the blastomeres. Division asynchrony between sister cells at different levels along the animal–vegetative egg axis and finally between corresponding cells at the same level would arise as a consequence of an unequal distribution of cytoplasmic substances.

INTERCELLULAR INTERACTIONS AND DIVISION ASYNCHRONY

So far, a possible influence of extrinsic factors determining division tempo and division asynchrony superimposed upon the intrinsic properties have been left out of consideration. Again for this aspect of molluscan development the experimental data are very scarce. For *Lymnaea* evidence has been obtained for the existence of intercellular interactions determining division tempo. As in other molluscs, the head region of *Lymnaea* arises from the first quartet of micromeres (Verdonk, 1965). Initially, the first quartet cells form a radially symmetrical cross-like figure (Fig. 2). The cell pattern becomes bilaterally symmetrical not only by means of a special position of the descendants of $1d$ (from which the dorsal arm of the molluscan cross is formed) but also of $1a$ and $1c$ (from which the lateral arms are formed). Thus, in contrast to the vegetative pole, the first indication of bilateral symmetry at the animal pole is not restricted to the D quadrant. In future exogastrulae bilateral symmetry in the first quartet of micromeres is repressed, whereas the appearance of bilateral symmetry at the vegetative pole remained unaffected (Verdonk, 1965, 1968).

On the analogy of the experiments of Clement (1952) with respect to the radializing effect of the removal of a vegetative part of the egg (the polar lobe at first cleavage) in *Ilyanassa*, Verdonk came to the following hypothesis. Bilaterality in the animal hemisphere of the *Lymnaea* egg is determined by a factor for bilateral symmetry located in the vegetative part of the D quadrant, presumably by means of an inductive influence. Strong evidence exists that this inductive influence is exerted at the 24-cell stage which, especially in many gastropods, is a resting stage in the development of the egg as for several hours no further divisions take place. Firstly, beyond the end of the resting stage it is no longer possible to obtain completely

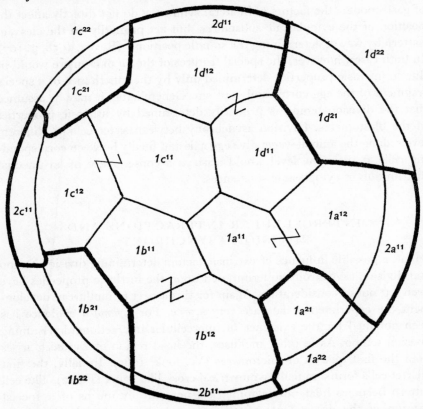

Fig. 2. First indication of the molluscan cross (outlined) at the animal side of the embryo; 49-cell stage. (After Verdonk, 1965.)

radialized embryos (Verdonk, manuscript in preparation). Secondly, in the 24-cell embryo of *Physa* (Wierzejski, 1905) and *Lymnaea* (Raven, 1946, 1970; Minganti, 1950) the cleavage cavity disappears and the micromeres of the animal hemisphere (the first quartet cells) come into contact with the inner side of the macromeres, especially with 3D. Moreover, in both species irregular complexes of RNA-containing granules, initially located around the cross furrow of the macromeres at the vegetative pole migrate along the innermost parts of the macromeres (Raven, 1970) and possibly pass into the micromeres (Wierzejski, 1905; Minganti, 1950). Apparently, the morphological requirements for an inductive influence of the vegetative macromeres, especially 3D, upon the animal micromeres are present. Finally, in *Lymnaea*, a transient indication of bilateral symmetry in the configuration of the first quartet cells can be observed at the animal pole at the end of the 24-cell stage (the presumed inductive period).

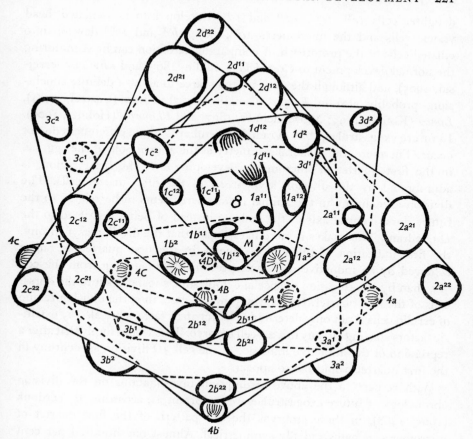

Fig. 3. 44-cell stage, at which division asynchrony in the first quartet of micromeres becomes visible. The cell $1d^1$ has just divided and the primary trochoblasts $1a^2$ and $1b^2$ are in metaphase, whereas the corresponding cells $1c^2$ and $1d^2$ are still in interphase. The primary mesoblast $4d$ is in interphase, whereas the cells $4a$, $4b$ and $4c$ are just formed. For convenience, the positions of the nuclei of corresponding cells in different quadrants are interconnected.

The descendants of the first quartet of micromeres, $1a^1$–$1d^1$ and $1a^2$–$1d^2$, do not divide synchronously. The cells that develop into the ventral ($1b^1$) and lateral arms ($1a^1$ and $1c^1$) of the molluscan cross (Fig. 2) divide about 30 min earlier than the cell $1d^1$ that develops into the dorsal arm (Figs. 1 and 3). On either side of the cell $1d^1$ the interradially located $1c^2$ and $1d^2$ can be found. These cells divide about 60 min later than the corresponding cells $1a^2$ and $1b^2$ which have an interradial position at the future ventral side of the embryo (Figs. 1 and 3). The division asynchrony of $1a^2$–$1d^2$ seems to be related with a difference in their final differentiation as the daughter cells $1c^{21}$, $1c^{22}$, $1d^{21}$ and $1d^{22}$ develop into non-ciliated head vesicle cells and the micromeres $1a^{21}$, $1a^{22}$, $1b^{21}$ and $1b^{22}$ develop into ciliated cells of the prototroch. A comparable situation can be seen during the normal development of *Crepidula* (Conklin, 1897) and *Physa* (Wierzejski, 1905), and although the data are too scarce to allow a definite conclusion, probably also in *Fulgur* (Conklin, 1907), *Fiona* (Casteel, 1904), *Limax* (Kofoid, 1895; Meisenheimer, 1897) and *Planorbis* (Holmes, 1900). In future exogastrulae, thus in radialized embryos, these differences do not occur. Therefore, the bilateral symmetry at the animal pole of the egg, i.e. in the first quartet of micromeres developing into the head region, is adumbrated by the division asynchrony of the first quartet cells. The division asynchrony in the animal hemisphere seems to be related to the future dorso-ventral axis of the embryo instead of being restricted to the D quadrant. This makes it more likely again that the division asynchrony, the first indication of bilateral symmetry in the first quartet, is rather achieved by an inductive influence, presumably emanating from the cell $3D$, than by a segregation of special cytoplasmic substances into the dorsal cells of the first quartet. If this would be the case, then the division tempo of certain cells is not only determined by intrinsic factors but also by extrinsic factors starting already in a 24-cell embryo. This would imply that after a repression of the inductive influence of the cell $3D$ division asynchrony in the first quartet would fail to appear.

With respect to the above assumption investigations on the division chronology of future exogastrulae are important as, according to Verdonk (1965, 1968), in these embryos the cell pattern of the first quartet of micromeres remains radially symmetrical. Almost one hundred per cent of exogastrulae can be obtained after a treatment with LiCl for two hours at 6 °C at the onset of the second cleavage (Geilenkirchen, 1967).

The results of an analysis of the division chronology of these future exogastrulae (van den Biggelaar, 1971c) is demonstrated in Fig. 4. The 16-cell stage is reached normally, but during further development the division chronology is disturbed. In the vegetative hemisphere the formation of the third quartet of micromeres and the first division of the second quartet cells is strongly delayed. Thus the formation of the cell $3D$, which by its central position and its precocious division normally marks the attainment of bilateral symmetry at the vegetative pole, is strongly delayed. During normal development the cleavage cavity disappears and the first quartet cells come into contact with the macromere $3D$ before bilaterality in this quartet is adumbrated by the division asynchrony between the cells $1a^1$–$1d^1$ and between the cells $1a^2$–$1d^2$. In future exogastrulae, as a consequence of the division delay in the vegetative hemisphere, the first quartet

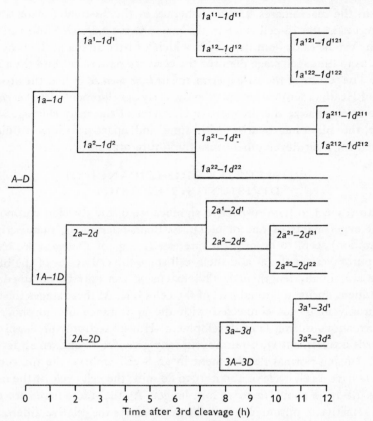

Fig. 4. Division chronology of eggs developing into exogastrulae after a treatment with LiCl at 6 °C for a period of two hours starting at the onset of the second cleavage. Note the division delay of the cells $2a$-$2d$ and $2A$-$2D$ and division synchrony in the first quartet of micromeres.

cells divide before bilaterality at the vegetative pole becomes apparent. This is associated with a repression of division asynchrony in the first quartet. The cells $1a^1$-$1d^1$ and similarly the cells $1a^2$-$1d^2$ now divide synchronously (Fig. 4).

Preliminary observations (J. A. M. van den Biggelaar) demonstrate that, in the Li-treated future exogastrulae, the division delay at the vegetative hemisphere is not associated with a delay in the disappearance of the cleavage cavity. In future exogastrulae the cleavage cavity disappears at the 16-cell stage and the cells of the first quartet come into contact with the vegetative macromeres $2A$-$2D$, whereas in normal development with $3A$-$3D$, especially with $3D$. The movement of the RNA-rich granules along the inner side of the macromeres normally takes place at the 24-cell stage in the macromeres *3A–3D*, whereas in the Li-treated exogastrulae this occurs at the 16-cell stage in the macromeres *2A–2D*. Unfortunately, nothing can be said about the way in which a Li-treatment at the transition of the 2- to the 4-cell stage disturbs the cleavage pattern beyond the 16-cell stage. The results of the Li-experiments in *Lymnaea* as well as the observations of Heath (1899) with respect to the early development of *Ischnochiton*: 'serve to emphasize the prospective character of the early cleavage stages where the blastomeres arise with time and space relations admirably adapted for their development into the future organism'.

PROSPECTIVE SIGNIFICANCE OF DIVISION CHRONOLOGY

With respect to the prospective significance of the division chronology, recent experiments of one of us (E. K. Boon-Niermeijer, manuscript in preparation) seem to be of great interest. If eggs of *Lymnaea* are treated with puromycin at the 2- and the 4-cell stage the cell cycles of the blastomeres are equally lengthened. This extension can only be obtained after a treatment during a limited part of the cell cycle. At these stages the effect of puromycin is not connected with the appearance of morphogenetic malformations during later development. However, from the 8-cell stage onwards a treatment with puromycin may give rise to abnormal development. During normal development in an 8-cell embryo the macromeres *1A–1D* have a cell cycle of about 75 to 80 min; the cell cycle of the micromeres *1a–1d* is about 12 to 15 min longer. As the micromeres are much more sensitive to puromycin than the macromeres the relative difference in the duration of the cycles is enlarged. In extreme cases it may be possible that the macromeres divide twice before the first quartet cells divide (Fig. 5).

Embryos with this altered division chronology nearly always exhibit an abnormal development. Therefore, it may be assumed that a reversible effect of puromycin in the form of an extension of the cell cycle may produce irreversible morphogenetic effects by means of a change in the division chronology. This agrees with the hypothesis that in a normal embryo the cell cycles in the diverse blastomere lineages are precisely adapted to each other and have a prospective significance indeed. From this point of view regulation of cell division becomes a major problem in the analysis of early molluscan development. It may be concluded that investigations on division chronology, on the duration of the different phases in the successive cell cycles and on the way in which a differentiation in the diverse cell lines is brought about may be of greater importance than is usually considered.

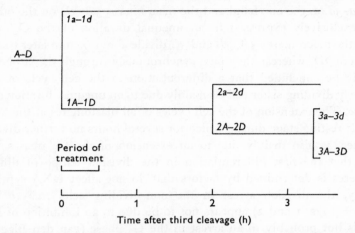

Fig. 5. Division chronology of highly sensitive eggs treated with puromycin at the onset of the 8-cell stage. The third quartet of micromeres is formed before the first division of the first quartet cells takes place.

ANALYSIS OF THE CELL CYCLES

At this moment *Lymnaea* is the only mollusc for which the cell cycles during the early cleavage stages have been analysed (van den Biggelaar, 1971*a*, *b*). Up to the 49-cell stage DNA synthesis starts at late telophase. As soon as the chromosomes swell into the karyomere vesicles incorporation of labelled thymidine into the nuclear DNA can be demonstrated. Thus a G_1 phase is absent. Up to and including the 4-cell stage the synchronously dividing blastomeres pass synchronously through the different phases of the cell cycle; mitosis takes about one half of the cycle, DNA synthesis and G_2 each about one quarter. From the 8-cell stage onwards a differentiation in the duration of the cycles in the diverse cell lines becomes apparent. The cell cycle of the macromeres *1A–1D* takes about 75 to 80 min, whereas the micromeres *1a–1d* divide about 12 to 15 min later. In both types of cells the DNA is reduplicated in the same period and the duration of mitosis remains constant. Consequently, the extension of the cell cycle of the micromeres must be due to an extension of the G_2 phase. Similarly, DNA synthesis in the micromeres of the second quartet, *2a–2d*, and in the macromeres *2A–2D* occurs almost synchronously and the difference in the length of the cell cycles between the two groups of cells is mainly due to an extension of the G_2 phase in the micromeres *2a–2d*. The same holds for the micromeres *3a–3d* and the macromeres *3A–3D*. Both types of cells pass almost synchronously through the S phase. The difference in the duration of interphase between the micro-

8

meres $3a-3d$ on the one hand and the macromeres $3A-3D$ on the other is almost exclusively expressed in an unequal duration of the G_2 phase. Finally, the macromeres $3A$, $3B$ and $3C$ divide about 90 min later than the macromere $3D$, whereas they pass synchronously through the S phase.

It may be concluded that a differentiation in the cell cycle of asynchronously dividing sister cells is mainly due to an unequal duration of the G_2 phase. The extension of the cell cycles of all blastomeres at the 24-cell stage, the resting stage during which for several hours no further divisions take place, is also mainly due to an extension of the G_2 phases. This implies that the first differentiation in the division tempo of different blastomeres is determined by factors that do not affect DNA synthesis. Similarly, the division arrest in the non-dividing cells $1a^{21}-1d^{21}$ and $1a^{22}-1d^{22}$ (Figs. 1 and 2) appears not to be due to an inhibition of DNA synthesis but probably to an arrest in the G_2 phase (van den Biggelaar, 1971b).

Summarizing it may be concluded that, in *Lymnaea*, a differentiation in the division rhythm is primarily due to an unequal duration of the G_2 phases. The factors determining division asynchrony probably determine the onset of mitosis. It remains to be demonstrated at which level this occurs. A first indication about the nature of the factors involved may be derived from the above-mentioned experiments with puromycin. The sensitivity of the blastomeres appears to be limited to a well defined period during the successive cell cycles, almost exactly coinciding with the DNA synthetic period, although DNA synthesis is not affected. The extension of the cell cycle in the puromycin-treated eggs is reflected in a lengthening of the G_2 phase only. This indicates that during the S phase puromycin exerts an influence on the factors that determine the duration of the G_2 phase, probably by an interference with the synthesis of specific proteins indispensable for the preparation of mitosis.

CONCLUDING REMARKS

It may be expected that experiments in which an influence is exerted on the duration of the cell cycles and on the division chronology will be of great importance in the search for the nature of the factors that determine when and why a cell will divide or stop dividing. Although the specific segregation of the ooplasm along the diverse cell lines will determine their future fate and their future division chronology, not only the question remains about 'the nature of the mechanism by which inheritance of a particular sector of egg cytoplasm can determine the future fate of an embryonic cell lineage' (Davidson, 1968, p. 109) but also the question about the nature of the mechanism by which inheritance of a particular sector of the egg cytoplasm and the egg cortex can determine the future division tempo in the different blastomere lineages and the way in which intercellular interactions can change the division chronology. With respect to these questions, comparative as well as experimental studies about the division chronology in a great variety of molluscs and studies of the division chronology of isolated blastomeres may be of great interest.

REFERENCES

BIGGELAAR, J. A. M. VAN DEN (1971a). Timing of the phases of the cell cycle with tritiated thymidine and Feulgen cytophotometry during the period of synchronous division in *Lymnaea*. *Journal of Embryology and Experimental Morphology*, 26, 351–66.

BIGGELAAR, J. A. M. VAN DEN (1971b). Timing of the phases of the cell cycle during the period of asynchronous division up to the 49-cell stage in *Lymnaea*. *Journal of Embryology and Experimental Morphology*, 26, 367–91.

BIGGELAAR, J. A. M. VAN DEN (1971c). Development of division asynchrony and bilateral symmetry in the first quartet of micromeres in eggs of *Lymnaea*. *Journal of Embryology and Experimental Morphology*, 26, 393–9.

CASTEEL, D. B. (1904). The cell-lineage and early larval development of *Fiona marina*, a nudibranch mollusk. *Proceedings of the Academy of Natural Sciences of Philadelphia*, 56, 325–405.

CLEMENT, A. C. (1952). Experimental studies on germinal localization in *Ilyanassa*. I. The role of the polar lobe in determination of the cleavage pattern and its influence in later development. *Journal of Experimental Zoology*, 121, 593–625.

CLEMENT, A. C. (1962). Development of *Ilyanassa* following removal of the D macromere at successive cleavage stages. *Journal of Experimental Zoology*, 149, 193–215.

CONKLIN, E. G. (1897). The embryology of *Crepidula*. A contribution to the cell lineage and early development of some marine gastropods. *Journal of Morphology*, 13, 1–226.

CONKLIN, E. G. (1907). The embryology of *Fulgur* : a study of the influence of yolk on development. *Proceedings of the Academy of Natural Sciences of Philadelphia*, 59, 320–59.

DAVIDSON, E. H. (1968). *Gene Activity in Early Development*. New York & London: Academic Press.

GEILENKIRCHEN, W. L. M. (1967). Programming of gastrulation during the second cleavage cycle in *Limnaea stagnalis* : a study with LiCl and actinomycin D. *Journal of Embryology and Experimental Morphology*, 17, 367–74.

HEATH, H. (1899). The development of *Ischnochiton*. *Zoologische Jahrbücher*, *Anatomie*, 12, 567–656.

HOLMES, S. J. (1900). The early development of *Planorbis*. *Journal of Morphology*, 16, 369–458.

KOFOID, C. A. (1895). On the early development of *Limax*. *Bulletin of the Museum of Comparative Zoology, Harvard College*, 27, 35–118.

MEISENHEIMER, J. (1897). Entwicklungsgeschichte von *Limax maximus* L. 1. Theil. Furchung und Keimblätterbildung. *Zeitschrift für wissenschaftliche Zoologie*, 62, 415–68.

MINGANTI, A. (1950). Acidi nucleici e fosfatasi nello sviluppo della *Limnaea*. *Rivista di Biologia*, 42, 295–319.

228 J. A. M. VAN DEN BIGGELAAR AND E. K. BOON-NIERMEIJER

RATTENBURY, J. C. & BERG, W. E. (1954). Embryonic segregation during early development of *Mytilus edulis*. *Journal of Morphology*, **95**, 393–414.

RAVEN, C. P. (1946). The development of the egg of *Limnaea stagnalis* L. from the first cleavage till the trochophore stage, with special reference to its 'chemical embryology'. *Archives Néerlandaises de Zoologie*, **7**, 353–434.

RAVEN, C. P. (1958). *Morphogenesis. The Analysis of Molluscan Development*. London, New York, Paris & Los Angeles: Pergamon.

RAVEN, C. P. (1970). The cortical and subcortical cytoplasm of the *Lymnaea* egg. *International Review of Cytology*, **28**, 1–44.

VERDONK, N. H. (1965). Morphogenesis of the head region in *Limnaea stagnalis* L. Thesis, University of Utrecht.

VERDONK, N. H. (1968). The determination of bilateral symmetry in the head region of *Limnaea stagnalis*. *Acta Embryologiae et Morphologiae Experimentalis*, **10**, 211–27.

WIERZEJSKI, A. (1905). Embryologie von *Physa fontinalis* L. *Zeitschrift für wissenschaftliche Zoologie*, **83**, 502–706.

WILSON, E. B. (1904a). Experimental studies on germinal localization. I. The germ-regions in the egg of *Dentalium*. *Journal of Experimental Zoology*, **1**, 1–72.

WILSON, E. B. (1904b). Experimental studies on germinal localization. II. Experiments on the cleavage-mosaic in *Patella* and *Dentalium*. *Journal of Experimental Zoology*, **1**, 197–268.

THE INFLUENCE OF THE CELL CYCLE ON THE RADIATION RESPONSE OF EARLY EMBRYOS

By L. HAMILTON

Department of Biology as Applied to Medicine, The Middlesex
Hospital Medical School, London WIP 6DB

INTRODUCTION

It is strange that studies of the effects of irradiation on synchronously dividing mammalian cells in tissue culture on the one hand and of the involvement of cell progression through the cell cycle with differentiation on the other are such popular subjects for research and yet the subject under discussion now has received so little attention.

One of the solutions of a problem facing radiation biologists, 'Where do we obtain a highly synchronous dividing cell population?', would continue to evade almost anyone but an embryologist. Tissue culturists at first avoided solving the problem by devising complex, timed labelling experiments and a great deal of microscopical work. In the early 1960s the technique of mitotic selection was devised in which rounded mitotic cells were washed off the culture dish and used to start a sub-culture. Although the washings might have contained 90% cells or more in mitosis there would only be about 50% synchrony at the next division, this synchrony declining in each subsequent cycle.

Sinclair & Morton (1965, 1966) used mitotically selected Chinese hamster cells to determine cell death or the lack of colony forming ability after X-irradiation at different stages of the cell cycle. Their results showed that the greatest sensitivity was at mitosis and the greatest resistance in late S. These workers also investigated the relationship between mitotic delay and stage of irradiation, as did Whitmore, Till & Gulyass (1967).

Other studies have concentrated on radiation-induced chromosomal aberrations with reference to the cell cycle.

These three well-used end points in radiation studies, cell death, delay in division and mutation, apply to metazoan embryos as well as tissue culture cells.

The embryologists' answer to the radiation biologists problem raised in the second paragraph is none other than 'the Embryo'.

In 1911 Bardeen published his results of X-irradiating frog eggs at different ages after fertilisation. He found that he obtained different levels of survival after treatment at different stages. Nowadays one might expect, and does find, that the one-celled, gastrula and neurula stages have different responses to X-rays. But at that time, with the inadequacies of technique under which Bardeen laboured, it is amazing that he was able to distinguish between survival levels within one- or two-celled stages. With a dose rate so low that it took up to half a cell cycle to deliver the dose, it is remarkable that Bardeen could pick up differences at all.

Half a century elapsed before research started again in earnest, on the responses of synchronously dividing eggs to insults of varying nature, irradiation being one.

Before proceeding to a more detailed examination of the effects of irradiation in embryos, it should be noted that there are two common sources of irradiation in use and three end points for the assessment of the effect of irradiation. These are ionising (e.g. X-, γ-rays) and non-ionising (e.g. UV) radiation; and cell death, mitotic delay and induced chromosomal aberration respectively.

IONISING RADIATION

Survival

X-rays have been the commonest source of ionising radiation, and survival or lethality are the commonest end point in experiments on dividing eggs, although some studies of division or cell cycle delays have been made.

In all cases to be described here, in contrast to tissue culture cells, irradiation does not lead to death of the cells or loss of reproductive ability within 3 or 4 cell cycles. Instead the cells continue to divide until, presumably, messenger RNA, which was laid down in the developing oocyte for 'cleavage', can no longer support development. In other words, gastrulation may be arrested in embryos which were irradiated early because they produce damaged or nonsense 'gastrulation' messenger RNA.

Neyfakh (1959) has investigated this aspect of X-ray damage to loach embryos. Although his experiments did not inquire into the cell cycle stage at irradiation it gave a relevant model of stage of death relative to developmental stage of irradiation. He found that irradiation of different stages of development led to a blockage at the stage of development next requiring a new or different type of protein or RNA.

Response of invertebrate embryos

Insects are the group of animals in which most studies of embryonic radiation sensitivity have been made. Notable amongst these are investigations using *Drosophila* (Würgler, Ulrich & Schneider-Minder, 1963)

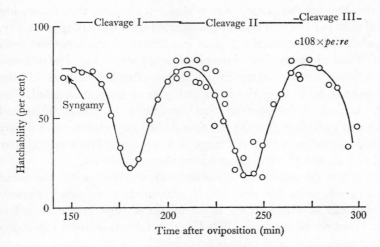

Fig. 1. Effect of X-rays on hatchability during early cleavage of silkworm eggs. The eggs used in this study were obtained from the F_1 eggs of a cross between females of c108 strain and males of marker strain *pe :re*. The sample eggs were collected every 5 min at 25 °C. These eggs were irradiated with 1000 R of X-rays (180 kVp, 25 mA, and 1.0 mm Al filter) at a dose-rate of 1000 R/min and irradiated at 5 min intervals. (From Murakami, 1969.)

Habrobracon (von Borstel & Amand, 1963) and *Bombyx* (Murakami, 1965).

Murakami's studies on the sensitivity of *Bombyx* embryos reflects their passage through many cell cycles with the same fluctuations each time. In each cycle sensitivity to X-rays was highest in mitosis and lowest at interphase. This extraordinary finding that increasing numbers of nuclei in each embryo made no difference to sensitivity at different divisions is remarkable. It may be that during this period prior to blastoderm formation, when the nuclei are separated by incomplete cell boundaries, the products from less damaged nuclei can spread into the shared cytoplasm and ameliorate the damage to other nuclei. The outcome of irradiation was measured at hatching when many embryos had died. Mortality varied from 30 % to 75 % depending on the nuclear stage of irradiation. When Murakami repeated the experiment (1969) with different strains and their hybrids, he obtained the results shown in Fig. 1 for the response of his 'standard' marker hybrid. He suggests that there is a maternal effect governing the sensitivity of different strains.

Similar findings of cyclical response to X-rays were found in *Drosophila* embryos, which have been used as synchronous 'cell cultures' at least since 1963. In their 1963 paper Würgler *et al.* produced results similar to those for *Bombyx* but there was a slight indication that increasing numbers of cells per embryo could be associated with increasing mortality. Again, low survival was obtained after irradiation of ana/telophase cells and high survival with late interphase/prophase cells. *Habrobracon* embryos behaved rather differently from the other two insects in that there appeared to be a sharp rise in mortality as cleavage proceeded. Von Borstel & Amand, 1963, showed that sensitivity to 500 R fluctuated during the cell cycle but that by the four-celled stage there were barely any survivors regardless of the cell stage at irradiation. These results may be typical of parasitic Hymenoptera where cleavage is total.

Geilenkierchen (unpublished observations) working with the snail *Lymnaea stagnalis* tells me that 600 R administered to synchronously dividing eggs had no effect on the development of eggs irradiated in late S and G_2. However embryos irradiated during mitosis showed increasing abnormality as the phases of mitosis proceeded.

Response of vertebrate embryos

Essentially the same findings have been obtained after irradiating vertebrate embryos. In one case (Hamilton, 1969) newly laid *Xenopus* eggs were collected at 5 min intervals. At known times after collection each batch was irradiated for one minute at 956 R/min (in all our experiments we employ this dose rate). Provided the room temperature remained constant all the embryos from one pair would be developing at the same rate. Gastrulation is the earliest stage of development at which X-irradiated embryos die, but it is so difficult then to judge whether an embryo is dead or alive that scoring of survival was delayed to the late tail bud stage (stages 27–29, Nieuwkoop & Faber, 1956). When this was done it could be seen that percentage survival ranged from 0% at early interphase to 100% at late interphase/prophase (Fig. 2). Graham (1966) has shown that DNA is being synthesised throughout interphase.

The explanation of these results seemed to the author to be that the amount of DNA in the nucleus determines the resistance of embryonic cells to irradiation. This seemed particularly likely in view of some preliminary experiments of exposing dividing haploid and diploid embryos to X-rays for 0.76 min (956 R/min). The results suggested that haploid survival at its greatest was the same as diploid survival at its lowest. At both these times the amount of DNA should be the same, i.e. twice 'n' in the haploid and once '2n' in the diploid.

It had previously been shown (Hamilton, 1967) that for *Xenopus* embryos irradiated at their most resistant early phase, which is syngamy, haploid embryos are twice as sensitive as diploid embryos. That is, the

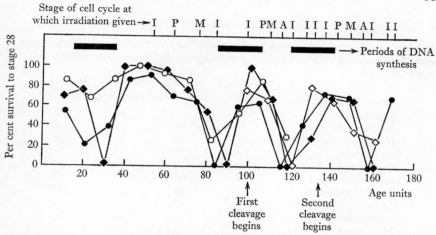

Fig. 2. Percentage of survival to stage 28 of embryos irradiated with 956 R of X-rays at different age-units after fertilisation. The extent of S and other periods of cell cycle are indicated at the appropriate age units. I = interphase, P = prophase, M = metaphase, A = anaphase and ▬ = synthetic phase. (Hamilton, 1969, © Academic Press.)

survival of diploid embryos after 1, 2 and 3 min was similar to the survival of haploid embryos after 0.5, 1 and 1.5 min exposure respectively.

An extension of the haploid cell cycle problem seemed to be an investigation of the cell cycle responses of haploid embryos to 478 R of X-rays as well as a reinvestigation of the effects of 720 R. Mr Brian Bennett has conducted these experiments as part of a University of London M.Sc. degree. He has found that haploid embryos irradiated with 478 R show the *same* survival pattern as diploid embryos irradiated with 956 R at successive stages in the first three cleavage cycles (Table 1).

If we now reconsider the three types of *Xenopus* cell cycle experiment, viz:

irradiating diploids for 1 min,
irradiating haploids for 0.5 min and
irradiating both haploid and diploid embryos for 0.76 min

we see that there is no clear-cut relationship between the amount of DNA present in each nucleus and the survival of the embryo to scoring. Haploid embryos do seem to react like diploid embryos do when exposed to X-rays at times during the first few cleavage cycles, but the dose must be half. Certainly the DNA content of their nuclei is also half. If DNA content were the sole explanation of differences in sensitivity then survival should only fluctuate by a factor of two within any cell cycle, be it haploid or diploid. This it does not do, so our observation that the haploid peak and

Table 1. *Effect of irradiation on* Xenopus *embryos*

Haploid embryos were irradiated with 478 R X-rays at different stages (column 1) and their survival and percentage survival recorded when the controls reached stage 28.

Stage	Number of nuclei	Number irradiated	Number surviving	Percentage survival
Prophase I	1	41	22	53
Anaphase I	1	66	5	8
Interphase (late) II	2	29	25	86
Metaphase II	2	76	53	70
Anaphase II	2	57	22	39
Interphase (late) III	4	96	84	86
Pro-metaphase III	4	34	30	88
Anaphase III	4	37	5	7
Interphase (late) IV	8	45	40	89
Metaphase IV	8	13	10	77
Anaphase IV	8	39	1	2
Inter-prophase V	16	32	28	78

the diploid low point coincide when both types of embryo are irradiated for 0.76 min is purely fortuitous. DNA content is surely the explanation of the differences observed between haploid and diploid embryos but not for differences between phases of the cell cycle. What substances in the cell might be responsible for changes in sensitivity during the cell cycle and conceal changes due to differences in nuclear DNA content?

Sulphydryl compounds and protection against ionising radiation

Sulphydryl compounds are known to exert a protective effect on cells that are X-irradiated (Révész & Bergstrand, 1963). It is thought that these compounds may work through their ability to 'scavenge' free radicals or repair electron-deficient sites produced by the interaction of ionising radiations and matter – in this case living matter. Evans postulated (1963) that since the amounts of these substances fluctuate in cells through the cell cycle they could be responsible for changes in sensitivity to X-rays.

The first report of treatment of synchronised cells with a sulphydryl compound during irradiation appeared in 1968 when Sinclair described his work on cultured Chinese hamster cells transferred to cysteamine during irradiation. Even though this compound is toxic it was possible to reduce the sensitivity of cells to X-rays at all stages of the cell cycle and, furthermore, to eliminate the differential sensitivity that is associated with stage. For instance, Chinese hamster cells were irradiated with 710 R in the presence of 5 mM cysteamine, or not. Survival of G_1, S and G_2 cells was

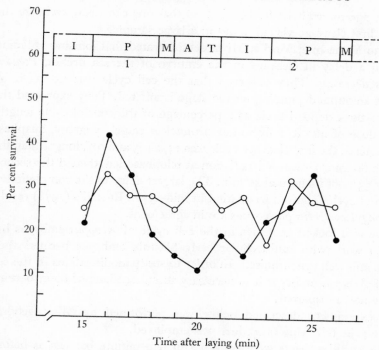

Fig. 3. *Drosophila* embryos were irradiated with 1000 R X-rays at different stages after syngamy, either in air (●) or hydrogen sulphide (○). The survival levels after H_2S alone are about 35%. (After Matter, 1970.)

about 0.06, 0.3 and 0.08 respectively in the absence of cysteamine and about 0.75 for all stages in the presence of cysteamine.

In 1970 Matter published the results of gassing *Drosophila* embryos with hydrogen sulphide during irradiation. At toxic doses it was possible to demonstrate that H_2S is a radioprotective agent that rescues ana/telophase stage embryos most. The results are plotted in Fig. 3 and show a flattening of the response curve which means that changes in survival after irradiation at different phases of the cell cycle can be modified by the presence of sulphydryl compounds.

Mr Bennett is now measuring the sulphydryl content of *Xenopus* embryos at different stages of the cell cycle and hopes to find fluctuations like those found for sea urchin eggs (Dan, 1966).

Delay in the cell cycle

In sea urchins and other small aquatic embryos it is possible to observe the early cleavages. It is also possible to obtain hundreds of embryos of the same age by artificial fertilisation, so that one can easily estimate the delay in first cleavage visually; and in a large sample.

In 1939 Yamashita, Mori & Miwa demonstrated that ionising radiation provoked a delay in cleavage in the embryo of the sea urchin, *Pseudocentrotus depressus*. They observed that the cell cycle was extended by different amounts depending on the stage irradiated. They expressed the length of the extended cycle as a percentage of the control cycle length. After a dose of 460 R delivered at pronuclear stage, syngamy, anaphase and telophase, the first cleavage cycle was 14, 24, 3 and 0 % longer than the control's 80 min. However irradiation at telophase lengthened the second cycle by 35 % of its normal 40 min. The largest delay in the completion of the second cycle followed irradiation in mid-cycle. Rustad's (1970) results for *Strongylocentrotus purpuratus* are in agreement.

I have also looked at delays in the cell cycle of *Xenopus* embryos but chose to work with multicellular early blastula embryos because their cells are still well synchronised. In order to study modifications in the cell cycle of *Xenopus* embryos it is necessary to cut sections of them because they are not transparent.

The experiment described below was performed by Miss Deborah Mayhew; the following procedure was employed.

Newly fertilised eggs were collected in five minute batches as before. This time they were allowed to divide six or seven times so that the animal pole contained between 32 and 64 well-synchronised cells. The vegetal pole cells were ignored in the analysis because they were often shattered during section cutting and because they were out of phase with the animal pole cells. The cycle time for the animal pole cells was about 22 min in this experiment. Four consecutive batches were selected, of which three were exposed to X-rays for 3 min (956 R/min, as before). The irradiated groups were spaced at 10 min age intervals and the stages of the cell cycle represented by these groups were ana/telophase, prophase and metaphase. Three embryos from each group were fixed at the time of irradiation and at subsequent five minutes. Control unirradiated embryos were fixed at equivalent ages. The embryos were prepared for microscopical examination and sections cut at 20 μm. The analysis which follows is based on the stage of each nucleus in the animal pole, where synchrony remains good.

The cells in the irradiated embryos also remained in good synchrony for at least half an hour after irradiation but they went through the cell cycle more slowly than the unirradiated controls (Fig. 4).

Delay in progression was greatest in those embryos irradiated at ana/telophase and least in those irradiated at prophase. Delays were manifested at prophase and metaphase since it was then that progression diverged

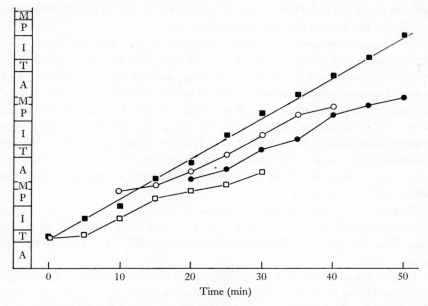

Fig. 4. Delay in progression through the cell cycle induced by irradiation of *Xenopus* blastula with 2868 R X-rays. ■, control; □, irradiated at telophase; ○, irradiated at inter/prophase; ●, irradiated at meta/anaphase. The times of fixation, after the start of the experiment, are plotted against the nuclear stage of the embryos.

from the rate observed in the controls. The slopes of all the graphs at interphase appeared to be parallel so that, on the criterion of nuclear morphology, we can say that interphase (S phase?) is not traversed more slowly in irradiated blastular cells.

Miss Mayhew also determined the survival of blastulae irradiated with 956 R. Her survival curves were similar to those previously described for younger embryos. 25 % of ana/telophase embryos survived to scoring when the controls were at stage 28 as did 95 % of the embryos irradiated at prophase/metaphase. There were approximately half as many vegetal pole cells as animal pole cells and they were almost at the same mitotic stage.

One may tentatively say that the cells most likely to be delayed in the 1st cell cycle after irradiation are similar in stage to those most likely to be killed by the X-rays.

Mutations

Mutation studies can only be carried out well in species that have been used for genetic research for many years. For that reason good evidence for specific mutation induction has been achieved with *Drosophila* (Petermann, 1968) and the mouse (Russell & Montgomery 1966).

Petermann was able to estimate recessive lethals induced in the X-chromosome and correlate the degree of loss with the stage in the cell cycle. The cell cycles he investigated were meiosis I and meiosis II. The sperm X-chromosome was much more sensitive to damage than the female X during this period even though it was not involved in meiotic divisions. The peak of mutation induction in the sperm coincided with early pro-nucleus formation, and reached 5.2%. It fell to about 1.6% when the two pronuclei approached each other. Even though the oocyte nucleus under-went two meiotic divisions induction of maternal recessive lethals was about 2.9% throughout.

During the period under investigation the total number of lethals produced falls from about 8 to 4% at the same time mortality rises *and* falls. The low point for both mutation and mortality is at the pronuclear stage.

In the mouse there is a very close correlation between the number of embryos failing to implant and the estimated number of chromosomes lost. These were calculated from the known frequency of X-chromosome loss (Russell & Montgomery, 1965).

NON-IONISING RADIATION

Although more investigations have been carried out with X-rays than ultraviolet light there is a considerable volume of work using non-ionising radiation.

It is well known that ultraviolet light of wave length 253.7 nm is selectively absorbed by DNA and brings about the formation of dimers, thus disturbing the helical structure of DNA.

Sinclair & Morton (1965) state that Chinese hamster cell cultures are particularly sensitive to the killing effect of UV in G_1 and early S. There is a rise in survival sometime during the S phase. Mitosis is not conspic-uously sensitive to UV-irradiation in comparison to X-irradiation. An investigation of delay in mitosis led them to the conclusion that cells only show delay because DNA synthesis was at a reduced level and prolonged. G_2 cells do not show mitotic delay. UV light has poor penetrating power so the embryos which have been treated with this irradiation have all been small and transparent. The embryos all contain large quantities of stable mRNA as well as DNA precursors.

Survival of embryos

Labordus (1970c) irradiated fertilised eggs of the snail *Lymnaea stagnalis* with 2×10^3 ergs/mm^2 at different times during the first four cleavage cycles. He found no evidence for stage-related mortality or abnormalities within each cell cycle. However, he did find a change in the *type* of abnormalities as the embryos got older. It is possible that a combination of the changes in position of cell constituents, and of the absorption of UV by the egg constituents in general is responsible for differential damage. For instance, if some organelle moves deeper into the embryo less UV will reach it than would have done on the surface. Labordus showed that the effects of UV also depended on which surface of the embryo was uppermost – i.e. which was nearest the light. Irradiation of the animal pole during the third and fourth cleavage cycles resulted in about 20% of embryos with head abnormalities, whereas irradiation of the vegetal pole cells at this age led to no head malformations but about 50% exo-gastrulae. Irradiation of the side of the embryo produced both these anomalies in lower proportions.

Delay in the cell cycle

Labordus (1970a) also considered the effects of ultraviolet light on the length of the cell cycle. In each of the first three cycles treatment before halfway through the DNA synthetic phase delayed the appearance of mitosis and cell division. Irradiation in the second half of S and in G_2 did not affect progression through the cycle of irradiation but prolonged the subsequent cycle. In a second paper Labordus (1970b) studied the effects of UV on other events in the cell cycle. He concluded that since both the S phase and prophase appeared to be of normal duration it must be G_2 that was lengthened. He suggested that this could be brought about if irradiation had led to abnormal messenger RNA and protein production.

Rustad (1960) had previously shown very similar results for the induction of mitotic delay in the sea urchin *Strongylocentrotus purpuratus*. During the first cycle there were four periods with distinct delay patterns. The first period, associated with the events of fertilisation, showed the greatest sensitivity declining to the second period. This was a plateau during which time delays of the same length were always produced. At the end of this period delays became shorter and finally could not be generated for the cell cycle in which irradiation took place. However irradiation towards the end of the first cell cycle led to a delay in the appearance of the second cell division.

DISCUSSION

The experiments described in this paper demonstrate a close similarity in the responses of embryos to irradiation despite large differences in the pattern and rate of cleavage.

The embryos described represent groups of animals in which oocytes mature at different times after fertilisation. The elimination of polar bodies must be completed before male and female pronuclei can fuse. Both polar bodies may already have been eliminated before sperm entry (sea urchins); or one (vertebrates) or two (snails and insects) still remain to be eliminated.

Correlation of survival with stage of the cell cycle at X-irradiation

In spite of these differences and those of absolute size the observed variations in response to irradiation during different phases of the cell cycle are remarkably constant. The dose of X-rays required to achieve a well defined age–response curve lies between 500 and 1000 R.

What do eggs and early embryos have in common that might explain the similarity in the results of irradiating them with X-rays? Most oocytes and zygotes are very large cells that contain reserves of nucleic acid precursors, amino acids, ribosomes, mitochondria etc. Their initial development is concerned with very rapid cleavages that, in most cases, cannot be halted. In *Xenopus*, for instance, it is thought that enough appropriate mRNA is present to programme and support cell division throughout the cleavage stages. After X-irradiation the majority of all embryos remain alive up to gastrulation. There is no immediate killing effect. Neyfakh (Neyfakh 1959; Neyfakh & Rott, 1968) has developed a technique for establishing the nuclear activity of embryos of different ages which is based on the concept of long lived embryonic messenger RNA. Although his investigations are not directly relevant to the results discussed here it is interesting to note that loach embryos cannot be killed immediately with 60 kR X-irradiation. The embryos will survive as long as their advance stocks of mRNA last.

Embryos are not as hardy as these experiments suppose, neither are much lower doses of irradiation without effect on the cytoplasm, as Sambuichi (1964) demonstrated. He irradiated blastulae or freshly ovulated eggs with 783 R γ-rays from a cobalt source and transplanted nuclei so that either the nucleus or cytoplasm of the transplant embryos had been irradiated. He found that only 3 % of X-irradiated eggs developed into normal swimming tadpoles as opposed to 19% of the control un-irradiated transplants. On the other hand unirradiated eggs never survived beyond the tail bud stage when injected with an irradiated blastula nucleus. One has to explain these results on the basis of irreparably damaged synthetic mechanisms in the irradiated cytoplasm that might lead to damage of the initially normal nucleus. If this line of thought is correct the embryos with unirradiated cytoplasm should cleave more normally than those with irradiated cytoplasm; but Sambuichi did not find this. Ulrich also found that heavy irradiation of that half of the zygote of *Drosophila* that contained no nucleus could nevertheless provoke mutations in the nucleus.

We do not yet know whether cytoplasm irradiated at different phases of the cell cycle will promote different effects in the nucleus of a certain phase but we do know that the survival of embryos will depend on the stage of the nucleus at irradiation. That is; sensitivity to killing is high from anaphase to early interphase and low from late interphase to prophase. These results have been repeated in many different embryos. In all cases the levels of synchrony are very high, both within any one embryo and between embryos of the same age and parentage. What embryologists have not yet investigated satisfactorily are the molecular events intervening between irradiation and the late effects on the embryo.

Alexandre (1970) found that control embryos and those irradiated with 1000 R at the 2–4-celled stage of a Urodele incorporated the same high level of [^{3}H]thymidine and low level of [^{3}H]uridine when they had reached the 16–32-celled stage. However, by the late blastula stage incorporation of [^{3}H]thymidine was only 19% of the control level; incorporation of [^{3}H]uridine was down to 35.8% of the unirradiated control level. This work should be repeated using synchronous cells for irradiation.

It might seem that mammalian cells in tissue culture provide a better system for such studies because it is easier to obtain sufficient cells for the required chemical analyses. The most thoroughly investigated cells have been those of the Chinese hamster. These enable us to reach some conclusions that may have a bearing on the effects of irradiation on embryos. Chinese hamster cells have a cycle time of about 12 h. When they are irradiated with 660 R (Sinclair & Morton, 1966) the survival levels are 1% for mitotic cells, 3% in G_1, up to 40% in late S and down to 2% for G_2 cells. Recently Bacchetti & Sinclair (1971) studied the effects of 710 rad on synchronised Chinese hamster cells with respect to the induction of DNA, RNA and protein synthesis. Synthesis of all these compounds was altered. DNA synthesis was heightened or prolonged in the irradiated cell cycle; and delayed in the subsequent cell cycle by a time corresponding to the length of the intervening, extended G_2. When they measured the uptake of tritiated uridine or leucine Bacchetti & Sinclair observed an immediate increase, on irradiation, in the synthesis of RNA and protein respectively. If the cells were treated with 50 μg/ml of cyclo-heximide in order to suppress DNA, RNA and protein synthesis their colony forming ability increased. Despite indications that excess synthesis of these compounds was inversely related to survival there was no reduction in mitotic delay observed after removing cycloheximide. Bacchetti and Sinclair (1971) therefore concluded that division delay and loss of colony-forming ability are brought about by different mechanisms.

Dewey et al. (1971), also working with Chinese hamster cells, compared cell death and chromosomal aberration frequency following irradiation at different phases of the cell cycle. They found a positive linear relationship between chromosomal aberrations and $\log_{10}$ cell death. After X-irradiation following 5-bromodeoxyuridine treatment, an enhancement of radiation sensitivity was found. Again, aberration frequency was linearly related to $\log_{10}$ cell death so Dewey et al. (1971) proposed that chromosomal aberrations were responsible for cell death.

How does the evidence for the mode of action of irradiation in mammalian cells affect our interpretation of the results of irradiating embryos?

It appears from mammalian cell studies that induction of chromosomal aberrations may be the cause of cell- or embryonic-death. This is indeed the conclusion that Russell & Montgomery (1966) reached when comparing loss of chromosomes with the number of embryos failing to implant. In other species it has been unusual for studies to be made on the radiation sensitivity of cell cycle stages with respect to both specific mutation induction and lethality. Petermann's (1969) example taken from X-irradiated Drosophila is not a clear positive correlation between mortality and mutation rate. However the sperm and oocyte pronuclei were unfused during most of the stages investigated. It also appears that the only studies combining mitotic delay and survival (Mayhew, 1971) have been those using Xenopus embryos.

The period immediately after fertilisation was not the best time for studying mitotic delay because the investigation had to be histological and there were then too few nuclei. However, early blastulae were still synchronous enough for the investigation. We did not realise how long a delayed cell cycle would be, so allowed barely enough time for one irradiated cycle to be completed. On that basis we could say that in the first cycle after irradiation delay was greatest following irradiation at ana/telophase. Mortality was also greatest at this stage. In mammalian cells delay in the cell cycle is greatest after irradiation in G_2 and least after irradiation in G_1. Embryonic cells have very short G_1 and G_2 which in our system probably correspond with ana/telophase and late inter/prophase, respectively. Mayhew & Hamilton have been able to trace this progress of embryonic *Xenopus* cells through the different phases of mitosis which lasts almost half of the cell cycle. Such histological evaluation of cell progress through the cell cycle leads us to state that delays are not apparent while cells pass from telophase to prophase, i.e. through S. However, cells are delayed both in prophase and metaphase. Delays at these stages have been reported for the plasmodium of *Physarum polycephalum* when treated with cycloheximide before or during mitosis (Cummins *et al.* 1966). I am not aware of such findings following X-irradiation.

The role of radiation protective substances on the effects of X-irradiation on embryos is most important. Révész & Bergstrand (1963) observed that the sulphydryl content of cells increased when they were incubated in cysteamine. The amount of increase corresponded with the degree of radiation protection afforded by these compounds, which are supposed to protect by scavenging free radicals. This means that the free radicals produced by the interaction of ionising radiations and living matter are 'inactivated' by reacting with sulphydryl compounds rather than by reacting with 'essential' cell components like nucleic acids.

In 1968 and 1969 Sinclair published results of X-irradiating Chinese hamster cells in the presence of cysteamine and found dramatic protection. In fact the protective effect of 5 mM cysteamine raised the survival level of all stages of the cell cycle to over 65 % when irradiated with 710 R. In 1970 Yu & Sinclair demonstrated that cysteamine also protects cells against damage expressed as mitotic delay and chromosomal aberrations but that the degree of protection was the same for all cell stages.

Two experiments with embryos suggest that sulphydryl compounds may be implicated in the cyclic responses of their cells to X-irradiation. Firstly, Dan (1966) demonstrated that naturally occurring sulphydryl compounds fluctuate in cleaving sea urchin embryos in such a way that levels are high at nuclear stages known to be resistant to X-rays. Secondly, Matter (1970) showed that application of hydrogen sulphide (H_2S) to *Drosophila* embryos abolished their stage sensitivity to X-rays. Hydrogen sulphide is toxic like all sulphydryl compounds at effective doses, but the results showed that even a highly toxic compound could reduce the mortality in irradiated embryos (Fig. 3).

It now seems likely that DNA – or chromosomes – are the most vital part of the cell and if these suffer damage the embryo may die. However, there is no indication that they are preferentially damaged. It is more likely that free radicals are produced at random and in the absence of 'scavengers' like sulphydryl compounds will dissipate their energy at random – and perhaps in the chromosomes and DNA.

Experiments with ultraviolet light

In mammalian cells there is good evidence that DNA is the target of ultra-violet light of wave length 253.7 nm. This is also accepted for embryos. Han *et al.* (1971) and Sinclair & Morton (1965) have demonstrated that the effects of UV on Chinese hamster cells may be explained on the basis of damaged nucleic acids especially DNA. Survival levels fluctuate but cell death is more common after irradiation in the first half of the cycle than in the second half. Delay in cleavage is larger than for X-rays if similar survival reduction is obtained. Many cells that are delayed at mitosis do not complete it so that loss of reproductive potential may be immediate. Cells that were irradiated in G_2 were not delayed, but many were not reproductive.

Embryonic cells were delayed in their first or second cleavage by UV irradiation at selected stages in the first or second cleavage cycles. Embryos in any one batch were delayed by a similar amount. Cells irradiated after late S in the first cycle were delayed in the second cycle only (Rustad, 1960; Labordus, 1970a). Labordus (1970c) found no evidence for stage sensitivity in the induction of embryonic death which would be contrary to the finding with Chinese hamster cells.

Labordus (1970b) suggested that UV might damage cortical, cytoplasmic, nuclear and mitochondrial RNA. All or any of these could cause division delay or other abnormalities of the development of *Lymnaea*.

A fairly clear-cut case for UV damage to embryonic RNA is afforded by Smith's (1966) experiments on irradiating the vegetal pole of frogs' eggs. The frogs that develop from such eggs are sterile, presumably because the RNA lying in the ventrally located germinal plasm at that time is destroyed. Smith's subsequent experiments of injecting unirradiated germinal plasm back into irradiated eggs allowed fully fertile frogs to develop.

SUMMARY AND CONCLUSIONS

The effects of X-rays and UV light on synchronously dividing embryos have been discussed. Although the effects of UV light on mammalian cells may be related to the state of synthesis of DNA the situation is complicated in embryos by the vaste reserves of DNA precursors and RNA. Ultra-violet damage is therefore rather aspecific. There is the further problem in working on the effects of ultraviolet light on even small embryos and that is its poor penetrating powers.

Much more consistent results have been obtained from X-irradiating synchronously dividing cells and embryos. It seems likely that in all cases there is a large dose modifying effect provided by endogenous sulphydryl, the amount of which is undergoing fluctuations similar to the stage related post-irradiation survival. Embryonic cells afford the opportunity of dissecting mitosis into its constituent phases for the purpose of investigating cyclic responses.

REFERENCES

ALEXANDRE, H. (1970). Etude autoradiographique de l'effet des rayons X sur les synthèses de DNA et de RNA au cours de la segmentation et de la gastulation de *Pleurodeles waltlii* Michah. *Archives de Biologie, Liège*, **81**, 139–62.

BACCHETTI, S. & SINCLAIR, W. K. (1971). The effects of X-rays on the synthesis of DNA, RNA and proteins in synchronised Chinese hamster cells. *Radiation Research*, **45**, 598–612.

BARDEEN, C. R. (1911). Further studies on the variation in susceptibility of amphibian ova to the X-rays at different stages of development. *American Journal of Anatomy*, **11**, 419–98.

VON BORSTEL, R. C. & AMAND, W. ST (1963). Stage sensitivity to X-radiation during meiosis and mitosis in the egg of the wasp *Habrobracon*. In *Repair from Genetic Radiation Damage*, ed. Sobels, pp. 87–97. London & Oxford: Pergamon.

CUMMINS, J. E., BLOMQUIST, J. C. & RUSCH, H. P. (1966). Anaphase delay after inhibition of protein synthesis between late prophase and prometaphase. *Science, New York*, **154**, 1343–4.

DAN, K. (1966). Behaviour of sulphydryl groups in synchronous division. In *Cell Synchrony*, ed. I. L. Cameron & G. M. Padilla. New York & London: Academic Press.

DEWEY, W. C., STONE, L. E., MILLER, H. H. & GIBLAK, R. E. (1971). Radiosensitisation with 5-bromodeoxyuridine of Chinese hamster cells X-irradiated during different phases of the cell cycle. *Radiation Research*, **47**, 672–87.

EVANS, H. J. (1963). Possible reasons for variation in chromosome radiosensitivity during mitotic and meiotic cycles. In *Repair from Genetic Radiation Damage*, ed. Sobels, pp. 31–44. London & Oxford: Pergamon Press.

GRAHAM, C. F. (1966). The regulation of DNA synthesis and mitosis in multinucleate frog eggs. *Journal of Cell Science*, **1**, 363–74.

HAMILTON, L. (1967). A comparison of the X-ray sensitivity of haploid and diploid zygotes of *Xenopus laevis*. *Radiation Research*, **30**, 248–60.

HAMILTON, L. (1969). Changes in survival after X-irradiation of *Xenopus* embryos at different phases of the cell cycle. *Radiation Research*, **37**, 173–80.

HAN, A., SINCLAIR, W. K. & YU, C. K. (1971). Ultraviolet light induced division delay in synchronised Chinese hamster cells. *Biophysical Journal*, **11**, 540–9.

LABORDUS, V. (1970a). The effect of ultraviolet light on developing eggs of *Lymnaea stagnalis* (Mollusca, Pulmonata). I. The pattern of the effect on mitotic cycles. *Proceedings of Koninkl. Nederland. Akademie van Wetenschappen-Amsterdam*, Series C, **73**, 382–96.

LABORDUS, V. (1970b). The effect of ultraviolet light on developing eggs of *Lymnaea stagnalis* (Mollusca, Pulmonata). III. Determination of the lengthened phase of the cell cycle. *Proceedings of Koninkl. Nederland. Akademie van Wetenschappen-Amsterdam*, Series C, **73**, 397–413.

LABORDUS, V. (1970c). The effect of ultraviolet light on developing eggs of *Lymnaea stagnalis* (Mollusca, Pulmonata). IV. The interference of irradiations with

246 L. HAMILTON

morphogenesis. *Proceedings of Koninkl. Nederland. Akademie van Weten-schappen-Amsterdam*, Series C, **73**, 477–93.

MATTER, B. E. (1970). Zur Ursache der unterschiedlichen Strahlenemfindlichkeit verschiedener Kernteilungsstadien in der frühen Furchung von *Drosophila melanogaster*. *Mutation Research*, **10**, 567–82.

MAYHEW, D. M. (1971). A study of mitotic delay and survival after X-irradiation of *Xenopus* embryos at different phases of the cell cycle. M.Sc. report, University of London.

MURAKAMI, A. (1965). Variation in radiosensitivity during the early developmental stage of the silkworm egg. *Annual Report No. 16, National Institute of Genetics, Japan*, 107–9.

MURAKAMI, A. (1969). Comparison of radiosensitivity among different silkworm strains with respect to the killing effect on the embryos. *Mutation Research*, **8**, 343–52.

NIEUWKOOP, P. D. & FABER, J. (1956). Normal table of *Xenopus laevis* (Daudin). Amsterdam: North Holland Publishing Co.

NEYFAKH, A. A. (1959). X-ray inactivation of nuclei as a method for studying their function in the early development of fishes. *Journal of Embryology and Experimental Morphology*, **7**, 173–92.

NEYFAKH, A. A. & ROTT, N. N. (1968). A quantitative approach to the detection of nuclear activity after differential damage to nucleus and cytoplasm in early development. *Journal of Embryology and Experimental Morphology*, **20**, 129–40.

PETERMANN, U. B. (1968). Mutationsraten und Sterblichkeiten nach Röntgen-bestrahlung früher Entwicklungsstadien von *Drosophila melanogaster*. *Mutation Research*, **5**, 397–410.

RÉVÉSZ, L. & BERGSTRAND, H. (1963). Radiation protection by cysteamine and cellular sulphydryl levels. *Nature, London*, **200**, 594–5.

RUSSELL, L. B. & MONTGOMERY, C. S. (1966). Radiation-sensitivity differences within cell-division cycles during mouse cleavage. *International Journal of Radiation Biology*, **10**, 151–64.

RUSTAD, R. C. (1960). Changes in the sensitivity to ultraviolet induced mitotic delay during the cell division cycle of the sea urchin egg. *Experimental Cell Research*, **21**, 596–602.

RUSTAD, R. C. (1970). Variation in the sensitivity to X-ray induced mitotic delay during the cell division cycle of the sea urchin. *Radiation Research*, **42**, 498–512.

SAMBUICHI, H. (1964). Effects of radiation on frog eggs. 1. The development of eggs transplanted with irradiated nuclei or cytoplasm. *Japanese Journal of Genetics*, **39**, 259–67.

SINCLAIR, W. K. (1968). Cysteamine: Differential X-ray protective effect on Chinese hamster cells during the cell cycle. *Science, New York*, **159**, 442–4.

SINCLAIR, W. (1969). Protection by cysteamine against lethal X-ray damage during the cell cycle of Chinese hamster cells. *Radiation Research*, **39**, 135–54.

SINCLAIR, W. K. & MORTON, R. A. (1965). X-ray and ultraviolet sensitivity of synchronised Chinese hamster cells at various stages of the cell cycle. *Biophysical Journal*, **5**, 1–25.

SINCLAIR, W. K. & MORTON, R. A. (1966). X-ray sensitivity during the cell generation cycle of cultured Chinese hamster cells. *Radiation Research*, **29**, 450–74.

SMITH, L. D. (1966). The role of a 'germinal plasm' in the formation of primordial germ cells in *Rana pipiens*. *Developmental Biology*, **14**, 330–47.

ULRICH, H. (1963). Partial irradiation of *Drosophila* zygotes by X-rays. *International Journal of Radiation Biology*, **6**, 381, abs.

WHITMORE, G. F., TILL, J. E. & GULYASS, S. (1967). Radiation induced mitotic delay in L cells. *Radiation Research*, **30**, 155–71.

WÜRGLER, F. E., ULRICH, H. & SCHNEIDER-MINDER, A. (1963). Variation of radiosensitivity during meiosis and early cleavage in newly laid eggs of *Drosophila melanogaster*. In *Repair from Genetic Radiation Damage*, ed. Sobels, pp. 101–4. London & Oxford: Pergamon Press.

YAMASHITA, H., MORI, K. & MIWA, M. (1939). The action of ionising rays on sea urchin. II. The effects of Roentgen, gamma and beta rays upon the fertilised eggs. *Gann*, **33**, 117–21.

YU, C. K. & SINCLAIR, W. K. (1970). Protection by cysteamine against mitotic delay and chromosomal aberrations induced by X-rays in synchronised Chinese hamster cells. *Radiation Research*, **43**, 357–71.

THE PATTERNS OF MITOSIS AND DNA
SYNTHESIS IN THE PRESUMPTIVE
NEURECTODERM OF
XENOPUS LAEVIS (DAUDIN)

By R. P. MALEYVAR and R. LOWERY

Department of Biological Sciences, Sir John Cass School of Science
and Technology, City of London Polytechnic, 31 Jewry Street,
London EC3N 2EY

INTRODUCTION

The emergence of systemic differences in an organism during development
or regeneration may be termed differentiation. More aptly perhaps, this
phenomenon ought to be described in terms of the molecular species and
the complete 'response repertory' of a cell (Weiss, 1968). It has become
increasingly clear that the changes which constitute differentiation reflect
a differential sequence of gene action leading to the appearance of new
RNAs and subsequent emergence of new proteins. Among others, Clayton
(1953), Deuchar (1958) and Stanisstreet & Deuchar (1972) have shown
that the antigenic make-up of neurectoderm changes during gastrulation
in amphibia.

Holtzer (1963) and Ishikawa, Bischoff & Holtzer (1968) have suggested
that a 'critical' or 'quantal' mitosis may be coupled in an obligatory
manner with the emergence of new proteins during differentiation;
the work of Mills & Topper (1970) and Bischoff & Holtzer (1970) has
further supported this idea. In the light of this suggestion and the previously cited work, it was felt that a survey of the temporal and spatial
patterns of mitosis and DNA synthesis in the presumptive neurectoderm
of *Xenopus laevis* gastrulae might provide information towards an understanding of primary induction.

MATERIAL AND METHODS

Embryos of *Xenopus laevis*, stages 10½ to 13 (Nieuwkoop & Faber, 1967),
were obtained by injecting chorionic gonadotrophin. In all cases jelly and
vitelline membrane were removed manually in Holtfreter's solution. For
the survey of mitotic pattern embryos were incubated in 10% Holtfreter's
solution with 0.0215 mg/ml colchicine for one hour before being fixed

[249]

in Smith's fixative. Thereafter, they were embedded in paraffin wax or
ester wax and sectioned at 9 μm in a sagittal plane.

In order to investigate DNA synthesis embryos were injected with
10^{-3} ml [³H]thymidine (5 Ci/mmole) and incubated in Holtfreter's solution for one hour before fixation. Subsequent procedures were based on
those described by Graham & Morgan (1966). To accentuate the nuclear
stain, which was not very outstanding after Feulgen's method, 0.1%
aqueous cresyl fast violet was used after developing the autoradiograms.

Nuclear count and statistical treatment

The nuclei in late prophase to telophase were counted as being in mitotic
phase. In the autoradiograms the mean background plus its standard
deviation was 5 grains per nucleus. 15 median 9 μm sections in case of
mitotic count and 30 median 5 μm sections for the labelled nuclei count
were considered. These median sections cover the region which gives rise
to the neural tube (Schroeder, 1970). The dorsal ectoderm was divided
into seven zones, each 0.2 mm across, using an eyepiece graticule. These
zones are shown in the inset diagrams of Fig. 1. The results are based on
counts made on one embryo of each stage. Compilation of data from the
counts of two or more embryos which were not precisely synchronised
could lead to a flattening of the curve resulting in a less clear picture. The
data have been computed to work out the analysis of variance. Tukey's Q
test (Snedecor, 1956) for the comparison of the means was applied to
ascertain the significance of the results.

RESULTS AND DISCUSSION

Distribution of mitosis

The mean percentages of mitosis in the seven zones of the neurectoderm
of gastrulae of *Xenopus laevis* are shown in Fig. 1. These show a distinct
peak in one zone at each stage of gastrulation examined. This peak is
invariably in contiguity with the mesodermal roof of the rostral tip of the
archenteron which progressively invaginates from stage $10\frac{1}{2}$ to stage 13.
The peak of mitotic activity moves from zone 2 to zone 7 in synchrony
with the invagination of the subjacent mesoderm. Although the invaginat-
ing tip of the archenteron does not extend up to the dorsal ectoderm at
stage $10\frac{1}{2}$, the rostral part of its mesodermal roof already lies under the
presumptive neurectoderm (Sudarawati & Nieuwkoop, 1971). The data
were subjected to an analysis of variance. In all zones the variance ratio was
significant at 0.1% level. Tukey's Q test was applied to ascertain which of

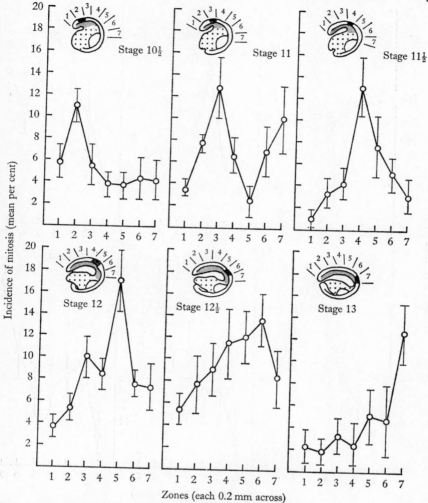

Fig. 1. Distribution of mitosis in the neurectoderm of *Xenopus laevis*. Zones showing maximum mitotic activity are shown black; standard deviations indicated by vertical bars.

the seven zones contributed the most to this variance. Table 1 shows the zones which contribute the most to this variance. These results are in agreement with those shown in Fig. 1.

Distribution of DNA synthesis

The labelled nuclei count, Fig. 2, shows a wave of DNA synthesis passing through the neurectoderm from zone 2 to zone 7 during the stages examined.

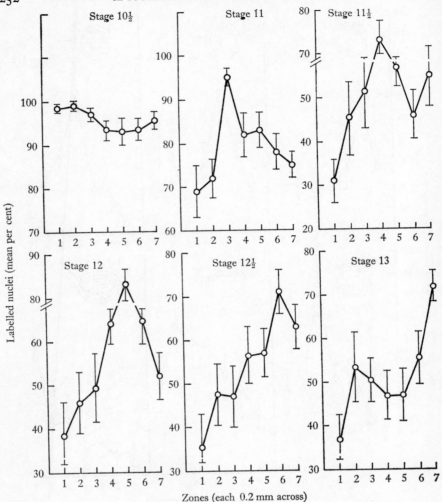

Fig. 2. Distribution of DNA synthesis in the neurectoderm of *Xenopus laevis*.
Standard deviation shown by vertical bars.

As in the mitotic study, the peak in this study also occurs near the mesodermal roof over the tip of the archenteron in all six stages. The fact that
the peak of DNA synthesis is not as pronounced at stage $10\frac{1}{2}$ as during the
succeeding stages could be due to the brevity of the cell cycle in the
neurectoderm at this time; this is indicated by more than 86 % of the
stage $10\frac{1}{2}$ nuclei being labelled. In contrast to this, about 52 % of the
nuclei are labelled during stages $11\frac{1}{2}$ to 13 which show more pronounced
peaks of DNA synthesis. Graham & Morgan (1966) have shown for the

Table 1. *Summary of Tukey's Q test*

Mitotic count

Value of Q for degrees of freedom = 7 and 98 upper 5% points = 4.24.

Stages of gastrula	$10\frac{1}{2}$	11	$11\frac{1}{2}$	12	$12\frac{1}{2}$	13
Most significant of the seven zones	2	3	4	5	6	7

Labelled nuclei count

Value of Q for degrees of freedom = 7 and 273 upper 5% points = 4.17.

Stages of gastrula	$10\frac{1}{2}$	11	$11\frac{1}{2}$	12	$12\frac{1}{2}$	13
Most significant of the seven zones	1, 2	3	4	5	6	7

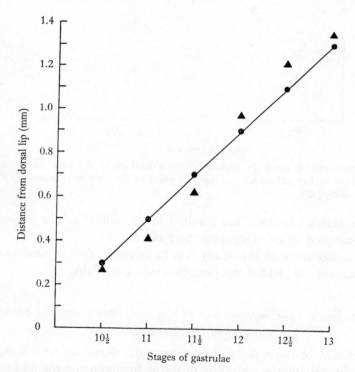

Fig. 3. Correlation between the distance from dorsal lip of the mesodermal roof over the tip of the archenteron (▲) and the mid point of the zone of maximum mitotic activity (●).

endoderm of *Xenopus laevis* that the length of the cell cycle increases as the embryo advances in age.

There is an apparent temporal and spatial coincidence of the waves of mitosis and DNA synthesis (Figs. 1 and 2). The brevity of the cell cycle

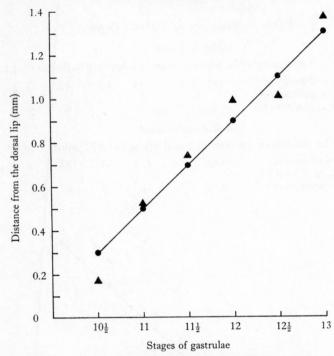

Fig. 4. Correlation between the distance from dorsal lip of the mesodermal roof of the tip of the archenteron (▲) and the mid point of the zone of maximum DNA synthetic activity (●).

during the stages examined has resulted in the failure of the counting technique adopted by us to separate these events.

Statistical treatment in this study was the same as that applied to the mitotic data and has yielded comparable results (see Table 1).

Correlation between the invagination of the archenteron and the peaks of mitosis and DNA synthesis

The correlation between the distances from the dorsal lip to the mesodermal roof overlying the rostral tip of the archenteron and the mid point of the zones of maximum mitotic and DNA synthetic activities is shown in Figs. 3 and 4 respectively. The spatial association between the mesodermal roof and the peaks of mitosis and DNA synthesis is quite significant. Bearing in mind the suggestion that a 'quantal' mitosis may be involved in differentiation (Ishikawa *et al.* 1968) and that primary induction occurs during gastrulation, it is our intention to further examine the possible significance for differentiation of these waves of mitosis and DNA synthesis.

SUMMARY

The spatial and temporal patterns of mitosis and DNA synthesis in the presumptive neurectoderm of *Xenopus laevis* gastrulae have been mapped. The count of the nuclei in mitotic phase after using colchicine blocking method and the count of nuclei in S phase by using [³H]thymidine labelling technique have made it evident that a wave of mitosis and a wave of DNA synthesis pass through the neurectoderm of *Xenopus* gastrulae. These events are correlated with the invagination of the mesodermal roof overlying the rostral tip of the archenteron. The possible significance of this phenomenon for the process of primary induction is discussed.

The authors are grateful to Mr P. B. Mordan of the Computer Centre, City of London Polytechnic for his assistance with the statistical analysis.

REFERENCES

BISCHOFF, R. & HOLTZER, H. (1970). Inhibition of myoblast fusion after one round of DNA synthesis in 5-bromodeoxyuridine. *Journal of Cell Biology*, **44**, 134.

CLAYTON, R. M. (1953). Antigens in the developing newt embryos. *Journal of Embryology and Experimental Morphology*, **1**, 25.

DEUCHAR, E. M. (1958). Regional differences in catheptic activity in *Xenopus laevis* embryos. *Journal of Embryology and Experimental Morphology*, **6**, 223.

GRAHAM, C. F. & MORGAN, R. W. (1966). Changes in the cell cycle during early amphibian development. *Developmental Biology*, **14**, 439.

HOLTZER, H. (1963). Comments on induction during cell differentiation. *XIII colloquim Gesellschaft für physiologische Chemie*, p. 127. Berlin: Springer-Verlag.

ISHIKAWA, H., BISCHOFF, R. & HOLTZER, H. (1968). Mitosis and intermediate-sized filaments in developing skeletal muscle. *Journal of Cell Biology*, **38**, 538.

MILLS, E. S. & TOPPER, Y. J. (1970). Some ultrastructural effects of insulin, hydrocortisone and prolactin on mammary gland explants. *Journal of Cell Biology*, **44**, 310.

NIEUWKOOP, P. D. & FABER, J. (1967). *Normal Table of Xenopus laevis (Daudin)*, 2nd edit. Amsterdam: North-Holland Publishing Co.

SCHROEDER, T. E. (1970). Neurulation in *Xenopus laevis*. An analysis and model based upon light and electron microscopy. *Journal of Embryology and Experimental Morphology*, **23**, 427.

SNEDECOR, G. W. (1956). *Statistical Methods*, 5th edit. Iowa State College Press.

STANISSTREET, M. & DEUCHAR, E. M. (1972). Appearance of antigenic material in gastrulae ectoderm after neural induction. *Cell Differentiation*, **1** (1), 15.

SUDARAWATI, S. & NIEUWKOOP, P. D. (1971). Mesoderm formation in the anuran *Xenopus laevis* (Daudin). *Wilhelm Roux' Archiv, EntwMech. Org.* **166**, 189.

WEISS, P. (1968). The problems of cellular differentiation. In *Dynamics of Development: Experiments and Inferences*. New York & London: Academic Press.

CELL PROLIFERATION IN LATE EMBRYOS AND YOUNG LARVAE OF THE NEWT *PLEURODELES WALTLII* MICHAH.

BY P. CHIBON

Laboratoire de Zoologie, Université scientifique et médicale de Grenoble, B.P. 53 Centre de Tri, 38-041 Grenoble Cedex, France

INTRODUCTION

Cellular proliferation studies using amphibians are much less numerous than studies made on mammals. Amphibians display, however, certain particular advantages: embryos are easily available and handled, and may be exposed to external medium variations, stimuli and/or factors.

Cell cycle methods of study, introduced by Howard & Pelc (1953) and Quastler & Sherman (1959), and applied to amphibian material, recently enabled Graham (1966), Graham & Morgan (1966), and Chulitskaya (1967) to demonstrate the non-existence of the G_1 phase and the short duration of G_2 phase during amphibian embryo cleavage. In the developing embryo, the cell cycle lengthens afterwards, for G_1 phase settles and becomes long, while other phases lengthen unequally. Simultaneously mitotic index diminishes unequally in the different primordia.

Other studies have demonstrated that cell cycles are much longer in adult amphibians than in mammals: the cell cycle lasts for about 75 h in adult newt liver and intestine (Grillo & Urso, 1968). For some regenerating organs it is of a relatively long duration, and the mitotic index and labelling index vary during the regeneration process (Eisenberg & Yamada, 1966; Zalik & Yamada, 1967; Mitashov, 1969).

Culture studies in-vitro showed a relatively long duration of cell cycle (Reddan & Rothstein, 1966; Malamud, 1967) and a mitotic index comparable to the value in-vivo (Simnett & Balls, 1969).

Our research project on amphibian cell tissue proliferation mainly concerned stages around hatching. This stage ends embryonic development and involves internal organ differentiation. It enabled us to concentrate on (1) normal cell proliferation in various organs; (2) the effect of temperature on cell proliferation; (3) the effect of thyroxine, a hormone known as cell-division-stimulating; (4) the relationship between cell proliferation and differentiation in each tissue.

MATERIALS AND METHODS

Experiments were made on the newt *Pleurodeles waltlii* Michah., kept according to Gallien's (1952) recommendations. The development table used was that of Gallien & Durocher (1957).

Using a microneedle, embryos or larvae are injected intraperitoneally with a size-variable volume of a tritiated thymidine solution (specific activity, 5 or 10 Ci/mmole; total activity 50 or 100 μCi/ml). Animals are fixed from 5 min to 100 h after injection. The labelled mitoses percentage, the labelling index and the mitotic index are determined with histo-autoradiographic techniques.

In order to determine the tritiated thymidine availability time, grains were counted in different tissues, according to the time elapsed between injection and fixation; DNA was extracted, dosed, and its radioactivity measured by liquid phase scintillation.

RESULTS

Duration of cell cycles at stage 38

For 12 different organs and tissues of 5-day-old (post hatching) *Pleurodeles* larvae kept at 21 °C, we first plotted labelled mitosis percentage variations, and calculated the duration of the cell cycle and its phases (Chibon, 1968).

Figs. 1 and 2 show seven of these graphs. Generation time or cell cycle time (T) values, and its phases S, G_2 and M were graphically determined, while G_1 phase obtained by subtraction.

Cell cycle duration ranges from 42 to 48 h; S phase is the longest phase and lasts from 27 to 34 h; G_2 phase varies from 2.5 h duration to 5 h, while the M phase lasts from 3.5 to 7 h. G_1 is the most variable phase: it may vary from 0 to 11 h, according to the tissue concerned.

These results mainly bring out the fact that generation time has an almost equal duration in different tissues of young larvae.

Effect of temperature on cellular proliferation

In poikilothermic animals it is well known that the rate of development varies according to numerous exogenous factors and particularly to temperature. In order to appreciate the effect of temperature on the cell cycle duration, and on its phases, cell proliferation was studied in three tissues of different embryonic origin in embryos at stage 34, kept at 12, 17, 23 and 26 °C.

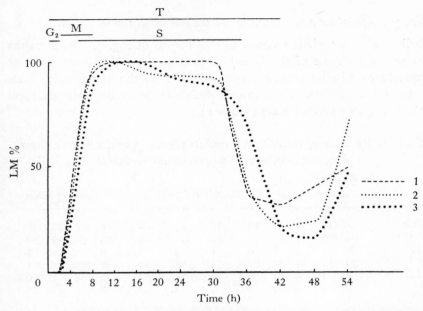

Fig. 1. Labelled mitoses percentage (LM) in the nervous system of *Pleurodeles* larvae (stage 38): 1, spinal ganglia; 2, telencephalon; 3, spinal cord. Durations of cell cycle and its phases are indicated on top.

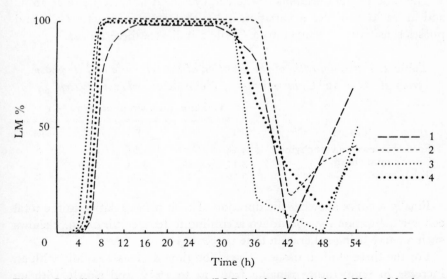

Fig. 2. Labelled mitoses percentage (LM) in the fore-limb of *Pleurodeles* larvae (stage 38): 1, limb tip epidermis; 2, cartilage; 3, limb tip mesenchyme; 4, limb basis mesenchyme.

Effect of temperature on the duration of the cell cycle and its phases

Labelled mitoses variation versus time is shown in Fig. 3. These graphs refer to mesenchyme of the fore-limb in larvae reared at four different temperatures. Similar curves were observable for the gastric epithelium and the telencephalon. These graphs enable us to measure the duration of the cell cycle and its phases (Table 1).

Table 1. *Duration of the cell cycle and its phases, graphically calculated from the labelled mitoses percentage variation*

Temperature (°C)	Fore-limb basis mesenchyme				Gastric epithelium				Telencephalon			
	12	17	23	26	12	17	23	26	12	17	23	26
T (h)	91	40	24	22	92	42	32	29	93	48	32	30
G_1	0	1	4	6	0	1	4	6.5	0	4	5	7.5
S	79	31	16	14	82.5	34	21	20	81	38	24	21
G_2	6	4	2	1	3.5	2.5	1.5	1	4	2	1	0.5
M	6	4	2	1	6	4.5	2.5	1.5	8	4	2	1

Graphs of the cell cycle duration and of its phases may also be drawn versus temperature: gastric epithelium curves are shown in Fig. 4. Similar curves have been obtained for both other tissues.

The ratio of the durations of the cell cycle and of each phase at 26 °C and at 12 °C provides a variation index that express specific cycle and phases sensitivity to temperature. Table 2 indicates these ratios.

Table 2. *Variation ratio of the duration of the cell cycle and its phases from 26 °C to 12 °C, in 3 tissues of* Pleurodeles *embryos at stage 34*

Tissues	Variation ratio (from 26 to 12 °C)				
	T	G_1	S	G_2	M
Fore-limb basis mesenchyme	4.14	∞	5.64	6	6
Gastric epithelium	3.17	∞	4.13	3.40	4
Telencephalon	3.10	∞	3.86	8	8

Finally a curve showing the duration of each phase relative to the total cell cycle duration may be drawn according to temperature. Fig. 5 shows such a curve using generation time percentage.

For the three studied tissues, generation time decreases rapidly with an increase in temperature ranging from 12 to 17 °C and slowly with an increase over 17 °C: thus Fig. 4 shows a hyperbolic curve. Table 2 shows that the mesenchyme of the fore-limb basis is the most sensitive to

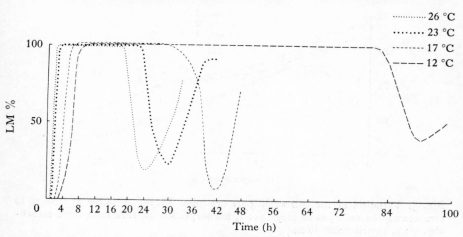

Fig. 3. Labelled mitoses percentage (LM) in the fore-limb basis mesenchyme of *Pleurodeles* embryos (stage 34), reared at four different temperatures.

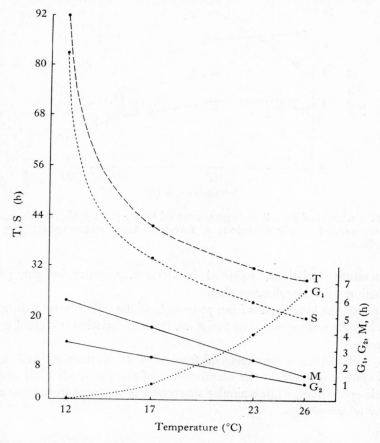

Fig. 4. Variation of the duration of the cell cycle (T) and its phases according to temperature, obtained in the gastric epithelium of *Pleurodeles* embryos at stage 34. Ordinates scale is four times smaller for T and S than for G_1, G_2 and M.

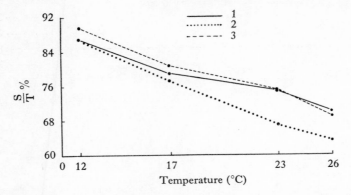

Fig. 5. Variation of the cell cycle part occupied by the S phase according to temperature in embryos of stage 34: 1, telencephalon; 2, fore-limb basis mesenchyme; 3, gastric epithelium.

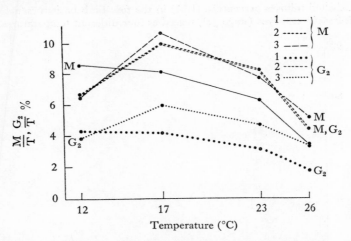

Fig. 6. Variation of the cell cycle part occupied by the G_2 and M phases according to temperature: 1, telencephalon; 2, fore-limb basis mesenchyme; 3, gastric epithelium.

temperature variations, followed in decreasing order by the gastric epithelium and the telencephalon.

S phase, the longest, fills from 70 to 90 % of the cell cycle total duration: thus it is the main cause of its variation. Its own variation is plotted hyperbolically.

G_2 and M phases are shortened from 12 to 26 °C proportionally to the temperature increase in the mesenchyme of the fore-limb basis and the gastric epithelium, while a similar shortening is observed from 17 to 26 °C in the telencephalon.

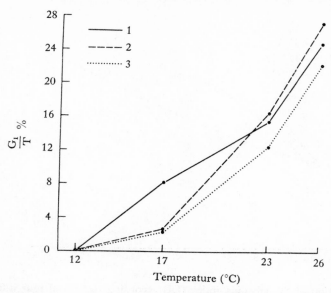

Fig. 7. Variation of the cell cycle part occupied by the G_1 phase according to temperature: 1, telencephalon; 2, fore-limb basis mesenchyme; 3, gastric epithelium.

Fig. 6 demonstrates that the cell cycle portion filled by the G_2 and M phases decreases from 17 to 26 °C while they increase or decrease from 12 to 17 °C, according to tissues. G_2 and M variation indices (Table 2) are close and indicate the fact that the range of variations of the G_2 and M phases is most important in the telencephalon.

G_1 phase varies counter to the three other phases: it is non-existent at 12 °C but rapidly lengthens, particularly over 23 °C: at 26 °C it fills a quarter of the generation time (Fig. 7). Thus, the variation index does not significantly indicate its increase. G_1 phase is the most sensitive to temperature.

Effect of temperature on labelling index

The labelling index was studied according to post-injection time in the three tissues already mentioned, in embryos reared at four different temperatures.

Gastric epithelium labelling index variations in embryos kept at 12, 16, 23 and 26 °C are shown in Fig. 8. The mesenchyme of the fore-limb basis and the telencephalon variation curves were similar. All these curves show a typical pattern of minima and maxima sequences. With the help of a theoretical graphical model we pointed out the fact that this characteristic

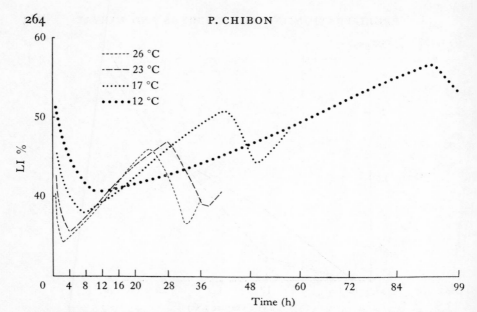

Fig. 8. Variation of the labelling index (LI) according to time at four different temperatures in the gastric epithelium of *Pleurodeles* embryos at stage 34. (In drawing these curves, nuclei with more than 8 grains were considered as labelled).

pattern corresponds to an asynchronous exponentially-growing cell population and to the labelled and non-labelled cells' alternating divisions (Brugal & Chibon, 1970).

In-vivo, cell populations show a certain cell cycle variability, and the amplitude of the labelling index variations gradually damps down, though keeping the original periodicity. Therefore these curves are used to calculate the duration of the cell cycle and its phases: T duration lies between two successive minima or maxima; S duration (including the tritiated thymidine availability time) is equal to the time corresponding to the ascending part of the curve; G_2 and M duration fill the time gap between injection and first minimum.

Labelling index variation curves and labelled mitoses percentage variation curves can be used to calculate the duration of the cell cycle and its phases; for the three studied tissues, values obtained from both methods are quite similar and corroborate the fact that these tissues have a typical exponentially growing asynchronous cell population.

Moreover, it is possible to calculate, according to the mathematical model of Brugal & Bertrandias (1970), the growth fraction, that means the proportion of cells in mitotic activity, in exponentially growing and asynchronous populations, if labelling index and durations of T, G_2 and S are known (this model assumes that no mortality and no cellular migration exist).

Fig. 8 shows that an increase of temperature induces a displacement to the left and a lowering of the curves, but they keep their characteristic shape. The fold results from the effect of temperature on the cell cycle and the duration of its phases, and the drop from the decrease of labelling index, which itself comes from the shortening of the relative duration of the S phase.

The three studied tissues present different values of the observed labelling index and different variation of it according to time: for a given temperature, the labelling index is higher, and its variations are greater in the fore-limb basis mesenchyme than in the gastric epithelium and in the telencephalon. Incidentally, these three tissues have very close theoretical labelling indices; the duration of their cell cycle and their ratios S/T are very similar: Brugal (1971a) has evaluated the growth fraction for each of them using the equation:

$$LI = \frac{S}{T} \log_e (P+1) \left[1 + \log_e (P+1) \frac{M+G_2+\frac{1}{2}S}{T} \right],$$

where S, M, G_2 = durations of the S, M, and G_2 phases,
T = generation time, P = growth fraction.

Table 3 shows that $\log_e (P+1)$ and consequently growth fraction P may be considered as equal for different temperatures in a given tissue.

Thus, as early as stage 34, the rate of cells in mitotic activity is very different in these three tissues, but is independent of temperature.

Effect of temperature on the mitotic index

Mitotic index has been calculated in the three tissues, for the four temperatures: in the telencephalon, the mitotic index decreases when temperature increases and more rapidly the higher the temperature; in the fore-limb basis mesenchyme, and in the gastric epithelium, it increases when temperatures increases from 12 to 17 °C, and decreases above 17 °C.

These variations are exactly parallel to the variation of the relative duration of the M phase (Fig. 6); the growth fraction does not depend on temperature, thus the equation (Brugal, 1971a)

$$MI = M/T . \log_e (P+1)$$

shows that in an exponentially growing cell population, mitotic index depends only on the relative duration of M. This equation enables us to calculate growth fraction if the durations of M and T, and mitotic index are known (Table 4).

Table 3. *Growth fraction (P) calculated from the labelling index, in 3 tissues of Pleurodeles embryos at stage 34*

	Fore-limb basis mesenchyme				Gastric epithelium				Telencephalon			
Temperature (°C)	12	17	23	26	12	17	23	26	12	17	23	26
Log_e (P+I)	0.604	0.586	0.619	0.613	0.537	0.539	0.560	0.541	0.317	0.302	0.295	0.296
Average of (log_e P+I)		0.606				0.544				0.303		
P+I		1.83				1.72				1.35		
Growth fraction P (%)		83				72				35		

Table 4. *Growth fraction (P) calculated from the mitotic index, in 3 tissues of Pleurodeles embryos at stage 34*

	Fore-limb basis mesenchyme				Gastric epithelium				Telencephalon			
Temperature (°C)	12	17	23	26	12	17	23	26	12	17	23	26
Mitotic index MI (‰)	39.7	62.4	54.2	28.8	38.2	61.1	46.6	25.3	27.9	25.2	19.6	10.3
M/T (‰)	65.9	100	83.3	45.5	65.2	107.1	78.1	51.7	86	83.3	62.5	33.3
MI.M/T = log_e (P+I)	0.602	0.624	0.651	0.633	0.586	0.570	0.597	0.489	0.324	0.303	0.314	0.309
Average log_e (log_e P+I)		0.628				0.561				0.313		
P+I		1.88				1.75				1.37		
Growth fraction P (%)		88				75				37		

Growth fraction values calculated with this method are in good agreement with those calculated from the labelling indices (Table 3): it corroborates that these three tissues have very different, but temperature independent, growth fractions.

Discussion and conclusion

The variation of generation time according to temperature observed here is comparable with the results obtained in young embryos of Amphibia and Fish by Chulitskaya (1967), Detlaff (1964), Ignatieva & Kostomorova (1966), and in meristematic plant cells by Evans & Savage (1959) and by Wimber (1966). Even at 26 °C, the generation time is much longer than in avian or mammalian embryos, probably on account of the higher temperature of these last named.

G_1 is the phase most sensitive to temperature; this fact has already been mentioned (Reddan & Rothstein, 1966), but here this phase varies inversely to the other phases, and that seems unexpected. At low temperature, if G_1 phase is absent, tissue differentiation should be late in comparison with proliferation; in fact, it has been observed by Decker & Kollros (1966) in Rana pipiens tadpoles, that animal size and cell number are greater if breeding temperature is lower. Furthermore, a drop of yolk utilization at low temperature has been reported (Lovtrup, 1959). Thus one may suppose that the S phase lengthening at low temperature is a consequence of a decrease in the production of enzymes and precursors necessary to DNA synthesis: these molecules are elaborated during G_1 phase, and then a shortening of G_1 phase involves a S phase lengthening.

The labelling index and chiefly the mitotic index are not reliable representations of the mitotic activity of any tissue, for mitotic rate increases according to temperature.

In fact, the mitotic index relies at one time on the growth fraction and on the relative duration of M phase. Moreover, in cases where duration of the cell cycle and its phases is known, these indices enable us to calculate the growth fraction, which is of great interest.

Effect of thyroxine on cellular proliferation

In different amphibian metamorphosing tissues, or in-vitro, several workers (Champy, 1922; May & Mugard, 1955; Ferguson, 1966; Reynolds, 1966; Pesetsky, 1969; Atkinson & Just, 1970; Fährmann, 1971) have pointed out the effect of thyroxine in increasing mitotic activity. Few data are reported concerning thyroxine effect on proliferation parameters: Chibon & Brugal (1969) observed a shortening of the generation time and of all phases, except for the G_1 phase which is shortened or lengthened according to the tissue considered. Burki & Tobias (1970) observed in in-vitro human cells a shortening of the cell cycle and of the G_1 phase, while Laguchev (1971) reports in the rat a shortening of the G_1 phase, without modification of the G_2 phase.

Thus, we have undertaken a study of cellular proliferation in the amphibian tadpole (*Bufo bufo*) before and during metamorphosis, in order to observe the possible modifications of the cell cycle and growth fraction during metamorphosis. Preliminary results indicate (Chibon & Dournon, 1972) that during the metamorphic climax a shortening of the generation time and G_1 phase takes place in intestinal epithelium and telencephalon, and an increase of the growth fraction occurs. These facts seem to be induced by the circulating thyroxine.

Table 5. *Duration of the cell cycle and its phases in 3 tissues of* Pleurodeles *embryos and larvae, at different stages of development* (*reared at* 17 °C)

Tissues	Stages	Duration (h)							
		T	S		G_2		M		G_1
Telence-	34	48	38	(79.2)	2	(4.2)	4	(8.3)	4	(8.3)
phalon	36	56	42	(75)	3	(5.4)	5	(8.9)	6	(10.7)
	38	64	46	(70.8)	4	(6.2)	6	(10.7)	8	(12.3)
	41	69	49	(71)	5	(7.2)	6	(8.7)	9	(13.1)
Spinal cord	34	41	35	(85.4)	2	(4.9)	3	(7.3)	1	(2.4)
	36	52	41	(78.8)	3	(5.8)	5	(9.6)	3	(5.8)
	38	72	52	(72.3)	5.5	(7.6)	7.5	(10.4)	7	(9.7)
	41	84	56	(66.7)	8	(9.5)	9	(10.7)	11	(13.1)
Gastric	34	42	34	(81)	2.5	(5.9)	4.5	(10.7)	1	(2.4)
epithelium	36	48	37	(77.1)	3.5	(7.3)	5	(10.4)	2.5	(5.2)
	38	56	40	(71.5)	5	(8.9)	5.5	(9.8)	5.5	(9.8)
	41	63	40	(63.3)	6	(9.5)	5	(7.9)	11	(17.3)

The part of the cell cycle (represented in percentage of the generation time) occupied by each phase is indicated in parentheses.

Variations of cellular proliferation in relation to age

Duration variability of the cell cycle and its phases

Brugal (1971b) investigated several proliferation parameters in embryos and larvae at stages 34, 36 (hatching), 38 and 41, reared at a constant temperature of 17 °C. Three tissues were considered: telencephalon. gastric epithelium, and spinal cord. Table 5 summarizes data graphically calculated from the labelled mitoses percentage variation curves, shown by

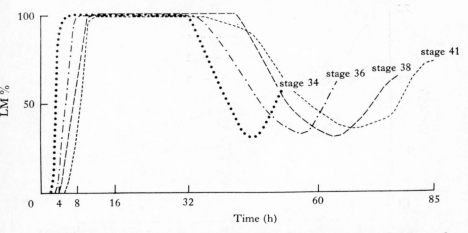

Fig. 9. Labelled mitoses (LM) percentage variation in telencephalon at 4 stages of development in *Pleurodeles* embryos and larvae reared at 17 °C. (From Brugal, 1971*b*, © Academic Press).

Fig. 9. It is obvious that for each tissue, cell cycle duration lengthens as larvae get older: this lengthening is greater at the beginning and becomes slower and slower in the successive stages. A variation index may be defined, representing the cycle and its phases lengthening from stage 34 to stage 41. Table 6 collects these calculated values and demonstrates that during this time interval, the cell cycle duration lengthens from 44% to 105% according to tissue, and that the most sensitive phase is G_1. G_2 phase is also significantly lengthened, while S and M phases are only slightly altered. The part of the cell cycle occupied by the G_1 phase therefore increases regularly and rapidly as the animal grows old. The part occupied by the G_2 phase increases slowly, whereas the S phase portion decreases.

Table 6. *Variation ratio of the duration of the cell cycle and its phases from stage 34 to stage 41*

Tissues	S	G_2	M	T	G_1
Telencephalon	1.29	2.50	1.50	1.44	2.25
Gastric epithelium	1.18	2.40	1.11	1.50	11
Spinal cord	1.60	4	3	2.05	11

The fact that G_1 phase gets more and more time-consuming is probably related to the progressing differentiation of the tissues. Its lengthening is unequal in the three tissues considered: it reaches the longest duration in gastric epithelium, and spinal cord, which acquire their functional activity at hatching.

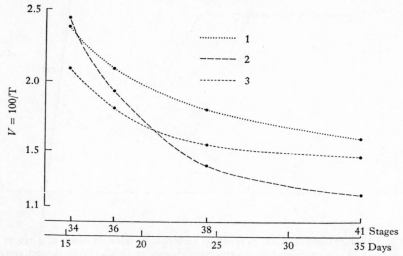

Fig. 10. Variation of the mitotic rate from stage 34 to stage 41 in 3 tissues of *Pleurodeles* embryos and larvae reared at 17 °C. The mitotic rate indicates the number of divisions undergone in a single cell during 100 h; it is inversely proportional to cell cycle duration, and is expressed by the equation $v = (100/T)$. 1, gastric epithelium; 2, spinal cord; 3, telencephalon. (From Brugal, 1971b, © Academic Press).

It is obvious that the mitotic rate is inversely proportional to generation time and can be expressed by the number of mitoses undergone by a cell in 100 h. Sequence of mitosis slows progressively in the three tissues as shown in Table 7 and Fig. 10.

Table 7. *Number of mitoses detected in* 100 *hours in cells of 3 tissues* (=*proliferation rate*) *in* Pleurodeles *embryos and larvae reared at* 17 °C

Tissues	34	36	38	41
Telencephalon	2.08	1.79	1.56	1.45
Spinal cord	2.44	1.92	1.39	1.19
Gastric epithelium	2.38	2.08	1.79	1.59

Stages

Variation of the labelling index, mitotic index and growth fraction

One hour after tritiated thymidine injection, labelling index and mitotic index have been evaluated: both indices decrease during development (Table 8).

Growth fraction values (P) have been calculated from both indices with the help of previously-used formulae; Table 8 and Fig. 11 show its variation. Both computation methods give quite similar results: growth

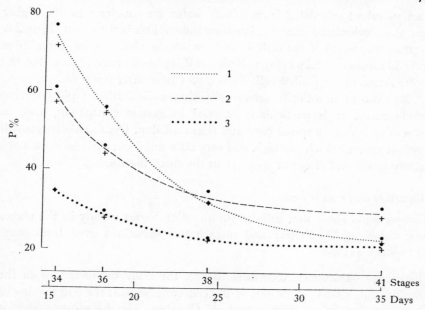

Fig. 11. Variation of the growth fraction (P) from stage 34 to stage 41 in 3 tissues of *Pleurodeles* embryos and larvae; growth fraction values are calculated from labelling index (+) and from mitotic index (●). 1, gastric epithelium; 2, spinal cord; 3, telencephalon. (From Brugal, 1971*b*, © Academic Press.)

Table 8. *Growth fraction* (P) *variation, calculated from the labelling index* (LI) *and from the mitotic index* (MI) *in 3 tissues of* Pleurodeles *embryos and larvae from stage 34 to stage 41*

Tissues	Stages	LI (%)	MI (‰)	P From LI	From MI
Telencephalon	34	27.7	25.2	35	35
	36	21.3	23.6	28	30
	38	15.7	24.4	22	23
	41	14.1	17	20	22
Spinal cord	34	47.6	35.1	57	61
	36	35.2	36.1	44	46
	38	23.1	31.6	32	35
	41	18.7	28.7	28	31
Gastric epithelium	34	57.1	61.1	71	77
	36	41.3	46.2	54	56
	38	22.9	27.6	32	32
	41	13.9	16.5	22	23

fraction values calculated from mitotic index are usually 3 to 6% higher than those calculated from the labelling index. This is the result (Brugal & Bertrandias, 1970) of the difference between the observed labelling index and real number of cells being in S phase: this phenomenon is caused by the proliferation of unlabelled cells from 0 to 1 hour after injection.

The ratio of mitotically active cells always diminishes rapidly during development. It is particularly marked in gastric epithelium, and, in decreasing order, in spinal cord and telencephalon. Consequently, growth fraction values, which are high and very different from each other at stage 34, are lower and closer at stage 41 in the three tissues.

Discussion and conclusions

These results show that cellular proliferation varies greatly in the stages near hatching. Qualitative and quantitative variations exist from stage 34 to stage 41.

Qualitative variations: According to age, there are modifications in the absolute and relative durations of the cell cycle phases: the lengthening of G_1 phase and, to a minor extent, of G_2 phase are the most important observations. They have to be correlated with the degree of histological differentiation of each tissue at the beginning and at the end of the period under study. Telencephalon shows at stage 34 an advanced differentiation, while the other organs, particularly the gastric epithelium, are less differentiated at stage 34, but functional at stage 41. When embryos reach stage 34, the tissue whose histogenesis has progressed the most has a minor lengthening of G_1 phase from stage 34 to stage 41.

Without any doubt, this G_1 phase lengthening in differentiating cells is related to the differentiation process itself. Plant meristematic cells (Clowes, 1965), tumour cells (Defendi & Manson, 1961; Baserga, 1963; Kissel *et al.* 1966) and sea urchin blastomeres (Hinegardner *et al.* 1964) have a very short G_1 phase or none at all. G_1 appearance or lengthening takes place after egg cleavage in the amphibian embryo (Graham & Morgan, 1966), during mouse neural tube histogenesis (Kaufmann, 1968), during chick leg muscle histogenesis (Marchok & Herrmann, 1967), and during growth of the chick wing bud (Janners, 1968).

A similar sequence of events occurs for G_2 phase, which occupies a progressively extending part in the cell cycle. While these two phases are going on, tissue-specific protein synthesis takes place. These are periods of intense synthetic activity. S and M phases relative durations are less variable: they do not seem to be directly involved in the differentiation process.

Quantitative variations. Growth fraction and mitotic rate are concerned in these variations; mitotic rate varies with the cell cycle duration.

The ratio of cells in mitotic activity and their mitotic rate decrease in the three studied tissues from stage 34 to stage 41. At stage 34, the gastric epithelium has the highest growth fraction (71 to 77%); from stage 34 to stage 38, during histogenesis, it decreases rapidly; from stage 39 to stage 41, it diminishes slowly. While modifications of growth fraction are going on, the cell proliferation rate decreases slightly but remains relatively high (2.38 to 1.59). Stage 38 corresponds to the beginning of feeding: from that moment, the gastric epithelium differentiation is completed. Cell division still goes on for a small number of cells, in order to ensure growth and cell renewal in the tissue.

At stage 34, contrary to gastric epithelium, the telencephalon shows a low (35%) and less decreasing growth fraction. Yet histological differentiation is well on, and cellular proliferation is slow.

Compared to both preceding tissues, the spinal cord has an intermediate developmental rate: growth fraction does not decrease below a certain level (28 to 31%) essential to larval growth.

We observe therefore that the improvement of histological differentiation and the decrease of cellular proliferation are synchronous, whereas according to other published results, a reduction of proliferation occurs before differentiation. Such a difference might be attributable to tissues or species considered, or may also come from the parameters chosen to define cellular differentiation, for cellular differentiation indeed has multiple facets and is a progressive phenomenon.

Recently, in order to verify our experimental procedure reliability and thus to correct if necessary the values expressed for S and G_1 phase duration (values calculated from the labelled mitoses percentage variation curves), we determined the tritiated thymidine availability time after injection into young larvae and young metamorphosed animals. We have come to the conclusion (Chibon & Brugal, 1973) that in larvae reared at 21 °C (stage 36) the availability time is 60 min and in young metamorphosed *Pleurodeles* 4 h.

CONCLUSIONS

A fall in temperature induces in the mitotically active cells a significant lengthening of S, G_2 and M phases; conversely, G_1 phase duration decreases, and at 12 °C disappears. Generation time, which is almost entirely subordinated to S phase duration, lengthens; at 12 °C it is three times longer than at 26 °C. During the 20 days development period we have investigated, generation time lengthens from 50 to 100% against time.

The high sensitivity of the G_1 phase, and to a lesser degree, of the G_2 phase, to temperature variations and to ageing, suggests that their durations are directly related to cellular metabolic activity; during these two phases, cells are engaged in tissue-specific differentiation processes.

Compared to G_1 and G_2 phases, the lengthening of S and M phases during development is less important. This lengthening seems more directly related to the intrinsic cell ageing than to differentiation processes.

Labelling index variation according to time enables us to set up a cell cycle and a growth fraction calculation method. The growth fraction calculation method is of a particular interest to embryologists; at a given stage, the growth fraction is different in the different tissues, while insensitive to temperature. Growth fraction is high in the less differentiated tissues, low in the more differentiated: in cells differentiated to a certain degree, this lowering indicates that cell division will soon cease. Consequently the G_0 phase, in which the cells are blocked, may be considered as an indefinitely prolonged G_1 phase.

The continuous growth fraction variation during development suggests that all cells forming a given tissue more or less control the number of cells which stop dividing after they have reached a certain degree of differentiation (Brugal, unpublished data). Moreover, comparison between the embryo's rate of development and the cell cycle durations suggests that each embryonic stage is reached after a definite number of cell divisions: this number is probably tissue-specific. We are inclined to believe that the beginning of differentiation in embryonic cells is closely related to the number of mitoses undergone by these cells.

In summary, at least three factors appear to influence the cell's mitotic behaviour: the number of previous mitoses, the cell's state of differentiation and the number of cells in the population it belongs to.

SUMMARY

The duration of the cell cycle and its phases are strictly subordinated to temperature; when temperature decreases, S, G_2 and M phases durations, and generation time (T) lengthen. On the contrary, G_1 phase duration shortens. Generation time lasts on for about 90 h at 12 °C and 30 h at 26 °C.

Labelling index variation curves against time correspond to theoretical curves calculated from a mathematical model; this shows that proliferation in the tissues under study is asynchronous and exponential.

If labelling index, generation time and S phase duration are known, or if mitotic index, generation time and M phase duration are known, one may calculate the growth fraction. This growth fraction is independent of temperature, and smaller as the tissue considered is more differentiated.

Cell cycle time and the duration of its phases increase with age: lengthening of G_1 and G_2 phases appears to be in relation to progressive cell differentiation, while the shorter lengthening of S and M phases seems to rely on intrinsic cellular ageing.

The growth fraction decreases significantly when tissue differentiation improves: after hatching, the growth fraction is almost equal in the different tissues studied.

The author wishes to thank Mrs Michèle Bélanger-Barbeau for her assistance with the translation of the manuscript.

REFERENCES

ATKINSON, B. G. & JUST, J. J. (1970). Cellular proliferation in bullfrog tadpole tissues in spontaneous and triiodothyronine induced metamorphosis. *Proceedings of the 54th annual meeting of the Federation of American Societies for Experimental Biology, Atlantic City*, Abstract no. 3455, 856.

BASERGA, R. A. (1963). Mitotic cycle of ascites tumor cells. *Archives of Pathology*, **75**, 156–61.

BRUGAL, G. (1970). Etude autoradiographique de la prolifération cellulaire, en fonction de l'âge et de la température, chez les embryons âgés et les jeunes larves de *Pleurodeles waltlii* Michah. (Amphibien Urodèle). Thèse 3° Cycle, Université scientifique et médicale de Grenoble.

BRUGAL, G. (1971a). Etude autoradiographique de l'influence de la température sur la prolifération cellulaire chez les embryons âgés de *Pleurodeles waltlii* Michah. (Amphibien Urodèle). *Wilhelm Roux' Archiv für Entwicklungsmechanik der Organismen*, **168**, 205–25.

BRUGAL, G. (1971b). Relations entre la prolifération et la différenciation cellulaires: étude autoradiographique chez les embryons et les jeunes larves de *Pleurodeles waltlii* Michah. (Amphibien Urodèle). *Developmental Biology*, **24**, 301–21.

BRUGAL, G. & BERTRANDIAS, J. P. (1970). Méthode mathématique d'évaluation du coefficient de prolifération dans les populations cellulaires embryonnaires en croissance exponentielle. *Comptes-rendus de l'Académie des Sciences, Paris*, série D, **270**, 1603–6.

BRUGAL, G. & CHIBON, P. (1970). Signification des variations périodiques de l'indice de marquage en fonction du temps dans les tissus embryonnaires. *Comptes-rendus de l'Académie des Sciences, Paris*, série D, **270**, 998–1001.

BURKI, H. J. & TOBIAS, C. A. (1970). Effect of thyroxine on the cell generation cycle parameters of cultured human cells. *Experimental Cell Research*, **60**, 445–8.

CHAMPY, C. (1922). L'action de l'extrait thyroïdien sur la multiplication cellulaire. *Archives de Morphogenèse générale et expérimentale*, **4**, 1–58.

CHIBON, P. (1968). Etude au moyen de la thymidine tritiée de la durée des cycles mitotiques dans la jeune larve de *Pleurodeles waltlii* Michah. (Amphibien Urodèle). *Comptes-rendus de l'Académie des Sciences, Paris*, série D, **267**, 203–5.

CHIBON, P. & BRUGAL, G. (1969). Etude autoradiographique de l'action de la température et de la thyroxine sur la durée des cycles mitotiques dans l'embryon âgé et la jeune larve de *Pleurodeles waltlii* Michah. (Amphibien Urodèle). *Comptes-rendus de l'Académie des Sciences, Paris*, série D, **269**, 70–3.

CHIBON, P. & BRUGAL, G. (1973). Durée de disponibilité de la thymidine exogène chez la larve et le jeune du triton, *Pleurodeles waltlii* Michah. *Annals d'Embryologie et de Morphogénèse*, in press.

CHIBON, P. & DOURNON, C. (1972). Etude autoradiographique de quelques paramètres de la prolifération cellulaire pendant la métamorphose naturelle du crapaud *Bufo bufo* L. *Comptes-rendus de l'Académie des Sciences, Paris*, série D, **275**, 255–8.

CHULITSKAYA, E. V. (1967). Onset of desynchronization and change in the rhythm of nuclear division in the cleavage period. *Doklady Akademii Nauk S.S.S.R.* **173**, 163–6.

CLOWES, F. A. (1965). The duration of the G1 phase of the mitotic cycle and its relation to radiosensitivity. *New Pathology*, **64**, 355–9.

DECKER, R. S. & KOLLROS, J. J. (1969). The effects of cold on hind-limb growth and lateral motor column development in *Rana pipiens*. *Journal of Embryology and Experimental Morphology*, **21**, 219–33.

DEFENDI, V. & MANSON, L. A. (1961). Studies on the relationships of DNA synthesis time to proliferative time in cultured mammalian cells. *La semaine des Hôpitaux : Pathologie et Biologie*, **9**, 525–8.

DETLAFF, T. A. (1964). Cell division, duration of interkinetic states and differentiation in early stages of embryonic development. In *Advances in Morphogenesis*, vol. 3, ed. M. Abercrombie & J. Brachet, pp. 323–62. New York & London: Academic Press.

EISENBERG, S. & YAMADA, T. (1966). A study of DNA synthesis during the transformation of the iris into lens in the lentectomized newt. *Journal of Experimental Zoology*, **162**, 353–68.

EVANS, H. J. & SAVAGE, J. R. K. (1959). The effect of temperature on mitosis and on the action of colchicine on root meristem cells of *Vicia faba*. *Experimental Cell Research*, **18**, 51–61.

FÄHRMANN, W. (1971). Die Morphodynamik der Epidermis des Axolotls (*Siredon mexicanum* Shaw) unter dem Einfluss von exogen applizierten Thyroxin. II. Die Epidermis während der Metamorphose. *Zeitschrift für mikroskopisch-anatomische Forschung*, **83**, 535–68.

FERGUSON, T. (1966). Thyroxine effects upon the mitotic activity of the medulla oblongata after unilateral excision in embryos of the frog. *General and Comparative Endocrinology*, **7**, 74–9.

GALLIEN, L. (1952). Elevage et comportement du pleurodèle au laboratoire. *Bulletin de la Société Zoologique de France*, **77**, 456–61.

GALLIEN, L. & DUROCHER, M. (1957). Table chronologique de développement chez *Pleurodeles waltlii*. *Bulletin biologique de la France et de la Belgique*, **91**, 97–114.

GRAHAM, C. F. (1966). The regulation of DNA synthesis and mitosis in multinucleate frog eggs. *Journal of Cell Science*, **1**, 363–74.

GRAHAM, C. F. & MORGAN, R. M. (1966). Changes in the cell cycle during the early embryonic development of *Xenopus laevis*. *Developmental Biology*, **14**, 439–60.

GRILLO, R. S. & URSO, P. (1968). An autoradiographic evaluation of the cell reproduction cycle in the newt *Triturus viridescens*. *Oncology*, **22**, 208–17.

HINEGARDNER, R. T., RAO, B. & FELDMAN, D. E. (1964). The DNA synthetic period during the early development of the sea urchin egg. *Experimental Cell Research*, **36**, 53–61.

HOWARD, A. & PELC, S. R. (1953). Synthesis of desoxyribonucleic acid in normal and irradiated cells and its relation to chromosome breakage. *Heredity*, supplement, 6, 261–73.

IGNATIEVA, G. M. & KOSTOMAROVA, A. A. (1966). Duration of the mitotic cycle in the period of synchronous cleavage divisions (t_0) and its relationship to temperature in the loach embryo. *Doklady Akademii Nauk S.S.S.R.* 168, 330–3.

JANNERS, M. N. Y. (1968). An autoradiographic study of the growth rate of the embryonic chick wing bud. Ph.D. Thesis, University of Virginia.

KAUFMANN, S. L. (1968). Lengthening of the generation cycle during embryonic differentiation of the mouse neural tube. *Experimental Cell Research*, 49, 420–4.

KISSEL, P., DUPREZ, A., BESSOT, M., SCHMITT, J. & DOLLANDER, A. (1966). Autoradiography *in vivo* of human cancers. *Nature, London*, 210, 274–6.

LAGUCHEV, S. S. (1971). The influence of L-thyroxine on the duration of periods of the mitotic cycle. *Biulleten eksperimentalnoi biologii meditsiny*, 71, 93–4.

LOVTRUP, S. (1959). Utilization of energy sources during amphibian embryogenesis at low temperatures. *Journal of Experimental Zoology*, 140, 383–94.

MALAMUD, D. (1967). DNA Synthesis and the mitotic cycle in frog kidney cells cultivated *in vitro*. *Experimental Cell Research*, 45, 277–80.

MARCHOK, A. C. & HERRMANN, H. (1967). Studies on muscle development. I. Changes in cell proliferation. *Developmental Biology*, 15, 129–55.

MAY, R. M. & MUGARD, H. (1955). Action de l'ingestion de poudre de thyroïde sur la multiplication cellulaire dans l'encéphale des têtards de *Rana temporaria*. *Annales d'Endocrinologie*, 16, 46–66.

MITASHOV, V. I. (1969). Dynamics of the DNA synthesis in pigment epithelium cells throughout the eye restitution after a surgical removal of the retina in adult newt, *Triturus cristatus*. *Tsitologyia*, 11, 434–46.

PESETSKY, I. (1969). Autoradiographic analysis of thyroxin-stimulated ependymal cell proliferation and migration in *Rana pipiens*. *American Zoologist*, 9, 1123.

QUASTLER, H. & SHERMAN, F. G. (1959). Cell population kinetics in the intestinal epithelium of the mouse. *Experimental Cell Research*, 17, 420–38.

REDDAN, J. R. & ROTHSTEIN, H. (1966). Growth dynamics of an amphibian tissue. *Journal of Cell Physiology*, 67, 307–18.

REYNOLDS, W. A. (1966). Mitotic activity in the lumbosacral spinal cord of *Rana pipiens* larvae after thyroxine or thiourea treatment. *General and Comparative Endocrinology*, 6, 453–65.

SIMNETT, J. D. & BALLS, M. (1969). Cell proliferation in *Xenopus* tissues: a comparison of mitotic incidence *in vivo* and in organ culture. *Journal of Morphology*, 127, 363–72.

WIMBER, D. E. (1966). Duration of the nuclear cycle in *Tradescantia* root tips at three temperatures measured with [H³]thymidine. *Americal Journal of Botany*, 53, 21–4.

ZALIK, S. E. & YAMADA, T. (1967). The cell cycle during lens regeneration. *Journal of Experimental Zoology*, 165, 385–94.

CYCLIC SYNTHESIS OF DNA IN POLYTENE CHROMOSOMES OF DIPTERA

By G. T. RUDKIN*

University of Nijmegen, Nijmegen, The Netherlands

The periodic nature of replication is obvious in chromosomes of mitotic cells where morphological features alternate between the doubled, compact state of metaphase and the disperse state of interphase – whether the nuclei divide (mitosis, meiosis, syncitia) or not (endopolyploidy) the chromosomes reappear at intervals, doubled in number, and the daughter chromosomes undergo a process of separation from one another. Polytene chromosomes, on the other hand, usually do not express recognizable, repeated changes in state during their development and in most studied forms the daughter elements remain closely paired with one another. It is, therefore, necessary to prove that synthesis is 'cyclic' and that replication is not in progress at all times somewhere in every growing polytene nucleus. This discussion is principally about the replication of DNA in polytene nuclei along with suggestions of areas in which further study would advance our understanding of chromosome replication in general. Much of the material has been recently reviewed (Rudkin, 1972; Berendes, 1973).

DEVELOPMENT OF POLYTENE NUCLEI

Polytene chromosomes have been reported in forms as diverse as plants (Nagl, 1969; Avanzi, Cionini & d'Amato, 1970), ciliates (Ammerman, 1971) and a wide variety of dipteran insects (see Beermann, 1962). The majority of studies have been carried out on a few species of the last named group in which polytene nuclei are characteristic of many larval and a few adult tissues. The largest chromosomes are usually found in salivary gland nuclei which have, therefore, received the most attention. The glands are derived from ectodermal invaginations in early embryogeny, their cell number is fixed shortly thereafter, and, at least in *Drosophila melanogaster*, replications leading to the polytene state have already begun before the embryo hatches (Rudkin, 1964).

* Permanent address: The Institute for Cancer Research, Philadelphia, Pennsylvania, 19111, U.S.A.

REPLICATION CYCLES

Nuclear cycles

The existence of a period of replication marked by the synthesis of DNA (S period) alternating with a non-replication period (called here G period) is clear from two kinds of evidence. The most convincing is that a substantial proportion of salivary gland nuclei are not labelled by a pulse of tritium-labelled thymidine ([³H]TdR) when scored by autoradiography at times of development when all cells are in a growth phase. A case in point is *D. hydei* larvae in which replication virtually ceases at the time of the last larval moult, then resumes after the moult to effect a four- to eight-fold increase in DNA content during the last instar. During the whole third instar, no more than half of the salivary gland nuclei are in an S period as judged by the autoradiographic test (Danieli & Rodino, 1967). Similar findings have been reported in most other studied forms (for example, Rodman, 1967a; Darrow & Clever, 1970). The fact that the DNA contents of individual nuclei are usually found to be distributed in discrete classes with mean DNA contents related by a factor of two further encourages the view that synthesis is carried out in complete rounds leading to successive doublings of DNA contents (Swift, 1962; Rudkin, 1969; Rodman, 1967a; Rasch, 1970).

The duration of the S period has not been accurately measured in all dipteran tissues. Neither close synchrony nor demonstrable random asynchrony are easily obtained. Maximum values for the doubling times have been estimated during development for *D. melanogaster* from nuclear DNA contents measured at intervals during the larval stage. It turns out that the number of replications achieved is proportional to the logarithm of developmental age, suggesting that the average duration of the total cell cycle $(S + G)$ is longer in the larger, older cells. The range is from approximately 6 h for the first replications in late embryonic stages to 20 h or more for the eighth replication at the end of the last larval instar (Rudkin, 1972). It is, thus, not surprising that Keyl & Pelling (1963) should have estimated that the S period alone in the fourth instar of *Chironomus thummi* may be 20 h long at approximately the 12th to 14th replication cycle. The reason for the extended S period in larger cells has not been investigated. Purely geometrical considerations show that the surface/volume ratio decreases as the spherical nuclei grow, and that the distances over which material would have to be transported from haemolymph to nucleus or from nuclear membrane to chromosome increase by a factor of ten in the course of polytenization from 2 C to 2048 C. On the other hand, the DNA content of the chromosomes themselves may prove a limitation. The

X-chromosome in male larvae begins in a haplo-state (the Y-chromosome is not replicated in polytene nuclei) and continues through polytene development with half the relative DNA content that is present in diplo-X female larvae. It appears to reach a discontinuous label pattern (see below) somewhat earlier than the (diplo-)autosomes in male, not in female, nuclei (Berendes, 1966; Lakhotia & Mukherjee, 1970). The issue is clouded by the facts that the structure of the X-chromosome is more compact in female larvae (Rudkin, 1964), that the question of the relative timing is not completely resolved (Rudkin, 1972), and that the assumption, implicit in the above discussion, that all chromatids are replicated in synchrony, has not been tested.

The relative length of the S and G phases appears to vary in step with metamorphic events. The gradual decline in frequency of labelled nuclei near the end of an instar implies that by the time the moult is accomplished, some nuclei will not have synthesized DNA for a much longer period than others. Such 'rest' periods vitiate calculated average cycle times and indicate the need for direct determination of the necessary parameters with respect to metamorphic events. A study in which microspectrophotometric measurements of DNA contents are being made on nuclei which are also scored with respect to S and G phases is under way in *D. hydei* larvae.

Control of replication cycles

The regular timing with respect to moulting cycles implies external control of the processes of initiation or maintenance of DNA synthesis in the polytene tissues so far studied. A prime candidate for a controlling molecular signal was the moulting hormone, ecdysone, which has long been known to affect chromosome activity by causing the induction of specific puffs (Clever & Karlson, 1960; see Ashburner, 1971, 1972; Berendes, 1972; Panitz, Wobus & Serfling, 1972 for reviews). Extensive studies on the correlation between DNA synthesis and the moulting process as well as on possible effects of the hormone itself have been carried out on insects which do not have polytene nuclei. The clear result, that there cannot be a one-to-one relationship between the secretion of the hormone and the suppression or initiation of DNA synthesis (Krishnakumaran, Berry, Oberlander & Schneiderman, 1967), appears to hold for polytene cells as well. Thus, Darrow & Clever (1970) observed no decrease in the frequency of nuclei in S following an injection of ecdysterone that induced premature pupation, although the normal pupation process is accompanied by a drop from 12.5 to 5.6% of nuclei labelled in the salivary gland nuclei of *Chironomus*. In a similar experiment on Sciarid larvae, Crouse (1968) reported an increase in the frequency of labelled nuclei during the preco-ciously induced metamorphic events, but the injections were made shortly before a similar increase occurred in the controls. An attempt to observe a direct affect of the hormone on nuclei of salivary glands of *D. hydei* in-vitro proved negative: the frequency of labelling neither increased nor decreased within 18 h of exposure to the hormone (Berendes & de Boer, personal communication). Thus, direct intervention of ecdysone in the biosynthetic pathways concerned with chromosome replication appears to be excluded and regulation co-ordinated with moulting must be mediated through other physiological factors.

Several lines of evidence point to the existence of extracellular control-ling factors implied in the above discussion. Removal of salivary glands from their normal *milieu* by transplantation into the haemocoel of adult hosts permits some cells to undergo more replications than are observed in-situ. 'Super-giant' chromosomes were reported by Staub (1969), while Berendes & Holt (1965) observe that proximal cells, which never achieve as high a grade of polyteny as their more distal neighbours, become fully as large in the transplants. The results could be formally explained in part by postulating the inhibition of DNA synthesis in late larval glands, perhaps by an inhibitor not found in adult flies, but other mechanisms must be adduced to explain why removal of the inhibition does not permit un-checked endoreduplication and to account for the absence of replication in polytene nuclei of the Malpighian tubules which persist, unchanged in size, from late larvae through pupal and adult life.

The analysis of the control of polytene replication could be assisted by studies of mutants in which it is affected. Two kinds of such mutants are known in *D. melanogaster* in one of which replication stops before the glands have reached normal size (lethals) and in the other of which the salivary gland nuclei reach unusually large dimensions (non-lethals). The early cessation in *lethal giant larva* and *lethal translucida* (Welch & Resch, 1968; Welch & Debault, 1968) are probably consequences of defects in the hormonal system controlling other metamorphic events which become critical just before the last few replications take place in salivary gland nuclei. The proximate cause of the effect is not known. The mutants *giant* and *tumorous-head* cause an extended larval period in some (but not all) homozygous larvae. Their 'super-giant' salivary gland nuclei are known to undergo at least one replication in the extension period of tumorous-head larvae, an observation that is compatible with the maintenance of a late-larval hormonal *milieu*. However, the *tu-h* mutation appears to affect the replication of chromosomes in dividing cells (imaginal anlage) which were found to contain more than 2 C amounts of DNA (Rodman, 1967*b*). The problem of finding the primary defect caused by a mutation detected by effects so generally distributed over a complex organism is a difficult one (see Weideli, 1971) but it must be solved before one can distinguish intrinsic from extrinsic effects on particular cell types.

Polytene replication cycles are thus seen to be subject to the co-ordinated pattern of growth of an organism in much the same manner as are their counterpart mitotic cycles. The analogy is further strengthened by observations that certain parasites have the property of releasing individual infected cells from the replication block that prevents growth beyond normal limits (see Pavan & da Cunha, 1969, for review). A gregarine parasite and a viral infection have the property of inducing polytene nuclei in intestinal caecum cells of *Rhynchosciara* species to undergo many more replication cycles than have been encountered in tissues of uninfected gnats, up to a million times the haploid genome DNA. The effect is restricted to some cells in an infected larva, implying that only those cells actually invaded by the virus are affected. The system appears to be a useful one for a study of the mechanisms by which parasites influence DNA synthesis without the added complication of the mitotic process.

PATTERNS OF CHROMOSOME SYNTHESIS

Intranuclear patterns of synthesis

A special feature of polytene chromosomes is the ease with which the existence of separate replication units can be detected and the possibility for study of their control within the chromosomes (Keyl & Pelling, 1963; Plaut, 1963). Autoradiographically, the nuclei labelled by pulses of [^{3}H]TdR fall into two general classes; those nuclei with chromosomes evenly covered with grains (so-called 'continuous' label) and other nuclei in which the label is unevenly distributed, some chromosome segments are unlabelled and other segments are marked by many grains (so-called 'discontinuous' label). Studies of the distribution of the locations of labelled spots have been carried out in most detail for *Drosophila melanogaster* (Howard & Plaut, 1968; Nash & Bell, 1968; Lakhotia & Mukherjee, 1970) and to a lesser degree for other forms (Mulder, van Duijn & Gloor, 1968; Hägele, 1970; Gabrusewycz-Garcia, 1964; Berendes, 1966; Swift, 1965). It turns out that for any two chromosome segments which are not always found labelled or unlabelled simultaneously, only one of the segments is labelled when the other is not; the reverse pattern is seldom found. The implication is that replication takes place at one of the spots for a longer period of time than at the other. The analysis has been extended to as many as 65 segments, 45 on the distal X-chromosome, 20 on the distal right arm of chromosome 2, each segment composed of 6 to 30 bands (out of a total of 5000 bands in the chromosone complement of *D. melanogaster*). When more than two segments are considered simultaneously, the analysis becomes more complicated but it turns out that most labelling patterns can be arranged in a scheme consistent with the view that replication proceeds uninterruptedly in any one segment but that some segments require a longer time to complete it than do others. Most data are consistent with the model, first proposed by Keyl & Pelling (1963) from doubly-labelled ([³H]TdR and [¹⁴C]TdR) chromosomes of *Chironomus*, that all segments begin replication essentially simultaneously and finish in a definite sequence. There are, however, in every data set 'exceptional patterns' which can be accommodated either by viewing them as trivial exceptions to Keyl & Pellings' simultaneous initiation model or by adopting another model which would propose that the S period begins with synthesis in a subset of one or a few spots, then engages all replicating units, and finally terminates in a different subset of spots (cascade initiation model, Howard & Plaut, 1968). Tiepolo & Laudani (1972) find that the salivary gland nuclei labelled immediately after the last moult in mosquito larvae display discontinuous patterns and that the frequency of continuously labelled nuclei then increases rapidly. The argument that their observation supports the cascade model includes the assumption that the earliest labelled nuclei were blocked in a G period, just before the initiation of a new S period. While that assumption is consistent with the frequent conclusion that an S period, once initiated, is likely to be completed (see, e.g. Rodman, 1968), the fact remains that continuity between individual nuclei observed at the end of the third instar and at the beginning of the fourth has not been established. A detailed analysis of the labelling patterns, difficult in such small chromosomes, should settle the question. Alternatively, determination of the DNA contents of labelled and unlabelled nuclei in the period preceding and following the moult (in progress for *Drosophila hydei*) would define the stage of the cell cycle at which arrest occurs.

While the patterns of synthesis at the beginning of the S period remain obscure, it seems well established that the number of active replicating units gradually decreases towards the end of S. Aside from the double labelling experiments of Keyl & Pelling (1963) (which are open to other interpretation) the inference is drawn from the fact that as the end of a metamorphic stage approaches (last larval instar for salivary glands or a specific time in the pupal stage for foot-pad cells), the frequency of nuclei in S decreases and the fraction of labelled nuclei which shows continuous label also falls, leaving only discontinuous patterns (Nash & Bell, 1968; Rodman, 1967*b*; Bultmann & Clever, 1969; Gabrusewycz-Garcia, 1964;

Hägele, 1970; Arcos-Terán, 1972; and others). Here, again, it is assumed that the termination of the developmental phase is marked by inhibition of initiation. The observed 'terminal' patterns are quite consistent within a species and tissue. They usually include, among the latest replicating sites, large, dense bands with larger than average DNA content, as well as centric and telomeric regions. Centric regions deserve mention because of the special place they occupy in the S period of dividing cells, entering and ceasing synthesis relatively late (see Lima-de-Faria, 1969). However, much of the most proximal chromatin appears not to be replicated in polytene chromosomes of *D. melanogaster* (Rudkin, 1969; Berendes & Keyl, 1967; Mulder, van Duijn & Gloor, 1968), so the homologies of the late replicating regions in mitotic (Barrigozzi *et al.* 1966; Berendes & Keyl, 1967) and polytene nuclei are not entirely clear. Centric and some telomeric regions are rich in repetitive sequences in polytene nuclei (Gall, Cohen & Polan, 1971; Jones & Robertson, 1970; Hennig, Hennig & Stein, 1970; Rudkin & Tartof, unpublished observations) but other late replicating segments appear not to be, for example, section 3 C in *D. melanogaster* (Arcos-Terán, 1972; Rudkin & Tartof, unpublished observations). Laird's (1971) observation that polytene nuclei are deficient with respect to repetitive sequences found in mitotic tissues supports the view that the repressed material has one or a few particular, repeated sequences but the relation, if any, of those sequences to the mechanism for inhibiting their replication remains obscure (see below).

Size of replication units

A convenient model of a polytene chromosome would propose that its structural organization is a reflection of its functional organization, that is, that the bands are units of function and activity as well as structure (Bridges, 1937; Beermann, 1962, 1972; Pelling, 1966), the so-called cytogenetic units of Rudkin (1965). The bands would then correspond to replicating units, a proposition to which a partial test has been applied by comparing the DNA contents of bands in *D. melanogaster* salivary gland chromosomes with the sizes of replicating units estimated by other methods for other eukaryotes (Huberman & Riggs, 1968; Taylor, 1968; Cairns, 1966), and with the time available in an S period (Mulder, van Duijn & Gloor, 1968; Rudkin, 1972). It turns out that the DNA contained in a single chromatid within the largest measured band is within the published range for 'replicons' and could be replicated within 8 h at a rate which is within those reported for eukaryotic chromosomal DNA. It could, then, be proposed that the relative labelling frequency of a band might be correlated with its DNA content, as implied above. That appears not to be true on the basis of visual judgements of band size (Howard & Plaut, 1968; Hägele, 1970). The resolving power of the autoradiographic method is not fine enough to resolve any but the largest bands, but it can be shown that the labelling frequencies of the segments actually scored are not significantly correlated with their DNA contents (see Table 1). A more critical test would be achieved by considering only the DNA contents of the largest band in each scored segment, but the data are not yet available.

Table 1. *Non-correlation between relative DNA content and frequency of pulse-labelling by* [³H]*TdR for segments of the X-chromosome of female* Drosophila melanogaster. *Spearman's rank correlation coefficient* (0.45) *is not significantly different from* 0.0 ($P = 0.1$)

DNA*			Label frequency†	
Segment	A_1	Rank	Rank	Segments scored
7 E	1.71	1	9	7 E
7 F	2.47	2	4.5	7 F
10 C	2.53	3	2	10 C
2 C–F	3.62	4	3	2 CD, 2 EF
7 D	3.88	5	4.5	7 D
10 D–F	4.40	6	1	10 D–F
1 C–F	4.52	7	8	1 C, 10–F
11 A	5.65	8	12	11 A
1 A	5.78	9	10.5	1 A
1 B	6.20	10	7	1 B
2 AB	7.86	11	6	2 AB
7 A–C	11.49	12	10.5	7 A–C

* Relative DNA contents are given as integrated absorbance (A_1) measured at 257 nm by microdensitometric scanning of calibrated photomicrographs (see Rudkin, Aronson, Hungerford & Schultz, 1955; Rudkin, 1961).

† Relative labelling frequencies are taken from the data of Lakhotia & Mukherjee (1970). Where two regions were scored for label within a single segment measured for DNA content, the region with the higher relative labelling frequency was used to determine rank. Absolute frequencies were not determined by Lakhotia & Mukherjee.

Independent control of replicating units

The nuclear cycles in growing polytene nuclei appear to be essentially the same as the cycles in dividing cells except for the omission of all processes connected with the act of division. Replication is under regulative controls as strict as those which govern mitotic activity during development, and equally obscure. Polytene chromosomes take on a special interest in cases where their special properties reveal phenomena that would be difficult to detect in the usual interphase nucleus. Thus, under-replication of the centric regions in *Drosophila* salivary glands may be directly connected with one of the essential properties of polytene chromosomes: the failure of kinetochore separation and, possibly, division. The amount of DNA in the repressed segments is enough to include tens of replicating units (Rudkin, 1972). On the other hand, the phenomenon of gene amplification, established for nucleolar DNA in nuclei of other types (Brown & Dawid, 1968; Gall, 1969), finds an analogy in the DNA puffs of the Sciaridae (Breuer & Pavan, 1955; Rudkin & Corlette, 1957; Ficq, Pavan & Brachet, 1958; Pavan & da Cunha, 1969, and others). The DNA content of a very short region of chromosome is replicated several-fold while the DNA of the rest of the chromosome is replicated only once (Rudkin & Corlette, 1957; Crouse & Keyl, 1968; Rasch, 1970). Unlike amplified nucleolar DNA, the puff DNA remains in-situ on the chromosome and, unlike the oocytes in which nucleolar DNA is amplified, the salivary gland cells with DNA puffs are in the terminal stage of their development: they will be broken down and resorbed in the pupa and never undergo another replication. It has recently been suggested that RNA-dependent DNA polymerase may be involved in the replication of the extra-chromosomal nucleolar DNA of oocytes (Crippa & Tocchini-Valentini, 1971; Ficq & Brachet, 1971), encouraging the suggestion that the relatively small amount of RNA in DNA puffs (as opposed to RNA puffs) may play a similar role (see Rudkin, 1972).

The fragmentary data relating to substances other than DNA in the replication cycles of polytene chromosomes have been omitted only because they do not yet contribute significantly to our understanding of the process. Techniques now available for obtaining insect organs in large quantity and for the analysis of proteins in small quantity promise to enable significant progress to be made in the near future.

CONCLUDING REMARKS

Polytene cells in dipteran insects undergo endoreduplication cycles which may be thought of as curtailed cell cycles in which the mechanics of chromosomal, nuclear and cell division are bypassed (see Fig. 1). The initiation of the cycles in any particular tissue is under the co-ordinated control of development which, in insects, is characterized by growth stages separated by ecdysis (moulting in the larval period). Replication is controlled by extracellular factors of unknown nature and origin, as well as by the competence of a given cell type to respond to those factors. The maximum number of replication cycles accomplished by a given cell type can be decreased by genetic factors and, on the other hand, can be increased not only by genetic factors but also by experimental alteration of the

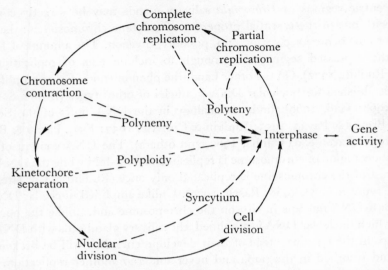

Fig. 1. Diagrammatic representation of some events in the mitotic cell cycle to illustrate the relationship between polytene nuclei and other deviants from the complete cycle. Polynemic chromosomes differ from polytenic chromosomes in that they retain the capability for undergoing a complete cycle after having completed at least one endoreduplication cycle. (From Rudkin, 1972.)

cellular *milieu* or by infection with intracellular parasites, including some viruses.

Polytene chromosomes are organized into replicating units, at least hundreds and probably thousands of which are in a genome size of the order of 10^{11} daltons DNA. It is likely that a chromomere (crossband) with an adjacent interband region contains the DNA of a single replicating unit. In most instances it appears that each active replicating unit undergoes a single round of synthesis in a nuclear period and that the order of initiation (and of completion) of synthesis is fixed for the different units within a species. However, replication controls acting at specific replicating units are demonstrably characteristic of certain cell types or as a response to genetic manipulation. In particular, repression of replication of centric DNA, demonstrated to date only in *Drosophila*, could be an essential feature of polytene nuclei whose kinetochores do not separate and may not be replicated.

Some of the work supported by National Institutes of Health (U.S.A.) grants CA-01613, CA-06927 and RR-05539; National Science Foundation (U.S.A.) grant GB-3525; and by an appropriation from the Commonwealth of Pennsylvania (U.S.A.).

The author is grateful to Prof. H. D. Berendes for permission to quote unpublished data and for critical reading of the manuscript.

REFERENCES

AMMERMAN, D. (1971). Morphology and development of the macronuclei of the ciliates *Stylonychia mytilus* and *Euplotes aediculatus*. *Chromosoma, Berlin*, **33**, 209–38.

ARCOS-TERÁN, L. (1972). DNS-Replikation und die Natur der spät replizierenden Orte im X-Chromosom von *Drosophila melanogaster*. *Chromosoma, Berlin*, **37**, 233–96.

ASHBURNER, M. (1971). Induction of puffs in polytene chromosomes of in vitro cultured salivary glands of *Drosophila melanogaster* by ecdysone and ecdysone analogues. *Nature, New Biology, London*, **230**, 222.

ASHBURNER, M. (1972). Puffing patterns in *Drosophila melanogaster*. In *Results and Problems in Cell Differentiation. 4. Developmental Studies with Giant Chromosomes*, ed. W. Beermann, H. Ursprung & J. Reinert, pp. 101–51. Berlin, Heidelberg & New York: Springer-Verlag.

AVANZI, S., CIONINI, P. G. & D'AMATO, F. (1970). Cytochemical and autoradiographic analysis on the embryo suspensor cells of *Phaseolus coccineus*. *Caryologia*, **23**, 605–36.

BARRIGOZZI, C., DOLFINI, S., FRACCARO, M., RAIMONDI, G. R. & TIEPOLO, L. (1966). *In vitro* study of the DNA replication patterns of somatic chromosomes of *Drosophila melanogaster*. *Experimental Cell Research*, **43**, 231–4.

BEERMANN, W. (1962). Riesenchromosomen. In *Protoplasmatologia, Handbuch der Protoplasmaforschung*, vol. 6D, ed. L. V. Heilbrun & F. Weber. Wien: Springer-Verlag.

BEERMANN, W. (1972). Chromosomes and genes. In *Results and Problems in Differentiation. 4. Developmental Studies with Giant Chromosomes*, ed. W. Beermann, H. Ursprung & J. Reinert, pp. 1–33. Berlin, Heidelberg & New York: Springer-Verlag.

BERENDES, H. D. (1966). Differential replication of male and female X-chromosomes in *Drosophila*. *Chromosoma, Berlin*, **20**, 32–43.

BERENDES, H. D. (1972). The control of puffing in *Drosophila hydei*. In *Results and Problems in Cell Differentiation. 4. Developmental Studies with Giant Chromosomes*, ed. W. Beermann, H. Ursprung & J. Reinert, pp. 181–207. Berlin, Heidelberg & New York: Springer-Verlag.

BERENDES, H. D. (1973). Synthetic activity of polytene chromosomes. *International Review of Cytology*, ed. G. H. Bourne & J. F. Danielli. New York: Academic Press. (In press.)

BERENDES, H. D. & HOLT, T. K. H. (1965). Differentiation of transplanted larval salivary glands of *Drosophila hydei* in adults of the same species. *Journal o Experimental Zoology*, **160**, 299–318.

BERENDES, H. D. & KEYL, H. G. (1967). Distribution of DNA in heterochromatin and euchromatin of polytene nuclei of *Drosophila hydei*. *Genetics*, **57**, 1–13.

BROWN, D. & DAWID, I. B. (1968). Specific gene amplification in oocytes. *Science, New York*, **160**, 272–80.

BREUER, M. & PAVAN, C. (1955). Behaviour of polytene chromosomes of *Rhynchosciara angelae* at different stages of larval development. *Chromosoma, Berlin*, **7**, 311–86.

BRIDGES, C. B. (1937). Correspondence between linkage maps and salivary chromosome structure as illustrated in the tip of chromosome 2R of *Drosophila melanogaster*. *Cytologia, Fujii Jubilee volume*, 745–55.

BULTMANN, H. & CLEVER, U. (1969). Chromosomal control of foot pad development in *Sarcophaga bullata*. I. The puffing pattern. *Chromosoma, Berlin*, 28, 120–35.

CAIRNS, J. (1966). Autoradiography of HeLa cell DNA. *Journal of Molecular Biology*, 15, 372–3.

CLEVER, U. & KARLSON, P. (1960). Induktion von Puff-Veränderungen in den Speicheldrüsenchromosomen von *Chironomus tentans* durch Ecdyson. *Experimental Cell Research*, 20, 623–6.

CRIPPA, M. & TOCCHINI-VALENTINI, G. P. (1971). Synthesis of amplified DNA codes for ribosomal RNA. *Proceedings of the National Academy of Sciences, U.S.A.* 68, 2769–73.

CROUSE, H. V. & KEYL, G. (1968). Extra replications in the 'DNA-Puffs' of *Sciara coprophila*. *Chromosoma, Berlin*, 25, 357–64.

CROUSE, H. V. (1968). The role of ecdysone in DNA-puff formation and DNA synthesis in the polytene chromosomes of *Sciara coprophila*. *Proceedings of the National Academy of Sciences, U.S.A.* 61, 971–8.

DANIELI, G. A. & RODINO, E. (1967). Larval moulting cycle and DNA synthesis in *Drosophila hydei* salivary glands. *Nature, London*, 213, 424–5.

DARROW, J. M. & CLEVER, U. (1970). Chromosome activity and cell function in polytene cells. III. Growth and replication. *Developmental Biology*, 21, 331–48.

FICQ, A. & BRACHET, J. (1971). RNA-dependent DNA polymerase: possible role in the amplification of ribosomal DNA in *Xenopus* oocytes. *Proceedings of the National Academy of Sciences, U.S.A.* 68, 2774–6.

FICQ, A., PAVAN, G. & BRACHET, J. (1958). Metabolic processes in chromosomes. *Experimental Cell Research*, 6, 105–14.

GABRUSEWYCZ-GARCIA, N. (1964). Cytological and autoradiographic studies in *Sciara coprophila* salivary gland chromosomes. *Chromosoma, Berlin*, 15, 312–44.

GALL, J. (1969). The genes for ribosomal RNA during oogeneses. *Genetics*, supplement, 61, 121–32.

GALL, J. G., COHEN, E. H. & POLAN, M. L. (1971). Repetitive sequences in *Drosophila* chromosomes. *Chromosoma, Berlin*, 33, 319–44.

HÄGELE, K. (1970). DNS-Replikationsmuster der Speicheldrüsenchromosomen von Chironomiden. *Chromosoma, Berlin*, 31, 91–138.

HENNIG, W., HENNIG, I. & STEIN, H. (1970). Repeated sequences in the DNA of *Drosophila* and their localization in giant chromosomes. *Chromosoma, Berlin*, 32, 31–63.

HOWARD, E. F. & PLAUT, W. (1968). Chromosomal DNA synthesis in *Drosophila melanogaster*. *Journal of Cell Biology*, 39, 415–29.

HUBERMAN, J. A. & RIGGS, A. D. (1968). On the mechanism of DNA replication in mammalian chromosomes. *Journal of Molecular Biology*, 32, 327–41.

JONES, K. W. & ROBERTSON, F. (1970). Localization of reiterated nucleotide sequences in *Drosophila* and mouse by *in situ* hybridization of complementary RNA. *Chromosoma, Berlin*, 31, 331–45.

KEYL, H. G. & PELLING, C. (1963). Differentielle DNS-Replikation in den Speicheldrüsenchromosomen von *Chironomus thummi*. *Chromosoma, Berlin*, 14, 347–59.

KRISHNAKUMARAN, A., BERRY, S. J., OBERLANDER, H. & SCHNEIDERMAN, H. A. (1967). Nucleic acid synthesis during insect development. II. Control of DNA synthesis in the Cecropia silkworm and other saturniid moths. *Journal of Insect Physiology*, 13, 1–57.

LAIRD, C. D. (1971). Chromatid structure: Relationship between DNA content and nucleotide sequence diversity. *Chromosoma, Berlin*, **32**, 378–406.

LAKHOTIA, S. C. & MUKHERJEE, A. S. (1970). Chromosomal basis of dosage compensation in *Drosophila*. III. Early completion of replication by the polytene X-chromosome in male: further evidence and its implications. *Journal of Cell Biology*, **47**, 18–33.

LIMA-DE-FARIA, A. (1969). DNA replication and gene amplification in heterochromatin. In *Handbook of Molecular Cytology*, ed. A. Lima-de-Faria, pp. 317–25. Amsterdam: North-Holland Publishing Co.

MULDER, M. P., VAN DUIJN, P. & GLOOR, J. H. (1968). The replicative organization of DNA in polytene chromosomes of *Drosophila hydei*. *Genetica*, **39**, 385–428.

NAGL, W. (1969). Banded polytene chromosomes in the legume *Phaseolus vulgaris*. *Nature, London*, **221**, 70–1.

NASH, D. & BELL, J. (1968). Larval age and the pattern of DNA synthesis in polytene chromosomes. *Canadian Journal of Genetics and Cytology*, **10**, 82–90.

PANITZ, R., WOBUS, U. & SERFLING, E. (1972). Effect of ecdysone and ecdysone analogues on two Balbiani rings of *Acricotopus lucidus*. *Experimental Cell Research*, **70**, 154–60.

PAVAN, C. & DA CUNHA, A. B. (1969). Chromosomal activities in *Rhynchosciara* and other Sciaridae. *Annual Review of Genetics*, **3**, 425–50.

PELLING, C. (1966). A replicative and synthetic chromosomal unit: the modern concept of the chromomere. *Proceedings of the Royal Society*, B, **164**, 279–89.

PLAUT, W. (1963). On the replicative organisation of DNA in the polytene chromosomes of *Drosophila melanogaster*. *Journal of Molecular Biology*, **7**, 632–5.

RASCH, E. M. (1970). DNA cytophotometry of salivary gland nuclei and other tissue systems in Dipteran larvae. In *Introduction to Quantitative Cytochemistry*, vol. II, ed. G. L. Wied & G. F. Bahr, pp. 335–56.

RODMAN, T. (1967a). DNA replication in salivary gland nuclei of *Drosophila melanogaster* at successive larval and prepupal stages. *Genetics*, **55**, 376–86.

RODMAN, T. (1967b). Control of polytenic replication in dipteran larvae. I. Increased number of cycles in a mutant strain of *Drosophila melanogaster*. *Journal of Cell Physiology*, **70**, 79–86.

RODMAN, T. (1968). Relationship of developmental stage to initiation of replication in polytene nuclei. *Chromosoma, Berlin*, **23**, 271–87.

RUDKIN, G. T. (1961). Cytochemistry in the ultraviolet. *Microchemical Journal Symposium*, series 1, 261–76.

RUDKIN, G. T. (1964). The structure and function of heterochromatin. *Genetics Today, Proceedings of the XI International Congress of Genetics*, The Hague, The Netherlands, 1963, **2**, 359–74.

RUDKIN, G. T. (1965). The relative mutabilities of DNA in regions of the X-chromosome of *Drosophila melanogaster*. *Genetics*, **52**, 665–81.

RUDKIN, G. T. (1969). Non replicating DNA in *Drosophila*. *Genetics*, supplement, **61**, 227–38.

RUDKIN, G. T. (1972). Replication in polytene chromosomes. In *Results and Problems in Cell Differentiation. 4. Developmental Studies with Giant Chromosomes*, ed. W. Beermann, H. Ursprung & J. Reinert, pp. 59–85. Berlin, Heidelberg & New York: Springer-Verlag.

RUDKIN, G. T., ARONSON, J. F., HUNGERFORD, D. A. & SCHULTZ, J. (1955). A comparison of the ultraviolet absorption of haploid and diploid salivary gland chromosomes. *Experimental Cell Research*, **9**, 193–211.

RUDKIN, G. T. & CORLETTE, S. L. (1957). Disproportionate synthesis of DNA in a polytene chromosome region. *Proceedings of the National Academy of Sciences, U.S.A.* **43**, 964–8.

STAUB, M. (1969). Veränderungen im Puffmuster und das Wachstum der Riesen-chromosomen in Speicheldrüsen von *Drosophila melanogaster* aus spätlarvalen und embryonalen Spendern nach Kultur *in vivo*. *Chromosoma, Berlin*, **26**, 76–104.

SWIFT, H. (1962). Nucleic acids and cell morphology in dipteran salivary glands. In *The Molecular Control of Cellular Activity*, ed. J. M. Allen, pp. 73–125. New York, Toronto & London: McGraw-Hill.

SWIFT, H. (1965). Molecular morphology of the chromosome. *In Vitro*, **1**, 26–49.

TAYLOR, J. H. (1968). Rates of chain growth and units of replication in DNA of mammalian chromosomes. *Journal of Molecular Biology*, **31**, 579–94.

TIEPOLO, L. & LAUDANI, U. (1972). DNA synthesis in polytenic chromosomes of *Anopheles atroparvus*. *Chromosoma, Berlin*, **36**, 305–12.

WEIDELI, H. (1971). RNS-Stoffwechsel und Proteinsynthese in einem zellfreien System des Wildtypus und der Lethalmutante '*1(3)tr*' von *Drosophila melanogaster*. *Molecular and General Genetics*, **112**, 167–96.

WELCH, R. M. & DEBAULT, L. E. (1968). Quantitative microspectrophotometry and microinterferometry of nucleic acids in salivary gland, proventriculus and ring gland of the lethal mutant *translucida* of *Drosophila melanogaster*. *Journal of the Royal Microscopical Society*, **88**, 85.

WELCH, R. M. & RESCH, K. (1968). A cytochemical analysis of deoxyribonucleic acid (DNA) and protein in salivary gland and gut of the lethal mutant *lgl* of *D. melanogaster*. *University of Texas Publication number 6818, Studies in Genetics*, **4**, 49–70.

THE CELL CYCLE DURING
MAMMALIAN DEVELOPMENT

By C. F. GRAHAM

Department of Zoology, University of Oxford, South Parks Road,
Oxford OX1 3PS

INTRODUCTION

The mammalian embryo is unusual in that its nucleic acid metabolism resembles that of an adult cell very early in development (reviewed by Graham, 1972). This review will deal with the following problems:

(1) The changes in the cell cycle during early development.

(2) The mechanism by which the trophoblast nuclei accumulate massive quantities of DNA.

(3) The extent to which the cell cycle is influenced by factors outside the embryo.

CHANGES IN THE CELL CYCLE

Methods of studying the cell cycle

Historically the first embryonic cell cycle to be described in any detail was that of the mouse. The early studies were on Feulgen-stained whole mounts and DNA quantities were measured by microdensitometry at particular intervals after the cleavage divisions (mouse – Alfert, 1950; mouse & rat – Dalcq & Pasteels, 1955) (Fig. 1). These cleavage divisions could be easily timed and observed in this slow-dividing, transparent embryo. However, the diffuse Feulgen-staining of the large nuclei of the early embryo made it particularly difficult to obtain microdensitometry readings which were comparable with those from adult cells and it was therefore important to confirm these earlier studies with labelling experiments.

Recent workers have studied the incorporation of tritiated thymidine into nuclear DNA in order to define the S phase of the cell cycle (terminology of Howard & Pelc, 1953). In the mouse blastocyst, tritiated thymidine is incorporated into both polysaccharide and DNA and it is therefore necessary to show that the radioactivity detected in autoradiographs is due to DNA synthesis (Piko, 1970).

In-vivo, it has proved difficult to label either DNA, RNA or proteins in the preimplantation embryo. It is known that the embryo can be labelled in-vitro during this stage of development (Silagi, 1963; Mintz,

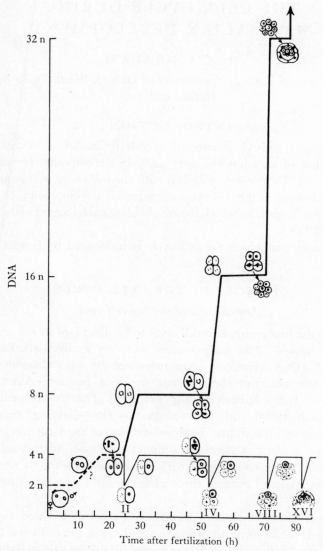

Fig. 1. DNA synthesis during cleavage in the eggs of rats. From the two cell stage on, two divergent graphs have been drawn. The thin line gives the microdensitometer readings of DNA amounts in one nucleus followed until its division, and then amounts in one of its daughter nuclei, and so on. Note that DNA is synthesized in the first few hours of each interphase. The bold line shows the total amount of DNA in all the nuclei of one embryo (assuming exactly synchronous division in the blastomeres). (From Dalcq, 1957.)

1964, 1965), and this problem is due either to the reproductive tract isolating the embryo from radioactive precursors injected into the peritoneum or to the impermeability of the embryo to these precursors when it is in the reproductive tract. The impermeability of the pre-morula is part of the problem, but it has recently been shown that it can be labelled by injecting thymidine into the oviduct. The embryo is not labelled when thymidine is injected intraperitoneally and so the oviduct must act as a barrier (Dyban, Samoshkina & Mystkowska, 1972). During and after implantation the embryo can be labelled by injecting radioactive thymidine into the peritoneum (Zavarzin, Samoshkina & Dondua, 1966), but during the pre-implantation period studies on DNA synthesis have had to be performed in culture. This introduces another difficulty because it is known that the cell cycle of the embryo is slowed by culture (Bowman & McLaren, 1970; Graham, 1971), and cell cycle times deduced from studies in-vitro should be interpreted with caution. Some workers inject thymidine directly into the uterus to avoid dilution of the label (Prasad, Dass & Mohla, 1968).

Rates of cell division

Cleavage rates in mammals are genetically determined (mouse – Whitten & Dagg, 1962), and any conclusions about the duration of particular phases of the cell cycle will only apply to the species or strain which has been studied. For instance, the embryos of a genetically heterogeneous Swiss albino stock contain, on average, twice as many cells as the embryos of the inbred C3H strain on the fourth day of pregnancy (first day of pregnancy = day on which copulation plug is found). In addition, the cell number of embryos from a particular strain or stock may also vary by a factor of four on this day of pregnancy (Bowman & McLaren, 1970; Barlow, Owen & Graham, 1972). The range of cell numbers is reduced if embryos are taken from the same mother, which suggests that part of this variation can be accounted for by differences in the time of mating and fertilization within the strain. It is usual to superovulate the female mice in order to obtain large numbers of eggs which were ovulated at a known time (Edwards & Gates, 1959; Donahue, 1972), and it is useful to record the time of the ovulating dose of human chorionic gonadotrophin (HCG) in order to standardize procedures.

The cell cycle during the first two cleavage divisions

The first two cell cycles are slow; in the mouse, the first lasts about 20 h and the second about 18 h. In all mammalian embryos which have been studied there is a post-meiotic pre-DNA synthesis phase (? G_1), a short S phase, and a long G_2 phase during the first cell cycle (mouse – Alfert, 1950; Sirlin & Edwards, 1959; Mintz, 1965; rat – Dalcq & Pasteels, 1955; hamster – Szollosi, 1964; rabbit – Oprescu & Thibault, 1965; Szollosi, 1966). The S phase may be complete before the pronuclei come together at the centre of the egg.

DNA synthesis begins again soon after the egg divides in two and a G_1 phase has not been detected in the second cell cycle (mouse and rat – Dalcq & Pasteels, 1955; rabbit – Oprescu & Thibault, 1965). Most of this cell cycle is occupied by a long G_2 phase (mouse – Gamow & Prescott, 1970).

The appearance of the G_1 phase

During division cycles 4 to 8 and 8 to 16 the G_1 phase must be short because DNA synthesis is complete in the first quarter of interphase (mouse and rat – Dalcq & Pasteels, 1955). Gamow & Prescott (1970) labelled mouse embryos in-vitro for 20 min during these early cleavage divisions and found that in division cycles 2 to 4, 4 to 8, 8 to 16, and 16 to 32, the nuclei in early interphase were always labelled with tritiated thymidine. These early interphase nuclei could be recognized because nuclear volume is approximately halved during each cleavage division (Alfert, 1950). They suggested that the G_1 phase was absent from the cell cycle up to the 32-cell stage. This suggestion contrasted with Samoshkina's deduction (1968) that the G_1 phase lasted 4–5 h during the 8- to 16-cell cycle. She arrived at this figure by studying the increase in labelled nuclei in mouse embryos cultured in continuous labelling conditions. Since this treatment might be expected to lengthen the cell cycle, it was necessary to reinvestigate the problem in conditions which reduced the slowing of the cell cycle by prolonged culture.

Barlow, Owen & Graham (1972) cultured embryos in tritiated thymidine within 5 min of the death of their mother. The embryos were fixed after 15 min in the label. Subsequently microdensitometry and autoradiography was performed on the air-dried embryos (Fig. 2). Unlabelled nuclei were found at the 5- to 16-cell cycles and these nuclei contained 2 C and 4 C DNA amounts. It was known that these nuclei were not unlabelled because the cells were dead; all the nuclei of these embryos become labelled under continuous labelling conditions (confirming the observation of Izquierdo & Roblero, 1965). It was known that these nuclei were not unlabelled because the label penetrated the embryos slowly; serial sections showed that the tritiated thymidine could significantly label the inside cells of the embryo after only 15 min in this radioactive precursor. It was therefore possible to conclude that these nuclei were unlabelled because they were not

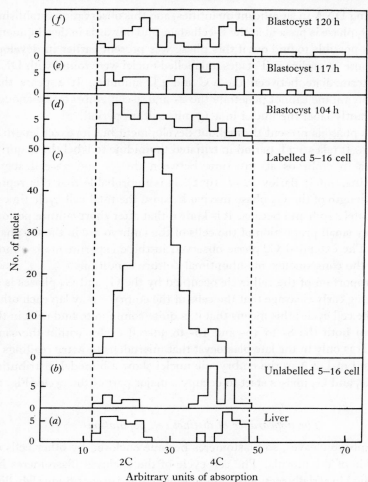

Fig. 2. The onset of polyploidy. Microdensitometry on morulae and blastocysts of the mouse.

(a) 49 liver nuclei used as absorbance standards of the 2 C and 4 C DNA amounts.

(b) Unlabelled nuclei in a sample of 324 nuclei from 5–16-cell embryos fixed at 69–72 h post-HCG after 15 min in tritiated thymidine. The distribution of these nuclei is bimodal and their DNA contents fall in the same range as the 2 C and 4 C amounts of the liver nuclei standard. This is evidence for the G_1 and the G_2 phases of the cell cycle being present at the 8-cell stage onwards.

(c) Labelled nuclei from the same sample as (b). The histogram has one mode and most of the nuclei have a DNA content intermediate between the 2 C and the 4 C amount.

(d) 81 nuclei from two blastocysts at 104 h post-HCG. Note the absence of nuclei with greater than the 4 C amount of DNA.

(e) 55 nuclei from four blastocysts at 117 h post-HCG. Note that some nuclei contain more DNA than the 4 C amount. These nuclei all incorporated tritiated thymidine in a 30 min labelling period before fixation. This shows that they are actively engaged in DNA synthesis.

(f) 89 nuclei from four blastocysts at 120 h post-HCG. Again note DNA contents in excess of the 4 C amount. (From Barlow et al. 1972.)

synthesizing DNA in significant quantities, and this observation establishes that the G_1 phase is present at the 8-cell stage and onwards in development. It was not possible to find out if this phase was present earlier in development because at earlier cell stages unlabelled nuclei were found with DNA values intermediate between the 2 C and 4 C amounts. It appears that tritiated thymidine cannot penetrate the 2- and 4-cell embryo fast enough to significantly label the nuclei in a 15 min culture period.

The G_2 phase is present throughout development but its exact length is not known. It takes a 7 h period in tritiated thymidine to label the majority of mitoses in embryos at any time between the 5- and 32-cell stages (Samoshkina, 1968; Barlow *et al.* 1972). It is improbable that this represents the length of the G_2 phase in-vivo because the total cell cycle time is approximately 10 h and because it is known that after short culture periods only a very small proportion of the cells of the embryo are in the G_2 phase (Fig. 2). The extended G_2 phase observed in these experiments is almost certainly the consequence of suboptimal culture conditions.

The proportion of the cell cycle occupied by the G_1 and G_2 phases is so short during early cleavage that the cells of the embryos may lap each other around the cell cycle; this means that it is quite common to find cells in the S phase of both the 8- to 16- and 16- to 32-cell cycles within the same embryo. It is only in the late blastocyst that microdensitometry readings of the amount of DNA in the embryonic nuclei show a bimodal distribution and the G_1 and G_2 phases start to occupy a major part of the cycle (Fig. 2).

The appearance of distinct cell populations

As the embryo cleaves, so blastomeres become enclosed by other cells on the outside of the morula. The cell cycle of the enclosed blastomeres has been studied in serially sectioned embryos maintained in continuous labelling conditions for various lengths of time (Barlow *et al.* 1972). Completely enclosed blastomeres are first found at the 8- to 16-cell stages and because they have small nuclei it is known that they are usually the daughter nuclei of the first four blastomeres to divide away from the 8-cell stage. These enclosed blastomeres have a significantly higher labelling index than the outside blastomeres at the time that they are first found and they continue to have a significantly higher labelling index when the embryo is exposed to radioactive thymidine for short periods at any stage up to the late blastocyst. The enclosed cells increase in number faster than the outside cells and in continuous labelling experiments they become 100 % labelled sooner than the outside cells. If it is assumed that there is little migration between the enclosed cells and the outside cells of the embryo then it is possible to conclude that the inside cells are derived from blastomeres with a short cell cycle at the 8-cell stage and they continue to have a short cell cycle throughout preimplantation development into a 128-cell blastocyst. The distinction between the cell cycles of the enclosed and outside cells is apparent at a stage of development two days before the formation of excess DNA amounts in the outside cells of the blastocyst.

In an earlier study, Samoshkina (1968) was unable to find a difference in the labelling index of cells with large nuclei and cells with small nuclei in the 16- to 32-cell cycle. This finding does not contradict the conclusions of Barlow and his associates who found cells with small nuclei in both the enclosed cells and outside cells of the morula; the distinction between these two cell populations is only apparent in serial sections.

However Zavarzin, Samoshkina & Dondua (1966) have reported that on the fourth and fifth days of pregnancy the outside cells of the blastocyst have a consistently higher labelling index than the inside cells. On the fourth day of pregnancy, 38% of the enclosed cell nuclei and 55% of the outside cell nuclei were labelled at short times after the injection of tritiated thymidine into peritoneum of the pregnant mother. On the fifth day, the low labelling index of the enclosed cells was still obvious; 70% of the enclosed cell nuclei and 90% of the outside cell nuclei were labelled under similar conditions. It is difficult to explain the discrepancy between these results and those of Barlow. One possibility is that in the experiments in-vivo of Zavarzin *et al.* the low grain counts and the slow penetration of the precursor from the wall of the uterus made it easier to detect labelled nuclei in the polyploid nuclei of the outside cells of the embryo.

Following implantation the cell derivatives of the enclosed blastomeres undergo organogenesis. This morphological differentiation is characterized by the development of distinct cell cycles in different regions of the embryo (e.g. mouse – Atlas & Bond, 1965; Zavarzin *et al.* 1966; Kauffmann, 1966, 1968; Solter & Skreb, 1968, 1971). The cell cycle is fastest in the ectoderm of seventh day embryos. It has been calculated that G_1 lasts 0.25 h, S lasts 4 h, G_2 lasts 0.7 h, and that mitosis lasts 1.5 h (Solter & Skreb, 1971). This is an instance of the unusual situation of cell cycles, for a brief period, becoming shorter as development proceeds. Close to parturition cell cycles begin to lengthen into those of the adult (18-day rat embryos – Wegener, Hollweg & Maurer, 1964).

The time of X-chromosome inactivation

Genetically inactive X-chromosomes usually form sex chromatin bodies in the interphase nucleus and these are characterized by the intense incorporation of tritiated thymidine late in the S phase (Lyon, 1962). The appearance of sex chromatin, of asynchronously replicating X-chromosomes, and of stably inactivated X-linked gene expression can therefore be taken as a guide to the stage of embryogenesis at which patterns of chromosome replication become fixed in development (reviewed by Lyon, 1972).

Sex chromatin appears in most species at about the time of blastocyst implantation (reviewed by Austin, 1966). It is not surprising that there is only one report of sex chromatin in earlier embryos (Syrian hamster – Lindmark & Melander, 1970); the nuclei of the cleaving embryo are large and diffuse which makes the detection of sex chromatin difficult.

However, asynchronously replicating X-chromosomes have been reported in earlier embryos. Asynchronous replication appears to be established in some cells of the 160–380-cell rabbit embryo at 96 h after mating (Issa, Blank & Atherton, 1969). In other species, such as *Misocricetus auratus* (Hill & Yunis, 1967), asynchronous replication does not appear to be established until after implantation.

Recently it has become possible to test the stability of X-chromosome inactivation during mouse development. The test consists of transferring a single cell from one embryo, which is heterozygous for a pair of X-linked coat colour markers, into another embryo which has a coat colour distinguishable from the two possible colours of the donor cell (Gardner & Lyon, 1972). If the donor cell can give rise to cells which express both types of the possible coat colour markers then stable X-chromosome inactivation has not occurred in the donor cells. Donor cells from blastocysts can usually produce cells which express both the coat colour types and these cells cannot have been stably inactivated at this stage of development.

It therefore appears that the pattern of chromosome replication is not fixed during early cleavage and the time at which it becomes stabilized varies from one mammalian species to another.

Nuclear DNA in excess of the 4 C amount

For a long time it has been known that the outside cells of the blastocyst and the trophoblast cells of later stages of development contain large nuclei (rat and mouse – Duval, 1891; Dickson, 1963). Nuclear DNA in excess of the 4 C amount is first found in the implanting blastocyst (mouse – Barlow *et al.* 1972; Barlow & Sherman, 1972; Sherman & Barlow, 1972). Subse-

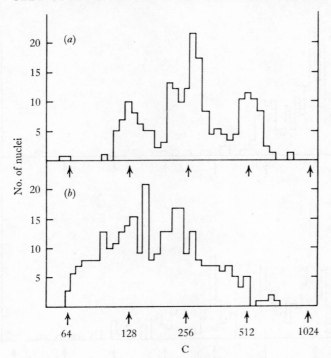

Fig. 3. Massive DNA amounts in the nuclei of mouse trophoblast on the 11th day of pregnancy. Microdensitometer readings on a selected sample of giant nuclei. (a) Nuclei released during dissection, (b) nuclei teased from the trophoblast. (From Barlow & Sherman, 1972.)

quently the amount of DNA in these giant nuclei may increase to enormous values; late in mouse development the trophoblast nuclei contain more than the 512 C amount of DNA (Fig. 3, Fig. 4; Barlow & Sherman, 1972), and in the rat placenta trophoblast nuclei may contain 4,096 C amounts of DNA (Nagl, 1972a).

The rodent embryo, like the bee embryo (Mittwoch, Kalmus & Webster, 1966), has the property of throwing off polyploid cells which rarely divide subsequently. In the case of the rodent these cells are located only in the trophoblast layer and the extra-embryonic membranes during embryogenesis. It has recently been pointed out that the suspensor tissue in plant ovaries has a similar function to that of the rodent trophoblast and it also contains highly polyploid cells (Nagl, 1972b).

302   C. F. GRAHAM

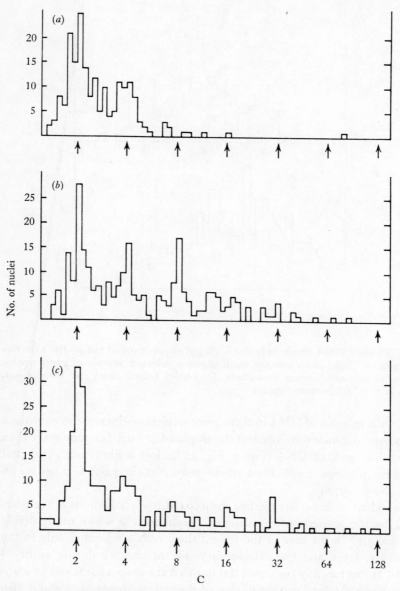

Fig. 4. The development of polyploid cells in the trophoblast. Microdensitometer
readings of the DNA contents in a random sample of trophoblast nuclei at different
stages of development. (a) 7th, (b) 8th, (c) 9th day of pregnancy. Note the continued presence of diploid cells in this tissue. (From Barlow & Sherman, 1972.)

ACCUMULATION OF DNA IN THE TROPHOBLAST NUCLEI

It has been suggested that DNA accumulation in the trophoblast nuclei might occur by three mechanisms:

Engulfment of maternal cells

Galassi (1967) found that if the DNA of the rat uterus was prelabelled with tritiated thymidine, then the trophoblast cells became labelled after implantation. A similar observation was made on mouse embryos transferred to prelabelled kidneys (Avery & Hunt, 1969). The nuclei of the trophoblast cells could either become labelled by incorporating fragments of DNA or by reutilizing tritiated thymidine from the breakdown of DNA in the uterus or kidney. It is not necessary to believe that the maternal nuclei are engulfed by the trophoblast nuclei in order to explain these observations and it is now known that if nuclear engulfment does occur, then it is either a rare event or the engulfed nucleus is not genetically active (see chimaera evidence below).

By cell fusion

It is known that many mammalian embryos have syncytial trophoblasts but it is not clear if these syncytia are formed by cell fusion or by the failure of mitosis. It has been suggested that this mechanism of DNA accumulation in one nucleus may be important in both the mouse and rat (rat – Jollie, 1964; mouse – Schlesinger & Koren, 1967; Saccoman, Morgan & Wells, 1967; Avery & Hunt, 1969).

However, there is now good evidence that the trophoblast nuclei rarely fuse either with each other or with cells of the uterus. In chimaeras between strains which are homozygous for different alleles of the same dimer enzyme locus, hybrid enzyme molecules are formed after fusion of myoblasts to form striated muscle (Mintz & Baker, 1967). However, in the trophoblast, such hybrid molecules are not formed and it is therefore likely that little cell fusion occurs (Chapman, Ansell & McLaren, 1972; Gearhart & Mintz, 1972; Hillman, Sherman & Graham, 1972). There are similar reasons for believing that the trophoblast nuclei do not fuse with genetically active maternal cell nuclei. Using this glucose phosphate isomerase-1 enzyme system, it is possible to detect hybrid enzyme if it represents 1 % or more of the total activity of the sample and it is therefore possible to conclude that cell fusion is a rare event in the trophoblast of mice.

By replication of the genome without cell division

It is a general observation that trophoblast nuclei incorporate tritiated thymidine but do not divide (rat – Alden, 1948; Dickson & Bulmer, 1960; Zybina, 1960; mouse – Atlas, Bond & Cronkite, 1960; Zavarzin et al. 1966; Saccoman et al. 1967). However, in the rat, tetraploid mitoses are observed during early trophoblast differentiation and it has been suggested that early DNA increases in the trophoblast nuclei are due to the failure of mitoses and the formation of binucleate cells. These are thought to give rise to cells with the 8 C amount of DNA by the two nuclei entering mitosis together to form a common tetraploid metaphase plate which divides into two tetraploid daughter cells (Zybina & Grischenko, 1970). Mitoses are not seen in nuclei which contain more than the 8 C amount of DNA and these nuclei are thought to increase their DNA content by endomitosis (Zybina, 1963).

There are several reasons for believing that the whole genome is replicated proportionately during the massive accumulation of DNA in the trophoblast nuclei. First, nuclei in the blastocyst which contain between 4 C and 8 C amounts of DNA are always labelled after short exposures to tritiated thymidine (mouse – Barlow et al. 1972). Secondly, the amount of DNA in the sex chromatin bodies and the total nuclear DNA increase in parallel during trophoblast growth in the rat. This shows that the inactive X-chromosome is replicated co-ordinately with the bulk of the genome (Zybina & Mos'yan, 1967). Co-ordinate replication of sex chromatin and total nuclear DNA also occurs in human amnion cells which may contain the 16 C amount of DNA (Klinger & Schwarzacker, 1958, 1961). And thirdly, the ratio of main band to satellite DNA remains constant during this DNA increase in the mouse (Sherman, McLaren & Walker, 1972).

It appears that the whole genome takes part in polyploidization and it has been claimed that polytene chromosome configurations can be seen in the trophoblast cells of the rat (Zybina, 1970). These have not been observed in the mouse (Barlow & Sherman, 1972) and if polyteny does occur, then it occurs in a different manner to the formation of salivary gland chromosomes in *Drosophila*; in this case the polytene chromosomes have a much reduced complement of satellite DNA (Gall, Cohen & Polan, 1971; Hennig & Meer, 1971; Dickson, Boyd & Laird, 1971). The observation that the ratio of main band to satellite DNA remains constant during this DNA increase in the mouse also excludes the possibility of large scale gene amplification (e.g. ribosomal cistrons in amphibia – Brown & Dawid, 1968; Perkowska, Macgregor & Birnsteil, 1968), unless the genes amplified fortuitously had the same ratio of main band to satellite DNA as the whole genome (Sherman et al. 1972).

EXTRINSIC CONTROL OF THE CELL CYCLE

At the blastocyst stage, the further development of the embryo comes under the control of the maternal physiology. Development and the cell cycle can be effectively blocked in three situations (reviewed by McLaren, 1972): (a) during lactation, (b) after ovariectomy, (c) in culture, in the absence of certain amino acids (mouse – Gwatkin, 1966a, b).

It is known that the onset of polyploidization is not under maternal control. Blastocysts grown in culture from the two cell stage will develop polyploid cells (Graham, 1971, Barlow et al. 1972; Sherman & Barlow, 1972) and polyploidy also develops in unhatched blastocysts whose cells have no contact with a substrate (Barlow & Sherman, 1972). Extensive polyploidy develops in blastocysts transferred to extra-uterine sites in male mice (Avery & Hunt, 1969; Hunt & Avery, 1971; Barlow & Sherman, 1972).

However, in lactational, ovariectomy or in-vitro delay, the embryo does not develop highly polyploid cells and does not develop further than the late blastocyst. Under these conditions the cell cycle stops (mouse – McLaren, 1968; Sherman & Barlow, 1972; rat – Gulyas & Daniel, 1969; Sanyal & Meyer, 1970, 1972).

The effect of these conditions on the cell cycle of the mouse differs (Sherman & Barlow, 1972). In ovariectomy delay, the majority of nuclei are stopped in the G_1 phase of the cell cycle. Blastocysts may be released from ovariectomy delay by an injection of oestrogen; after this injection many nuclei enter S phase at about the same time (rat – Prasad et al. 1968; Dass, Mohla & Prasad, 1969; Sanyal & Meyer, 1972). In contrast, during delay in-vitro the nuclei appear to be blocked in both the G_1 and the G_2 phases of the cell cycle and the effect of this treatment is quite different from that of ovariectomy (mouse – Sherman & Barlow, 1972). Again, following release from this delay, there is a burst of DNA synthesis (mouse – Ellem & Gwatkin, 1968). In lactational, ovariectomy, and in-vitro delay the cells of the embryo are known to be viable because following release from delay normal mice develop. The blockage at various stages of the cell cycle is therefore not a consequence of cell death.

CONCLUSIONS

The cell cycle of mammalian embryos is unusual in several ways. During early cleavage the cell cycle is long and it progressively shortens in some tissues as the embryo makes contact with the nutrient supplies in the uterine wall. Secondly the embryo forms the trophoblast cells in which massive DNA accumulation by polyploidization occurs very early in development. And thirdly, the cell cycle of the embryo can be reversibly blocked by changing the physiological conditions outside the embryo.

Many of the ideas expressed in this paper arose from discussions with Peter Barlow, David Owen, and Mike Sherman when we were working together in the Sir William Dunn School of Pathology in Oxford with the support of the M.R.C.

REFERENCES

ALDEN, R. H. (1948). Implantation of the rat egg. III. Origin and development of primary trophoblast giant cells. *American Journal of Anatomy*, **83**, 143–82.

ALFERT, M. (1950). A cytochemical study of oogenesis and cleavage in the mouse. *Journal of Cellular and Comparative Physiology*, **36**, 381–409.

ATLAS, M. & BOND, V. P. (1965). The cell generation cycle of the eleven-day mouse embryo. *Journal of Cell Biology*, **26**, 19–24.

ATLAS, M., BOND, V. P. & CRONKITE, E. (1960). Deoxyribonucleic acid synthesis in the developing mouse embryo studied with tritiated thymidine. *Journal of Histochemistry and Cytochemistry*, **8**, 171–81.

AUSTIN, C. R. (1966). Sex chromatin in embryonic and fetal tissues. In *Sex chromatin*, ed. K. L. Moore, pp. 241–54. Philadelphia: W. B. Saunders.

AVERY, G. B. & HUNT, C. V. (1969). The differentiation of trophoblast giant cells in the mouse studied in kidney capsule grafts. *Transplantation Proceedings*, **1**, 61–6.

BARLOW, P. W. & SHERMAN, M. I. (1972). The biochemistry of differentiation of mouse trophoblast: studies on polyploidy. *Journal of Embryology and Experimental Morphology*, **27**, 447–65.

BARLOW, P. W., OWEN, D. A. J. & GRAHAM, C. F. (1972). DNA synthesis in the preimplantation mouse embryo. *Journal of Embryology and Experimental Morphology*, **27**, 431–45.

BOWMAN, P. & McLAREN, A. (1970). Cleavage rate of mouse embryos *in vivo* and *in vitro*. *Journal of Embryology and Experimental Morphology*, **24**, 203–7.

BROWN, D. D. & DAWID, I. B. (1968). Specific gene amplification in oocytes. *Science, New York*, **160**, 272–80.

CHAPMAN, V. M., ANSELL, J. D. & McLAREN, A. (1972). Trophoblast giant cell differentiation in the mouse: Expression of glucose phosphate isomerase (GPI-1) electrophoretic variants in transferred and chimeric embryos. *Developmental Biology*, **29**, 48–54.

DALCQ, A. M. (1957). *Introduction to general embryology*. Oxford: Clarendon Press.

DALCQ, A. M. & PASTEELS, J. (1955). Détermination photométrique de le tenour relative en DNA des noyaux dans les œufs en segmentation du rat et de la souris. *Experimental Cell Research*, supplement, **3**, 72–97.

DASS, C. M. S., MOHLA, S. & PRASAD, M. N. (1969). Time sequence of action of oestrogen on nucleic acid and protein synthesis in the uterus and blastocyst during delayed implantation in the rat. *Endocrinology*, **85**, 528–36.

DICKSON, A. D. (1963). Trophoblast giant cell transformation of mouse blastocysts. *Journal of Reproduction and Fertility*, **6**, 465–6.

DICKSON, A. D. & BULMER, D. (1960). Observations on the placental giant cells of the rat. *Journal of Anatomy*, **94**, 418–24.

DICKSON, E., BOYD, J. B. & LAIRD, C. D. (1971). Sequence diversity of polytene chromosome DNA from *Drosophila hydei*. *Journal of Molecular Biology*, **61**, 615–27.

DONAHUE, R. P. (1972). Fertilization of mouse oocytes: sequence and timing of nuclear progression to the two-cell stage. *Journal of Experimental Zoology*, **180**, 305–18.

DUVAL, M. (1891). Le placenta des rongeurs: le placenta de la souris et du rat. *Journal de l'anatomie et de physiologie normales et pathologiques de l'homme et des animaux*, **27**, 24–106.

DYBAN, A. P., SAMOSHKINA, N. A. & MYSTKOWSKA, E. B. (1972). The oviduct as a barrier to exogenous thymidine in the development of the mouse embryo. *Journal of Embryology and Experimental Morphology*, **27**, 163–6.

EDWARDS, R. G. & GATES, A. H. (1959). Timing of the stages of the maturation divisions, fertilization and the first cleavage of eggs of adult mice treated with gonadotrophins. *Journal of Endocrinology*, **18**, 292–304.

ELLEM, K. A. O. & GWATKIN, R. B. L. (1968). Patterns of nucleic acid synthesis in the early mouse embryo. *Developmental Biology*, **18**, 311–30.

GALASSI, L. (1967). Reutilization of maternal nuclear material by embryonic and trophoblastic cells in the rat for synthesis of deoxyribonucleic acid. *Journal of Histochemistry and Cytochemistry*, **15**, 573–79.

GALL, J. G., COHEN, E. H. & POLAN, M. L. (1971). Repetitive DNA sequences in *Drosophila*. *Chromosoma, Berlin*, **33**, 319–44.

GAMOW, E. I. & PRESCOTT, D. M. (1970). The cell cycle during early embryogenesis of the mouse. *Experimental Cell Research*, **59**, 117–23.

GARDNER, R. L. & LYON, M. F. (1972). X-chromosome inactivation studied by injection of a single cell into the mouse blastocyst. *Nature, New Biology, London*, **231**, 381–3.

GEARHART, J. D. & MINTZ, B. (1972). Glucose phosphate subunit reassociation test for maternal–fetal and fetal–fetal cell fusion in the mouse placenta. *Developmental Biology*, **29**, 55–64.

GRAHAM, C. F. (1971). Virus assisted fusion of embryonic cells. In *In vitro methods in Reproductive Biology, Karolinska Symposium on Research Methods in Reproductive Biology*, ed. E. Diczfalusy, **3**, 154–67.

GRAHAM, C. F. (1972). Nucleic acid metabolism during early mammalian development. In *Regulation of mammalian reproduction*, ed. R. Crozier & P. Corfmann, National Institute of Health Monograph. Springfield: C. C. Thomas. (In press.)

GULYAS, B. C. & DANIEL, J. C. (1969). Incorporation of labeled nucleic acid and protein precursors by diapausing and non-diapausing blastocysts. *Biology of Reproduction*, **1**, 11–20.

GWATKIN, R. B. L. (1966a). Defined media and development of mammalian eggs in vitro. *Annals of the New York Academy of Sciences*, **139**, 79–80.

GWATKIN, R. B. L. (1966b). Amino acid requirements for attachment and outgrowth of the mouse blastocyst in vitro. *Journal of Cellular Physiology*, **68**, 335–44.

HENNIG, W. & MEER, B. (1971). Reduced polyteny of ribosomal RNA cistrons in giant chromosomes of *Drosophila hydei*. *Nature, New Biology, London*, **233**, 70–2.

HILL, R. N. & UNIS, J. J. (1967). Mammalian X-chromosomes: changes in pattern of DNA replication during embryogenesis. *Science, New York*, **155**, 1120–1.

HILLMAN, N., SHERMAN, M. I. & GRAHAM, C. F. (1972). The effect of spatial arrangement on cell determination during mouse development. *Journal of Embryology and Experimental Morphology*, **28**, 263–78.

HOWARD, A. & PELC, S. R. (1953). Synthesis of deoxyribonucleic acid in normal and irradiated cells and its relation to chromosome breakage. *Heredity*, supplement, 6, 261–73.

HUNT, C. V. & AVERY, G. B. (1971). Increased levels of deoxyribonucleic acid during trophoblast giant-cell formation in mice. *Journal of Reproduction and Fertility*, 25, 85–91.

ISSA, M., BLANK, C. E. & ATHERTON, G. W. (1969). The temporal appearance of sex chromatin and of late replicating X-chromosome in blastocysts of the domestic rabbit. *Cytogenetics*, 9, 219–37.

IZQUIERDO, L. & ROBLERO, L. (1965). The incorporation of labelled nucleosides by mouse morulae. *Experientia*, 21, 532–3.

JOLLIE, W. P. (1964). Radioautographic observations on variations in deoxyribonucleic acid synthesis in the rat placenta with increasing gestational age. *American Journal of Anatomy*, 114, 161–72.

KAUFFMANN, S. L. (1966). An autoradiographic study of the generation cycle in the ten-day mouse embryo neural tube. *Experimental Cell Research*, 42, 67–73.

KAUFFMANN, S. L. (1968). Lengthening of the generation cycle during embryonic differentiation of the mouse neural tube. *Experimental Cell Research*, 49, 420–4.

KLINGER, H. P. & SCHWARZACHER, H. G. (1958). Amount of sex chromatin in female tissues is correlated with degree of tissue ploidy. *Nature, London*, 181, 1150–2.

KLINGER, H. P. & SCHWARZACHER, H. G. (1960). The sex chromatin and heterochromatin bodies in human diploid and polyploid cells. *Journal of Biophysical and Biochemical Cytology*, 8, 345–65.

LINDMARK, G. & MELANDER, Y. (1970). Sex chromatin in pre-implantation embryos of the Syrian hamster. *Hereditas*, 64, 128–31.

LYON, M. F. (1962). Sex chromatin and gene action in the mammalian X-chromosome. *American Journal of Human Genetics*, 14, 135–48.

LYON, M. F. (1972). X-chromosome inactivation and developmental patterns in mammals. *Biological Reviews*, 47, 1–35.

McLAREN, A. (1968). A study of blastocysts during delay and subsequent implantation in lactating mice. *Journal of Endocrinology*, 42, 453–63.

McLAREN, A. (1972). Blastocyst activation. In *The regulation of mammalian reproduction*, ed. R. Crozier & P. Corfman. National Institute of Health Monograph. Springfield: C. C. Thomas. (In press.)

MINTZ, B. (1964). Synthetic processes and early development in the mammalian egg. *Journal of Experimental Zoology*, 157, 85–100.

MINTZ, B. (1965). Nucleic acid and protein synthesis in the developing mouse embryo. In *Preimplantation stages of pregnancy*, ed. G. E. W. Wolstenholme & M. O'Connor, pp. 145–61. London: J. & A. Churchill.

MINTZ, B. & BAKER, W. W. (1967). Normal mammalian muscle differentiation and gene control of isocitrate dehydrogenase synthesis. *Proceedings of the National Academy of Sciences, U.S.A.* 58, 592–8.

MITTWOCH, U., KALMUS, H. & WEBSTER, W. S. (1966). Deoxyribonucleic acid values in dividing and non-dividing cells of male and female larvae of the honey bee. *Nature, London*, 210, 264–6.

OPRESCU, SR. & THIBAULT, C. (1965). Duplication of DNA in the egg of the rabbit after fertilization. *Annales de biologie animale, biochimie et biophysique*, 5, 151–6.

NAGL, W. (1972a). Giant sex chromatin in endopolyploid trophoblast nuclei of the rat. *Experientia*, 28, 217–18.

NAGL, W. (1972b). The angiosperm suspensor and the mammalian trophoblast; organs with similar structure and function? *Bulletin Société botanique de France. Mémoir*, in press.

PERKOWSKA, E., MACGREGOR, H. C. & BIRNSTEIL, M. L. (1968). Gene amplification in the oocyte nucleus of mutant and wild-type *Xenopus laevis. Nature, London,* **217**, 649–50.

PIKO, L. (1970). Synthesis of macromolecules in early mouse embryos cultured *in vitro :* RNA, DNA, and polysaccharide component. *Developmental Biology,* **21**, 257–74.

PRASAD, M. N. R., DASS, C. M. S. & MOHLA, S. (1968). Action of oestrogen on the blastocysts and uterus in delayed implantation: an autoradiographic study. *Journal of Reproduction and Fertility,* **16**, 97–104.

SACCOMAN, F. M., MORGAN, C. F. & WELLS, L. J. (1967). Radioautographic studies of DNA synthesis in the developing extra embryonic membranes of the mouse. *Anatomical Record,* **158**, 197–206.

SAMOSHKINA, N. A. (1968). A study of DNA synthesis in the period of ovicell division in mice (*in vitro* experiments). *Tsitologiya,* **10**, 856–64. In Russian.

SANYAL, M. K. & MEYER, R. K. (1970). Effect of estrone on DNA synthesis in preimplantation blastocysts of gonadotrophin-treated immature rats. *Endocrinology,* **86**, 976–81.

SANYAL, M. K. & MEYER, R. K. (1972). Deoxyribonucleic acid synthesis *in vitro* in normal and delayed nidation preimplantation blastocysts of adult rats. *Journal of Reproduction and Fertility,* **29**, 439–42.

SCHLESINGER, M. & KOREN, Z. (1967). Mouse trophoblast cells in tissue culture. *Fertility and Sterility,* **18**, 95–101.

SHERMAN, M. I. & BARLOW, P. W. (1972). Deoxyribonucleic acid content in delayed mouse blastocysts. *Journal of Reproduction and Fertility,* **29**, 123–6.

SHERMAN, M. I., McLAREN, A. & WALKER, P. M. B. (1972). Studies on the mechanism of DNA accumulation in giant cells of mouse trophoblast: satellite DNA content. *Nature, New Biology,* **238**, 175–6.

SILAGI, S. (1963). Some aspects of the relationship of RNA metabolism to development in normal and mutant mouse embryos cultivated *in vitro. Experimental Cell Research,* **32**, 149–52.

SIRLIN, J. I. & EDWARDS, R. G. (1959). Timing of DNA synthesis in ovarian oocyte nuclei and pronuclei of the mouse. *Experimental Cell Research,* **18**, 190–4.

SOLTER, D. & SKREB, N. (1968). La durée des phases du cycle mitotique dans différentes régions du cylindre-oeuf de la souris. *Comptes-rendus de l'Académie des Sciences, Paris,* série D, **267**, 659–61.

SOLTER, D. & SKREB, N. (1971). The cell cycle analysis in the mouse egg-cylinder. *Experimental Cell Research,* **64**, 331–4.

SZOLLOSI, D. (1964). The time of DNA synthesis in mammalian eggs after sperm penetration. *Journal of Cell Biology,* **23**, 92A.

SZOLLOSI, D. (1966). The time and duration of DNA synthesis in rabbit eggs after sperm penetration. *Anatomical Record,* **154**, 209–12.

WEGENER, K., HOLLWEG, S. & MAURER, W. (1964). Autoradiographische Bestimmung der DNS-Verdopplungszeit und anderer Teil-phasen des Zell-zyklus bei fetalen Zellarten der Ratte. *Zeitschrift für Zellforschung und mikroskopische Anatomie,* **63**, 309–26.

WHITTEN, W. K. & DAGG, C. P. (1962). Influence of spermatozoa on the cleavage rate of mouse eggs. *Journal of Experimental Zoology,* **148**, 173–83.

ZAVARZIN, A. A., SAMOSHKINA, N. A. & DONDUA, A. K. (1966). Synthesis of DNA and kinetics of cells in early embryogenesis of mice. *Zhurnal obscheĭ biologii,* **27**, 697–709. In Russian.

ZYBINA, E. V. (1960). Sex chromatin in the trophoblast of early white rat embryos. *Dokladý Akademii nauk S.S.S.R.* **130**, 633–5. In Russian.

ZYBINA, E. V. (1963). Cytophotometric determination of DNA content in nuclei of trophoblast giant cells. *Doklady Akademii nauk S.S.S.R.* **153**, 1428–31. In Russian.

ZYBINA, E. V. (1970). Anomalies of polyploidisation of the cells of the trophoblast. *Tsitologiya*, **12**, 1081–93. In Russian.

ZYBINA, E. V. & MOS'YAN, I. A. (1967). Sex chromatin bodies during endomitotic polyploidization of trophoblast cells. *Tsitologiya*, **9**, 265–72. In Russian.

ZYBINA, E. V. & GRISCHENKO, T. A. (1970). Polyploid cells of the trophoblast in different regions of the white rat placenta. *Tsitologiya*, **12**, 585–95. In Russian.

THE DIFFERENTIAL EFFECT OF [³H]-THYMIDINE UPON TWO POPULATIONS OF CELLS IN PRE-IMPLANTATION MOUSE EMBRYOS

By M. H. L. SNOW

Institute of Animal Genetics, West Mains Road, Edinburgh

INTRODUCTION

The first major morphological differentiation that can be clearly defined in the developing mouse embryo is that of blastulation. In general terms the cavitation of the morula produces a structure with two well defined regions – the trophoblast, being the one cell thick outer cell layer, and the inner cell mass, being all those cells contained within the trophoblast sphere. Graham (1971, and in this Symposium) has described and discussed this process of differentiation in some detail. Although it is possible, and perhaps even advisable, to further designate regions of the blastocyst according to position, e.g. that trophoblast in contact with the inner cell mass and that not in contact, or that part of the inner cell mass totally surrounded by other cells and that in contact with the blastocoel, I shall not use such fine distinctions here.

In recent years micromanipulation techniques have been applied to cultured mammalian embryos, and Gardner (1968, 1971) has ably demonstrated that it is possible to surgically separate trophoblast and inner cell mass and to recombine them in various ways. As an aid in such embryo reconstruction experiments the ability to label parts of an embryo is of high priority. For the study of cell mixing and movement Wilson, Bolton & Cuttler (1972), using oil droplets, and Graham (1971), using isotopic labelling, have already indicated some of the applications for such a technique. It is apparent that considerable advantage attaches to the achievement of heavy isotopic labelling in a rapidly developing system such as the embryo. In 1966, in this laboratory, Caroline Naysmith (in unpublished work) investigated the possibility of labelling cleaving mouse embryos with high concentrations of [³H]thymidine ([³H]TdR). Using a high specific activity [³H]TdR she found that concentrations of 0.1 μCi/ml and above prevented 2-cell and 4-cell mouse embryos from cleaving more than once. In her culture system these embryos could have completed only one S phase. At concentrations between 0.025 and 0.05 μCi/ml about 75% of

[311]

these embryos produced morphologically normal morulae but after 70 h in culture only 15 % had cavitated to form blastocysts. No attempt was made to examine these later stage embryos for subsequent viability, nor was any attempt made to investigate cell number and cleavage rate.

It was known at that time that thymidine *per se* could inhibit cell growth in-vitro by increasing the pool of thymidine triphosphate which in turn inhibited the conversion of cytosine monophosphate to deoxycytosine monophosphate. Thus DNA synthesis is inhibited and cell growth slows and eventually ceases (Reichard, Canellakis & Canellakis, 1960, 1961). The concentration of thymidine necessary to cause this interference in growth was generally between 10^{-5} and 10^{-4} M (see Cleaver, 1967, for further discussion). Naysmith's labelling experiments involved thymidine concentrations of around 10^{-9} M, whereas she found that mouse embryos will cleave, albeit slowly, in 10^{-2} M thymidine and apparently normally in 10^{-5} M thymidine. In the light of more recent reports it seems most likely that the effect Naysmith observed was due to radiation damage. Disintegration of tritium atoms in-situ are now known to cause growth delay in-vivo in liver, ileal and spleen cells (Post & Hoffman, 1965, 1967, 1968, 1971) and cell death (Whitmore & Gulyas, 1966; Apelgot, 1966). Chromosome aberration in both plant and animal cells (Wimber, 1959; Brewen & Olivieri, 1966), mutation (Bateman & Chandley, 1962) and chromosome 'pulverisation' (Ikeuchi, Weinfeld & Sandberg, 1972) have also been ascribed to ^{3}H disintegrations.

The first part of the work described below is a re-examination of Naysmith's findings using an improved culture system and carrying the analysis further to obtain data upon embryo cell number and cleavage rate. The results of this work indicated that about half of the cells that would form a blastocyst are more sensitive to radiation damage than the remainder. Microscopic examination of the damaged blastocysts suggested that the radiosensitive cells would normally be located in the inner cell mass and that the resistant cells constituted the trophoblast. The second part of the data is the result of experiments designed to confirm or reject this opinion.

The most obvious property of trophoblast in the mouse (and many other mammals) is its ability to invade other tissue (see Billington, 1971). In-utero the mouse trophoblast erodes the uterine epithelium and migrates into the uterine stroma, which reacts by forming the deciduoma (see Finn, 1971), and in this way effects implantation. It seems that one function of the decidual tissue is to limit the degree of trophoblast invasion. If blastocysts are implanted in ectopic sites such as kidney or testis trophoblast invasion is much more extensive but there is little or no decidual cell reaction by the surrounding tissue. By transferring experimentally treated blastocysts either into the uterus of a pseudo-pregnant female mouse, or into an ectopic site it is possible to assess to some extent the growth potential of parts of the blastocyst. In-utero, trophoblast development may be limited but inner cell mass development is maximal and, furthermore, the only real test of normality of the blastocyst is to produce viable, normal young after a normal gestation period. In ectopic sites trophoblast growth is more extensive but foetal differentiation is limited, although endoderm, ectoderm and amnion are formed in their correct sequence. During trophoblast invasion in both these sites the cells enlarge and become highly polyploid.

Mouse blastocysts will grow further in-vitro if the culture medium is supplemented with a few per cent of foetal calf serum. In these conditions the blastocyst will hatch from its zona pellucida and stick to the surface of a plastic culture dish. Subsequently, the blastocyst collapses and the trophoblast cells will grow out laterally as a monolayer. They undergo a 'giant cell transformation' similar to that observed in-utero and in ectopic sites. Inner cell mass differentiation is poor and definitely abnormal in-vitro.

MATERIALS AND METHODS

All mice belonged to the randomly bred Q strain. 2-cell embryos were flushed from the oviducts at the appropriate time after mating. They were cultured in 0.1 ml droplets of Brinster's (1965) medium in which the albumin concentration was raised from 1.0 to 3.0 mg/ml (Bowman & McLaren, 1970). They were incubated at 36 °C with a gas phase of 10% CO_2 in air. [³H]TdR, obtained from the Radiochemical Centre, Amersham and diluted with phosphate buffered saline to a stock solution of 10 μCi/ml, was added to the medium to give a range of tritium concentration. The maximum dilution of the culture medium with PBS was 1%, which does not affect embryo growth in any discernible way. Unlabelled thymidine (Sigma Chemical Co.) in powder form was dissolved in the culture medium directly.

The effects of experimental conditions were measured in two ways. Firstly, the number of 2-cell embryos forming expanded blastocysts was scored and secondly, as an indication of their normality, the number of viable cells in the blastocyst was ascertained by counting undamaged nuclei in preparations fixed with 3:1 ethanol/acetic acid and air dried according to Tarkowski's (1966) technique. Such preparations were stained either with lacto-acetic orcein or Giemsa.

Further development of the blastocyst was monitored in the ways described in the introduction. For in-vitro outgrowth the culture medium was supplemented with 5 % foetal calf serum and one blastocyst was placed in each drop. For ectopic outgrowth groups of 5 or 6 blastocysts were placed either under the capsule of the kidney or in the cryptorchid testis of mature male Q mice. These operations were performed by J. D. Ansell. Finally, groups of 3 to 6 blastocysts (corresponding to $3\frac{1}{2}$ days development in-vivo) were transferred to the uterine horns of $2\frac{1}{2}$-day pseudo-pregnant Q females (obtained from a mating with a vasectomised male) according to the procedures established by McLaren & Michie (1956). Autopsies were made at various intervals and material for histological examination was fixed in Sanfelices' fluid and wax-embedded prior to sectioning.

RESULTS

TdR-6-T(n), specific activity 5.0 Ci/mmole, and methyl-T-TdR, specific activity 17.4 Ci/mmole, were used to corroborate Naysmith's data. Each precursor was tested in 5 replicate experiments comparing development in medium containing 0, 0.01, 0.025, 0.05 and 0.1 μCi/ml [^{3}H]TdR. At least ten 2-cell embryos were placed in each droplet and the mean number of blastocysts produced was expressed as an angular response (Biggers & Brinster, 1965). These results are shown in Table 1.

Table 1. *The effect of TdR-6-T(n) (sp. act. 5.0 Ci/mmole) and methyl-T-TdR (sp.act. 17.4 Ci/mmole) upon the development of blastocysts from 2-cell embryos. Data pooled from 5 replicate experiments*

Treatment (μCi/ml)	Thymidine-6-T(n)		Methyl-T-Thymidine	
	Mean angular response	Blastocyst mean cell no.	Mean angular response	Blastocyst mean cell no.
0	58.9 ± 8.2	51.7 ± 5.7	67.5 ± 6.5	61.9 ± 6.4
0.01	51.3 ± 3.9	32.9 ± 3.2	69.0 ± 3.2	40.3 ± 4.0
0.025	43.6 ± 12.4	21.0 ± 3.4	68.7 ± 6.1	33.1 ± 3.9
0.05	38.0 ± 2.4	16.0 ± 0.98	78.4 ± 6.0	30.3 ± 2.5
0.1	13.7 ± 5.8	12.0 ± 2.4	60.7 ± 3.6	16.9 ± 2.5

It is clear from these data that both forms of [^{3}H]TdR, when used in this range of concentrations, exert a deleterious effect upon the development of 2-cell embryos. Fewer embryos expand to form blastocysts in TdR-6-T(n) and those forming blastocysts have a reduced cell number. The regressions for both criteria against concentration are highly significant ($P < 0.001$). Methyl-T-TdR does not interfere with blastocyst formation but the regression of blastocyst cell number against concentration is again highly significant ($P < 0.001$). There is a significant difference between the effects of methyl-T-TdR and TdR-6-T(n) on blastocyst formation ($P < 0.01$) but not for the effects upon cell number.

A possible cause for the reduction in blastocyst formation in TdR-6-T(n) lay in the difference in the specific activities of the two forms of [³H]TdR. This possibility was tested by varying the specific activity of the nucleoside and also by testing the effect of unlabelled thymidine in this culture system. The results of these investigations are shown in Table 2. It is immediately clear that lowering the specific activity of methyl-T-TdR does not significantly reduce the number of blastocysts produced from 2-cell embryos. From Table 2 it can be seen that high concentrations of thymidine do not interfere with blastocyst formation, but do result in depression in cell number. Since thymidine significantly interferes with cell number only at concentrations above 10^{-6} M it seems fair to conclude that the effect of TdR-6-T(n) upon blastocyst formation is not due to thymidine itself but to some contaminant.

Table 2. *The effect of specific activity of methyl-T-TdR and of unlabelled thymidine upon the development of blastocysts from 2-cell embryos. See also Fig. 1*

Tdr conc. (M)	Methyl-T-TdR (0.05 μCi/ml) (6 replicates)			Unlabelled thymidine (3 replicates)	
	Sp.act. mCi/mmole	Mean L'r response	Mean cell no. of blastocysts	Mean L'r response	Mean cell no. of blastocysts
0	—	73.8 ± 5.5	60.3 ± 7.4	71.1 ± 6.2	41.8 ± 8.9
2 × 10⁻⁹	27000	62.4 ± 4.2	37.4 ± 3.2	—	—
2 × 10⁻⁸	2700.0	76.0 ± 4.7	40.6 ± 1.4	45.0 ± 4.3	39.0 ± 1.9
2 × 10⁻⁷	270.0	72.6 ± 9.0	43.0 ± 2.0	—	—
2 × 10⁻⁶	27.0	70.1 ± 1.5	48.2 ± 5.2	69.8 ± 5.1	39.5 ± 2.7
2 × 10⁻⁵	2.70	63.5 ± 3.7	35.6 ± 2.6	71.5 ± 3.7	30.2 ± 2.8
2 × 10⁻⁴	0.27	68.8 ± 3.3	30.5 ± 0.7	72.4 ± 2.7	23.3 ± 1.6
2 × 10⁻³	0.027	62.1 ± 4.6	25.1 ± 0.5	74.3 ± 2.7	25.2 ± 1.6

The data in Table 2 also indicate that specific activity has an important bearing upon the depression in cell number, which suggests that the mechanism causing the reduction in cell number is associated with cell-bound or incorporated [³H]TdR. Taking into account Naysmith's unpublished work using unlabelled thymidine, it is interesting to note that lowering the specific activity of [³H]TdR causes a lessening in the radiation toxicity until a point is reached where thymidine poisoning becomes the major consideration. This is clearly shown in Fig. 1 where these data are plotted graphically.

The air-dried preparations of [³H]TdR-treated blastocysts show some

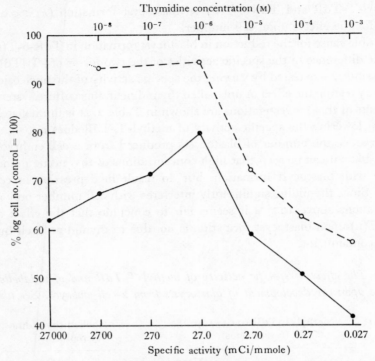

Fig. 1. The effect of varying the specific activity of methyl-T-Tdr, at a concentration of 0.05 μCi/ml (●—●); and of unlabelled thymidine (○--○) upon blastocyst cell number.

abnormal, misshapen nuclei, nuclear fragments and broken chromosomes. Chromatid exchanges have also been found. Blastocysts grown in unlabelled thymidine do not yield such damaged nuclei and show no evidence of cell death beyond that encountered in untreated embryos.

In order to examine cleavage rate and possibly the manner in which cell number is reduced, 2-cell embryos were grown in 0.05 μCi/ml methyl-T-TdR (a concentration that effectively reduces the cell number by 50%). Samples of embryos were taken at intervals and their cell numbers determined. Analysis of ten experiments indicates that development under control and experimental conditions is synchronous up to the 16-cell stage, but at that point embryos in [³H]TdR fail to increase their cell number for some 10–12 h. It is during this period of delay that fragmented and damaged nuclei (such as are visible in Plate 1a) can first be detected cytologically. When normal embryos are at the 32-cell stage these experimentally damaged embryos contain between 16 and 18 cells. Thereafter the experimental embryos cleave at a rate very similar to controls. It is as if

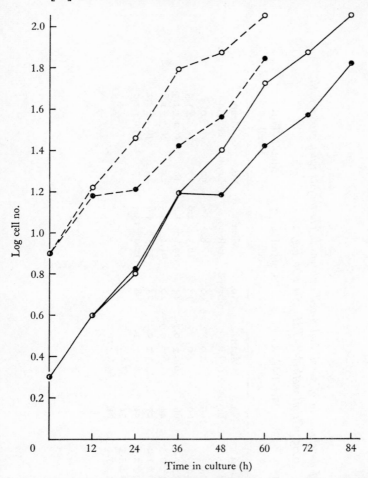

Fig. 2. The data in Table 3 plotted graphically. ○, control, ●, 0.05 μCi/ml
methyl-T-TdR, ———, 2-cell embryos, –––, 8-cell embryos. The radiation damage
is quite clearly restricted to the 16–32 cell stage.

half of the 16 blastomeres fail to develop any further but that the surviving
blastomeres cleave normally. These results are shown in Table 3 and Fig. 2,
which also include the results of eight experiments in which 8-cell
embryos were incubated with 0.05 μCi/ml methyl-T-TdR. These embryos
show the same marked reduction in cell number at the 16–32 cell stage
despite having undergone one less DNA replication than embryos grown
from the late 2-cell stage. Accumulation of [³H]thymidine in the cellular
DNA is therefore of little significance in explaining why the radiation
damage manifests itself particularly at the 16-cell stage.

Table 3. *The progress of development of 2-cell and 8-cell embryos to blastocysts when grown with 0.05 μCi/ml methyl-T-TdR. See also Fig. 2*

Time in culture (h)	Control 2-cell		Met-T-TdR 2-cell		Control 8-cell		Met-T-TdR 8-cell	
	No. embryos	Mean cell no.	No. embryos	Mean cell no.	No. embryos	Mean cell no.	No. embryos	Mean cell no.
0	165	2.0	199	2.0	90	8.0	90	8.0
12	161	4.0	160	4.0	12	16.5 ± 2.0	17	15.2 ± 1.5
24	42	6.3 ± 0.8	66	6.6 ± 0.7	14	29.1 ± 1.2	16	16.1 ± 1.5
36	48	15.4 ± 0.7	44	15.6 ± 0.9	20	61.1 ± 4.5	24	26.3 ± 1.8
48	34	24.6 ± 1.0	27	15.3 ± 0.4	23	73.5 ± 4.6	15	36.6 ± 2.4
60	61	52.0 ± 2.9	51	26.6 ± 1.8	8	112.0 ± 15.4	7	69.9 ± 7.0
72	22	73.4 ± 3.5	51	37.4 ± 3.7				
84	20	113.3 ± 7.1	21	65.6 ± 10.2				

Viewed with a dissecting microscope most of the blastocysts (about 80%) grown in 0.05 μCi/ml appear not to contain an inner cell mass. To test this possibility [³H]TdR damaged blastocysts were transferred into pseudo-pregnant foster mothers. Autopsies were carried out on the 4th, 5th, 6th or 7th day of pregnancy. At these times the decidual swellings at the sites of implantation can be detected and histological examination made of a localised area.

From 53 control blastocysts transferred into 10 females 36 (66.6%) implantation sites were obtained. Of 83 blastocysts grown in 0.05 μCi/ml methyl-T-TdR, transferred to 14 females, 55 (65.5%) gave rise to decidual cell reactions. Histological examination shows that the controls undergo a normal development (Plates 1b, 2a, c), although during this early post-implantation period they are often a few hours retarded, in comparison with embryos in-vivo. [³H]thymidine-treated embryos produce a normal decidual cell reaction but implantation is not normal. Trophoblast giant cell transformation is similar to controls and the uterine epithelium is eroded, but the trophoblast cells do not then penetrate through the epithelial basement membrane in an organised way (Plates 1c, 2b). At 7½ days of pregnancy individual giant cells can still be observed in the cavity that would normally be occupied by the developing foetus (Plate 2d). At this stage normal primary trophoblast should have migrated some way into the decidual tissue. In no case has any evidence of an inner cell mass been found in these [³H]TdR-treated blastocysts and it seems very probable that the cells lost as a result of radiation damage represent this tissue. It also is the case that these results could indicate that the trophoblast is abnormal and unable to fulfil its invasive role.

To further test this hypothesis, blastocysts were allowed to outgrow in-vitro and were also transferred to ectopic sites. In-vitro, 42/51 (84%) of control blastocysts but only 14/40 (35%) of those grown in 0.05 μCi/ml [³H]TdR hatch and outgrow. All those that succeed in hatching from their zona pellucida outgrow, and the extent of growth is similar in both cases. In ectopic sites groups of blastocysts of each class grow well, both in kidneys and in testes. Autopsies were performed 5, 8 or 10 days after transfer of the blastocysts. The ectopic growth of [³H]TdR-damaged blastocysts is vigorous and even after 10 days shows no sign of abating. In the trophoblast nodules produced there is no evidence for tissue derived from inner cell mass cells although this is very noticeable in the controls (Plate 3a, b). It is also clear that the trophoblast giant cells from the experimental blastocysts are considerably larger than those in the controls (Plate 3c, d). It is not known for how long this ectopic trophoblast will survive.

CONCLUSIONS AND DISCUSSION

It seems fair to conclude from the data above that the reduction in cell number observed in blastocysts grown in [³H]thymidine is the result of cell death, occurring principally at the 16-cell stage, and not to a slowing of cleavage rate. The nature of the damage and its relationship to tritium concentration and specific activity indicate that radiation from ³H disintegration is the causative factor. The investigations of the further development of the blastocysts grown in 0.05 µCi/ml methyl-T-thymidine confirm the early suspicion that the radiosensitive cells of the blastocyst are, or would be, located in the inner cell mass. It is interesting to note that it is at the 16–32-cell stage that the radio-resistant cells become distinct. It could be that at this stage inner cell mass differentiates from trophoblast. Other factors also point to the 16-cell stage as the time at which this differentiation begins. Graham (1971) and Wilson *et al.* (1972) have shown that there is a marked tendency for those cells, or regions of cells towards the outside of an 8- or 16-cell embryo to become trophoblast and those on the inside to become inner cell mass. Enclosed cells are first encountered at the 16-cell stage. Tight junctional complexes between the cells are characteristic of trophoblast and it has been demonstrated that these junctions become established at the early 16-cell stage, at least in the rat. (Enders & Schlafke, 1965; Schlafke & Enders, 1967). There is also a large increase in RNA synthesis in the 16-cell mouse embryo which could also be associated with differentiation and blastulation (Monesi & Salfi, 1967; Pikó, 1970).

The growth in-vitro and in ectopic sites of [³H]TdR-damaged blastocysts show that the surviving cells will continue to grow and differentiate but only as invasive trophoblast. The data from ectopic transfers indicating that [³H]TdR-treated trophoblast invades more extensively than controls suggest that the inner cell mass may exert some controlling influence over the proliferation of the trophoblast. This is also apparent in normal implantations in-utero where the trophoblast overlying the inner cell mass does not proliferate or become invasive until some 36 h after implantation commences. The reason for the failure of the [³H]TdR-treated trophoblast to invade in-utero is not clear. The absence of an inner cell mass may be an important factor but it is also likely that the uterus constitutes a harsher environment than the male kidney or testis and that the trophoblast is overcome by a reaction from the decidual cells. It is the case that when trophoblast is transferred to a non-pregnant uterus proliferation and invasion can be extensive but that the pregnant or pseudo-pregnant uterus, with its ability to form decidual tissue, is considerably less susceptible to

PLATE I

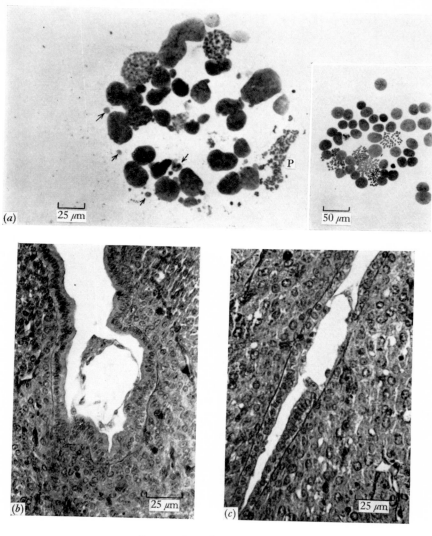

(a)   25 μm   P   50 μm

(b)   25 μm

(c)   25 μm

For explanation see p. 323

PLATE 2

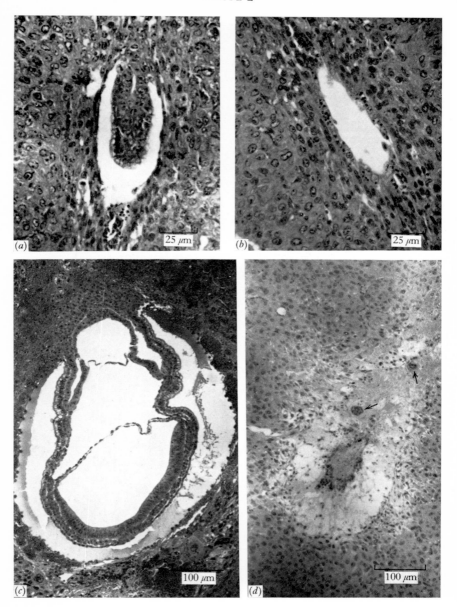

For explanation see p. 323

PLATE 3

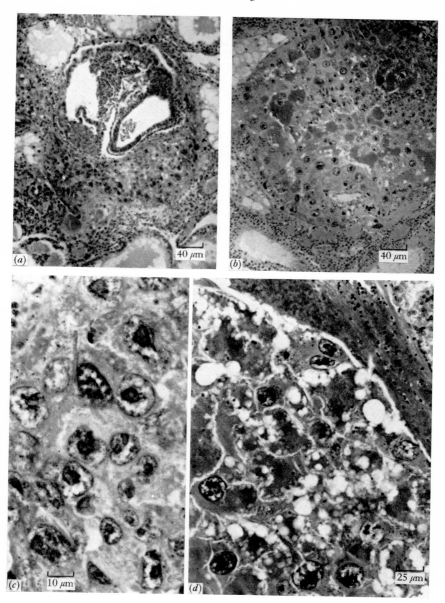

For explanation see pp. 323–4

such erosion (Kirby & Cowell, 1968). A better test of the viability of the [³H]thymidine-treated trophoblast would be to provide it with an un-damaged inner cell mass and place that reconstructed embryo in a pseudo-pregnant uterus. The techniques for doing this work are available and Richard Gardner has demonstrated that it can be done.

Why should the prospective inner cell mass be more susceptible to radiation damage? The fact that those cells that first become enclosed by other cells tend to form the inner cell mass may indicate a possible explana-tion. The cells on the inside of a morula (16–32-cell stage) are subject not only to the beta radiation from [³H]thymidine contained within those cells, but also to a certain fraction of the radiation emanating from the cells surrounding them. The actual concentration of beta particles within these 'inside' cells will therefore be higher than that in the 'outside' cells, and presumably radiation effects will be increased. It is doubtful if this is the major factor in the differential effect of [³H]TdR for two reasons. First, the early indications of radiation damage are localised in the nucleus (fragmen-tation, chromosome breakage, etc.) and whole-cell functions such as cleav-age and cell–cell contact are not impaired, even when a [³H]TdR-damaged blastocyst, with no inner cell mass, contains between 40 and 50 cells. In these cases, prior to cavitation, when the morula contains 30–40 cells it is not possible for all cells to be 'outside', and yet, there is little evidence for cell death at this later stage. Secondly, the energy of tritium-derived beta particles is very low, such that in an aqueous environment particles seldom travel more than 1 μm. Hence there would be very little inter-cell radiation and what there was could only be effective very near the cell periphery, and certainly could not contribute to the nuclear damage observed.

A more likely explanation is discussed by Graham (this Symposium). It has been observed that certain of the cells of a developing mouse embryo start to cleave more rapidly at about the 16-cell stage. These cells apparently become localised in the inner cell mass. The shortening of the cell cycle is associated with a decrease in the duration of DNA replication. This can be achieved in two ways, either the rate of DNA synthesis increases, or there is an increase in the rate at which replicons become available for synthesis. Either of these mechanisms would allow for the exposure of radiation-induced single-strand breaks in the DNA at a greater rate than previously and perhaps in a way that renders the DNA repair mechanism inadequate or ineffective. The accumulation of 'broken' chromosomes may then result in nuclear fragmentation and in extreme cases in chromosome pulverisation such as that reported by Ikeuchi et al. (1972), and also observed in the above study.

REFERENCES

APELGOT, S. (1966). Actions léthales comparés des rayons X et des β du tritium sur *E. coli. International Journal of Radiation Biology*, **10**, 495–508.

BATEMAN, A. J. & CHANDLEY, A. C. (1962). Mutations induced in the mouse with tritiated thymidine. *Nature, London*, **193**, 705–6.

BIGGERS, J. D. & BRINSTER, R. L. (1965). Biometrical problems in the study of early mammalian embryos *in vitro. Journal of Experimental Zoology*, **158**, 39–48.

BILLINGTON, W. D. (1971). Biology of the Trophoblast. In *Advances in Reproductive Physiology*, ed. M. W. H. Bishop, pp. 27–66. London: Logos Press.

BOWMAN, P. & McLAREN, A. (1970). Viability and growth of mouse embryos after *in vitro* culture and fusion. *Journal of Embryology and Experimental Morphology*, **23**, 693–704.

BREWEN, J. G. & OLIVIERI, G. (1966). The kinetics of chromatid aberrations induced in Chinese hamster cells by tritium-labelled thymidine. *Radiation Research*, **28**, 779–92.

BRINSTER, R. L. (1965). Studies on the development of mouse embryos *in vitro*. IV. Interaction of energy sources. *Journal of Reproduction and Fertility*, **10**, 227–40.

CLEAVER, J. E. (1967). *Thymidine Metabolism and Cell Kinetics*. Amsterdam: North-Holland Publishing Co.

ENDERS, A. C. & SCHLAFKE, S. (1965). The fine structure of the blastocyst: some comparative studies. Ciba Foundation Symposium, *Preimplantation Stages of Pregnancy*, ed. G. E. W. Wolstenholme & M. O'Connor, pp. 29–59. London: J. & A. Churchill.

FINN, C. A. (1971). The biology of decidual cells. In *Advances in Reproductive Physiology*, vol. v, ed. M. W. H. Bishop, pp. 1–26. London: Logos Press.

GARDNER, R. L. (1968). Mouse chimaeras obtained by the injection of cells into the blastocyst. *Nature, London*, **220**, 596–7.

GARDNER, R. L. (1971). Manipulations on the blastocyst. In *Advances in Biosciences*, 6: *Intrinsic and Extrinsic Factors in Early Mammalian Development*, ed. G. Raspé, pp. 279–96. Edinburgh: Pergamon Vieweg.

GRAHAM, C. F. (1971). The design of the mouse blastocyst. In *Control Mechanisms of Growth and Differentiation, Symposia of the Society for Experimental Biology*, **25**, 371–8.

IKEUCHI, T., WEINFELD, H. & SANDBERG, A. A. (1972). Chromosome pulverization in micronuclei induced by tritiated thymidine. *Journal of Cell Biology*, **52**, 97–104.

KIRBY, D. R. S. & COWELL, T. P. (1968). Trophoblast-host interactions. In *Epithelial-Mesenchymal Interaction*, ed. R. Fleischmajer & R. E. Billingham, pp. 64–77. Baltimore: Williams & Wilkins.

McLAREN, A. & MICHIE, D. (1956). Studies on the transfer of fertilized mouse eggs to uterine foster-mothers. 1. Factors affecting the implantation and survival of native and transferred eggs. *Journal of Experimental Biology*, **33**, 394–416.

MONESI, V. & SALFI, V. (1967). Macromolecular synthesis during early development in the mouse embryo. *Experimental Cell Research*, **46**, 632–5.

PIKÓ, L. (1970). Synthesis of macromolecules in early mouse embryos cultured *in vitro* : RNA, DNA and a polysaccharide component. *Developmental Biology*, **21**, 257–79.

POST, J. & HOFFMAN, J. (1965). Dose dependent effects of [H^3]TdR as a DNA label on the replications of liver cells in the growing rat. *Radiation Research*, **26**, 422–41.

POST, J. & HOFFMAN, J. (1967). Late effects of [H³]TdR as a DNA label on liver cell replication. *Radiation Research*, **30**, 748–58.

POST, J. & HOFFMAN, J. (1968). Early and late effects of [H³]TdR-labelled DNA on ileal cell replication *in vivo*. *Radiation Research*, **34**, 570–82.

POST, J. & HOFFMAN, J. (1971). Early and late effects of tritiated thymidine upon the replication of splenic lymphocytes. *Radiation Research*, **45**, 335–48.

REICHARD, P., CANELLAKIS, Z. N. & CANELLAKIS, E. S. (1960). Regulatory mechanisms in the synthesis of deoxyribonucleic acid *in vitro*. *Biochimica et Biophysica Acta*, **41**, 558–9.

REICHARD, P., CANELLAKIS, Z. N. & CANELLAKIS, E. S. (1961). Studies on a possible regulatory mechanism for the biosynthesis of deoxyribonucleic acid. *Journal of Biological Chemistry*, **236**, 2514–19.

SCHLAFKE, S. & ENDERS, A. C. (1967). Cytological changes during cleavage and blastocyst formation in the rat. *Journal of Anatomy*, **102**, 13–32.

TARKOWSKI, A. K. (1966). An air-drying method for chromosome preparations from mouse eggs. *Cytogenetics*, **5**, 394–400.

WHITMORE, G. F. & GULYAS, S. (1966). Synchronization of mammalian cells with tritiated thymidine. *Science, New York*, **151**, 691–4.

WILSON, I. B., BOLTON, E. & CUTTLER, R. H. (1972). Preimplantation differentiation in the mouse egg as revealed by microinjection of vital markers. *Journal of Embryology and Experimental Morphology*, **27**, 467–79.

WIMBER, D. E. (1959). Chromosome breakage produced by tritium-labelled thymidine in *Tradescantia paludosa*. *Proceedings of the National Academy of Sciences, U.S.A.* **45**, 839–46.

EXPLANATION OF PLATES

PLATE 1

(a) An air-dried preparation of a morula grown in 0.05 μCi/ml methyl-T-TdR. Misshapen, lobed nuclei, nuclear fragments (arrows) and a 'pulverised' metaphase plate (P) are clearly visible. The inset shows part of a control embryo of the same stage.

(b) A section through a control 4½-day embryo transferred in-utero. The inner cell mass is very distinct.

(c) A section through a 4½-day embryo grown in 0.05 μCi/ml methyl-T-TdR. Note the absence of the inner cell mass.

PLATE 2

(a) A section through a control embryo at 5½ days, in-utero.

(b) A section through a 5½-day embryo from a blastocyst grown in 0.05 μCi/ml methyl-T-TdR. Note the trophoblast cells and the absence of the inner cell mass.

(c) A section through a 7½-day control embryo.

(d) A section through the deciduoma produced at 7½ days by a blastocyst grown in 0.05 μC/ml methyl-T-TdR. There is clearly no embryonic tissue present except for two giant trophoblast cells (arrows).

PLATE 3

(a) A section through an 8-day-old ectopic outgrowth containing two control blastocysts. Embryonic differentiation has occurred to some extent and trophoblast growth seems limited.

(b) A section through an 8-day-old ectopic outgrowth produced from several (probably 5) blastocysts grown in 0.05 μCi/ml methyl-T-TdR. The trophoblast outgrowth has been vigorous but there is no trace of inner cell mass tissue.

(c) A group of the largest giant trophoblast cells visible in (a). Compare with (b) and (d).

(d) The edge of a large ectopic outgrowth produced 8 days after 5 blastocysts grown in 0.05 μCi/ml methyl-T-TdR were transferred to the kidney of a male mouse.

A CELL CYCLE IN BONE MINERALIZATION

By J. E. AARON and F. G. E. PAUTARD

MRC Mineral Metabolism Unit, The General Infirmary, Leeds

INTRODUCTION

Bone develops from the mesenchyme by the differentiation of osteopro-genitor cells which first form osteoid and then somehow allow the inter-cellular spaces to fill with calcium phosphate. While the morphogenesis of bone is now widely accepted, the steps which take place when the tissue calcifies have always been regarded as extracellular events. The prevailing view is that the fabric of bone is first laid down by the cellular synthesis of collagen, which is extruded to form, alone or in association with other substances, stereotactic sites upon which the calcium salts accumulate by crystallization or precipitation from the serum fluids.

There is, however, optical (Bohatirchuk, 1965; Kashiwa, 1970; Rolle, 1969) and chemical (Hirschman & Nichols, 1972) evidence that calcium phosphate is present within bone cells during mineralization, as is the case in many other calcified tissues (Pautard, 1966). The possibility that mineral might be present within the osteoblast or osteocyte has been emphasized by two recent observations. The first is that the absence of any calcium salts in electron micrographs of bone cells might be the result of the loss of soluble calcium phosphate complexes in the fixatives and stains conventionally used for ultrastructural research (Termine, 1972; Pautard, 1972). The second observation is that phosphate, at least, can be visualized in the electron microscope, and here the ultrastructural image corresponds to the optical image (Aaron & Pautard, 1972).

We have observed recently (Aaron & Pautard, manuscript in prepara-tion) that calcium and phosphorus are not present at the same time, or in the same place, in the cell types which have been described during the development of the mouse calvarium (Aaron, 1972). In this preliminary communication, we present additional evidence based upon staining with tetracycline, in support of the hypothesis that calcium phosphate may not only be formed inside bone cells during mineralization, but that its accumulation and organization into characteristic structures may be the result of a cycle of events involving a specific pathway through the cell.

MATERIALS AND METHODS

Throughout these experiments, 7-day-old mice were used. The preparation and staining of the calvaria for calcium with glyoxal bis-(2-hydroxyanil) (GBHA) and for phosphate with silver have been described elsewhere (Aaron & Pautard, 1972). Details of selected sites and cell types with time relative to this study have already been reported (Aaron, 1972). Calvaria were also treated with tetracycline (*Tetracyn*, Pfizer Ltd., England) under various conditions and viewed either whole, or in thin (1 μm) sections, under ultraviolet light, filtered (Zeiss excitor filter I, extinction filters 53, 0, 44) for tetracycline fluorescence. The natural autofluorescence of bone was reduced by examination in water where necessary.

RESULTS

Freshly-removed calvaria, fixed or unfixed, took up tetracycline rapidly. A striking feature is the strong staining of the cells in the whole mount (Plate 1a), the fluorescence being located, in most instances, principally in the region of the nuclei (Plate 1b). In section (Plate 1c) osteoblasts and cartilage cells also show pronounced staining in the region of the nucleus, the fluorescence often appearing as patches at the highest magnifications (Plate 1d). The staining characteristics of cells and matrix under a variety of conditions are set out in Table 1.

When tetracycline was used in combination with the GBHA stain for calcium, regions of fluorescence appeared in some cells not stained with GBHA. On the other hand, when tetracycline was used in combination with the silver stain for phosphate, the fluorescence was obscured (Plate 1f). The silver image (Plate 1e) occupied the same, and different, regions from the fluorescent image which was restored (Plate 1g) when the silver was removed by a saturated solution of potassium ferricyanide for 18 h.

DISCUSSION

The use of tetracycline as a direct stain, in contrast to the more usual practice of labelling over periods of time in-vivo adds more evidence to support the possibility that calcium phosphate may be organized within the cell and exported to the calcification front. Hitherto, the evidence for the *transport* of calcium phosphate (as opposed to the presence of the salt) has been absent, although the pattern of calcium with time inside bone cells has suggested a 'loading' and 'unloading' process (Aaron, 1972).

The distribution of fluorescent areas characteristic of tetracycline within bone cells immediately after removal from the animals, or after periods in

Table 1. *Some staining characteristics of tetracycline*

Tetracycline treatment	Tissue	Fluorescence	
1% in ion-free water	None	Solution negative; dry crystals positive but soluble in water	
1% with Ca^{2+} ions	None	Solution negative. Dry salt positive but soluble in water	
1% with amorphous calcium phosphate (Eanes, 1970)	None	Precipitate strongly positive in water; insoluble	
		'Nucleus'	'Matrix'
0.8%, pH 6.8, 30 min	Whole calvarium in tissue culture (Biggers *et al.* 1961)		
	Zero time	Positive	Weak
	3 h	Strongly positive some in cytoplasm	Weak
	Whole calvarium fixed 1.5% glut.	Positive	Weak
	+10% EDTA, 18 h	Absent	Weak
	Whole calvarium unfixed	Positive	Weak
	+10% EDTA, 18 h	Absent	Positive
	Section, early stage, site 5*a*	Strongly positive cartilage osteoblast	Weak
	Section, late stage, site 2*a*	Cartilage absent. Strongly positive osteocytes	Weak

tissue culture, suggests that the primary site for binding is in the vicinity of the nucleus. The loss of fluorescence after treatment with EDTA, together with the return of fluorescence after the removal of silver from the phosphate, suggests that calcium phosphate (or some complex of the salt) is responsible for the binding. The position of the fluorescent regions in relation to the nucleus, together with the structural and ultrastructural evidence (Aaron & Pautard, 1972) prompt us to propose that the Golgi apparatus (the juxtanuclear vacuole seen in the optical microscopy of the osteoblast and osteocyte) may be the organelle involved in the production of substances, and possibly structures, containing calcium phosphate.

Since it has been well established that tetracycline incorporates into bones and teeth in-vivo, either by chelation with calcium salts or some part of the matrix, or both (see Kaitila, 1971 for references) the presence of tetracycline within bone cells in fresh or fixed tissue immediately suggests transport of complexes from within the cell. In-vivo, the toxicity of tetracycline precludes the use of heavy doses and hence any fluorescence within the cells would be small in comparison to the fluorescence in the calcification zone outside the cells, where the tetracycline would be steadily

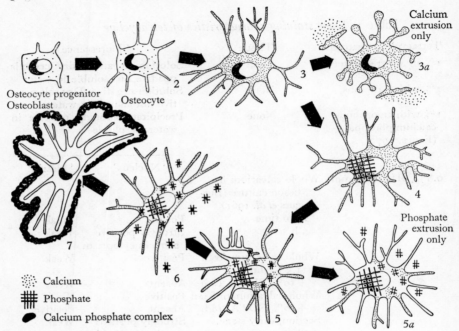

Fig. 1. Proposed cell cycle for bone mineralization.

accumulated during the labelling period. The absence of strong fluorescence in the matrix in stained fresh tissue indicates that the mature calcified tissue is not readily permeated by tetracycline, a fact well recorded for the behaviours of stains for calcium and phosphorus.

The evidence that some tetracycline labelled regions do not stain for calcium while the phosphate areas obscure all the tetracycline fluorescence (and probably parts not fluorescent) indicates that a certain proportion of the calcium and phosphorus inside the cell may not be associated together, or with the complex which stains strongly for tetracycline. The staining of both osteoblast and osteocyte with tetracycline, the variation in intensity from cell to cell and the diminution of the osteocytes in regions of the calvarium where the bone is thicker and more mature all illustrate the cell-by-cell nature of development and the changes that take place with time.

While all histological procedures should be treated with some caution and little is known at present about the ultrastructural and molecular features of binding with tetracycline, the evidence presented above, taken in conjunction with the present knowledge of the distribution of calcium and phosphorus within cells during the formation of selected regions of the calvarium, allow us to propose a cell cycle for bone mineralization, illustrated diagrammatically in Fig. 1. Here, the osteoblast (and also the cartilage cell, which appears to calcify before resorption) show calcium phosphate incorporation into the juxtanuclear vacuole, but to a lesser extent than the developing osteocyte. The osteocyte, during maturation, becomes progressively loaded with calcium, with an active Golgi apparatus beginning to synthesize precursors. Later, phosphate appears in the juxtanuclear region and spreads throughout the cell, the calcium diminishing. Finally, calcium and phosphorus in complexed form appear outside the cell and the Golgi activity diminishes as the osteocytes are surrounded by mature, fully calcified matrix. The observation (Pautard, 1972) that some cells, containing only calcium, appear to distribute their contents into the surrounding seam spaces, suggests a mechanism of calcium extrusion independent of the Golgi/phosphate cycle. Phosphate alone may be extruded in a similar way.

The scheme proposed in Fig. 1 lacks the ultrastructural evidence now being sought. The optical information clearly indicates that a cell cycle may be present and it is unfortunate that, with the exception of phosphate, the present methods of preparing specimens for the electron microscope do not allow us to see the ultrastructural details at present.

REFERENCES

AARON, J. E. (1972). Osteocyte types in the developing mouse calvarium. *Calcified Tissue Research*, in press.

AARON, J. E. & PAUTARD, F. G. E. (1972). Ultrastructural features of phosphate in developing bone cells. *Israel Journal of Medical Science*, **8**, 625–9.

BIGGERS, J. D., GWATKIN, R. B. L. & HEYNOR, S. (1961). Growth of embryonic avian and mammalian tibiae on a relatively simple chemically defined medium. *Experimental Cell Research*, **25**, 41–58.

BOHATIRCHUK, F. (1965). The study of calcification of mammalian cartilage in norm and pathology by stain historadiography. *American Journal of Anatomy*, **117**, 287–309.

EANES, E. D. (1970). Thermochemical studies on amorphous calcium phosphate. *Calcified Tissue Research*, **5**, 133–45.

HIRSCHMAN, P. N. & NICHOLS, G. (1972). The isolation and partial characterization of a calcium-rich particulate fraction from bone cells. *Calcified Tissue Research*, **9**, 67–79.

KAITILA, I. (1971). Effect of tetracycline on mineralization in cycloheximide-treated bones *in vitro. Calcified Tissue Research*, **7**, 46–57.

KASHIWA, H. K. (1970). Calcium phosphate in osteogenic cells. *Clinical Orth. and Related Research*, **70**, 200–11.

PAUTARD, F. G. E. (1966). A biomolecular survey of calcification. In *Calcified Tissues* 1965, ed. H. Fleisch, H. J. J. Blackwood & M. Owen, pp. 108–21, Berlin: Springer.

PAUTARD, F. G. E. (1972). Contribution to discussion. In *Comparative Biology of Extracellular Matrices*, ed. H. Slavkin. New York & London: Academic Press. (In press.)

ROLLE, G. K. (1969). The distribution of calcium in normal and tetracycline-modified bones of developing chick embryos. *Calcified Tissue Research*, 3, 142–150.

TERMINE, J. D. (1972). Contribution to discussion. In *Comparative Biology of Extracellular Matrices*, ed. H. Slavkin. New York & London: Academic Press. (In Press.)

EXPLANATION OF PLATE

PLATE I

Various features of staining the parietal plate of 7-day-old mouse calvaria with 0.8 % tetracycline, viewed optically.

(*a*) Tissue culture for 3 h. The nuclear region of the cells shows heavy labelling, × 200.

(*b*) Detail of a similar region to (*a*) above stained immediately on removal from the animal. The nuclear regions of the osteocytes show varying amounts of labelling. There is a pale fluorescence in the cytoplasm. Two clearly defined juxtanuclear areas are arrowed, × 800.

(*c*) Section from site 5*a* in the peripheral region showing immature bone. The cartilage cells C show positive staining in the nuclei, with less stain in the nuclei of the osteoprogenitor layer O. The thin table of bone contains osteocytes with dense nuclei (arrowed), × 200.

(*d*) Detail from the osteoprogenitor region O in (*c*) above. The developing osteocytes show staining in the nuclear region, often in patches (arrowed). The cytoplasm is not stained, × 800.

(*e*) A wedge from the central region stained with tetracycline and the von Kossa reagent. Viewed in bright field, the cells stain in the immature area and the extra-cellular spaces stain in the mature area, × 110.

(*f*) The same specimen as (*e*) above viewed in ultraviolet light. Only a faint fluorescence is visible at the edge, × 110.

(*g*) The specimen in (*e*) after removal of the silver with saturated potassium ferricyanide solution. The tissue now shows labelling in the cells, × 110.

PLATE I

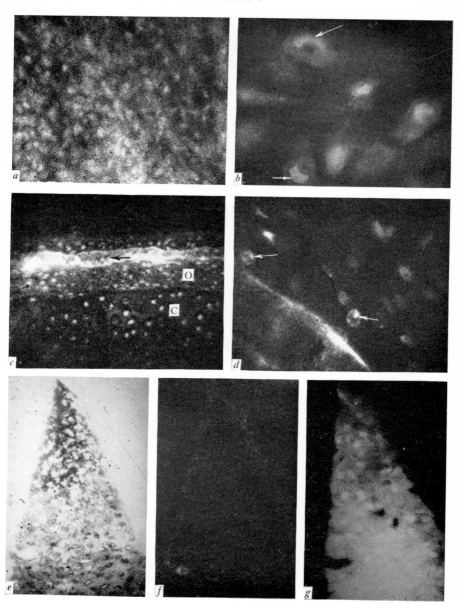

THE EMBRYOSPECIFIC ACTION OF STATHMOKINETIC AGENTS

By J. P. G. WILLIAMS

Department of Growth and Development, Institute of Child Health,
30 Guilford Street, London WC1N 1EH

INTRODUCTION

A stathmokinetic agent is a compound which will inhibit cell division in metaphase (Eigsti & Dustin, 1955). The tropolone derivatives colchicine and colcemid are well known examples. The ability to cause metaphase arrest is also found in several structurally dissimilar compounds such as the vinca alkaloids, vinblastine and vincristine, the antifungal agent griseo-fulvin and podophyllin, an extract of the resin of *Podophyllum* (Dustin, 1963). The stathmokinetic effect is due to disruption of the mitotic spindle preventing separation of the chromosomes at metaphase. The disruption or failure of assembly of the spindle fibres is due to the binding of the stathmokinetic agents to the protein tubulin (Borisey & Taylor, 1967). Under normal conditions tubulin polymerises to form the microtubules which are the component parts of the spindle fibres. The interaction of colchicine and vinblastine with microtubules accounts for several of the other properties of these drugs. Cell locomotion (Bhisey & Freed, 1971) platelet aggregation (Soppitt & Mitchell, 1969) cell adhesion (Kolodny, 1972) and hormone secretion (Williams & Wolff, 1970) are among the activities altered by these compounds which are attributed to their action on microtubules.

The ability of stathmokinetic agents to arrest and thus accumulate cells in metaphase has been used to determine the rates of cellular proliferation in many tissues. The conclusions drawn from experiments on cell pro-liferation are dependent on the drug acting at metaphase and at no other time in the cell cycle (Hooper, 1961). Several reports have stated that colchicine and colcemid do not affect cells during interphase (Taylor, 1965; Kleinfeld & Sisken, 1966). The examination of the effects of colchicine and colcemid in mammalian embryos in-utero however has revealed an effect of stathmokinetic agents which cannot as yet be attributed to the action of these compounds on microtubules.

OBSERVATIONS ON THE EFFECT OF COLCEMID, COLCHICINE AND VINBLASTINE ON METAPHASE ACCUMULATION AND UPTAKE OF [³H]THYMIDINE

In experiments to determine the effect of adult nephrectomy on foetal growth it was planned to use colcemid to accumulate cells in metaphase in foetal rats in-utero. To determine the optimum amount of colcemid to administer to pregnant rats, a dose response experiment was conducted. Rats at 12–13th day of gestation were injected with different concentrations (1–8 mg kg⁻¹) of colcemid; after four hours the animals were killed and the gut of the adults and the brain of the foetuses were fixed and sectioned and the number of metaphase figures counted. The adult gut and the foetal brain were examined because they were the most mitotically active tissues in the respective organisms. In the gut of the adult the same mitotic index was found at all of the dose levels employed, indicating that the lowest dose level was the optimum and was causing a total metaphase block. In the embryos the lowest dose produced the greatest number of metaphase figures and as the dose increased so the number of cells blocked in metaphase diminished (Fig. 1). Both the inverse dose relationship and the difference in adult and foetus were unexpected. Counts of cells in the various stages of mitosis indicated that a complete metaphase block was produced at all dose levels in both foetus and adult. Thus the results could not be attributed to a failure of the drug to act. The reduction of the number of cells in metaphase could only be explained by some prior event reducing the number of cells passing from G_2 to M. Knowing that metaphase arrest was complete, and using figure of 1.5 h for G_2 in foetal rat tissue (Wegener, Hollweg & Maurer, 1963) it was apparent that the block was prior to the onset of G_2. Since the G_1 population did not enter the sample it was concluded that a block to the passage of cells through the cell cycle was occurring in S or at the entry to G_2. This effect occurred only in the foetal tissue and not in the adult tissue (Williams & Carpentieri, 1967).

The effect of colcemid on the incorporation of radioactivity from tritiated thymidine into foetal and adult DNA was examined. Mice at the 12–13th day of gestation were injected with two dose levels of colcemid (1–8 mg kg⁻¹); one hour later [³H]thymidine was administered, and one hour after that the animals were killed. The incorporation of label into DNA was examined. In several adult tissues no significant difference was found between the treated and untreated groups. In the foetus the lower dose level produced a 50% inhibition of DNA synthesis and at the higher dose level the degree of inhibition was greater than 99%. In these experiments the different responses to the drug of adult and foetus were again noted but in

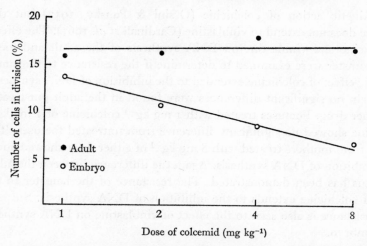

Fig. 1. The effect of 4 h treatment with various dosages of colcemid on the number of cells observed in mitosis.

this experiment DNA synthesis was inhibited in a dose-dependent manner. Thus the effect of colcemid on the accumulation of cells in metaphase in embryos in-utero can be explained by the reduction of DNA synthesis causing the cells to be retained in S (Williams, 1968). Similar results have been found with colchicine.

Vinblastine is another stathmokinetic agent which has a very different structure to colcemid (Dustin, 1963). For instance, the molecular weight of colcemid is 366 while that for vinblastine is 856. Colcemid is a troplone derivative containing two seven-member rings; vinblastine is a polycyclic derivative of indole.

The effect of vinblastine (1 mg kg^{-1} and 8 mg kg^{-1}) on the mammalian embryo has been examined and effects parallel to those of colcemid and colchicine have been found. No significant difference was found in mitotic indices in the adult gut but the number of cells found in metaphase was reduced at high levels of vinblastine in the foetuses. The indices found were lower than was anticipated from labelling experiments and did not correlate with the mitotic indices found with colcemid and colchicine (Carpentieri & Williams, 1968). The incorporation of [³H]thymidine into DNA in the adult gut is not altered by vinblastine but the incorporation into foetal DNA was inhibited by 40 % at 1 mg kg^{-1} and by 96 % at 8 mg kg^{-1} (Williams, 1970). The embryo specific inhibition of DNA synthesis appears to be related to the stathmokinetic property of these compounds.

The Syrian hamster *Mesocricetus aurateus* has a unique resistance to the stathmokinetic action of colchicine (Orsini & Pansky, 1952) but this resistance does not extend to vinblastine (Cardinali *et al.* 1961). The effects of colchicine and vinblastine on DNA synthesis of the adult and foetal Syrian hamster were examined to determine if the resistance to the stathmokinetic effect of colchicine extended to the inhibition of DNA synthesis. Once again no significant differences were found in the adult gut treated with either drug. Foetuses treated with 1 mg kg^{-1} colchicine or 1 mg kg^{-1} vinblastine showed no significant difference from untreated foetuses. The foetuses from mothers treated with 8 mg kg^{-1} of either drug showed about 50% inhibition of DNA synthesis. Again the difference between the adult and foetus has been demonstrated. The resistance of the hamster to the action of colchicine extends to the inhibition of DNA synthesis. Surprisingly, resistance is also seen to the effect of vinblastine on DNA synthesis in the embryo.

DISCUSSION

There are several reports of colchicine affecting DNA synthesis (Sartorelli & Creasey, 1969). Most of these studies are from tissue culture or examination of tissue in-vitro and many suffer from the same fault as do the fewer studies in-vivo. It has been shown above that an inhibition of synthesis in one part of the cell cycle (S) leads to a reduction of the number of cells appearing in another part (M). Colchicine accumulates cells by preventing them going through cell division, thus there will eventually be fewer cells arriving in S and this will lead to a diminution of DNA synthesis. Experiments examining the effect of stathmokinetic agents on DNA synthesis must take less time than the cell spends in G_1, similarly experiments examining metaphase must take less time than the cell spends in G_2 if an effect in S is suspected. With this reasoning in mind it may be said that inhibition of DNA synthesis by colchicine or colcemid has not been demonstrated in-vivo. Studies in-vitro such as those of Hell & Cox (1963) indicate that the DNA synthetic machinery in adult mammals can be influenced by these agents. A recent report by Mizel & Wilson (1971) indicates that inhibition seen in-vitro may be due to an inhibition of thymidine transport. Whatever the mechanism may be concerning the inhibition of DNA synthesis the most difficult phenomenon to explain is its embryo specific nature.

REFERENCES

BHISEY, A. N. & FREED, J. J. (1971). Ameboid movement induced in cultural macrophages by colchicine or vinblastine. *Experimental Cell Research*, **64**, 419–29.

BORISEY, G. G. & TAYLOR, E. W. (1967). The mechanism of action of colchicine. *Journal of Cell Biology*, **34**, 525–33.

CARDINALI, G., CARDINALI, G., HANDLEY, A. H. & AGRIFOGLIO, M. F. (1961). Comparative effects of colchicine and vincaleukoblastine on bone marrow mitotic activity in Syrian hamsters. *Proceedings Society Experimental Biology and Medicine*, **107**, 891–2.

CARPENTIERI, U. & WILLIAMS, J. P. G. (1968). Stathmokinetic agents and the mitotic index in embryo and adult rats. *Currents in Modern Biology*, **2**, 4–6.

DUSTIN, P. JR. (1963). New aspects of the pharmacology of antimitotic agents. *Pharmacological Reviews*, **15**, 449–80.

EIGSTI, O. J. & DUSTIN, P. JR. (1955). *Colchicine in Agriculture, Medicine, Biology and Chemistry*. Ames, Iowa State College Press.

HELL, E. & COX, D. G. (1963). Effect of colchicine and colcemid on synthesis of deoxyribose nucleic acid in skin of the guinea-pig ear *in vitro*. *Nature, London*, **197**, 287–8.

HOOPER, C. E. S. (1961). Use of colchicine for the measurement of mitotic rate in the intestinal epithelium. *Journal of Anatomy*, **108**, 231–44.

KLEINFELD, R. G. & SISKEN, J. E. (1966). Morphological and kinetic aspects of mitotic arrest by, and recovery from colcemid. *Journal of Cell Biology*, **31**, 369–79.

KOLODNY, G. M. (1972). Effect of various inhibitors on readhesion of trypsinised cells in culture. *Experimental Cell Research*, **70**, 196–202.

MIZEL, S. B. & WILSON, L. (1971). Inhibition of thymidine and uridine transport in HeLa cells by colchicine. *Abstracts 11th Annual Meeting of the American Society for Cell Biology*, p. 195.

ORSINI, M. W. & PANSKY, B. (1952). The natural resistance of the golden hamster to colchicine. *Science, New York*, **115**, 88–9.

SARTORELLI, A. C. & CREASEY, W. A. (1969). Cancer chemotherapy. In *Annual Review of Pharmacology*, vol. 9, pp. 51–72. Palo Alto: Annual Reviews Inc.

SOPPITT, G. D. & MITCHELL, J. R. A. (1969). Effect of colchicine on human platelet behaviour. *Journal of Atherosclerosis Research*, **10**, 247–52.

TAYLOR, E. W. (1965). The mechanisms of colchicine inhibition of mitosis. *Journal of Cell Biology*, **25**, 145–60.

WEGNER, K., HOLLWEG, S. & MAURER, W. (1963). Autoradiographische Bestimmung der Dauerder DNS – Verdopplung und der Generationszeit bei fetalen Zellen der Ratte. *Naturwissenschaften*, **50**, 738–9.

WILLIAMS, J. A. & WOLFF, J. (1970). Possible role of microtubules in thyroid secretion. *Proceedings of the National Academy of Sciences, U.S.A.* **67**, 1901–8.

WILLIAMS, J. P. G. (1968). Inhibition of embryonic deoxyribonucleic acid synthesis by colcemid. *European Journal of Pharmacology*, **3**, 337–40.

WILLIAMS, J. P. G. (1970). Selective inhibition of embryonic deoxyribonucleic acid synthesis by vinblastine. *Cell and Tissue Kinetics*, **3**, 155–9.

WILLIAMS, J. P. G. & CARPENTIERI, U. (1967). The different response of embryonic and adult rats to demecolcin. *Life Sciences*, **6**, 2613–20.

TESTOSTERONE-INDUCED CELL PROLIFERATION AND DIFFERENTIATION IN THE MOUSE SEMINAL VESICLE. AN EXPERIMENTAL AND COMPUTER SIMULATION STUDY

BY A. R. MORLEY, N. A. WRIGHT AND D. APPLETON

Department of Pathology, The Medical School,
University of Newcastle upon Tyne

INTRODUCTION

The prostatic complex of the castrated male rodent responds to androgen stimulation by cell proliferation. The major features of this response have been described by Burkhardt (1940) and Allen (1958). Following androgen treatment there is a latent period of 15–20 h, then an increase in DNA synthesis followed by a wave of mitotic activity which reaches a peak at 40–50 h and thereafter declines to near-castrate levels. This response is typical of the models of induced DNA synthesis recently reviewed by Baserga (1971). Such experimental models are valuable because they allow examination of the events preceding cell division. These changes have been extensively reviewed by Cooper (1971). The cell cycle parameters of induced DNA synthesis in the castrate rodent prostate complex have been examined by Tuohimaa & Niemi (1968) who showed that the duration of DNA synthesis was considerably shortened after testosterone treatment. Morley & Wright (1972) have suggested that the fall in mitotic and labelling indices in the latter half of the proliferation response is a result of cells leaving the proliferative cycle. In the present work this concept is incorporated in a computer simulation model. The results of the simulation are consistent with the experimental data, and suggest further investigations to verify the theoretical model.

MATERIALS AND METHODS

Male Balb/c mice aged 3 months were castrated, and used two weeks later. Daily injections of 250 μg testosterone propionate in sesame oil were administered for up to 5 days. Mitotic indices (I_M) were based on a count of 2000 interphase cells. Labelling indices (I_L) were obtained in the same manner. Animals were killed 1 h after the intraperitoneal injection of

[337]

$1 \mu Ci/g$ of tritiated thymidine. Fraction labelled mitoses (FLM) curves were started 24, 48, and 72 h after the first injection of testosterone propionate.

A stochastic model incorporating the following features was developed for an IBM 360/67 computer: (1) The flow of cells from an undifferentiated or G_0 compartment followed an Erlangian distribution; (2) Cells passed through repeated cell cycles composed of G_1, S, G_2, and M compartments; (3) Following mitosis (M) a decycling probability was introduced determining the likelihood of daughter cells remaining in the proliferative compartment. The decycling probability could be fixed or made proportional to either the total number of cells, or the number of differentiated cells. Graphical print-out of simulated values of I_M, I_L (Figs. 1 and 2), growth fraction, FLM curves, I_L curves with continuous labelling and total cell number was obtained.

RESULTS

After a period of 20 h I_L increased to 25% at 35–45 h; and then gradually diminished to 5% by 90 h (see Fig. 1). The increase in I_M occurred later beginning at 30 h, and rising to 1.8% by 48 h. Thereafter I_M fell gradually to 0.4% at 90 h (see Fig. 2). FLM curves at 24, 48, and 72 h gave values of t_2 ($t_{G_2} + \frac{1}{2}t_M$) of 1.6, 2.0, and 1.8 h and t_S of 10.5, 8.0, and 8.0 h respectively. The duration of the cell cycle (T_c) was not measured by the FLM curve at 24 h since the no second peak was apparent, probably a result of dilution by unlabelled mitoses. T_c at 48 h was 17.5 h. Calculation of growth fraction at 48 h gave a value of 0.64. It was not possible to measure T_c from the FLM curve at 72 h, the absence of a second peak can be attributed to the small proportion of cells remaining in the proliferative cycle after 90 h. Since t_2 and t_S appear to remain largely constant throughout the proliferative response it is probable that the fall in I_L and I_M indicates a diminution in the number of cells entering DNA synthesis and mitosis.

The parameters obtained from FLM curves were incorporated in the computer simulation model. Constrained in this way the model produces values of I_M and I_L which are close to those obtained experimentally (continuous lines Figs. 1 and 2). The plateau seen in the experimental I_M and I_L curves is reproduced by the simulation model in which it is postulated that cells pass through successive waves of division (continuous line, Figs. 1 and 2).

The gradual fall in I_M and I_L in the model is a result of postulating that after mitosis each cell has a probability of leaving the proliferative cycle, and this probability rose from 0.75 to 1.0 as cells entered the differentiated compartment. This probability is the same as that described by Bresciani

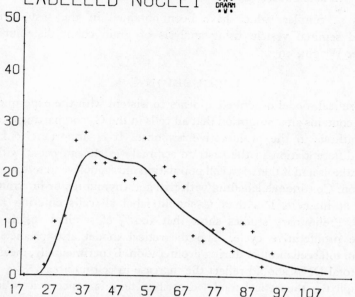

Fig. 1. Percentage labelled cells in the seminal vesicle of the castrate mouse during testosterone propionate treatment at 0, 24, 48, 72 and 96 h. +, experimental observations; −, computer simulation model.

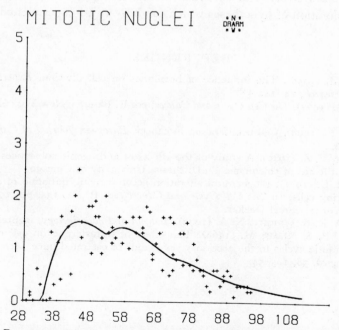

Fig. 2. Percentage mitotic cells in the seminal vesicle of the castrate mouse during testosterone propionate treatment at 0, 24, 48, 72 and 96 h. +, experimental observations; −, computer simulation model.

(1968) as $\overset{\rightarrow}{\eta}$. Similar values have been obtained in the testosterone-stimulated seminal vesicle using analysis of grain count distributions. (Morley & Wright, 1972).

DISCUSSION

The theoretical model described appears consistent with the experimental data, but contains an assumption that all cells in the G_0 compartment enter DNA synthesis in the proliferative response. It is important to know whether differentiation in the castrate seminal vesicle can occur without DNA synthesis and if there is a cell population unresponsive to testosterone stimulation. Continuous labelling with tritiated thymidine, or intermittent labelling at intervals less than t_S should label all cells entering DNA synthesis. Preliminary studies show that 100% of seminal vesicle cells enter the proliferative cycle. The theoretical model also predicts the increase in total cell number during proliferation. Experimentally, measurement of total DNA should reflect the increase in cell number.

Although the theoretical model described here is concerned with the seminal vesicle, preliminary results with the castrate mouse coagulating gland are encouraging. We believe that the model is flexible enough for use in the study of induced DNA synthesis in organs such as the partially hepatectomised liver. The computer model is rapid in operation and allows exploration of hypotheses which can be verified experimentally.

REFERENCES

ALLEN, J. M. (1958). The influence of hormones on cell division. *Experimental Cell Research*, **14**, 142–58.

BASERGA, R. (1971). *The Cell Cycle and Cancer*, ed. R. Baserga. New York: Marcel Dekker.

BRESCIANI, F. (1968). Cell proliferation in cancer. *European Journal of Cancer*, **4**, 343–66.

BURKHARDT, E. Z. (1940). A study of the effects of androgenic substances in the rat by the aid of colchicine. Ph.D. thesis, University of Chicago.

COOPER, H. L. (1971). Biochemical alteration accompanying initiation of growth in resting cells. In *The Cell Cycle and Cancer*, ed. R. Baserga, pp. 197–226. New York: Marcel Dekker.

MORLEY, A. R. & WRIGHT, N. A. (1972). *Journal of Endocrinology*, in press.

TUOHIMAA, P. & NIEMI, M. (1968). The effect of testosterone on cell renewal and mitotic cycles in the accessory sex glands of castrated mice. *Acta Endocrinologica*, **58**, 145–54.

ROLE OF CELL DIVISION AND NUCLEIC ACID SYNTHESIS IN ERYTHROPOIETIN-INDUCED MATURATION OF FOETAL LIVER CELLS IN-VITRO

BY P. R. HARRISON, D. CONKIE AND J. PAUL

Beatson Institute for Cancer Research,
132 Hill Street, Glasgow G3 6UD

Development of mammalian erythroid cells involves a continuous process from an immature precursor cell without detectable haemoglobin to a mature non-nucleated erythrocyte in which most of the cellular protein is haemoglobin. In mammals, the site of erythroid cell differentiation changes as foetal development proceeds: initially, erythroid cells develop as a synchronous cohort in the yolk sac, later as a non-synchronous population in the liver and spleen or, finally, in the bone marrow. In rodents, erythroid cell differentiation in the yolk-sac is associated with a special type of haemoglobin (foetal haemoglobin); in the remaining sites of foetal erythropoiesis only adult haemoglobin is synthesised. This change in type of haemoglobin synthesised is also associated with a dependence of erythroid cell development on the hormone erythropoietin. Thus in many respects, the processes of erythroid cell development in rodent foetal liver or spleen and adult marrow appear to be equivalent.

There is much indirect evidence to show that all recognisable erythroid cells are derived from multi-potential 'stem' cells (colony-forming cells, CFC) which are also the precursors of the granulocytic and thrombocytic cell series (for a recent review, see Lajtha, 1970). These stem cells are mainly in the resting state. There is further evidence (reviewed by McCulloch, 1970; Trentin, 1970; Hodgson, 1970) that in the appropriate environment the CFC may develop into a second, actively-proliferating cell which is committed to differentiate into recognisable erythroid cells in the presence of erythropoietin (erythropoietin-sensitive cell, ESC). Differentiation of ESC in the presence of erythropoietin may occur in G_1 (Kretchmar, 1966). The evidence for CFC and ESC is derived mainly from bone marrow cells. However, Stephenson & Axelrad (1971) and Stephenson, Axelrad, McLeod & Shreeve (1971) have provided clear evidence for the existence of both CFC and ESC in foetal liver. About three-quarters of the cells present in 13.5-day foetal liver are represented by recognisable erythroid cells, of which about 35 % are immature erythroblasts (Paul, Conkie & Freshney, 1969; Tarbutt & Cole, 1970; Chui, Djaldetti, Marks & Rifkind, 1971). By contrast, the proportion of recognisable erythroid cells in bone marrow is much lower; of these, immature erythroblasts represent only about 20%. Thus it is clear that both marrow and foetal liver comprise a very complex series of erythroid cell types which, under the influence of erythropoietin and other factors, develop to produce mature erythrocytes. The situation in marrow is further complicated by the presence of cells of the granulocytic and thrombocytic series.

For many purposes, it is desirable to obtain viable cultures of erythroid cells in-vitro. Techniques for obtaining such cultures have been devised in the case of foetal liver (Cole & Paul, 1966) and bone marrow (Krantz, Gallien-Lartigue & Goldwasser, 1963). It is the intention of this paper to discuss certain experiments performed with such cultures of erythroid cells, relating in particular to the mechanism by which erythropoietin induces differentiation of immature erythroid cells. However, it is imperative to examine critically any conclusions so obtained with regard to their possible relevance to the analogous situation in-vivo.

BIOCHEMICAL EVENTS INDUCED BY ERYTHROPOIETIN

Synthesis of RNA, DNA and haemoglobin

When foetal liver or marrow cells are placed in culture, the rates of synthesis of RNA, DNA and haemoglobin gradually decay. However, addition of erythropoietin allows foetal liver or marrow cells to maintain (or possibly increase) their rates of synthesis of RNA and haemoglobin in culture (Cole & Paul, 1966; Paul & Hunter, 1968, 1969; Krantz & Goldwasser, 1965a, b; Gallien-Lartigue & Goldwasser, 1965). The increase in synthesis of RNA is genuine and not due to increased incorporation into nucleotide pools (Nicol, Conkie, Lanyon, Drewienkiewicz, Williamson & Paul, 1972). Krantz & Goldwasser (1965b) made the interesting observation that the relationship between cell number and the rate of haemoglobin synthesis at the various levels of erythropoietin is consistent with a co-operative effect among the erythropoietin-sensitive cells. The uptake of labelled glucosamine into stroma by marrow cells is also stimulated by erythropoietin (Dukes & Goldwasser, 1965).

Paul & Hunter (1968, 1969) and Gross & Goldwasser (1970) also found that addition of erythropoietin to foetal liver or marrow cultures increased the rate of incorporation of thymidine into DNA. Under conditions identical to those employed by Paul & Hunter (1969) to measure the incorporation of labelled thymidine into DNA (i.e. in the presence of 10^{-4} M FUdR and 10^{-4} M thymidine, when de-novo synthesis of DNA

nucleotides is suppressed), the stimulation by erythropoietin of incorporation of radioactivity into either total acid-soluble pools or nucleotide pools
(isolated according to the method described by Harrison (1968) by
chromatography on DEAE–cellulose) represents only about 30% of the
extent of stimulation of thymidine incorporation into DNA (Conkie &
Harrison, unpublished results). Therefore, under these conditions stimulation of synthesis of DNA by erythropoietin is genuine. However, Djaldetti,
Preisler, Marks & Rifkind (1972) have not been able to confirm this effect
of erythropoietin on DNA synthesis on foetal liver cultures during the
first six hours of culture. Nevertheless, the same group of workers have
found that erythropoietin prevents the decay in rate of DNA synthesis
during culture for 10–24 h which occurred in its absence (Chiu *et al.* 1971).

Sequence of events after erythropoietin stimulation

Paul & Hunter (1969) observed that the erythropoietin-induced increase
in DNA synthesis in their foetal liver cultures preceded a stimulation of
synthesis of total RNA and haemoglobin, which occurred at similar times.
This led to the idea that DNA synthesis might be a prerequisite for
haemoglobin synthesis in these cultures. This was supported by the finding
that addition of 5′-fluorodeoxyuridine (FUdR) to the cultures prevented
the erythropoietin-induced increase in haemoglobin synthesis 6 h later,
and that the presence of thymidine reversed this effect of FUdR (Paul &
Hunter, 1968, 1969). The latter result suggested that the effect of FUdR
was due to inhibition of DNA synthesis *per se*. This correlation between
erythropoietin-induced DNA and haemoglobin syntheses has been confirmed with other inhibitors of DNA synthesis (Paul & Conkie, manuscript
submitted for publication). In the case of marrow cultures, Gross &
Goldwasser (1970) and Ortega & Dukes (1970) were able to demonstrate
inhibition by various DNA inhibitors of erythropoietin-induced haemoglobin synthesis after 24 h but not within the first four hours of treatment.
Some of the results may be due to cells being damaged during the DNA
synthetic phase by the addition of certain inhibitors (particularly hydroxy
urea), rather than due to inhibition of DNA synthesis *per se*. Nevertheless,
the results taken as a whole indicate that erythropoietin-induced DNA
synthesis (or passage of cells through S phase) is a prerequisite for induced
haemoglobin synthesis.

A choice between these alternative interpretations might be possible on
the basis of whether cell division is required in order to allow induction of
haemoglobin synthesis by erythropoietin. Paul & Hunter (1969) were able
to demonstrate that, under conditions in which colchicine reduced the

344 P. R. HARRISON, D. CONKIE AND J. PAUL

erythropoietin-induced increase in numbers of foetal liver cells, no reduction in induced haemoglobin synthesis was detected after 24 h, when the rate of induced haemoglobin synthesis was maximal. These results would suggest that cell division was not required for induced haemoglobin synthesis. On the other hand, Gross & Goldwasser (1970) were able to demonstrate a large inhibition by colchicine of erythropoietin-induced haemoglobin synthesis in marrow cultures after 24 h. However, after treatment for four hours with colchicine, these authors could not detect any effect of colchicine on erythropoietin-induced haemoglobin synthesis. Thus the difference between these results and those of Paul & Hunter may simply reflect the difference in cell cycle parameters in the two culture conditions. It is interesting to note that Dukes & Goldwasser (1965) could not demonstrate any effect of colchicine on erythropoietin-induced incorporation of labelled glucosamine into stroma under the conditions in which induced-haemoglobin synthesis was inhibited.

Paul & Hunter (1969) also found that addition of FUdR to foetal liver cultures inhibited totally the erythropoietin-induced increase in total RNA synthesis. More recent work (Nicol *et al.* 1972) has shown that this effect is variable, depending on the proportion of hepatocytes in culture and other factors. Paul & Conkie (manuscript submitted for publication) have evidence obtained with other inhibitors that erythropoietin-induced synthesis of RNA is partially inhibited when DNA synthesis is inhibited completely. Gross & Goldwasser (1970) and Djaldetti *et al.* (1972) have not been able to confirm this dependence of erythropoietin-induced increase in RNA synthesis on DNA synthesis either in marrow or foetal liver cultures, using a variety of inhibitors of DNA synthesis. This discrepancy has been discussed in some detail in a previous report by Paul, Freshney, Conkie & Burgos (1972).

The dependence of haemoglobin synthesis on DNA synthesis (or passage of cells through S phase) is apparently a link in a sequence of events induced by erythropoietin. Both in foetal liver cultures (Paul & Hunter 1969) and marrow culture (Gross & Goldwasser, 1970), the acceleration in erythropoietin-induced DNA synthesis is inhibited by puromycin and actinomycin D. Haemoglobin synthesis in both untreated and erythropoietin-stimulated cultures is also inhibited by actinomycin D in the case of 13.5-day foetal liver cultures (Paul & Hunter, 1969; Djaldetti, Chui, Marks & Rifkind, 1970) and marrow cultures (Gallien-Lartigue & Goldwasser, 1965). These effects of actinomycin D are only observed if it is added prior to erythropoietin (Paul & Hunter, 1969; Gallien-Lartigue & Goldwasser, 1965). This suggests that in these cultures the initial event after addition of erythropoietin is the transcription of minor components of RNA followed by early protein synthesis. This contrasts with the situation in 15-day foetal liver cultures, in which haemoglobin synthesis is resistant to actinomycin D. This change in stability of the haemoglobin synthetic apparatus appears not to be due to different proportions of individual erythroid types in the two cultures, but rather to some change in the foetal environment (Djaldetti et al. 1970). Freshney & Paul (1971) have concluded that three enzymes which may have a regulatory role in haem synthesis are not rate-limiting steps during the increase in haemoglobin synthesis between 14 and 15 days. Tarbutt & Cole (1970) have suggested, on the basis of kinetic data, that there are two distinct populations of erythroblasts in the liver, one superseding the other at about the 15th day.

Some direct evidence for early transcription of RNA after erythropoietin treatment of marrow or 13.5-day foetal liver cultures is available (Krantz, & Goldwasser, 1965a; Gross & Goldwasser, 1970; Djaldetti et al. 1972). Provided degradation of RNA during isolation is eliminated, synthesis of heterogeneous RNA with very high sedimentation coefficient (150S) is induced within 15 min after erythropoietin treatment (Gross & Goldwasser, 1969). This RNA component is considered not to be due to aggregation of small RNA species and may therefore represent the early transcription product whose existence had been inferred from the work with inhibitors discussed previously. At later times after addition of erythropoietin to both foetal liver (Nicol et al. 1972) and marrow culture (Gross & Goldwasser, 1969) the synthesis of a whole series of RNA species is stimulated, including 9S RNA (Gross & Goldwasser, 1971). This 9S RNA component may be the messenger RNA specifying globin or another unrelated protein (e.g. histone), as is the case with 9S RNA isolated from yolk-sac erythroid cells (Terada, Banks & Marks, 1971).

Thus, to conclude this section of the discussion, all the available evidence from both early foetal liver and marrow cultures indicates the following sequence of events in induction of haemoglobin synthesis by erythropoietin: (1) early transcription of heterogeneous RNA; (2) translation of necessary protein; (3) DNA synthesis; (4) 9 S RNA synthesis, possibly the messenger RNA for globin; (5) haemoglobin and stroma synthesis. Associated with this sequence of events is a general increase in cellular RNA synthesis and cell numbers.

CELL POPULATION STUDIES

A clear difficulty in interpreting all the data presented previously is that of associating given biochemical changes with the precise cell types in which they occur. Thus events which appear to be causally related in the whole culture may in fact be due to unrelated and independent events in separate cell types. It is possible to avoid this difficulty by cytological studies combined with autoradiography in order to measure the rates of synthesis of macromolecules in recognisable cells.

Plate 1 illustrates the various types of erythroid cells present in foetal liver cultures. After incubation for 24 h in the presence of erythropoietin, certain atypical mature erythroid cells are formed. These atypical cells are similar to those observed in-vivo in macrocytic (Wintrobe, 1961) or megaloblastic (Killmann, 1970) anaemia. Polychromatic macroerythroblasts resemble basophilic erythroblasts in size and chromatin condensation, but polychromatic erythroblasts in nucleus to cytoplasm ratio and haemoglobinisation of cytoplasm. Orthochromatic macroerythroblasts may also be as large as basophilic erythroblasts but otherwise resemble orthochromatic erythroblasts in nucleus to cytoplasm ratio and haemoglobinisation of cytoplasm. Nuclear condensation in orthochromatic macroerythroblasts is rarely as complete as in orthochromatic erythroblasts. After extrusion of the nucleus, the orthochromatic macroerythroblast becomes a macrocyte of up to 12 μm in diameter, as compared with about 8 μm for the normal hepatogenic reticulocyte. The formation of these atypical cells is observed with either relatively impure human urinary erythropoietin (0.6 u/ml) or highly purified sheep erythropoietin (Step III, 0.2 u/ml), and when the culture medium is supplemented with vitamin B_{12}, folic acid, folinate and ferric chloride (either alone, or bound to homologous transferrin).

The changes in distribution of the major groups of foetal liver cells during culture are shown in Fig. 1. During culture for 24 h, the total number of cells remains approximately constant; there is slight maturation of immature cells which leads to a slight accumulation of orthochromatic erythroblasts, reticulocytes and erythrocytes. It should be noted that these estimates of the number of reticulocytes and erythrocytes formed in culture are likely to be underestimates, in view of the fragility of these cells. Few atypical macroerythroblasts are detected. However, in the presence of erythropoietin, the number of cells increases by about 50 %; a large number of macroerythroblasts are formed (often about 30 % of the total population after 24 h), accompanied by a reduction in proerythroblasts and an increase in basophilic and polychromatic erythroblasts. Chui et al. (1971) have made similar measurements of foetal liver cells in culture. These authors have not reported the formation of the atypical macroerythroblasts in cultures containing erythropoietin. However, the overall increase in cell numbers and the general maturation of immature cells in the presence of erythropoietin are in agreement with the present results.

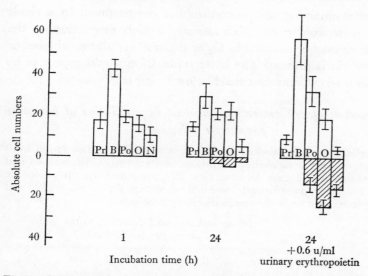

Fig. 1. Cell population changes in total foetal liver cell cultures. Pr, proerythroblasts; B, basophilic erythroblasts; Po, polychromatic compartment; O, orthochromatic compartment; N, non-nucleated cells. Unshaded regions indicate typical cells; shaded regions, atypical macroerythroblasts or macrocytes.

Kinetics of erythroid cell maturation

The DNA labelling index of proerythroblasts is maintained for at least eight hours after being placed in culture at about 85 %, as compared with a value of 60 % for all erythroblasts. After culture in the presence of erythropoietin, the corresponding labelling indices of proerythroblasts and total erythroblasts are increased to about 95 % and 70 % respectively. Chui et al. (1971) have reported similar high labelling indices for proerythroblasts and erythroblasts in foetal liver cell cultures, although the slight increase in labelling index in the presence of erythropoietin was not observed. Tarbutt & Cole (1970) have reported labelling indices of the combined proerythroblast and basophilic erythroblast compartment in 13–16-day foetal mouse liver in-vivo to be about 75 %. In rat marrow cells in-vivo, Blackett (1968) has shown that the labelling indices of proerythroblasts increases from 65 % in a normal animal to a value of 80 % in protracted anaemia. Thus the results obtained in cultured systems appear to be very similar in this respect to those reported for the living animal.

Studies with cine time-lapse photography have shown that shortly after addition of erythropoietin, a considerable fraction of the proerythroblasts in G_2 or S phase is induced to enter mitosis (Burgos, 1971; Paul et al. 1971). Thereafter, the mitotic index falls to zero. This may be due to the difficulties of maintaining the proerythroblast compartment in a viable state under these conditions. Alternatively, a high proportion of the proerythroblasts may be arrested in G_2 or S phase in culture, whether or not erythropoietin is present. The latter explanation would appear to be unlikely in view of evidence discussed below.

Table 1. *Numbers of cell generations involved in maturation of immature foetal liver cells*

Foetal liver cells were pulse-labelled for 30 min with [³H]thymidine (10 μCi/ml) in the presence of 10^{-4} M unlabelled thymidine and 10^{-4} M FUdR, washed and incubated in normal medium for the given time. Pr, proerythroblasts; B, basophilic erythroblast; Po and O, polychromatic and orthochromatic erythroblasts; PM and OM, polychromatic and orthochromatic macroerythroblasts.

Duration of culture (h)	Treatment	Mean and standard deviation of the grain count per labelled cell					
		Pr	B	Po	PoM	O	OM
1	Control	53 ± 19	18 ± 9	16 ± 9	—	—	—
	Erythropoietin	53 ± 16	19 ± 10	19 ± 7	—	—	—
4	Control	36 ± 11	18 ± 11	13 ± 6	—	—	—
	Erythropoietin	37 ± 16	20 ± 10	11 ± 7	—	—	—
6	Control	36 ± 15	19 ± 11	16 ± 6	—	—	—
	Erythropoietin	27 ± 16	18 ± 13	12 ± 4	—	—	—
24	Control	20 ± 13	15 ± 7	13 ± 6	20 ± 5	7 ± 6	14 ± 12
	Erythropoietin	20 ± 9	18 ± 8	13 ± 6	18 ± 5	5 ± 3	13 ± 1
48	Control	—	17 ± 7	11 ± 4	15 ± 3	6 ± 4	8 ± 4
	Erythropoietin	—	8 ± 3	12 ± 5	17 ± 5	10 ± 6	15 ± 6

Attempts to elucidate the kinetics of erythroid cell population changes have been made in the following way (Paul, Conkie & Burgos, manuscript submitted for publication). Immature cells in the initial culture were first pulse-labelled with thymidine; after resuspending the labelled cells in medium containing only unlabelled thymidine, the effect of erythropoietin on the maturation of the labelled immature cells was monitored in subsequent culture. The results of such experiments are given in Table 1. From the reduction in grain count of the proerythroblast compartment, it can be deduced that at least a fraction of these cells proliferate in culture, and that this proliferation is stimulated by the addition of erythropoietin. This fact is readily apparent when the frequency distribution of grain counts in proerythroblasts is considered (Fig. 2). This shows clearly that incubation of proerythroblasts in the presence of erythropoietin causes accumulation of proerythroblasts with very low grain counts. These results can only be explained on the assumption that erythropoietin causes at least 50% of the

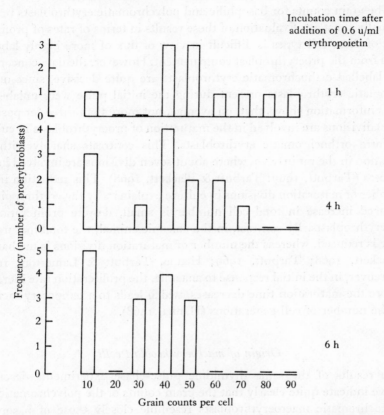

Fig. 2. Proliferation of the proerythroblast compartment in total foetal liver cultures. Data given in Fig. 1 are analysed regarding the change in grain count distribution of proerythroblasts during culture in the presence of 0.6 u/ml erythropoietin.

proerythroblasts to proliferate rapidly during the first six hours in culture. About 10 % of the proerythroblast compartment appears to undergo two divisions within the first six hours in culture in the presence of erythropoietin. This implies an extremely short cell cycle time in this fraction of the proerythroblast compartment. Thereafter, the rate of proliferation appears to decrease. This is thought unlikely to be due to radiation damage, since the specific activity of the labelled thymidine was very low (100 μCi/μmole). The increase in proliferation of proerythroblasts in-vitro in the presence of erythropoietin is similar to the situation observed in rat marrow during protracted anaemia (Blackett, 1968), and in spleens of polycythemic mice after erythropoietin treatment (Orlic, Gordon & Rhodin, 1965).

The grain counts for basophilic and polychromatic erythroblasts remain virtually constant. Evaluation of these results in terms of rates of proliferation of these cell types is difficult in view of flux of more highly labelled cells from the proerythroblast compartment. However, the data concerning the labelled orthochromatic erythroblasts are quite decisive, since orthochromatic erythroblasts present during the initial pulse were unlabelled. This information shows that, on average, not more than three (or possibly four) divisions are involved in the maturation of proerythroblasts in culture to form orthochromatic erythroblasts. This contrasts sharply with the situation in the rat in-vivo, where about seven divisions are involved in this process (Tarbutt, 1967; Tarbutt & Blackett, 1968). This reduction in the number of maturation divisions in culture explains why the erythropoietin-induced increase in total cell number is small, despite proliferation of proerythroblasts. By contrast, in the anaemic animal, the total maturation time is reduced, whereas the number of maturation divisions is unchanged (Blackett, 1968; Tarbutt, 1969; Hanna, Tarbutt & Lamerton, 1969). Moreover, in the initial response to anaemia, the proliferation rate increases before the maturation time decreases, which leads to a temporary increase in the number of cell generations (Hanna, 1968).

Origin of macroerythroblast cells

The results of the same thymidine pulse-chase experiments described above indicate quite clearly that the grain counts of the polychromatic and orthochromatic macroerythroblasts resemble closely those of basophilic and polychromatic erythroblasts respectively (Table 1). This suggests that orthochromatic and polychromatic macroerythroblasts represent the same generation of cells as the polychromatic and basophilic erythroblasts respectively. It would also be consistent with the above grain count data if the polychromatic macroerythroblasts were related to the component of the proerythroblast compartment which has undergone rapid proliferation. Although not yet proven, it is likely that the macroerythroblast cells contain tetraploid amounts of DNA. This could be due to the fact that the macroerythroblast cells are formed from immature cells arrested in late S or G_2. This would imply that maturation of the haemoglobin synthesising mechanism in these cells could occur without cell proliferation. This would not be consistent with the observation that formation of macroerythroblast cells is inhibited by inhibitors of DNA synthesis. Alternatively, it is possible, as suggested by Killmann (1970), that the acceleration of maturation by erythropoietin relative to flow through the cell cycle leads to a certain proportion of basophilic or polychromatic erythroblasts containing polychromatic or orthochromatic characteristics before the next mitosis is reached. This interpretation implies that at least one division has been omitted along the macroerythroblast pathway relative to the typical maturation. In view of evidence discussed previously, this means that not more than two generations elapse in formation of the orthochromatic macroerythroblast from the proerythroblast in culture.

Biochemical events in recognisable cell types

Synthesis of nucleic acids

One of the initial events which may be observed after addition of erythropoietin to foetal liver cultures is an increase in incorporation of thymidine into DNA (Paul, Conkie & Burgos, manuscript submitted for publication) or of uridine into RNA after a 15 or 5 min pulse-labelling (Djaldetti *et al.* 1972; Conkie & Harrison, unpublished data) in the proerythroblast compartment. However, with longer labelling times, stimulation of incorporation of uridine into RNA is only detected at longer intervals after erythropoietin addition, when the labelling of all immature cells is increased (Nicol *et al.* 1972). It should be emphasised that it is at present difficult to measure the effects of erythropoietin on incorporation of labelled nucleosides into the nucleotide pools of individual cell types. Therefore these studies on incorporation of thymidine or uridine into individual cell types should be interpreted with caution in this respect.

Rate of haemoglobin synthesis

A crucial question in elucidating the mechanism of action of erythropoietin is to determine whether erythropoietin affects the rate of synthesis of haemoglobin in individual cells or rather the numbers of haemoglobin-synthesising cells. In the total cell population, an erythropoietin-induced increase in total cell number of about 50–60% is to be compared with a 150–250% increase in haemoglobin synthesis, measured either by incorporation of ^{59}Fe into haem (Paul, Conkie & Burgos, manuscript submitted for publication) or incorporation of labelled leucine into isolated globin chains (Chui *et al.* 1971). Both these groups of workers have demonstrated by autoradiographic methods that the incorporation of labelled leucine into polychromatic or orthochromatic erythroblasts is not increased in erythropoietin-treated cultures. Thus, neglecting amino acid pool effects and assuming that all the leucine incorporation into protein in these cells represents globin synthesis, these results show that erythropoietin does not increase the rate of globin synthesis in individual typical cells. However, the incorporation of leucine into polychromatic or orthochromatic macroerythroblasts is double that into the corresponding erythroblasts. This is compatible with the concept that these macroerythroblast cells are formed by omission of one division during development. By combining these results with those concerning cell population changes discussed earlier, it can be shown that the erythropoietin-induced increase in leucine incorporation into the total polychromatic and orthochromatic compartments (about 140%) is similar to the increase in haemoglobin synthesis in the whole culture, as measured by leucine incorporation into isolated globin chains (about 150% – Chui *et al.* 1971), or by measurement of ^{59}Fe into protein-bound haem (175% – Paul, Conkie & Burgos).

STUDIES WITH ISOLATED ERYTHROID CELL TYPES

It has become increasingly clear that there is a limit to the biochemical information which may be obtained concerning differentiation of foetal liver cells in culture by classical autoradiography. Thus we have attempted to separate certain erythroid cell types by unit gravity sedimentation, according to the method developed by Miller & Phillips (1969). The results of a typical cell separation are shown in Fig. 3. It is evident that, as would be predicted on theoretical grounds, the method separates the various immature cell types better than the more mature members of the series. A further important point is that any colony-forming or erythropoietin-sensitive 'stem' cells would probably sediment in the region of basophilic erythroblasts and proerythroblasts respectively (Stephenson & Axelrad, 1971; McCool, Miller, Painter & Bruce, 1970; Worton, McCulloch & Till, 1969). However, a considerable fraction of the total foetal liver cells (30–40%) routinely sediment through the gradient very rapidly due to aggregation; these cells contain an enriched proportion of proerythroblasts. Attempts to eliminate disaggregation have so far been unsuccessful, despite changes in sedimentation fluid, temperature and serum. It may reflect some form of functional co-operation among immature cells in-vivo (Orlic, Gordon & Rhodin, 1965).

The role of cell division

The first question which has been studied using this technique has been whether the macroerythroblast cells (in particular the polychromatic macroerythroblasts) arise without division from a specific typical cell type. Fractions of the unit-sedimentation gradient were pooled so as to give a proerythroblast or basophilic erythroblast-rich fraction, with few poly-chromatic erythroblasts and virtually no cells of mature type. Aliquots of these fractions were then pre-treated with colchicine or colcemid in an

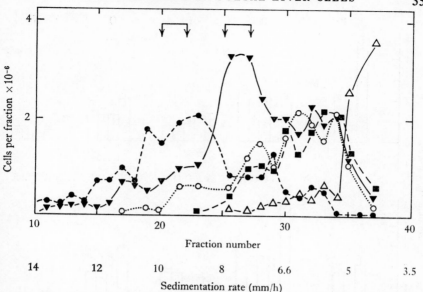

Fig. 3. Separation of foetal liver cells by unit gravity sedimentation. The method described by Miller & Phillips (1969) was used, except that the sedimentation was performed at 20° in a 14.7 cm diam. chamber. About 3×10^8 cells were loaded in 50–100 ml medium containing 3 % foetal bovine serum under a 50 ml Hanks balanced-salt solution and above a steep, convex 50 ml 5–15 % serum gradient followed by a linear 1200 ml 15–30 % serum gradient. Cells were allowed to sediment for 4 h, before the chamber was unloaded. The proerythroblast-rich and basophilic erythroblast-rich fractions (arrows) were pooled, sedimented and resuspended in complete Waymouth's medium. Specimens for cytology were then taken. For routine work, the required cell fractions were chosen on the basis of measurements of cell volume made with a Coulter size-distribution plotter, and then checked subsequently by cytology. ●, proerythroblasts, ▼, basophilic erythroblasts; ○, polychromatic erythroblasts; ■, orthochromatic erythroblasts; △, reticulocytes and erythrocytes.

attempt to inhibit cell division, before incubating for 24 h either with or without erythropoietin. The intention was to discover whether macroery-throblasts were formed from either of the purified immature cell fractions without division. The combined results of three such experiments are summarised in Fig. 4. The following conclusions may be made:

(1) Proerythroblasts disappear in culture but are maintained to a large extent by the presence of erythropoietin, unless colchicine or colcemid is also present. Moreover, in the colcemid-treated cultures the proportion of cells blocked in metaphase (1–2 %) is substantially lower than in the erythropoietin and colcemid-treated cultures (6–13 %). These facts are very strong evidence that erythropoietin maintains a considerable proportion of the proerythroblast population by enhancing its proliferation rate.

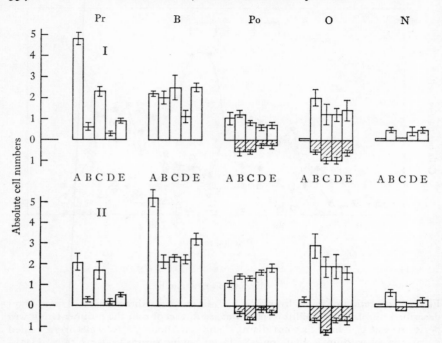

Fig. 4. Effect of colchicine or colcemid on maturation of proerythroblast-rich (I) or basophilic erythroblast-rich (II) fractions. Cell fractions were obtained as described in Fig. 3. Key for cell types is as in Fig. 1. A, initial culture; B, 24 h culture; C, 24 h culture + erythropoietin (human urinary, 0.6 u/ml); D, 24 h culture + colchicine or colcemid (10^{-4} M); E, 24 h culture + erythropoietin and colchicine/colcemid. Colchicine (2 experiments) or colcemid (one experiment) was added 1 h before erythropoietin.

(2) Immature cells mature to an identical extent whether erythropoietin is present or not, and this typical maturation is only slightly affected by the presence of the inhibitors. This may be due to inability of colchicine or colcemid to block division in mature erythroid cells.

The loss of proerythroblasts and the extent of maturation of immature cells in culture in the absence of erythropoietin is somewhat greater than that observed in total cell cultures (cf. Fig. 1). Various explanations of this difference may be suggested: (a) Slightly different culture conditions obtain (the total cell cultures were buffered with bicarbonate, in contrast to the use of HEPES in the cultures of isolated cell fractions); (b) In total cell cultures, mature cells may inhibit the proliferation and/or maturation of immature cells; in the isolated cell fractions the proportion of mature cells is very low and this influence would not operate; (c) Immature cells may be 'triggered' to mature by erythropoietin in the serum used to form the gradient; this is probably an unlikely possibility since the same phenomenon occurs when the sedimentation is performed at 4°; (*d*) The proerythroblast compartment in total cultures is maintained by differentiation of stem cells; in isolated cell fractions, stem cells may be removed and therefore maintenance would not occur.

(3) Formation of both the atypical polychromatic and orthochromatic macroerythroblasts is enhanced to about the same extent by erythropoietin; moreover, this enhancement is inhibited by colchicine or colcemid. Thus the effects of erythropoietin and colchicine on proliferation of proerythroblasts and formation of macroerythroblasts are very similar. This correlation, which will be referred to again in connection with studies with 5'-bromodeoxyuridine (BUdR), is an unexpected result since macroerythroblasts appear to be derived from immature erythroid cells by fewer divisions than is necessary in the case of typical mature cells. Possibly inhibition of the proliferation of proerythroblasts reduces the precursors of the atypical macroerythroblasts; maturation of these precursors may not be inhibited by colchicine or colcemid, as appears to be the case with the typical cell series.

(4) Macroerythroblasts appear to be formed to more or less equal extents from both proerythroblast-rich and basophilic erythroblast-rich fractions. This would suggest that at least basophilic erythroblasts are involved in atypical maturation. It is possible that a proportion of the basophilic erythroblasts proliferate in the presence of erythropoietin (as do proerythroblasts), but that this is undetected in these experiments owing to the flow of proerythroblasts and polychromatic erythroblasts into and out of the basophilic erythroblast compartment.

Haemoglobin synthesis and cell maturation

In both proerythroblast- and basophilic erythroblast-rich fractions, culture for 24 h in the presence of erythropoietin enhances considerably the rate of incorporation of ^{59}Fe into protein-bound haem. Furthermore, addition of colchicine to the culture eliminates this effect of erythropoietin in the case of the proerythroblast-rich fraction; in the basophilic erythroblast-rich fraction, the effect of colchicine is not as marked (Table 2). These results are identical to those obtained by McCool *et al.* (1970) who studied the effects of erythropoietin and vinblastine on bone marrow cell fractions separated by unit-gravity sedimentation. The simplest interpretation of these results is that a greater number of cell divisions are required during maturation of proerythroblasts than of basophilic erythroblasts.

An attempt has been made to correlate these experimental results on rate of haemoglobin synthesis (strictly haemoprotein synthesis) with those

Table 2. *Observed and predicted rates of synthesis of haemoglobin in erythro-poietin- and colchicine-treated foetal liver cell fractions*

Observed results are those obtained for incorporation of ^{59}Fe into acid/butanone-extracted haem. Predicted values are calculated as described in text. Results relating to the three experiments described in Fig. 3 are pooled. Results are normalised relative to the 24 h control value.

		Proerythroblast-rich fraction		Basophilic erythroblast-rich fraction	
		Experimental	Predicted	Experimental	Predicted
Initial		115 ± 30	25 ± 10	175 ± 60	28 ± 1
24 h	—	100	100	100	100
	+ colchicine	84 ± 3	105 ± 20	79 ± 4	90 ± 10
	+ erythropoietin	245 ± 35	77 ± 3	185 ± 35	104 ± 3
	+ colchicine and erythropoietin	93 ± 5	75 ± 6	135 ± 25	100 ± 12

predicted on the basis of the observed population changes (Fig. 3) and the rates of synthesis of globin chains in individual mature erythroid cell types (measured autoradiographically by leucine incorporation, and described previously). These predicted values assume that there is no haemoglobin synthesis in proerythroblasts and basophilic erythroblasts. It is clear from Table 2 that the predicted and observed values are not in good agreement in the case of initial and erythropoietin-treated cultures. These discrepancies are readily explained on the following bases: (1) immature cells (particularly proerythroblasts) may incorporate ^{59}Fe in the absence of protein synthesis, due to the pool of unsaturated apoferritin in these cells (Shepp, Toff, Yamada & Gabuzda, 1972); (2) a pool of free globin chains may exist in mature cells, as in reticulocytes (Felicetti, Colombo & Baglioni, 1966); alternatively, haem–haem exchange may occur (Bunn & Jandl, 1966). In view of these considerations interpretation of the effects of colchicine on erythropoietin-stimulated incorporation of ^{59}Fe into haem must be interpreted with caution.

Maturation and DNA synthesis

An interesting observation discussed previously is that inhibition of DNA synthesis in total cultures of foetal liver cells prevents the erythropoietin-induced increase in haemoglobin synthesis. It is therefore important to identify the cell type in which DNA synthesis is required. Consequently an investigation has been made of the effect of cytosine arabinoside (which is a very effective inhibitor of DNA synthesis under the conditions used) on maturation of proerythroblast-rich and basophilic erythroblast-rich

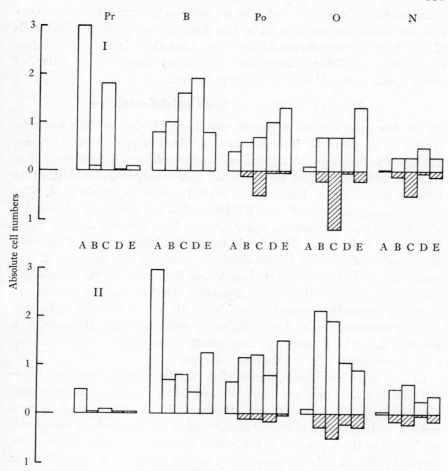

Fig. 5. Effect of cytosine arabinoside on maturation of proerythroblast-rich (I) or basophilic erythroblast-rich (II) fractions. Details as in Fig. 4. A, initial culture; B, 24 h culture; C, 24 h culture + erythropoietin; D, 24 h culture + cytosine arabinoside (40 μg/ml); E, 24 h culture + erythropoietin and cytosine arabinoside. Cytosine arabinoside was added 1 h before erythropoietin.

fractions of foetal liver cells. The design of such experiments was analogous to those described in the case of colchicine. The results (Fig. 5) show clearly that maintenance of the proerythroblast compartment and formation of atypical mature cells in the presence of erythropoietin are both abolished by cytosine arabinoside. It is not clear whether typical maturation is significantly reduced in the presence of the drug. On the basis of the evidence for the involvement of cell division in typical maturation in culture (Table 1), some inhibition of typical maturation by cytosine arabinoside would be expected. The present results with cytosine arabinoside are somewhat complicated by the distorted morphology of the cells after culture in the presence of the drug. Thus there is some uncertainty regarding quantitative (rather than qualitative) changes in numbers of recognisable erythroid cells in this experiment.

Effects of BUdR on erythroid cell differentiation

BUdR has been shown to inhibit erythropoietin-induced haemoglobin synthesis, measured by incorporation of ^{59}Fe into protein-bound haem, and also the formation of atypical macroerythroblasts in total foetal liver cultures. However it has no effect on erythropoietin-induced RNA or DNA synthesis or on formation of typical mature erythroid cells. This effect is analogous to the effect of BUdR in inhibiting specifically expression of differentiated functions in other systems (e.g. Koyama & Ono, 1971; Coleman & Coleman, 1968; Mayne, Sanger & Holtzer, 1971). This raises the question of whether BUdR exerts its effect on any specific cell type during erythroid cell differentiation. Autoradiographic studies have shown that tritium-labelled BUdR is incorporated into immature erythroid cells in exactly the same manner as thymidine with respect to labelling indices, rates of incorporation and stimulation by erythropoietin. However, it is possible that BUdR, although incorporated into all immature erythroid cells, nevertheless affects a specific cell type.

In fact, there is evidence, derived from studies with isolated foetal liver cell fractions, that this is the case (Fig. 6). Many features of these experiments (e.g. proliferation of proerythroblasts in the presence of erythropoietin, and formation of typical mature cells in its absence) are completely analogous to those previously described. In these particular experiments, atypical maturation in the presence of erythropoietin was more advanced in the basophilic erythroblast-rich fraction. Moreover, in the latter fraction of cells, atypical maturation was not markedly erythropoietin-dependent, in contrast to the proerythroblast-rich fraction. This difference is not usually observed (cf. Fig. 4); it may be detected on account of the particularly effective separation of the two cell fractions achieved in this experiment.

Of particular interest was the finding that BUdR prevented the erythropoietin-induced maintenance of proerythroblasts. BUdR also reduced the formation of macroerythroblasts from the proerythroblast-rich fraction, but not from the basophilic erythroblast-rich fraction. These effects of BUdR were fully apparent at a concentration of 10^{-4} M; at 10^{-5} M BUdR was only partly effective. In contrast, formation of typical mature cells was not affected significantly by BUdR. This would suggest that the observed effects of BUdR were not due to its toxicity.

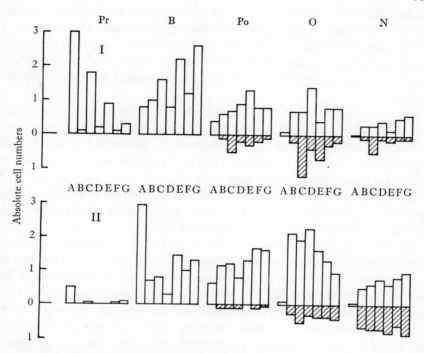

Fig. 6. Effect of BUdR on maturation of proerythroblast-rich (I) or basophilic erythroblast-rich (II) fractions. Details as Fig. 4. A, initial culture; B, 24 h culture; C, 24 h + erythropoietin; D, 24 h culture + 10^{-5} M BUdR; E, 24 h culture + erythropoietin and 10^{-5} M BUdR; F, 24 h culture + 10^{-4} M BUdR; G, 24 h culture + erythropoietin and 10^{-4} M BUdR. BUdR was added 1 h before erythropoietin.

CONCLUSIONS AND FUTURE POSSIBILITIES

From the various results discussed in this chapter, certain general conclusions emerge:

(1) Erythropoietin induces proliferation of the proerythroblast compartment. This is inhibited by colchicine, colcemid, cytosine arabinoside and BUdR. The way in which BUdR acts remains an intriguing mystery.

(2) Erythropoietin also induces an increased rate of haemoglobinisation relative to the frequency of maturation divisions. This leads to the formation of an atypical erythroid cell series in which a given extent of haemoglobinisation is apparent one generation earlier than in the typical series. This erythropoietin-induced increase in rate of haemoglobinisation is inhibited by colchicine, cytosine arabinoside and BUdR. Thus, the correlation of the effects of erythropoietin and the above-mentioned drugs on proliferation of the proerythroblast compartment and on the rate of

haemoglobinisation argues for a causal relationship between them. Possibly erythropoietin has a dual action: by inducing proliferation of a component of the proerythroblast compartment, a greater number of proerythroblasts may pass through a phase in the cell cycle in which, under a secondary action of erythropoietin, they may become committed to become haemoglobinised.

(3) This would explain why, in the absence of erythropoietin, proerythroblasts are not maintained and the rate of haemoglobinisation is less, with the result that mainly typical mature cells are formed. It would also explain why formation of typical cells in culture is not as sensitive to inhibition by the drugs mentioned.

(4) There is evidence to suggest that this maturation of immature cells in culture without erythropoietin may be inhibited by the presence of a substantial proportion of mature erythroid cells.

(5) There is no evidence in these studies that erythropoietin affects directly mature members of the erythroid cell series. Certainly, the rate of haemoglobin synthesis in specific mature cell types is not increased by culture with erythropoietin. The increase in haemoglobin synthesis which occurs in total foetal liver cultures after incubation with erythropoietin is therefore due to increased *numbers* of mature erythroid cells, which in turn are due to the direct effect of erythropoietin on proliferation of proerythroblasts.

The relationship between this situation and that obtaining in-vivo is far from clear. The effect of erythropoietin on proliferation of the proerythroblast compartment is totally analogous to the situation which obtains in the anaemic state in-vivo. Formation of atypical cells very similar to those observed in the present work is also observed in certain anaemias in-vivo. However, a clear distinction between the situation obtaining in culture and in-vivo concerns the number of maturation divisions involved. A plausible explanation of this difference is that the greater numbers of maturation divisions in-vivo are involved in producing sufficient numbers of red blood cells, rather than in the haemoglobinisation process *per se*. These amplification divisions may not be possible in culture, perhaps due to lack of some essential metabolite.

A vital piece of knowledge which we require in order to elucidate further the mechanism of action of erythropoietin is the stage in the maturation process at which the messenger sequences for globin are first transcribed in the nucleus, and at which messenger RNA for globin appears in the cytoplasm. These facts would enable important conclusions to be made concerning the mode of action by erythropoietin. If messenger sequences are not transcribed in proerythroblasts, then the activation of the genes specifying globin may be activated under the influence of erythropoietin in culture. On the other hand, if messenger sequences exist in the nucleus but not in the cytoplasm of proerythroblasts, then this would argue for an effect of erythropoietin on 'processing' of heterogeneous nuclear RNA. Yet again, if messenger RNA sequences are present in the cytoplasm of proerythroblasts, the role of erythropoietin would appear to be in regulating translation of these sequences or of activating the messenger RNA in some way.

Elucidation of this problem clearly demands a very sensitive probe for messenger sequences in erythroid cells. We believe such a sensitive probe exists in the form of a DNA-copy of the messenger RNA for globin which we have been able to prepare, using the reverse transcriptase from avian myeloblastosis virus (Harrison, Hell & Paul, 1972; Harrison, Hell, Birnie & Paul, 1972). By hybridisation of this DNA-copy with RNA from immature erythroid cells, either in-situ or by conventional methods, it should be possible to answer this crucial question.

These investigations were supported by the Medical Research Council and Cancer Research Campaign. Erythropoietin was obtained through the offices of the Committee on Erythropoietin of the National Heart Institute, U.S.A. The authors are grateful to Dr Anna Hell for drawing the figures.

REFERENCES

BLACKETT, N. M. (1968). Changes in proliferation rate and maturation time of erythroid precursors in response to anaemia and ionising radiation. In *Effects of Radiation on Cellular Proliferation and Differentiation*, pp. 235–46. Vienna: International Atomic Energy Agency.

BUNN, H. F. & JANDL, J. H. (1966). Exchange of heme among hemoglobin molecules. *Proceedings of the National Academy of Sciences, U.S.A.* **56**, 974–8.

BURGOS, H. (1971). Studies on erythroid cell maturation. Ph.D. Thesis, University of Glasgow.

CHUI, D. H. K., DJALDETTI, M., MARKS, P. A. & RIFKIND, R. A. (1971). Erythropoietin effects on fetal mouse erythroid cells. I. Cell population and hemoglobin synthesis. *Journal of Cell Biology*, **51**, 585–95.

COLE, R. J. & PAUL, J. (1966). The effects of erythropoietin on haem synthesis in mouse yolk sac and cultured foetal liver cells. *Journal of Embryology and Experimental Morphology*, **15**, 245–60.

COLEMAN, J. R. & COLEMAN, A. W. (1968). Muscle differentiation and macromolecular synthesis. *Journal of Cell Physiology*, **22** (Supplement 1), 19–34.

DJALDETTI, M., CHUI, D., MARKS, P. A. & RIFKIND, R. A. (1970). Erythroid cell development in fetal mice: stabilization of the hemoglobin synthetic capacity. *Journal of Molecular Biology*, **50**, 345–58.

DJALDETTI, M., PREISLER, H., MARKS, P. A. & RIFKIND, R. A. (1972). Erythropoietin effects on fetal mouse erythroid cells. II. Nucleic acid synthesis and the erythropoietin-sensitive cell. *Journal of Biological Chemistry*, **247**, 731–5.

DUKES, P. P. & GOLDWASSER, E. (1965). On the mechanism of erythropoietin-induced differentiation. III. The nature of erythropoietin action on ^{14}C glucosamine incorporation by marrow cells in culture. *Biochimica et Biophysica Acta*, **108**, 447–54.

FELICETTI, L., COLOMBO, B. & BAGLIONI, C. (1966). Assembly of hemoglobin. *Biochimica et Biophysica Acta*, **129**, 380–94.

FRESHNEY, R. I. & PAUL, J. (1971). The activities of three enzymes of haem synthesis during hepatic erythropoiesis in the mouse embryo. *Journal of Embryology and Experimental Morphology*, **26**, 313–22.

GALLIEN-LARTIGUE, O. & GOLDWASSER, E. (1965). On the mechanism of erythropoietin-induced differentiation. I. The effects of specific inhibitors on hemoglobin synthesis. *Biochimica et Biophysica Acta*, **103**, 319–24.

GROSS, M. & GOLDWASSER, E. (1969). On the mechanism of erythropoietin-induced differentiation. V. Characterization of the ribonucleic acid formed as a result of erythropoietin action. *Biochemistry*, **8**, 1795–805.

GROSS, M. & GOLDWASSER, E. (1970). On the mechanism of erythropoietin-induced differentiation. VII. The relationship between stimulated deoxyribonucleic acid synthesis and ribonucleic acid synthesis. *Journal of Biological Chemistry*, **245**, 1632–6.

GROSS, M. & GOLDWASSER, E. (1971). On the mechanism of erythropoietin-induced differentiation. IX. Induced synthesis of 9S ribonucleic acid and of hemoglobin. *Journal of Biological Chemistry*, **246**, 2480–6.

HANNA, I. R. A. (1968). An early response of the morphologically recognisable erythroid precursors to bleeding. *Cell and Tissue Kinetics*, **1**, 91–9.

HANNA, I. R. A., TARBUTT, R. C. & LAMERTON, L. F. (1969). Shortening of the cell cycle time of erythroid precursors in response to anaemia. *British Journal of Haematology*, **16**, 381–7.

HARRISON, P. R. (1968). Studies of potential radiosensitizing agents. Inhibition of nucleic acid synthesis by Synkavit (2-methyl-1,4-naphthaquinol bis disodium phosphate) in Ehrlich mouse ascites tumour cells. *British Journal of Cancer*, **22**, 274–89.

HARRISON, P. R., HELL, A. & PAUL, J. (1972). Inhibition of reverse transcriptase by high concentrations of tritium-labelled substrates. *FEBS Letters*, in press.

HARRISON, P. R., HELL, A., BIRNIE, G. D. & PAUL, J. (1972). Evidence for single copies of globin genes in the mouse genome. *Nature, London*, in press.

HODGSON, G. (1970). Mechanism of action of erythropoietin. In *Regulation of Hematopoiesis*, ed. A. S. Gordon, vol. 1, chapt. 20, pp. 459–69. New York: Appleton-Century-Crofts.

KILLMANN, S.-A. (1970). Cell classification and kinetic aspects of normoblastic and megaloblastic erythropoiesis. *Cell and Tissue Kinetics*, **3**, 217–28.

KOYAMA, H. & ONO, T. (1971). Effect of 5'-bromodeoxyuridine on hyaluronic acid synthesis of a clonal line of mouse and Chinese hamster in culture. *Journal of Cell Physiology*, **78**, 265–72.

KRANTZ, S. B., GALLIEN-LARTIGUE, O. & GOLDWASSER, E. (1963). The effect of erythropoietin upon heme synthesis by marrow cells *in vitro*. *Journal of Biological Chemistry*, **238**, 4085–90.

KRANTZ, S. B. & GOLDWASSER, E. (1965a). On the mechanism of erythropoietin-induced differentiation. II. The effect on RNA synthesis. *Biochimica et Biophysica Acta*, **103**, 325–32.

KRANTZ, S. B. & GOLDWASSER, E. (1965b). On the mechanism of erythropoietin-induced differentiation. IV. Some characteristics of erythropoietin action on hemoglobin synthesis in marrow cell culture. *Biochimica et Biophysica Acta*, **108**, 455–62.

KRETCHMAR, A. L. (1966). Erythropoietin: Hypothesis of action tested by analog computer. *Science, New York*, **152**, 367–70.

LAJTHA, L. G. (1970). Stem cell kinetics. In *Regulation of Hematopoiesis*, ed. A. S. Gordon, vol. 1, chapt. 6, pp. 111–31. New York: Appleton-Century-Crofts.

McCOOL, D., MILLER, R. J., PAINTER, R. H. & BRUCE, W. R. (1970). Erythropoietin sensitivity of rat bone marrow cells separated by velocity sedimentation. *Cell and Tissue Kinetics*, **3**, 55–65.

McCULLOCH, E. A. (1970). Control of hematopoiesis at the cellular level. In *Regulation of Hematopoiesis*, ed. A. S. Gordon, vol. 1, chapt. 7, pp. 132–59. New York: Appleton-Century-Crofts.

MAYNE, R., SANGER, J. W. & HOLTZER, H. (1971). Inhibition of mucopolysaccharide synthesis by 5'-bromodeoxyuridine in cultures of chick amnion cells. *Developmental Biology*, **25**, 547–67.

MILLER, R. G. & PHILLIPS, R. A. (1969). Separation of cells by velocity sedimentation. *Journal of Cellular Physiology*, **73**, 191–202.

NICOL, A. G., CONKIE, D., LANYON, W. G., DREWIENKIEWICZ, C. E., WILLIAMSON, R. & PAUL, J. (1972). Characteristics of erythropoietin-induced RNA from foetal mouse liver erythropoietic cell cultures and the effects of FUdR. *Biochimica et Biophysica Acta*, in press.

ORLIC, D., GORDON, A. S. & RHODIN, J. A. G. (1965). An ultrastructural study of erythropoietin-induced red cell formation in mouse spleen. *Journal of Ultrastructure Research*, **13**, 516–42.

ORTEGA, J. A. & DUKES, P. P. (1970). Relationship between erythropoietin effect and reduced DNA synthesis in marrow cell cultures. *Biochimica et Biophysica Acta*, **204**, 334–9.

PAUL, J. & CONKIE, D. (1972). Effects of inhibitors of DNA synthesis on the stimulation of mouse erythroid cells by erythropoietin. *Experimental Cell Research*. (In press.)

PAUL, J., CONKIE, D. & FRESHNEY, R. I. (1969). Erythropoietic cell population changes during the hepatic phase of erythropoiesis in the foetal mouse. *Cell and Tissue Kinetics*, **2**, 283–94.

PAUL, J., FRESHNEY, R. I., CONKIE, D. & BURGOS, H. (1971). Biochemical aspects of foetal erythropoiesis. *Proceedings of International Conference on Erythropoiesis, Capri*, in press.

PAUL, J. & HUNTER, J. A. (1968). DNA synthesis is essential for increased haemoglobin synthesis in response to erythropoietin. *Nature, London*, **219**, 1362–3.

PAUL, J. & HUNTER, J. A. (1969). Synthesis of macromolecules during induction of haemoglobin synthesis by erythropoietin. *Journal of Molecular Biology*, **42**, 31–41.

SHEPP, M., TOFF, H., YAMADA, H. & GABUZDA, T. G. (1972). Heterogeneous metabolism of marrow ferritins during erythroid cell maturation. *British Journal of Haematology*, **22**, 377–82.

STEPHENSON, J. R. & AXELRAD, A. A. (1971). Separation of erythropoietin-sensitive cells from hemopoietic spleen colony-forming stem cells of mouse fetal liver by unit gravity sedimentation. *Blood*, **37**, 417–27.

STEPHENSON, J. R., AXELRAD, A. A., McLEOD, D. L. & SHREEVE, M. M. (1971). Induction of colonies of hemoglobin-synthesizing cells by erythropoietin *in vitro*. *Proceedings of the National Academy of Sciences, U.S.A.* **68**, 1542–6.

TARBUTT, R. G. (1967). A study of erythropoiesis in the rat. *Experimental Cell Research*, **48**, 473–83.

TARBUTT, R. G. (1969). Cell population kinetics of the erythroid system in the rat. The response to protracted anaemia and to continuous γ-irradiation. *British Journal of Haematology*, **16**, 9–24.

TARBUTT, R. G. & BLACKETT, N. M. (1968). Cell population kinetics of the recognisable erythroid cells in the rat. *Cell and Tissue Kinetics*, **1**, 65–80.

TARBUTT, R. G. & COLE, R. J. (1970). Cell population kinetics of erythroid tissue in the liver of foetal mice. *Journal of Embryology and Experimental Morphology*, **24**, 429–46.

TERADA, M., BANKS, J. & MARKS, P. A. (1971). RNA synthesized during differentiation of yolk sac erythroid cells. *Journal of Molecular Biology*, **62**, 347–60.

TRENTIN, J. J. (1970). Influence of hematopoietic organ stroma (hematopoietic inductive microenvironments) on stem cell differentiation. In *Regulation of Hematopoiesis*, ed. A. S. Gordon, vol. 1, chapt. 8, pp. 159–86. New York: Appleton-Century-Crofts.

WINTROBE, M. M. (1961). *Clinical Haematology*, 5th edit. London: Kempton.

WORTON, R. G., McCULLOCH, E. A. & TILL, J. E. (1969). Physical separation of hemopoietic stem cells differing in their capacity for self-renewal. *Journal of Experimental Medicine*, **130**, 91–103.

EXPLANATION OF PLATE

PLATE 1

Typical and atypical cells derived from foetal liver cultures. Pr, proerythroblast; B, basophilic erythroblast; Po, polychromatic erythroblast; O, orthochromatic erythroblast; R, reticulocyte; PM, polychromatic macroerythroblast; OM, orthochromatic macroerythroblast; M, macrocyte. All cells are magnified to the same extent.

PLATE I

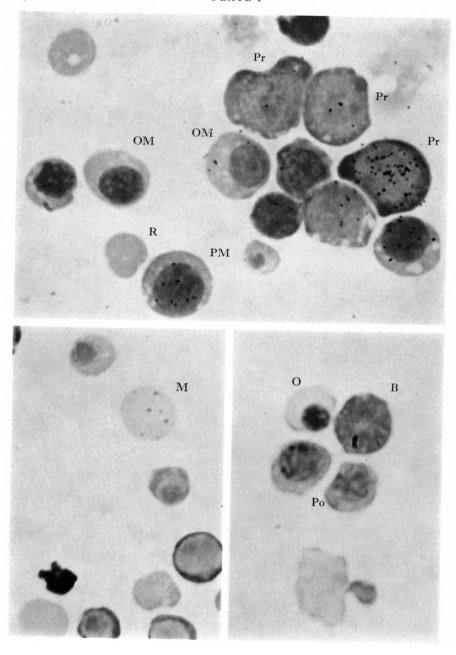

KINETICS OF CELL MULTIPLICATION AND DIFFERENTIATION DURING ADULT AND PRENATAL HAEMOPOIESIS

BY R. J. COLE AND R. G. TARBUTT

School of Biological Sciences,
University of Sussex, Falmer, Brighton

INTRODUCTION

The cells of the blood are highly differentiated and, except for the lymphocytes, are incapable of division. Moreover, these mature cells have lifespans that are short compared to that of the whole animal, and consequently they have to be continually replaced. In the adult, this replenishment is achieved by proliferation and maturation of precursor cells in the bone-marrow, spleen and lymphoid tissue; in the foetus, the precursors may be located in the yolk-sac, liver, spleen or bone-marrow, according to the stage of development of the animal.

The different types of blood cells (erythrocytes, lymphocytes, granulocytes, monocytes and platelets) are morphologically and functionally very distinct. Nevertheless, the erythrocytes and granulocytes undoubtedly arise from a common progenitor cell type which is normally active for the lifespan of the animal, and it seems almost certain that the other haemopoietic cells also arise from this common stem cell (Wu, Till, Siminovitch & McCulloch, 1967, 1968). Thus within the haemopoietic system (both foetal and adult) we have an example of an 'initial differentiation' involving choice between one of several different pathways, followed by 'continued differentiation' along the selected pathway. In this paper we have attempted to describe some of the events that occur during these differentiations, and have tried to relate them to changes in cell cycles of the differentiating cells. The discussion covers adult and foetal haemopoiesis, and also considers some aspects of haemopoiesis in genetically anaemic mice.

The common progenitor (stem) cell is a prime target for studies of cytodifferentiation, since it is at the stem-cell level that the 'choice' between different pathways is made. Unfortunately the morphological identity of this cell remains a matter of controversy; at one extreme are those who suggest that the large reticulum cell of the bone-marrow is the stem cell, and at the other extreme are those who campaign for the small bone-marrow lymphocyte. The biggest barrier to positive identification is the very small number of stem cells in haemopoietic tissue – estimated by

Blackett (1972) to be about 0.2% of the cell population in adult bone marrow. Recently-introduced techniques of cell separation (or at least concentration) offer some hope of solving the problem, but have so far proved disappointing in this respect. Clearly, then, techniques based on morphological or biochemical properties are inappropriate to the study of haemopoietic stem cells, and other techniques have to be devised. One such technique is the *spleen colony technique*; this technique has proved so useful in studies of the regulation of haematopoiesis that a fairly detailed description of the technique and its applications will serve as a useful introduction to the current concept of the haemopoietic system.

SPLEEN COLONY ASSAY FOR HAEMOPOIETIC STEM CELLS

The spleen colony technique (Till & McCulloch, 1961) may be compared to the clonal techniques used in microbial systems, and indeed haemopoietic tissue has often been compared to such systems in that it consists of a series of free-living cells, apparently acting independently of each other. While such a comparison has proved useful in suggesting many elegant experimental designs, it is now becoming more and more evident that this is a gross over-simplification.

The spleen colony assay for haemopoietic stem cells depends upon the observation that when a suspension of haemopoietic cells is injected into a lethally irradiated mouse, discrete nodules appear on the surface of the recipient's spleen, and that these nodules contain erythroid, granulocytic or megakaryocytic cells. The number of nodules is directly proportional to the number of cells injected, suggesting that each nodule is formed from a single entity, and also that the nodules develop independently of each other. It is tempting to suggest that each nodule arises from a single haemopoietic stem cell.

That each nodule did, in fact, arise from a single cell was demonstrated by Becker, McCulloch & Till (1963), who showed that whenever radiation-induced chromosomal abnormalities were seen in the cells of a nodule, the abnormality was present in at least 95% of the metaphases of the nodule, showing that at least 95% of the cells within a nodule were derived from a single cell. Hence each nodule may reasonably be regarded as a clone. In order to demonstrate that the original cell of each clone was a haemopoietic stem cell, it was necessary to show that this cell had the characteristic properties associated with stem cells: these have been listed by McCulloch (1971) as extensive proliferative potential, including self renewal; the capacity to differentiate; and responsiveness to control. The most convincing demonstration of their proliferative potential is to measure the rate of growth of spleen colonies, which may exceed 10^6 cells in 10 days, all arising from a single cell. As the great majority of these cells consist of mature haemopoietic elements, it is also clear that the original cell has the capacity to differentiate. Furthermore, by re-transplanting cells from original colonies it can be shown that the number of cells in a clone capable of clone-formation increases as the colony grows, so that colony-forming-cells have the self-renewal property of stem cells. Finally McCulloch (1971) has pointed out that colony-forming cells must be subject to very precise controlling factors, for if their proliferative and differentiative capacities were not under control a condition of leukaemia would result. Thus colony-forming cells possess all of the essential properties of haemopoietic stem cells, and so the spleen colony technique may reasonably be regarded as an assay for haemopoietic stem cells.

The composition of individual spleen colonies may be studied by microscopic examination of spleen sections or of smears from isolated colonies. The colonies are usually found to contain only one cell type during the first 9–10 days of growth, but later on the majority of colonies contain both erythroid and granulocytic cells (Curry & Trentin, 1967). During the early stages (i.e. while the colonies contain only one differentiated cell type) the ratio of erythroid:granulocyte colonies is approximately 66:22, the remainder being undifferentiated (Curry & Trentin, 1967). If individual colonies containing only one differentiated cell line are dissected out and retransplanted into secondary recipients, then it is found that the colonies so formed are not all of the same type, but more or less reflect the usual ratio of erythroid: granulocyte: megakaryocyte colonies, perhaps with a slight shift towards the 'mother colony' type This provides strong evidence for the pluripotent nature of the colony-forming stem cell.

Why do spleen colonies become 'mixed' only after about 11 days of growth if the stem cell is pluripotent? It is possible that each colony has its own definitive cell line, perhaps determined by the local environment, and that the second cell line is a contaminant caused by cell immigration. However, experiments with mixtures of chromosomally-distinguishable cells have shown unequivocally that the second line is *not* a contaminant, but arose from within the same clone (Trentin et al. 1969). But why should it take 10–11 days for this second line to appear? Curry & Trentin (1967) have suggested that the haemopoietic tissue contains many different microenvironments, each being 'good' for just one line of cell differentiation, and operating over a volume equivalent to 10 days-worth of colony proliferation, or 10^6 cells in the case of erythroid colonies. When a stem cell 'seeds' in one of these microenvironments, it and its progeny can only differentiate along the pathway governed by that microenvironment. However, as the colony increases in size it will eventually outgrow its original microenvironment, and spill over into the next microenvironment. If this happens to correspond to a different line of differentiation, then a second line of differentiated cells will appear. Thus practically all colonies will eventually tend to become mixed. Strong support for this hypothesis comes from experiments involving transplantation of marrow fragments into the spleens of irradiated recipients (for review, see Trentin, 1971).

HUMORAL FACTORS IN HAEMOPOIESIS. ERYTHROPOIETIN

It is well known that the haemopoietic system is responsive to certain humoral factors, the best known being *erythropoietin*, which is a glycoprotein found in small quantities in the plasma of normal animals. Its concentration increases under certain conditions of hypoxia, and is practically abolished if an animal is made polycythemic, for example by hypertransfusion. When a large amount of erythropoietin is injected into a normal animal, the rate of erythropoiesis is increased. When the erythropoietin concentration is severely reduced by hypertransfusion or by anti-erythropoietin injections erythropoiesis ceases. If small amounts of erythropoietin are injected into hypertransfused (or anti-erythropoietin treated) mice, a wave of erythropoiesis is induced, whose size depends upon the amount of erythropoietin injected. This latter finding provides the basis for the most commonly used assay for erythropoietin, in which hypertransfused mice are injected with the preparation under test, and the amount of erythropoiesis so induced is measured by the uptake of ^{59}Fe into haem. The ^{59}Fe uptake is then compared to that obtained with standard erythropoietin preparations.

Erythropoietin-sensitive cells

In the erythropoietin assay the erythropoietin is injected into recipients whose haemopoietic tissues have been severely depleted of morphologically identifiable erythroblasts by hypertransfusion or similar treatment. It is therefore natural to suggest that erythropoietin must act directly on the unidentified, multipotent haemopoietic stem cell, causing it to differentiate along the erythropoietic pathway, and that in the absence of erythropoietin this differentiation does not take place. However, this suggestion is apparently in conflict with the idea that it is the local environment that determines the line of differentiation of a stem cell. But if erythropoietin does *not* cause stem cells to differentiate, and there are virtually no recognisable erythroblasts in the hypertransfused animals into which the erythropoietin was injected, where does the erythropoietin act? One way out of the dilemma is to postulate the existence of a second type of morphologically unidentified precursor, intermediate between stem cells and proerythroblasts, which is responsive to erythropoietin. The experimental distinction between stem cells and 'erythropoietin sensitive cells' has now been demonstrated several times (Bruce & McCulloch, 1964; Schooley, 1966; Till, Siminovitch & McCulloch, 1966; Fried, Martinson, Weisman & Gurney, 1966). Although it is not our purpose to review all of the available evidence, it will be useful to outline one particular experimental design ('thymidine suicide', Becker et al. 1965), since it will be necessary to refer to this again later. In their adaptation of the 'suicide' technique, Lajtha, Pozzi, Schofield & Fox (1969) injected mice either with very large doses of tritiated thymidine ([^{3}H]TdR), sufficient to kill every cell actively incorporating thymidine into DNA, or with saline. Some mice from each group were killed 2 h later, and a spleen colony assay performed on their bone marrow. This showed that less than 10% of the colony-forming cells were killed by the high dose of tritiated thymidine, i.e. were in DNA synthesis when the [^{3}H]TdR was injected. The remaining mice in each group were made polycythemic, and later injected with standard doses of erythropoietin. The wave of erythropoiesis so produced was then assayed by measuring the uptake of ^{59}Fe into erythrocytes. The size of the wave was assumed to be proportional to the number of erythropoietin-sensitive cells (standard dose of erythropoietin), and consequently it was possible to deduce the number of erythropoietin-responsive cells killed by the [^{3}H]TdR. This was found to be approximately 70%, compared to less than 10% for colony-forming cells. Thus erythropoietin-response cells have cell cycle characteristics that are markedly different from those of colony-forming stem cells, and may therefore be regarded as comprising a distinct population.

So far we have reviewed evidence suggesting that the multipotent haemopoietic stem cell is induced by local environmental factors to differentiate into a cell committed to a particular line of differentiation, and that in the case of the erythroid system this morphologically unidentified cell is responsive to erythropoietin, which causes its further differentiation into a proerythroblast. In making the stem cell/erythropoietin-sensitive cell transition the cells undergo a change in their proliferative states, so that in the normal adult a change from a slow rate of proliferation to a much faster rate accompanies the differentiation of a stem cell. Experiments on adult rats suggest that the further differentiation as a result of interaction with erythropoietin is accompanied by a second increase in proliferation rate (Hanna, 1968a). Thus in this system there appears to be a tendency for the proliferation rate to increase as differentiation proceeds, probably reaching a maximum at the proerythroblast stage.

MORPHOLOGICALLY IDENTIFIABLE ERYTHROBLASTS

Once an erythropoietin-sensitive cell has properly interacted with a sufficiently high concentration of erythropoietin it begins to mature through the erythroblast stages of the system. This maturation is irreversible, and the cells will continue to proliferate and mature normally to erythrocytes even in the absence of erythropoietin (Tarbutt, 1969a). The maturation time (i.e. the interval between transition to proerythroblast and entry into the circulation) appears to be determined by the concentration of erythropoietin present at the time of interaction between erythropoietin and erythropoietin-sensitive cell (Tarbutt, 1969b; Hanna, 1968b).

Despite the fact that erythroblasts continue to mature and proliferate in the absence of erythropoietin, their proliferation rate does depend, in part, upon the concentration of erythropoietin, for when this is increased *above* the normal level during anaemia, the cycle time of erythroblasts is reduced (Fig. 1; Hanna, Tarbutt & Lamerton, 1969). As the cycle time can be reduced by elevated levels of erythropoietin *after* differentiation into proerythroblasts, but the maturation time is fixed *before* differentiation, the erythroblasts will initially undergo more divisions during maturation, but later on, when cells with shortened maturation times have reached the circulation, these extra divisions are lost. The important physiological implications of this sequence of events are outside the scope of this paper (see Hanna, 1968b; Tarbutt, 1969a).

From the foregoing discussion it will be evident that the recognisable erythroblasts as well as a class of morphologically unidentified precursors are responsive to erythropoietin. The term 'erythropoietin-sensitive cell' is, however, usually reserved for the unidentified precursor, and unless otherwise stated we will follow this usage. It should not, however, be forgotten that the recognisable erythroblasts are also sensitive to the hormone.

RESPONSE OF THE ADULT ERYTHROID SYSTEM
TO CONTINUOUS STRESS

The manner in which the haemopoietic system responds to stress depends not only upon the nature of the stress, but also on its duration. We will consider two types of stress, namely continuous whole-body irradiation at

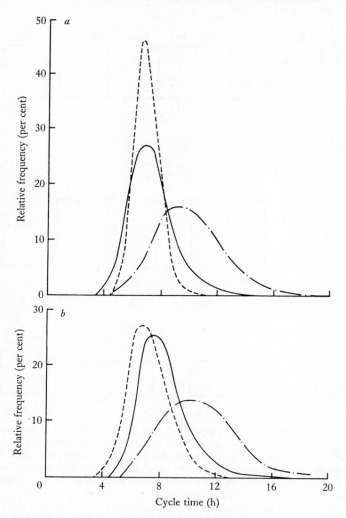

Fig. 1. The distribution of erythroblast cell cycle times in normal (—·—); bled (——); phenylhydrazine-treated (--) rats. (a) Pro- and basophilic erythroblasts; (b) polychromatic erythroblasts. (From Hanna *et al.* 1969).

a dose consistent with normal red cell production in the rat (45 rad/day), and long-term haemolytic anaemia, induced by phenylhydrazine treatment. In the anaemic rats, red cell production was approximately five times normal.

It is not intended to describe the experimental methods here, as they have been described fully by Tarbutt (1969*a*), Blackett (1968) and Hanna (1967). The measurements made to characterise the response were the

Normal adult rats

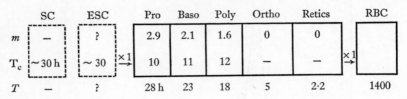

Protracted anaemia

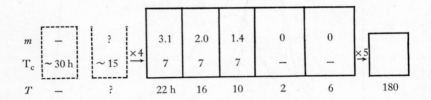

Continuous irradiation (45 rads/day)

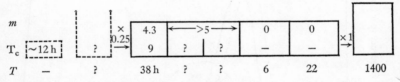

Fig. 2. Cell kinetic parameters of erythroid system in normal rats and in rats subjected to protracted anaemia or to continuous γ irradiation at 45 rad/day. SC: stem cells. ESC: erythropoietin sensitive cells. Pro: proerythroblasts. Baso: basophilic erythroblasts. Poly: polychromatic erythroblasts. Ortho: Orthochromatic erythroblasts. Retics: reticulocytes. RBC: red blood cells. m: number of divisions. T_c: cycle time. T: transit time. The heights of the boxes are approximately proportional to the numbers of precursors in normal, irradiated and anaemic rats, taking the values for normal rats as one unit of height. The numbers above the arrows refer to cell flow-rates, again relative to those in normal rats (one unit).

number of cells of different classes, their cell cycle times, maturation times through the different stages and the number of divisions undergone during maturation. The results are shown diagrammatically in Fig. 2. It must be emphasised that the estimates shown refer only to the new steady state achieved during the stress, and not to the early changes that occur before the steady state is achieved.

No difference was detected from normal in the stem cell compartment of the anaemic rats, both the number and proliferation rate being unchanged.

Since the turnover of erythropoietin is very high during phenylhydrazine anaemia, it is therefore apparent that the stem cells do not respond directly to erythropoietin, as shown in an earlier section. In contrast the stem cells in the irradiated rats respond by a very marked increase in the proliferation rate of the cells surviving the irradiation. Thus the stem cells appear to respond to a decrease in their number by increasing their proliferation rate (note, however, that the stem cell compartment is still only about one-tenth its normal size in the irradiated animals, despite this increased proliferation).

As one would expect, the erythropoietin-responsive cells respond to the anaemia by an increased proliferation rate, as the result of which the rate of cell-flow into the proerythroblast compartment is increased five-fold. This increased flow is due entirely to the increased proliferation rate of the erythropoietin-sensitive cells, and is in no way due to changes in the stem cell compartment. Good estimates for the proliferation rate of erythropoietin-sensitive cells in continuously irradiated rats have not yet been obtained, but since the animals are not anaemic (normal red cell production) it seems unlikely that these cells will respond. Some preliminary results on acutely-irradiated rats support this view (Blackett & Hanna, personal communication).

The rate at which cells flow into the proerythroblast compartment is markedly different in the two stress situations, for in the anaemic rats it is 4–5 times normal, whereas in the irradiated rats it is $\frac{1}{4}$-normal (despite the increased proliferation by stem cells). Thus the recognisable precursor compartment is presented with flow rates that are markedly different from normal. How does this compartment deal with these different flow rates? One way of approaching this problem is to regard the recognisable precursor compartment as a cell-flow amplifier, whose gain is the ratio of the cell output to the cell input. In the case of the protracted anaemia the gain is normal, since both the input and output are increased five-fold. However, in the case of the continuously irradiated rats, the gain is increased by a factor 4, since the input is $\frac{1}{4}$-normal and the output is normal. This implies that the cells undergo two extra divisions during maturation in the irradiated rats, in order to ensure a normal output of cells from a depleted precursor pool. These findings are confirmed by detailed calculations of the number of cell divisions undergone in each of the morphologically identifiable compartments (Fig. 2).

Although the 'amplification' is normal during protracted anaemia, the kinetics of the recognisable precursor compartment do, in fact, change, for there is a considerable reduction in the cell cycle time (Fig. 1). However, because the maturation time is also decreased by about the same proportion, the number of divisions does not change. As pointed out previously, this is not the case during the early stages of anaemia, since the proliferation rate changes *before* the maturation rate (Hanna, 1968b).

FOETAL ERYTHROPOIESIS

The above presentation of some of the more important features of the haemopoietic system provides a background for a discussion of the developmental changes that occur in the mouse erythropoietic system.

Foetal erythropoiesis occurs first in the yolk-sac, followed by the liver, spleen and bone marrow. In this paper we will consider only the hepatic phase of erythropoiesis, which in the mouse occupies the period from the tenth day of gestation until birth. The pattern of hepatic erythropoiesis differs somewhat from one strain of mouse to another; in this part of the paper we will refer only to results for the random-bred Swiss albino mouse.

The foetal haemopoietic system bears some resemblance to the adult system under continuous irradiation, since in both cases the stem cells must attempt to increase their numbers, in order to meet the demands of radiation-induced depletion or of growth. Thus we may expect a considerably enhanced rate of stem cell proliferation compared to that in the adult, and also possibly more divisions during maturation. The foetal system is also somewhat like the adult system under protracted anaemia, for the foetus is considerably anaemic compared to the adult. This may suggest that the rate of proliferation of erythropoietin sensitive cells and recognisable erythroblasts will be increased in the foetus relative to the adult.

Cell kinetics of recognisable erythroblasts in mouse foetal livers

During the hepatic period of erythropoiesis the embryo increases in weight by a factor of 20, and the liver produces about 500 million red cells. This massive production of red cells does not occur at a constant rate in the Swiss albino mouse, but in two rather distinct phases (Paul, Conkie & Freshney, 1969). The first phase lasts until the 15th day of gestation, and corresponds to an 8 h doubling time for the red cell mass. The second phase lasts from the 15th day until birth, and corresponds to a doubling time of 48 h. These results are reproduced in Fig. 3 (Paul *et al.* 1969). This change in doubling time correlates with the change in sensitivity to erythropoietin of cultured foetal liver cells, to be discussed later.

The doubling time of a population is partially dependent upon the cell cycle time of the proliferative cells in the population, and it is possible that the change in doubling time of the red cell mass is related to a change in the

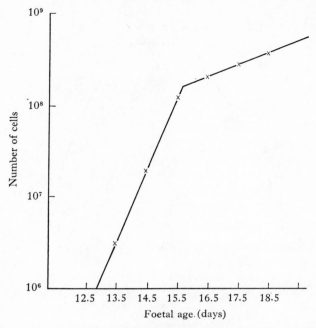

Fig. 3. Total numbers of hepatogenic red cells and reticulocytes in mouse foetuses.
(From Paul *et al.* 1969.)

Table 1. *Cell cycle parameters of erythroblasts in foetal mice* (*Swiss albino*)

Phase	13 days' gestation		16 days' gestation	
	Early erythroblasts	Late erythroblasts	Early erythroblasts	Late erythroblasts
G_2 (h)	0.7	1.0	1.0	1.5
G_1 (h)	0.4	1.5	0–0.5	0.5
S (h)	4.5	5.5	4.0–4.5	~4.0
T_C (h)	5.6	8.0	5.5	5.0–6.0

cell cycle time of the recognisable precursor population. In order to check
the possibility, direct measurements of the cell cycles of erythroblasts in
13-day foetuses were compared with those in 16-day foetuses, using the
labelled mitoses technique. The results are shown in Table 1. It is clear
that there are no major cycle-time differences between 13- and 16-day
foetuses, and that in fact the total cycle time is somewhat shorter at 16 days
than at 13 days, which is the reverse of the much larger difference in
doubling time. Thus the change in doubling time (and sensitivity to
erythropoietin, see later) is not correlated with a change in cell cycle time.
Other factors that may affect the doubling time of the red cell mass are

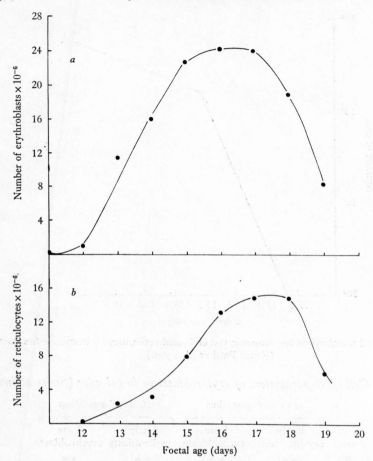

Fig. 4. (*a*) The total number of erythroblasts in mouse foetal livers. (*b*) The total number of reticulocytes in mouse foetal livers. (From Tarbutt & Cole, 1970.)

the total number of precursors in the liver, and the proportion of them that are capable of dividing. Destruction of red cells is not considered as a possible contributory factor, as the lifespan of a red cell is greater than the time elapsed from conception to birth. In measuring the total number of erythroid precursors in the liver, it was assumed that the number of un-identified precursors is negligible compared to the morphologically recognisable precursors. The results are shown in Fig. 4, where it can be seen that the number of erythroblasts increases until the 15th day of gestation, after which it levels off and then declines. Thus until the 15th day of gestation we have an increasing precursor population serving the expanding circulation, but after day 15 we have a static or declining

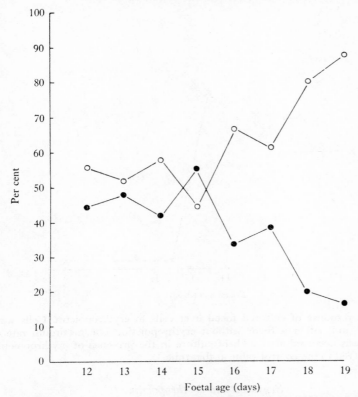

Fig. 5. Relative numbers of early (●) and late (○) erythroblasts in Swiss mouse foetal livers.

population. This alone would be expected to lead to a change in the rate of red cell production. The relative numbers of early and late erythroblasts also change rather abruptly on day 15 (Fig. 5). There are more or less equal numbers of early and late erythroblasts up to the 15th day, but thereafter the proportion of late erythroblasts increases at the expense of the early cells. Since all of the early erythroblasts but only the less mature late erythroblasts can proliferate, this reflects a change towards a non-proliferating population of cells, starting from the 15th day of gestation. This will also lead to a decrease in the rate of red cell production around about day 15.

Calculations based upon the number of erythroblasts in the liver, their cell cycle times and the proportion of them that are capable of proliferation have shown that the changes in cell numbers and proliferative fraction discussed above can completely account for the observed bi-phasic pattern of growth of the red cell mass (Tarbutt & Cole, 1970). We will indicate later on how these changes may come about.

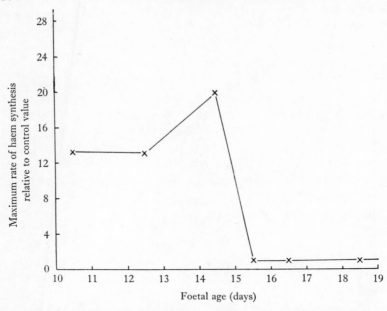

Fig. 6. The response of cultured foetal liver cells to erythropoietin. Cells were cultured for 30 h either with or without erythropoietin. The maximum rate of haem synthesis occurred after 28 h of culture in the presence of erythropoietin, and is compared to the control value at that time.

Sensitivity to erythropoietin

The sensitivity of foetal liver cells to erythropoietin cannot be measured in-vivo, but satisfactory methods exist for cells in suspension cultures. We do not at this point wish to review the many changes that have been observed in cultured cells responding to erythropoietin (see for example, Paul & Hunter, 1969; Bateman *et al.* 1972; and a later section of this paper), and the results to be presented will be discussed only in terms of the presence or absence of cells capable of responding to erythropoietin by increased haem synthesis. The assay system employs disaggregated foetal liver cells suspended in Waymouth's medium supplemented with foetal calf serum and adult mouse serum. Erythropoietin is added to half of the cultures at the start of the incubation, and the rate of haem synthesis estimated after about 28 h of incubation, by measuring the rate of incorporation of ^{59}Fe into haem. Results of experiments of this type are shown in Fig. 6, which demonstrates clearly the biphasic nature of the response as mentioned earlier.

Responses of up to twenty times the control level are obtained prior to the 15th day of gestation, but thereafter there is virtually no response to erythropoietin. The interpretation of this sort of experiment is not as straightforward as we would wish, for the erythropoietin added to the culture may act upon more than one cell type, since the cultures are derived from whole livers. However, we would like to suggest that after the comparatively long (28 h) culture period, the response is due primarily to cells that have differentiated from the erythropoietin-sensitive cell compartment, under the action of erythropoietin. This being the case, the experiment measures essentially the size of the erythropoietin-sensitive cell compartment, and therefore suggests that this compartment empties rather suddenly just before the 15th day of gestation. This conclusion could in principle be confirmed by separating out the recognisable erythroblasts from the remainder of the cell suspension.

Stem cells

The mouse foetal liver produces very few granulocytes, megakaryocytes and lymphocytes, but very large numbers of erythrocytes. Is the haemopoietic stem cell in foetal liver capable of producing spleen colonies in irradiated adults, and if so, are the colonies all erythroid? A fairly detailed analysis (Silini, Pons & Pozzi, 1968) has shown that foetal liver cells do produce spleen colonies in irradiated adult recipients, and that these colonies are erythroid, granulocytic, megakaryocytic or mixed. The proportion of mixed colonies is rather greater than in colonies arising from adult bone marrow, but this may simply reflect the somewhat faster growth-rate of foetal-derived colonies (see next section), so that the colonies more quickly outgrow their original microenvironment and spill over into a different microenvironment. What is needed to confirm this is an analysis of colonies at earlier times after transplantation. However, the important point is that foetal liver haemopoietic stem cells have the capacity to form granulocytes and megakaryocytes as well as erythrocytes. Thus we may conclude that the foetal liver offers a predominantly 'erythropoietic environment' to multipotent stem cells.

Although at birth haemopoiesis has almost ceased in the foetal liver, this organ still contains a large number of colony forming cells – more, for example, than at 14 or 15 days of gestation, when hepatic haemopoiesis is at a peak (Silini, Pozzi & Pons, 1967). Moreover, these colony-forming cells are in an active state of cell division for a large proportion of them are killed by high doses of [³H]TdR (Becker *et al.* 1965). Thus although during the later stages of gestation the liver supports the proliferation of a large population of haemopoietic stem cells, these cells do not complete their differentiation into recognisable erythroblasts. We will suggest later that the 'block' is at the stem cell/erythropoietin-sensitive cell transition.

Are foetal haemopoietic cells identical to adult haemopoietic cells? Are they regulated by the same factors?

The most convincing demonstration of a difference between foetal and adult haemopoietic cells has come from the recent experiments of Micklem *et al.* (1972), which showed that when chromosomally distinguishable foetal and adult cells are transplanted into the same irradiated adult, the foetal cells eventually outgrow the adult cells, even though they started off at a disadvantage in terms of cell numbers. The foetal liver-derived cells continued to increase relative to the bone-marrow-derived cells for approximately two months after transplantation, at which time a balance was achieved between the two populations, lasting for at least 400 days (the duration of the experiments). The authors prefer to explain the eventual stabilization of the foetal liver/adult bone-marrow ratio by supposing that the daughters of the original foetal-liver-derived stem cells gradually acquire the properties of adult stem cells, the transition taking two months. An alternative explanation, also suggested by Micklem *et al.* is that as a new steady state is reached after the period of post-radiation regeneration in the recipient mouse, most stem cells would probably be out of cycle so that any inherent proliferative advantage of foetal-liver-derived cells would be less readily expressed.

Table 2. ^{59}Fe *uptakes in irradiated Swiss mice injected with* 5×10^6 *foetal liver or bone-marrow cells. Effect of hypertransfusion*

Source of cells	^{59}Fe uptakes in non-transfused recipients	^{59}Fe uptakes in transfused recipients	Apparent suppression
12-day f/L*	27.7 ± 5.4 %	0.16 ± 0.08 %	99.5 ± 0.9 %
12-day f/L†	6.5 ± 1.8 %	0.16 ± 0.17 %	97.5 ± 1.1 %
13-day f/L*	26.0 ± 2.1 %	1.12 ± 0.22 %	95.7 ± 0.7 %
16-day f/L*	30.0 ± 3.4 %	0.67 ± 0.35 %	97.8 ± 0.8 %
19-day f/L*	22.4 ± 4.0 %	0.40 ± 0.34 %	98.0 ± 0.8 %
Adult b/m*	25.2 ± 1.0 %	0.16 ± 0.04 %	99.5 ± 0.5 %

f/L = foetal liver. b/m = bone marrow.
* Male recipients. † Female recipients.

Previous attempts to demonstrate difference in the proliferative capacities of foetal and adult stem cells have yielded controversial results. Thus Schofield (1970) reported a faster proliferation rate in foetal compared to adult colony-forming cells growing in an irradiated adult, but Vogel *et al.* (1970) reported that foetal and adult cells have similar generation times. Micklem's experiments appear to have settled this particular controversy.

Some authors (Latsinik *et al.* 1971; Bleiberg & Feldman, 1969) claim to have shown that foetal-liver-derived erythroid spleen colonies grow independently of erythropoietin, in contrast to adult-bone-marrow-derived colonies. Kubanek, Rencricca, Porcellini & Stohlman (1971) were unable to substantiate this claim, and our own data (Table 2) suggest that foetal-liver-derived erythropoiesis is abolished if the irradiated recipients are hypertransfused, so that we also were unable to find a difference between foetal and adult cells in terms of erythropoietin sensitivity.

Other differences between foetal and adult haemopoetic cells include differences in cell size and lifespan. However, these are probably simply an expression of the response to differing environmental conditions, since changes in adult cell size and lifespan can be induced by subjecting the animal to haemopoietic stress.

Regulation

(i) *Environmental factors.* As stated previously, when foetal haemopoietic tissue is transplanted into an irradiated adult, it produces roughly the same proportions of erythroid, granulocytic and megakaryocytic colonies as does adult bone marrow. This suggests that foetal and adult stem cells respond to similar differentiative stimuli, presumably derived from the microenvironment. Since the mouse foetal liver produces many erythroblasts but few non-erythroid haemopoietic cells, the foetal liver must provide a predominantly 'erythropoietic' environment.

The factors regulating stem cell proliferation are unknown, but probably also reside in the microenvironment (see section on genetic anaemias). The fact that foetal-liver-derived cells eventually outgrow bone-marrow-derived cells living in the same environment (Micklem *et al.* 1972) may suggest that foetal liver cells are more responsive to these stimuli. There is no evidence for a fundamentally different sort of control mechanism.

(ii) *Erythropoietin.* The role of erythropoietin in adult haemopoiesis has already been discussed, and is a comparatively non-controversial subject, although its action at the molecular level is not understood. However, its role in foetal and neonatal haemopoiesis is not well understood, and is a matter of controversy.

That mouse foetal liver cells can respond to erythropoietin is shown by the experiments in-vitro of Cole & Paul (1968), and on this point there is general agreement. However, the extent to which erythropoietin is responsible for controlling foetal and neonatal erythropoiesis under physiological conditions is not clear. We have already mentioned the controversy that exists over the requirement for erythropoietin by foetal liver cells growing in irradiated adults. Similar controversy exists concerning erythropoietin in neonatal animals. Erythropoietin is undoubtedly present in neonatal rat plasma (Table 3), but early experiments indicated that hypertransfusion, nephrectomy or starvation all apparently only partially inhibited erythropoiesis, whereas in the adult erythropoiesis is practically abolished by these procedures, which all lower the erythropoietin concentration (for review, see Stohlman, 1970). More recent experiments have shown that nephrectomy does not, in fact, reduce the concentration of erythropoiesis in neonatal rats (Carmena, Howard & Stohlman, 1968), and that the rate of erythropoiesis in neonatal starved rats is probably much less than was originally suggested by the high reticulocyte and erythroblast counts in these animals. The reason for the continued erythropoiesis in hypertransfused newborn rats remains unexplained, but it could be that erythropoietin production is not suppressed in these animals. Injection of anti-erythropoietin into newborn rats was reported by Schooley, Garcia, Cantor & Havens (1968) to abolish erythropoiesis in these animals, but even these results have been criticised by Stohlman (1970), who suggests that the observed change in haematocrit was not consistent with the complete cessation of erythropoiesis.

Table 3.

Substance injected*	^{59}Fe uptake
Saline	0.63 %
Ep.	
0.1 u	3.64 %
0.5 u	8.02 %
Starved serum	
0.5 ml	0.76 %
1.0 ml	0.64 %
2.0 ml	0.84 %
Normal serum	
0.5 ml	1.03 %
1.0 ml	1.80 %
2.0 ml	2.29 %

* The volumes given are the total amounts *before dilution*. Injections were fractionated into 4 equal parts, and the substances diluted with saline so that the volume administered at each injection was 0.5 ml.

We are of the opinion that the evidence for a partially erythropoietin-independent erythropoiesis during foetal and neonatal life is becoming less and less convincing, and that the role of erythropoietin during these stages of development is similar to, and of the same importance as during adult life. Further discussion of the factors controlling haemopoiesis in both adult and foetal tissue is presented in the next section of the paper.

GENETIC DISORDERS OF MOUSE HAEMOPOIESIS

A large number of genetic defects affecting the mouse haemopoietic system have been isolated (see reviews by Russell, 1970; Pinkerton & Bannerman, 1968). Of these only three have been studied in any detail, namely 'W', 'Steel' (Sl) and 'flexed' (f), all of which produce severe anaemias. Haemopoiesis is a genetically-controlled process, so that genetic defects that upset the regulation of haemopoiesis may give some clue to the nature of the underlying processes.

We shall first present a brief summary of some of the more important results relating to W and Sl, and then present our results on foetal 'flexed' mice.

'W' and 'Sl' anaemias

Several mutations of the 'W' locus are known, all producing anaemia, pigmentation defects and sterility in the homozygote. The heterozygote is affected to a much smaller degree, and is not sterile. The most important 'W' mutations are known as W and W^v, the first (W) of which is usually lethal in the homozygote, the second (W^v) being less severe and producing viable homozygotes. Crosses yielding WW^v mice have an anaemia intermediate in severity between WW and W^vW^v, and are viable.

The most common mutations at the 'Steel' locus are Sl (lethal) and Sl^d (viable). Homozygotes are phenotypically similar to W homozygotes, as they lack coat colour, are sterile and severely anaemic. Crosses yielding $SlSl^d$ mice are viable, with anaemia of intermediate severity.

In the following discussion 'W' may generally be taken to mean any mouse with one or other of the 'W' series mutations on both chromosomes, e.g. WW, WW^v etc., and 'Sl' to mean Sl^dSl^d, $SlSl^d$ etc.

'W' and 'Sl' mice lack coat pigmentation, are sterile, extremely radiosensitive and suffer from severe macrocytic anaemia. Thus at first sight W and Sl mutations appear to produce identical abnormalities, and might be thought to affect the same regulatory process. The most striking evidence that this is not the case comes from transplantation studies. If bone marrow cells from Sl mice are transplanted into coisogenic W mice, then the W mice are permanently cured of their anaemias (Russell, Smith & Lawson, 1956). Transplantation of haemopoietic cells into Sl mice has no beneficial effect, but haemopoietic W tissue (as opposed to cell) transplants are effective in curing the anaemia of Sl mice (Bernstein, 1970). Transplants of normal cells into W mice or normal tissues into Sl mice are also effective.

These results indicate that the defect in W mice is intrinsic to the haemopoietic cells themselves; injection of normal cells cures the anaemia

because these 'good cells' are able to take over from the host cells. *Sl*
haemopoietic cells are presumably 'good', since these can also cure *W*. The
fact that injections of normal cells do not cure *Sl*, whereas organ transplants
are effective, suggests that in *Sl* the haemopoietic cells are 'good', but their
environment is defective. This is borne out by the fact that *Sl* is also cured
by *W* organ transplants, presumably because 'good' *Sl* cells are able to
seed in the 'good' environment offered by the *W* organ.

Further insight into the defects in *W* and *Sl* mice comes from spleen
colony experiments. Early experiments on the colony-forming ability of *W*
marrow suggested a severe deficiency in the number of colony-forming
cells (McCulloch, Siminovitch & Till, 1964). Later experiments (Lewis *et
al.* 1967) showed that colonies are produced, but that they are extremely
small in size, and can be detected only by microscopic examination. This
suggested a failure of *W* haemopoietic cells to proliferate normally.

W mice are also very defective in their response to erythropoietin, since
to achieve a given response in *W* mice it is necessary to use about 150 times
the dose of erythropoietin used in normal mice (Keighley, Lowy, Russell &
Thompson, 1966). Nevertheless, injections of anti-erythropoietin effectively abolish erythropoiesis in *W* mice (Schooley *et al.* 1968), showing that
the hormone is necessary for erythropoiesis in these animals. Moreover, *W*
mice show a more or less normal response to hypoxia, and show markedly
elevated levels of erythropoietin under these conditions, even compared to
normal hypoxic animals (Russell & Keighley, 1972).

Haemopoietic cells from *Sl* mice show a normal capacity to produce
spleen colonies in normal irradiated recipients. However, normal haemopoietic cells transplanted into irradiated *Sl* mice show very low proliferative capacity (McCulloch, Siminovitch, Till, Russell & Bernstein, 1965),
particularly those located in the spleen (Sutherland *et al.* 1970). These
results fully support the idea that *Sl* haemopoietic cells are normal (cf. *W*
haemopoietic cells) but that the environment in which they proliferate is
deficient in some essential factor. The factor is not a humoral one, since
parabiosis with normal mice has no effect on the growth of colony-forming
cells in the *Sl* parabiont, although those in the normal partner proliferate
satisfactorily (McCulloch *et al.* 1965).

Sl mice are quite unresponsive to erythropoietin – doses of 1000-fold the
'minimum dose' effective in normal mice produced no effect in *Sl* mice
(Bernstein, Russell & Keighley, 1968). The response to hypoxia is satisfactory, provided that the mice are slowly acclimatised to low oxygen
tension (few *Sl* mice survive sudden exposure to hypoxia, and the survivors
show only slightly increased haematocrits (Bernstein *et al.* 1968)).

Metcalf & Moore (1971) have suggested that haemopoietic cells have on their surfaces receptor sites that are complementary to receptor sites on the surface of cells making up the stroma of the haemopoietic organs. Unless there is proper interaction between the haemopoietic and stroma cell receptors, the haemopoietic cells are unable to proliferate normally. In *W*, the receptor sites on the haemopoietic cells were supposed to be defective, and in *Sl* the stroma cells have ineffective receptors. Thus in both animals there will be inefficient haemopoietic cell proliferation. This hypothesis would predict that the *W* and *Sl* defects are not additive, and this has indeed been found to be the case (Sutherland, Till & McCulloch, 1970). Similar receptor sites must exist on pigment cells and germinal cells and their respective stroma, for these organs are also affected in *W* and *Sl*.

These two mutations have helped focus attention on the regulation of haemopoiesis via short-range (cell–cell?) interactions, the importance of which has recently been neglected in favour of humoral regulatory processes in haemopoiesis.

Flexed-tailed mice

The anaemia of the flexed tailed (*f/f*) mouse is normally present only during foetal and neonatal life, although its effects may reappear in an adult whose haemopoietic system has been stressed. The effect of the *f* gene is detectable in homozygotes (*f/f*) by the 11th day of gestation as a reduction in the haemoglobin concentration (Bateman, Cole, Regan & Tarbutt, 1972), showing that the yolk-sac period of erythropoiesis is affected as well as liver, spleen and bone-marrow erythropoiesis. During liver erythropoiesis, the number of erythrocytes produced is somewhat less than normal, and they contain considerably less haemoglobin than normal hepatogenic erythrocytes (Tarbutt & Cole, 1972). Erythrocytes in *f/f* foetuses and newborns contain many iron-containing siderotic granules, which may be detected by the Prussian Blue reaction (Grüneberg, 1942). This iron accumulation is suggestive of a derangement of haem synthesis (see later).

Fig. 7a shows the proportion of erythroblasts in *f/f* foetal livers, plotted against foetal age. Comparison is made with heterozygotes (*f/+*), which are haematologically normal. The relative proportions of early and late erythroblasts are shown in Fig. 7b.

In both cases the curves for *f/f* livers tend to lag behind those for *f/+* livers, suggesting a delay in the initiation of hepatic erythropoiesis. A similar lag is seen in the arrival of hepatogenic red cells in the circulation (Tarbutt & Cole, 1972).

The capacity for haem synthesis is more or less normal in *f/f* hepatic erythroblasts (Bateman *et al.* 1972). This is in contrast to the situation in *f/f* foetal reticulocytes, which synthesise haem at only one-quarter of the

13

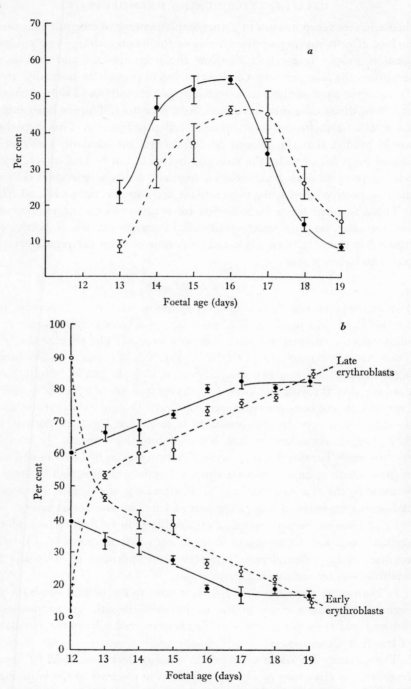

Fig. 7. (*a*) The number of nucleated erythroblasts as a percentage of the total number of nucleated cells in *f/f* (○) and *f/+* (●) foetal livers.

(*b*) Relative proportions of early and late erythroblasts in *f/f* (○) and *f/+* (●) foetal livers. (From Bateman *et al.* 1972.)

normal rate (Cole, Regan & Tarbutt, 1972). This striking difference between f/f erythroblasts (normal haem synthesis) and f/f reticulocytes (defective synthesis) may provide an important clue to the nature of the defect in flexed mice.

Linked to the defective haem synthesis in f/f reticulocytes is an abnormality in globin synthesis, which shows up as an imbalance between the synthesis of the α-chain and that of the β-chain. This defect is 'cured' if haem is added to reticulocyte cultures, but is not influenced by the addition of protoporphyrin (Cole, Garlick & Tarbutt, 1972).

Experiments in-vitro on the intrinsic rates of DNA synthesis in f/f and normal hepatic erythroblasts indicated that the rate was approximately 50% higher in f/f erythroblasts (Bateman et al. 1972). Experiments in-vivo on the cell cycles of f/f and normal foetal erythroblasts have shown that this difference probably represents a failure of the cell cycle time to increase during maturation of f/f erythroblasts (Table 4).

Table 4. *Cell cycle parameters for* f/f *and* f/+ *14-day foetal erythroblasts*

Phase	f/+ foetuses		f/f foetuses	
	Early erythroblasts	Late erythroblasts	Early erythroblasts	Late erythroblasts*
G_2 (h)	1.0 ± 0.2	1.3 ± 0.3	1.0 ± 0.3	$1\frac{1}{4}$
S (h)	4.0 ± 0.8	5.5 ± 1.3	4.9 ± 0.8	$5\frac{1}{4}$
G_1 (h)	0.3 ± 0.5	0.5 ± 1.0	0.3 ± 1.0	$\frac{1}{4}$
T_c (h)	5.3 ± 1.0	7.3 ± 1.7	6.2 ± 1.3	$6\frac{1}{2}$

* Approximate mean values

Small differences between cells at different stages of maturation within a genotype can be established more accurately than small differences *between* genotypes, since the comparisons are made on the same set of slides in the former case but between sets in the latter case. Thus we can say with some certainty, for example, that the cell cycle time in *early f/+* erythroblasts is about 50% shorter than in *late f/+* erythroblasts (Table 4), but the apparent small difference between f/f and $f/+$ early erythroblasts is unlikely to be significant.

The number of colony-forming units in post-12-day foetuses is only slightly reduced in f/f compared to $f/+$ foetal livers, and when colony-forming units are expressed in terms of the number of recognisable erythroblasts, there is little difference between the two genotypes (Bateman & Cole, 1972). The difference may be greater prior to the 13th day of gestation (one-quarter normal?), but is difficult to quantitate in these rapidly-growing early foetuses. The spleen colonies originating from f/f

cells are somewhat smaller than those originating from $f/+$ cells (Thompson et al. 1966), and there is a smaller proportion of erythroid colonies than occur in spleens of animals injected with $f/+$ cells (McCulloch et al. 1965). Thus it appears that the number of haemopoietic stem cells in f/f foetal livers is similar to that in $f/+$ foetal livers, although their eventual differentiation into recognisable erythroblasts may be impaired.

The major effects of the f mutation in foetal erythropoiesis thus appear to be to delay the appearance of recognisable erythroblasts in the liver, and to produce a derangement of haem synthesis and globin synthesis in reticulocytes, but not in erythroblasts. The abnormal globin synthesis is secondary to the derangement of haem synthesis, and can be 'cured' invitro by addition of haem to cultures of reticulocytes (Cole, Garlick & Tarbutt, 1972). There is also a small change in the cell cycle of erythroblasts, probably representing an attempt to compensate for the anaemia. These apparently diverse effects may be explained if it is assumed that the defect results primarily in inefficient synthesis or activity of an early haemsynthesising enzyme, which is produced in excess in erythroblasts. This will lead directly to a delay in the initiation of erythropoiesis, normal haem synthesis in erythroblasts and defective synthesis in reticulocytes (Tarbutt & Cole, 1972). The defect in globin synthesis in reticulocytes then follows directly from the defect in haem synthesis, and the reduced cycle time is a response to stress.

Response of f/f foetal erythropoiesis to erythropoietin

Foetal liver cultures from f/f and $f/+$ foetuses at certain stages of development are responsive to erythropoietin when this is measured in terms of the rate of haem synthesis 28 h after explanation. However, the changes in the response during foetal development are quite different from those described previously for Swiss mice, and there is also a difference between $f/+$ and f/f livers. Livers from $f/+$ foetuses give a high response at day 13 of gestation, but the response gradually diminishes as the animal develops, until on day 17 no response can be detected. Livers from f/f animals behave similarly, except that the response lasts until the 18th day of gestation. This difference correlates with the lag in haemopoiesis that occurs in f/f relative to $f/+$ livers. There may also be a correlation with the time taken to reach a particular concentration of red cells in the circulation; thus the three genotypes discussed in this paper (f/f, $f/+$ and Swiss) achieve a red cell concentration of 2.4×10^9 cells/ml on days 18, 17 and 15 respectively, which coincide with the days on which the livers first become refractory to erythropoietin. The pattern of change in red cell concentra-

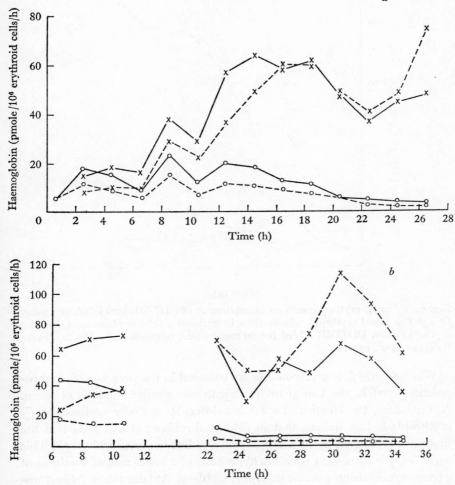

Fig. 8. (a) Effect of erythropoietin on haem synthesis in 13-day f/f and $f/+$ foetal livers in-vitro. Data expressed in terms of 10^6 nucleated erythroid cells present in the cultures at explanation. ^{59}Fe added for 1 h pulses. ×—×, $f/+$, + Ep; o—o, $f/+$, − Ep; ×--×, f/f, + Ep; o----o f/f, − Ep.

(b) As (a), but up to 35 h. (From Bateman et al. 1972.)

tion is also similar to the pattern of erythropoietin responsiveness: in Swiss mice, there is an abrupt change in doubling time, occurring on day 15 and the change from erythropoietin responsiveness to unresponsiveness also occurs abruptly on day 15; in f/f and $f/+$ foetuses, the red cell count increases more gradually, and the loss of responsiveness is also a gradual process.

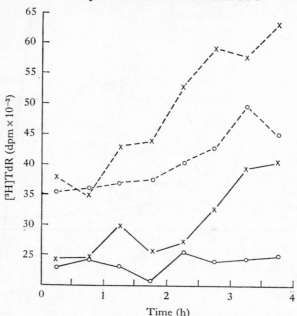

Fig. 9. Effect of erythropoietin on incorporation of [³H]TdR into DNA of 13-day *f*/*f* and *f*/+ foetal erythroid cells in-vitro, in terms of a 10⁶ complement of dividing erythroid cells. [³H]TdR added for 30 min pulses. Symbols as in Fig. 8. (From Bateman *et al.* 1972.)

When *f*/*f* and *f*/+ erythroblasts are cultured in the presence of erythropoietin for 36 h, the two genotypes synthesise similar amounts of haem. Nevertheless, the kinetics of synthesis differ, for initially synthesis in *f*/*f* erythroblasts lags behind that in *f*/+ erythroblasts (Fig. 8), so that the first peak is reached by 12 h in *f*/+ cultures but not until 18 h in *f*/*f*. This early delay is consistent with the hypothesis of a lower rate of synthesis of a haem-synthesising enzyme in *f*/*f* erythroblasts. At later times, *f*/*f* erythroblasts tend to synthesise haem more rapidly than *f*/+ erythroblasts; this later response cannot easily be interpreted, especially as we do not know the detailed population changes taking place during the response.

Foetal liver cultures from both *f*/*f* and *f*/+ mice can synthesise both DNA and RNA in-vitro, and respond to erythropoietin by increased synthesis of these molecules. As already pointed out, the initial rate of DNA synthesis is greater in *f*/*f* than in *f*/+ cultures, probably reflecting a slightly shorter cell cycle time in late *f*/*f* erythroblasts. The DNA response to erythropoietin is similar in the two genotypes (Fig. 9), and is apparent within the first hour of culture. This response almost certainly occurs in already-differentiated erythroid cells, since these make up the bulk of the

DNA-synthesising cells in the culture. As for RNA synthesis, the response is unfortunately complicated by the rapid disappearance from the culture of RNA-synthesising non-erythroid cells. However, there is little doubt that RNA synthesis in-vitro reaches higher levels in f/f than in $f/+$ cultures in the absence of erythropoietin, and that erythropoietin elicits similar responses in the two genotypes (Bateman *et al.* 1972). The greater rate of RNA synthesis in f/f compared to $f/+$ cultures may reflect the greater proliferative activity in-vitro of f/f cells, since RNA synthesis is closely related to DNA synthesis in this system.

A HYPOTHESIS CONCERNING THE SEQUENCE OF EVENTS OCCURRING DURING HEPATIC HAEMOPOIESIS IN THE MOUSE

The following account of some of the more important events of hepatic haemopoiesis is mainly based on the experiments described in this paper.

Hepatic haemopoiesis is probably initiated by immigrant haemopoietic stem cells of yolk-sac origin (Metcalf & Moore, 1971). On arrival in the liver bud, the stem cells interact with the stroma cells, which influence their proliferation. If this interaction is inefficient, defective haemopoiesis may result (*Sl* and *W* anaemias). The liver environment in 10–14-day foetuses is also able to induce stem cells to differentiate, predominantly into erythropoietin-sensitive cells, which can be detected in the liver bud within half a day of its appearance (Bateman, 1971). The corresponding induction in normal adult tissue is associated with a marked reduction in cycle time, but this probably does not occur in the foetus, where the cycle time of stem cells may already be maximal. After induction, the cells become responsive to erythropoietin, which, according to a hypothesis of Kretchmar (1966), enters the cells during part of the G_1 phase, but is only effective in initiating further differentiation during the S period. Kretchmar proposed that control of differentiation could be achieved by varying the length of G_1, so that, for example, if G_1 was prolonged, the erythropoietin that had entered the cell during that cycle would have decayed before the cell reached the critical period in the S phase, and so the cell could not differentiate. There is evidence that the erythropoietin-dependent differentiation step is also accompanied by an increase in proliferation rate, but once again this may not occur in the already rapidly-proliferating cells of the foetus.

The 'inductive microenvironment' (stem cell/erythropoietin-sensitive cell transition) is probably maintained until about the 14th day of gestation in Swiss mice. On day 14, no new erythropoietin-sensitive cells are formed from stem cells, and the erythropoietin-sensitive cell compartment rapidly empties, so that by day 15 the livers are refractory to erythropoietin. Although not inducive to stem cell differentiation, the stem cell/stroma cell interaction is still effective in allowing stem cells to proliferate. Some of these stem cells migrate to the spleen, to initiate haemopoiesis in that organ.

The 'cut-off' of the stem cell supply leads in turn to a cut-off in the supply of cells differentiating into proerythroblasts, so that there is a more or less orderly decrease in the numbers of recognisable precursors at the various stages of maturation (Tarbutt & Cole, 1970). This in turn leads to a slowing down in the output of erythrocytes, and calculations show that the expected red cell output is similar to the biphasic pattern observed.

In f/f and $f/+$ foetuses the loss of the inductive environment is apparently a slower process than in Swiss mice, as reflected by the gradual decline in erythropoietin sensitivity. The defect in f/f haemopoiesis is probably located near to the time of differentiation into proerythroblasts, and delays this process as well as upsetting haem synthesis in reticulocytes.

REFERENCES

BATEMAN, A. E. (1971). Foetal erythropoiesis. Ph.D. Thesis, University of Sussex.
BATEMAN, A. E. & COLE, R. J. (1972). Colony forming cells in the livers of prenatal flexed (f/f) anaemic mice. *Cell and Tissue Kinetics*, **5**, 165–73.
BATEMAN, A. E., COLE, R. J., REGAN, T. & TARBUTT, R. G. (1972). The role of erythropoietin in prenatal erythropoiesis of congenitally anaemic flexed-tailed (f/f) mice. *British Journal of Haematology*, **22**, 415–27.
BECKER, A. J., McCULLOCH, E. A., SIMINOVITCH, L. & TILL, J. E. (1965). The effect of differing demands for blood cell production on DNA synthesis by hemopoietic colony-forming cells of mice. *Blood*, **26**, 296–308.
BECKER, A. J., McCULLOCH, E. A. & TILL, J. E. (1963). Cytological demonstration of the clonal nature of spleen colonies derived from transplanted mouse marrow cells. *Nature, London*, **197**, 452–4.
BERNSTEIN, S. E. (1970). Tissue transplantation as an analytic and therapeutic tool in the treatment of hereditary anaemias. *American Journal of Surgery*, **119**, 448–51.
BERNSTEIN, S. E., RUSSELL, E. S. & KEIGHLEY, G. (1968). Two hereditary mouse anaemias (Sl/Sl^d and W/W^v) deficient in response to erythropoietin. *Annals of the New York Academy of Science*, **149**, 475–85.
BLACKETT, N. M. (1968). Investigation of bone-marrow stem cell proliferation in normal, anaemic and irradiated rats, using methotrexate and tritiated thymidine. *Journal of the National Cancer Institute*, **41**, 909–18.
BLACKETT, N. M. (1972). The proliferation and maturation of hemopoietic cells. In *The Cell Cycle and Cancer*, ed. R. Baserga, chapt. 2, pp. 27–53. New York: Marcel Dekker.
BLEIBERG, I. & FELDMAN, M. (1969). On the regulation of hemopoietic spleen colonies produced by embryonic and adult cells. *Developmental Biology*, **19**, 566–80.
BRUCE, W. R. & McCULLOCH, E. A. (1964). The effect of erythropoietic stimulation on the hemopoietic colony-forming cells of mice. *Blood*, **23**, 216–32.

CARMENA, A. O., HOWARD, D. & STOHLMAN, F. JR. (1968). Regulation of erythro-poiesis. XXII. Erythropoietin production in the newborn animal. *Blood*, **32**, 376–84.

COLE, R. J., GARLICK, J. & TARBUTT, R. G. (1973). Disturbed haem and globin synthesis in reticulocytes of prenatal (*f/f*) anaemic mice. *Genetical Research*, in press.

COLE, R. J. & PAUL, J. (1966). The effects of erythropoietin on haem synthesis in mouse yolk sac and cultured foetal liver cells. *Journal of Embryology and Experimental Morphology*, **15**, 245–60.

COLE, R. J., REGAN, T. & TARBUTT, R. G. (1972). Haemoglobin synthesis in reticulocytes of prenatal (*f/f*) anaemic mice. *British Journal of Haematology*, **23**, 443–52.

CURRY, J. L. & TRENTIN, J. J. (1967). Hemopoietic spleen colony studies. I. Growth and differentiation. *Developmental Biology*, **15**, 395–413.

FRIED, W., MARTINSON, D., WEISMAN, M. & GURNEY, C. W. (1966). Effect of hypoxia on colony-forming units. *Experimental Hematology*, **10**, 22–29.

GRÜNEBERG, H. (1942). The anaemia of flexed-tailed mice (*Mus musculus* L.). II. siderocytes. *Journal of Genetics*, **44**, 246–71.

HANNA, I. R. A. (1967). Response of early erythroid precursors to bleeding. *Nature, London*, **211**, 355–7.

HANNA, I. R. A. (1968*a*). Ph.D. Thesis, University of London.

HANNA, I. R. A. (1968*b*). An early response of the morphologically recognisable erythroid precursors to bleeding. *Cell and Tissue Kinetics*, **1**, 91–8.

HANNA, I. R. A., TARBUTT, R. G. & LAMERTON, L. F. (1969). Shortening of the cell cycle time of erythroid precursors in response to anaemia. *British Journal of Haematology*, **16**, 381–7.

KEIGHLEY, G. H., LOWY, P., RUSSELL, E. S. & THOMPSON, M. W. (1966). Analysis of erythroid homeostatic mechanisms in normal and genetically anaemic mice. *British Journal of Haematology*, **12**, 461–77.

KRETCHMAR, A. L. (1966). Erythropoietin: Hypothesis of action tested by analogue computer. *Science, New York*, **152**, 367–70.

KUBANEK, B., RENCRICCA, N., PORCELLINI, A. & STOHLMAN, F. JR. (1971). The effect of plethora and erythropoietin on erythropoiesis in heavily irradiated recipients receiving foetal liver cells. In *The Regulation of Erythropoiesis and Haemoglobin Synthesis*, ed. T. Travnicek & J. Neuwirt, pp. 157–64. Prague: Universita Karlova.

LAJTHA, L. G., POZZI, L. V., SCHOFIELD, R. & FOX, M. (1969). Kinetic properties of hemopoietic stem cells. *Cell and Tissue Kinetics*, **2**, 39–49.

LATSINIK, N. V., SAMOYLINA, N. L. & CHERTKOV, L. (1971). Susceptibility to polycythemia of hemopoietic spleen colonies produced by cultured embryonal liver cells. *Journal of Cellular Physiology*, **78**, 405–10.

LEWIS, J. P., O'GRADY, L. F., BERNSTEIN, S. E., RUSSELL, E. S. & TROBOUGH, F. E. (1967). Growth and differentiation of transplanted *W/W* marrow. *Blood*, **30**, 601–16.

MCCULLOCH, E. A. (1971). Control of hematopoiesis at the cellular level. In *Regulation of Hematopoiesis*, vol. 1, ed. A. S. Gordon, pp. 133–59. New York: Appleton-Century-Crofts.

MCCULLOCH, E. A., SIMINOVITCH, L. & TILL, J. E. (1964). Spleen colony formation in anaemic mice of genotype *W/W^v*. *Science, New York*, **144**, 844–6.

MCCULLOCH, E. A., SIMINOVITCH, L., TILL, J. E., RUSSELL, E. S. & BERNSTEIN, S. E. (1965). The cellular basis of the genetically determined hemopoietic defect in anaemic mice of genotype *Sl/Sl^d*. *Blood*, **26**, 399–40.

METCALF, D. & MOORE, M. A. S. (1971). *Haemopoietic Cells*, chapt. 10, pp. 488–98. Amsterdam: North-Holland Publishing Co.

MICKLEM, H. S., FORD, C. E., EVANS, E. P., OGDEN, D. A. & PAPWORTH, D. S. (1972). Competitive *in vivo* proliferation of foetal and adult haematopoietic cells in lethally irradiated mice. *Journal of Cellular Physiology*, **79** (2), 293–8.

PAUL, J., CONKIE, D. & FRESHNEY, R. I. (1969). Erythropoietic cell population changes during the hepatic phase of erythropoiesis in the foetal mouse. *Cell and Tissue Kinetics*, **2**, 283–94.

PAUL, J. & HUNTER, J. A. (1969). Synthesis of macromolecules during induction of haemoglobin synthesis by erythropoietin. *Journal of Molecular Biology*, **42**, 31–40.

PINKERTON, P. H. & BANNERMAN, R. M. (1968). The hereditary anaemias of mice. *Hematological Review*, **1**, 119–92.

RUSSELL, E. S. (1970). Abnormalities of erythropoiesis associated with mutant genes in mice. In *Regulation of Hematopoiesis*, vol. 1, ed. A. S. Gordon, pp. 649–75. New York: Appleton-Century-Crofts.

RUSSELL, E. S. & KEIGHLEY, G. (1972). The relation between erythropoiesis and plasma erythropoietin levels in normal and genetically anaemic mice during prolonged hypoxia or after whole body irradiation. *British Journal of Haematology*, **22**, 437–52.

RUSSELL, E. S., SMITH, L. J. & LAWSON, F. A. (1956). Implantation of normal blood-forming tissues in radiated genetically anaemic hosts. *Science, New York*, **124**, 1076–7.

SCHOFIELD, R. (1970). A comparative study of the repopulating potential of grafts from various haemopoietic sources. *Cell and Tissue Kinetics*, **3**, 119–30.

SCHOOLEY, J. C. (1966). The effect of erythropoietin on the growth and development of spleen colony-forming cells. *Journal of Cellular Physiology*, **68**, 249–62.

SCHOOLEY, J. C., GARCIA, J. F. M., CANTOR, L. N. & HAVENS, V. W. (1968). A summary of some studies on erythropoiesis using antierythropoietin immune serum. *Annals of the New York Academy of Science*, **149**, 266–80.

SILINI, G., PONS, S. & POZZI, L. V. (1968). Quantitative histology of spleen colonies in irradiated mice. *British Journal of Haematology*, **14**, 489–500.

SILINI, G., POZZI, L. V. & PONS, S. (1967). Studies on the haemopoietic stem cells of mouse foetal liver. *Journal of Embryology and Experimental Morphology*, **17**, 303–18.

STOHLMAN, F. JR. (1970). Fetal erythropoiesis. In *Regulation of Hematopoiesis*, vol. 1, ed. A. S. Gordon. pp. 471–85. New York: Appleton-Century-Crofts.

SUTHERLAND, D. J. A., TILL, J. E. & McCULLOCH, E. A. (1970). A kinetic study of the genetic control of haemopoietic progenitor cells assayed in culture and *in vivo*. *Journal of Cellular Physiology*, **75**, 267–74.

TARBUTT, R. G. (1969*a*). Cell population kinetics of the erythroid system in the rat; the response to protracted anaemia and to continuous γ-irradiation. *British Journal of Haematology*, **16**, 9–24.

TARBUTT, R. G. (1969*b*). Erythroid cell proliferation in hypertransfused rats. *British Journal of Haematology*, **17**, 191–8.

TARBUTT, R. G. & COLE, R. J. (1970). Cell population kinetics of erythroid tissue in the liver of foetal mice. *Journal of Embryology and Experimental Morphology*, **24**, 429–46.

TARBUTT, R. G. & COLE, R. J. (1972). Foetal erythropoiesis in genetically anaemic flexed-tailed (*f/f*) mice. *Cell and Tissue Kinetics*, **5**, 491–503.

THOMPSON, M. W., McCULLOCH, E. A., SIMINOVITCH, L. & TILL, J. E. (1966). The cellular basis for the defect in haemopoiesis in flexed-tailed mice. *British Journal of Haematology*, **12**, 152–60.

TILL, J. E. & McCULLOCH, E. A. (1961). A direct measurement of the radiation sensitivity of normal mouse bone marrow cells. *Radiation Research*, **14**, 213–22.

TILL, J. E., SIMINOVITCH, L. & McCULLOCH, E. A. (1966). Growth and differentiation of marrow cells transplanted in anemic and plethoric mice of genotype *W/W^v*. *Experimental Hematology*, **9**, 59.

TRENTIN, J. J. (1971). Influence of hematopoietic organ stroma on stem cell differentiation. In *Regulation of Hematopoiesis*, vol. 1, ed. A. S. Gordon, pp. 161–86. New York: Appleton-Century-Crofts.

TRENTIN, J. J., BRAATEN, B. A., AMEND, N., PRASAD, N., WOLF, N. S. & JENKINS, V. K. (1969). Intra-clonal origin of the second hemopoietic cell line in spleen colonies. *Federation Proceedings*, **28**, 295–304.

VOGEL, H., HADJIK, I., SULTANIAN, I. & MATIOLI, G. (1970). Growth kinetics of hemopoietic fetal stem cells. *Journal of Cellular Physiology*, **76**, 117–26.

WU, A. M., TILL, J. E., SIMINOVITCH, L. & McCULLOCH, E. A. (1967). A cytological study of the capacity for differentiation of normal hemopoietic colony-forming cells. *Journal of Cellular Physiology*, **69**, 177–84.

WU, A. M., TILL, J. E., SIMINOVITCH, L. & McCULLOCH, E. A. (1968). Cytological evidence for a relationship between normal hematopoietic colony-forming cells and cells of the lymphoid system. *Journal of Experimental Medicine*, **127**, 455–64.

THE MITOTIC ACTIVATION OF
LYMPHOCYTES – BIOCHEMICAL
AND IMMUNOLOGICAL CONSEQUENCES

BY D. A. HARDY AND N. R. LING

Department of Experimental Pathology,
University of Birmingham Medical School, Birmingham B15 2TJ

INTRODUCTION

It is widely accepted that the lymphocyte is the cell responsible for the specific recognition of, and response to, foreign material entering a mammal. When such antigenic material is introduced into the body, changes are seen in the lymph node draining the site. Lymphocytes become activated by antigen into lymphoblastoid cells which are intensely basophilic and some of them appear in the efferent lymph.

The activation of lymphocytes by specific antigen can also take place in tissue culture. Lymphocytes separated from blood, lymph or various lymphoid organs from a large number of species have all been shown to respond, in-vitro, to antigenic challenge by blastogenesis, characteristic morphological changes and mitosis (see Ling, 1968, for references). Although this can be shown to be a highly specific stimulation, obtainable only with antigens to which the animal has been previously immunised, it is possible to obtain a phenomenon, which is at least superficially similar, with various non-specific stimulants. The first of these to be discovered (Nowell, 1960), and still the best known, is the extract of the bean *Phaseolus vulgaris*, phytohaemagglutinin (PHA). There are also a whole range of other substances such as staphylococcal filtrate (SF), anti-lymphocyte serum (ALS) and concanavalin A (con A) which are potent lymphocyte mitogens and also some heavy metal ions. Lymphocytes from another individual of the same species, differing in antigenic make up, are also capable of inducing a similar reaction (Bain, Vas & Lowenstein, 1964). This so called mixed lymphocyte reaction (MLR) may well have significance in the body as a defence against malignant cells (see Hardy & Steel, 1971 and section 3c).

This paper deals with three aspects of lymphocyte activation: (1) the early changes in the lymphocyte following the addition of mitogen; (2) the cell cycle of activated lymphocytes; (3) the properties of activated lymphocytes.

We shall attempt to discuss some of the evidence on these topics with special reference to the mechanism of the activation of the cells, the possible changes in expression of the genome following activation and the likely biological significance of the process.

(1) THE EARLY CHANGES IN THE LYMPHOCYTE FOLLOWING ACTIVATION

(a) Effects of stimulant on the membrane

It is generally accepted that stimulants such as PHA act by binding to the cell. Cytological techniques using fluorescein-conjugated (Michalowski, Jasinska, Brzosko & Nowoslawski, 1964; Razavi, 1966) or radio-labelled PHA (Rieke, 1965; Conard, 1967; Stanley, Frenster & Rigas, 1971) or indirect techniques using conjugated antisera to PHA (Michalowski, Jasinska, Brzosko & Nowoslawski, 1965; Byrd, Hare, Finley & Finley, 1967) have been employed to determine the intracellular localisation of the PHA. Most workers agree that labelled material appears very rapidly in the cytoplasm of the cell or in the nucleus (Michalowski *et al.* 1964; Conard, 1967; Stanley *et al.* 1971) or next to the cytoplasmic membrane (Rieke, 1965; Michalowski *et al.* 1965). These studies do not do more than trace the fate of the (usually impure) PHA during the course of the activation and appear to be secondary to the initiatory events occurring at the cell surface.

Evidence that the activation is dependent upon surface membrane events comes from a number of studies. We and others (see Hardy & Ling, 1969; Hardy, Knight & Ling, 1970) have shown that for effective stimulation in the MLR two viable intact cell populations are required. It is difficult to imagine how activation could occur in this case other than by a direct interaction of the surface membranes of the cells involved. Stimulation with insoluble antigens such as sheep red cells (see Knight, Walker & Ling, 1971) would also make surface stimulation highly probable, as would the fact that ALS, containing antibodies principally to surface antigens, is an effective stimulant (Grasbeck, Nordman & De La Chapelle, 1964). Recently Greaves & Bauminger (1972) have observed stimulation with PHA made insoluble by covalently coupling it to sepharose. This confirms the earlier studies of Kay (1971) who found that PHA bound to cell membrane was an effective stimulant.

Enzymes, such as trypsin or neuraminidase, known to act at the cell surface, suppressed the response of lymphocytes to PHA (Lindahl-Kiessling & Peterson, 1969a). However, the cells were capable of recovering responsiveness after some hours, suggesting that surface regeneration must occur before the PHA response can be initiated. Interestingly the enzyme-treated cells were able to respond to allogeneic lymphocytes in a MLR, indicating that the activation process in this system is somewhat different (Lindahl-Kiessling & Peterson, 1969b). In the case of PHA, it has been shown that the binding of the mitogen to the membrane is not dependent upon active metabolism of the cells, for it is not inhibited by drugs which block glycolysis or the citric acid cycle (Lindahl-Kiessling & Mattsson, 1971). This strongly suggests that PHA exerts its effect on the outer cell membrane and that endocytosis of the molecule is not a prerequisite for initiation of proliferation.

All these observations, as well as the fact that the mitogenic action of PHA can be stopped by the addition of anti-PHA antibody, which presumably acts at the cell surface (see below), provides very strong evidence that the initiatory event of activation occurs at the cell surface.

The length of time that lymphocytes have to be in contact with stimulant for subsequent activation to occur has been extensively studied. The usual approach involves washing the lymphocytes free from PHA after various times and measuring the subsequent activation. Results varying from maximal activation after 5 min exposure to PHA (Richter et al. 1966) to little or no activation following incubation with PHA for as long as 12 or 24 h (Tormey & Mueller, 1965; Yamamoto, 1966) have been reported, which probably reflect the difficulties involved in effectively removing the stimulant without damaging the cells.

The fact that binding of the mitogen con A to lymphocytes can be reversed by the sugar methyl α-D-mannoside has been used by Powell & Leon (1970) and Lindahl-Kiessling (1972) to study the length of time that the stimulant must remain in contact with the lymphocytes for activation to occur. They showed that exposure of the cells to con A for periods of even 19–20 h did not appear to commit the cells irrevocably to incorporate thymidine. Antiserum to PHA has also been used as a specific 'stop' reagent for lymphocytes stimulated with PHA and washed. Using this device it has been found that the PHA needs to be left on the lymphocyte surface for some hours for activation to proceed normally. This time has has been reported to be 6 h (Kay & Oppenheim, 1971) or at least 24 h (Hausen & Stein, 1968). Thus, although binding of stimulants such as PHA to the lymphocyte membrane occurs rapidly, being complete within 2 h (Kay, 1969, 1971) its presence on the cells is required for much longer periods in order for the population to reach maximal mitotic activity. The crucial question of whether stimulant is required continuously for the activated cell to progress through one or more cell cycles is difficult to answer with certainty, but it would seem more likely that once the cell has reached a certain point in the activation process it will continue through it without the subsequent involvement of stimulating moiety. The question of whether an activated cell can proceed in the reverse direction in the stimulation sequence and can return to a non-dividing small lymphocyte is considered in Section 2(d). However, it is important to note that there is a distinction between the experiments discussed here, where the cell is *stopped* at a point in the activation process and those discussed in Section 2(d) where an activated cell *reverses* to become a resting small lymphocyte once more. That the cell does not revert to its original state as a consequence of treatment with the PHA antiserum is shown by the fact that removal of the antiserum by washing, followed by the addition of fresh PHA, resulted in a larger and more rapid incorporation of uridine into the cells compared with that found during the primary induction with PHA (Hausen & Stein, 1968). This indicates that the cells remain in some form of activated state throughout the treatment.

 Changes in the biochemistry and physiology of the cell membrane occur soon after stimulation with PHA or other stimulant. Increased incorporation of radiolabelled precursors such as phosphate ions, inositol, choline and acetate into phospholipids can be detected in lymphocytes within minutes or hours following the addition of stimulant (Fisher & Mueller, 1968, 1969, 1971; Kay, 1968a; Huber *et al.* 1968; Lucas, 1970; Lucas, Shohet & Merler, 1971). Fisher & Mueller (1969, 1971) consider there to be two phases in this increased synthesis – an early effect on phosphatidyl inositol which is independent of protein synthesis and a later synthesis of other phospholipids which is sensitive to inhibitors of protein synthesis. The latter may represent the production of new membranes in preparation for cell division. The former is considered to be an important event in the activation process. Hokin & Hokin (1965), studying phospholipid synthesis in pancreas slices, suggest that accelerated phosphatidyl inositol synthesis is associated with a stimulation of transmembranal transport. Increased membrane transport is seen in stimulated lymphocytes (see later) and the two phenomena may be closely linked. One study on the antigen-induced stimulation of human lymphocytes with tetanus toxoid, however, may argue against this simple interpretation. Lucas (1970) and Lucas, Shohet & Merler (1971) have shown that although phosphatidyl choline and phosphatidyl ethanolamine synthesis were increased, that of phosphatidyl inositol was not. Thus it appears that the initial phospholipid changes in the membrane may vary with different stimulants, although the subsequent blast transformation and DNA synthesis would appear to be the same.

A marked acceleration of the transport of low molecular weight substances across the cell membrane occurs in lymphocytes soon after the addition of stimulant. Ions such as K^+ (Quastel, Dow & Kaplan, 1970) or Ca^{2+} (Allwood, Asherson, Davey & Goodford, 1971) have been shown to enter the cells very rapidly and an increased efflux of Na^+ has been reported (Dent, unpublished – quoted in Dent, 1971). The accelerated uptake of nucleosides such as uridine (Kay & Handmaker, 1970; Peters & Hausen, 1971a), amino acids such as α-aminoisobutyric (Mendelsohn, Skinner & Kornfeld, 1971) and sugars such as 3-O-methyl-glucose (Peters & Hausen, 1971b) indicate that this is a general phenomenon. The fact that such uptake is not affected by inhibitors of RNA and protein synthesis (Kay & Handmaker, 1970; Mendelsohn et al. 1971; Peters & Hausen, 1971a) suggests that PHA is having a direct effect on the cell membrane.

Membrane poisons have a marked effect on the incorporation of substances into the cell. Quastel & Kaplan (1968) have extensively studied the inhibitory effects of ouabain on the incorporation of labelled amino acids, uridine and thymidine into control and stimulated lymphocytes. The inhibitory effects have been confirmed by Kay (1972) and by ourselves (Fig. 1). Quastel & Kaplan (1970) consider that the effect of the inhibitor is on the lymphocyte transformation process itself. However, the facts that the lymphocytes are equally sensitive to ouabain no matter how soon after stimulation the inhibitor is added, that the effect is reversible, that all cells of the population, not only those stimulated are affected, suggest that it is more likely to be a non-specific inhibition of isotope incorporation and not primarily an effect on the transformation process.

It may be, as Kay (1972) suggests, that protein synthesis in stimulated lymphocytes is more sensitive to variations in the K^+ environment of the cells than non-stimulated cells and the drug in blocking this appears to have a differential effect on stimulation. However, it is not established whether the increase in K^+ uptake causes the increase in protein synthesis or is merely required for it.

The inhibitor dipyridamole also blocks the incorporation of nucleotides into stimulated lymphocytes (Peters & Hausen, 1971a) but this drug does not markedly inhibit amino acid incorporation (Fig. 2). Activation of the cells can occur (as measured by protein synthesis) even though nucleoside uptake is blocked.

The effects of membrane poisons on lymphocytes need to be interpreted with caution. The drugs, in blocking membrane transport, are bound to interfere with the metabolism of the cell, not to mention direct effects on the incorporation of labelled precursors which may be used for the assay system. It is thus difficult to distinguish between any general inhibitory effect and a primary action of the drugs on the activation mechanism of the cells.

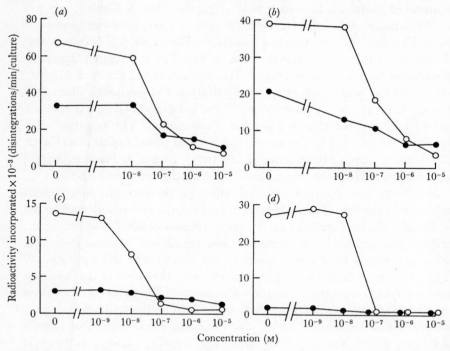

Fig. 1. The effect of ouabain on the incorporation of radiolabelled amino acids and nucleosides by stimulated and non-stimulated lymphocytes. Lymphocytes were separated from human cord blood (*a* and *b*) or from pig blood (*c* and *d*), after gelatin sedimentation of the red cells, by a ficoll–triosil gradient technique (see Mackintosh *et al.* 1972). Human cells, after washing, were resuspended to 1×10^6/ml in 20 % pooled human serum gelatin/Eagles MEM without leucine and 1 ml volumes dispensed into $3 \times \frac{1}{2}$ inch capped tubes. Ouabain was added to give the final concentration shown. PHA (2 µg; Burroughs Wellcome partially purified batch X 5 – kindly donated by Dr B. A. L. Hurn) was added to the stimulated (open circles) cultures. Controls having no PHA are shown as closed circles. [³H]4,5-L-leucine (1 µCi; 52000 mCi/mmole) (Fig. 1*a*) or [³H]uridine (1 µCi; 5000 mCi/mmole) (Fig. 1*b*) was added at time 0 and the cells harvested 18 h later. Pig cells were resuspended to 2×10^6/ml in 10 % foetal calf serum/RPMI 1640 medium. Ouabain was added as before. Burroughs Wellcome crude PHA (30 µg/ml) was used as stimulant (open circles) and controls contained no PHA (closed circles). [³H]leucine (1 µCi; 29000 mCi/mmole) was added 24 h after initiating the cultures (Fig. 1*c*) and [³H]thymidine after 48 h (Fig. 1*d*). Cells were harvested 24 h after the addition of isotope. Radioactivity in the TCA insoluble material was estimated by a liquid scintillation technique (Ling & Holt, 1967). Values in all experiments are the means of triplicate cultures. Ouabain at the higher concentrations was toxic. In the experiment with pig lymphocytes 25 % and 50 % of the number of cells in the control culture were viable at day 3, as judged by a trypan blue dye exclusion test, at ouabain concentrations of 10^{-5} and 10^{-6} respectively.

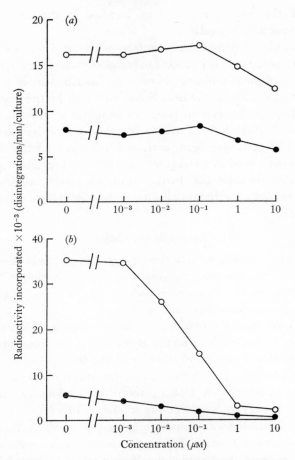

Fig. 2. The effect of dipyridamole on the incorporation of [³H]leucine or [³H]thy-midine by stimulated and non-stimulated pig lymphocytes. A suspension of pig blood lymphocytes was prepared as for Fig. 1c and d at 2 × 10⁶/ml and dispensed into 1 ml cultures. Dipyridamole was added to give the concentration shown. PHA (crude 30 μg/culture) was added to stimulated cultures (open circles). Controls had no PHA (closed circles). [³H]leucine (1 μCi; 29 000 mCi/mmole) (Fig. 2a) was added at time 0 and [³H]thymidine (Fig. 2b) after 24 h. Cells were harvested 24 h after the addition of isotope. Values are means of triplicate cultures. Dipyridamole was not toxic even at 10 μM as judged by trypan blue dye exclusion counts on the cells performed up to 96 h.

Numerous other effects of PHA on lymphocyte membrane phenomena during activation have been reported. For example, Hirschhorn, Brittinger, Hirschhorn & Weissman (1968) have shown an enhanced uptake of neutral red dye and Robineaux, Bona, Anteunio & Orme-Rosselli (1969) have found an increased pinocytosis following stimulation. Smith, Steiner,

Newberry & Parker (1971) reported an increase in the activity of the membrane enzyme adenyl cyclase.

Taken as a whole all these observations strongly implicate a stimulant–membrane event as the primary process in lymphocyte activation. However, it is unlikely that all the membrane phenomena described are important for the subsequent activation of the cell. Some may well be the result of the stimulant binding to the cell surface. An example of this may be the similar efflux of Na^+ ions from mouse lymphocytes or lymphoma cells treated with PHA, the former being activated while in the latter thymidine incorporation is depressed by PHA (Dent, unpublished results – quoted in Dent, 1971). It would seem that further study on membrane phenomena in relation to lymphocyte activation is vital if the stimulation mechanism is to be understood.

(b) Changes in the nucleus

Many authors have considered that the changes occurring in the lymphocyte following stimulation are a reflection of gene activation of the cells. However, attempts to demonstrate gene activation in the early stages of the transformation process have been inconclusive.

Pogo, Allfrey & Mirsky (1966, 1967) showed that acetylation of pre-existing histones in the nucleus occurred soon after the addition of PHA and, in fact, preceded a very rapid increase in the incorporation of uridine into RNA. The observed acetylation of histones has since been confirmed by others (e.g. Mukherjee & Cohen, 1969; Darżynkiewicz, Bolund & Ringertz, 1969). However, MacGillivray & Monjardino (1968) found that, although it was possible to demonstrate histone acetylation, this was not related to lymphocyte activation. PHA preparations which were incapable of causing activation of the lymphocytes nevertheless produced increased acetylation of histone, whereas other purified PHA preparations which contained mitogenic activity caused a depression in histone acetylation compared with controls. Thus it would appear that histone acetylation is more a reflection of some membrane effect of the PHA rather than its mitogenicity but the true significance of this is not understood.

An increased incorporation of [^{32}P]phosphate into nuclear phosphoproteins following stimulation of lymphocytes by PHA has also been reported (Kleinsmith, Allfrey & Mirsky, 1966). However, Cross & Ord (1970) demonstrated that although uptake of [^{32}P]phosphate occurred immediately after stimulation there was no rise in the net phosphate content of the histones until the period of DNA synthesis. Subsequently they observed (Cross & Ord, 1971) that either PHA or 6-N, 2'-O-dibutyryl-adenosine 3':5'-cyclic monosphosphate (which gave similar lymphocyte stimulation) resulted in an increase in the size and specific activity of the intracellular phosphate pool. The activity of the enzymes histone kinase and histone phosphatase increased soon after stimulation with PHA and then decreased again. These observations could suggest that the phosphorylation of histones observed is a reflection of an apparent increase in the turnover of phosphate in the histones due to changes in the intracellular pool of phosphate rather than evidence for gene activation.

Other chemical and physical changes in the deoxyribonucleic acid protein (DNP) complex of lymphocyte nuclei have been observed soon after the addition of PHA. The binding of dyes such as acridine orange (Killander & Rigler, 1969) and alkaline bromophenol blue (Zetterberg & Auer, 1969) or of the drug actinomycin D (Ringertz, Darżynkiewicz & Bolund, 1969) is increased. The thermal sensitivity of the DNA following PHA addition has also been reported to be altered (Rigler & Killander, 1969; Rigler, Killander, Bolund & Ringertz, 1969). All these events were considered to reflect different stages of dissociation of the nucleoprotein complex and were assumed to be part of the 'multi step preparatory mechanism for initiation of transcription' (Darżynkiewicz, Bolund & Ringertz, 1969). However, more recent observations have made such an interpretation unlikely. Auer, Zetterberg & Killander (1970) and Bolund, Darżynkiewicz & Ringertz (1970), have shown in the case of acridine orange at least, that the increased binding of the dye to the DNP is related to a 'crowding effect' caused by the PHA agglutinating the cells and resulting in greater numbers of them adhering to the glass of the culture vessel and is not related to the preparation of the cell for mitotic activation.

Thus the observations on the changes in the lymphocyte nuclear DNP complex following the addition of stimulant cannot be taken as proof of gene activation. Indeed it appears that many, if not all, of the phenomena reported are in some way connected with changes in the cell membrane brought about by the stimulant and occur whether or not the cell is activated into mitotic division. Gene activation may well be involved later in the transformation process but the evidence here and in the next section suggests that this is not an early effect of stimulant on the cells.

(c) Changes in RNA and protein synthesis

It is well established that RNA synthesis increases following lymphocyte activation (see Ling, 1968; Cooper, 1970a, for references). For the measurement of RNA synthesis many workers have used the incorporation of radio-labelled uridine into RNA. This is open to the severe criticism that it is not an accurate measure of the rate of RNA synthesis but is dependent upon the uptake of uridine into various pools and changes in the activity of the enzyme uridine kinase (Lucas, 1967; Hausen & Stein, 1968; Kay & Handmaker, 1970). This means that the many studies using uridine and reporting an early increase in incorporation (e.g. Pogo *et al.* 1966) cannot be interpreted as quantitatively reflecting new RNA synthesis. Using other techniques, increases in the *net* RNA synthesis have been reported to begin about 12 h after stimulation (Killander & Rigler, 1965; Cooper & Rubin, 1965).

Perhaps the most conclusive evidence for gene activation would be the demonstration of new species of messenger RNA in the stimulated cell which are not present in the control cells. However, it is now generally thought that ribosomal RNA forms the bulk of the new RNA synthesised in activated cells (e.g. Kay, 1968*b*; Cooper, 1969*a*, 1969*b*, 1970*b*; Neiman & Henry, 1971). Kay, Ahern & Atkins (1971), however, found that although there was a large increase in the rate of ribosomal RNA synthesis, as measured by uridine incorporation, the increase in the number of ribosomes was less dramatic, suggesting that turnover rather than net synthesis was being measured. In these experiments the uridine was added from day 2 to 3 and apart from ribosomal RNA (80–85 % of the total label incorporated) the remainder was predominantly transfer RNA. Other labelled species of RNA (e.g. messenger RNA) could have comprised only a very small percentage of the total. Attempts to demonstrate the presence of new messenger RNA by hybridisation techniques have been unsuccessful. Neither Torelli, Henry & Weissman (1968) nor Neiman & MacDonnell (1970) were able to show any differences between control and 24 h PHA-stimulated lymphocytes.

Lymphocytes when incubated with various stimulants increase their rate of synthesis of protein (Sell, Rowe & Gell, 1965; Kay & Korner, 1966; Kay, 1968*b*). The increase, which is normally measured by the incorporation of radiolabelled amino acids into the cell, begins within a few hours after the addition of stimulant (Fig. 3), and reaches a maximum after about 48 h. It can be inhibited by drugs which are known to inhibit protein synthesis in other systems (e.g. cycloheximide – Kay & Korner, 1966) – proving that genuine protein synthesis is being measured. The question of whether new proteins which are not found in small lymphocytes are synthesised by activated lymphocytes is considered in Section 3*b*. New proteins, if they could be shown to be present in the activated cells, would be evidence for altered gene expression.

Thus it may be concluded that the early biochemical events observed following the addition of stimulant to lymphocytes in culture give no definite evidence for changes in the qualitative expression of the genome

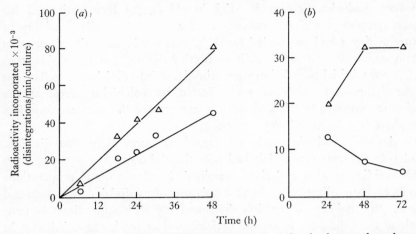

Fig. 3. Stimulation by PHA of [³H]leucine incorporation by human lymphocytes. Suspensions of human lymphocytes were prepared from adult blood (Fig. 3a) or from cord blood (Fig. 3b) as described in Fig. 1. Adult cells were cultured in 20 % pooled human serum gelatin/Eagles MEM without leucine at 1 × 10⁶/ml. PHA (1 μg/ml; Batch X 5) was added to stimulated cultures (open triangles). No PHA was added to controls (open circles). [³H]leucine (1 μCi; 19 000 mCi/mmole) was added at time o and the cells harvested at the times shown. Cord blood lymphocytes were cultured at 2 × 10⁶/ml in 20 % pooled human serum gelatin/normal Eagles MEM. PHA (2 μg/ml; Batch X 5) was added to stimulated cultures (open triangles). No PHA was added to controls (open circles). [³H]leucine (1 μCi; 19 000 mCi/ mmole) was added 24 h before harvesting. Cells were harvested at the time shown. Values in both experiments are the means of triplicate cultures.

but seem more likely to reflect an acceleration of existing metabolic processes. Studies on the turnover of ribosomal RNA, synthesis and wastage, led Cooper (1970a) to the conclusion that the addition of PHA results, amongst other things, in alterations at the level of RNA transport and survival and also at the level of translation. Thus it appears that one of the modes of action of PHA may involve the activation of pre-existing messenger RNA or the enhancement of ribosomal function. Such a mechanism would operate at a cytoplasmic level, and thus not necessarily involve gene activation.

(2) THE CELL CYCLE IN ACTIVATED LYMPHOCYTES

(a) Kinetics of lymphocyte activation

Lymphocytes activated with PHA do not reach mitosis until about 40 h after the addition of stimulant (Bender & Prescott, 1962). Not all of the cells in a culture respond to stimulation. It is not possible, with current techniques, to quantitate exactly the proportion of cells which are activated.

However, Jasinska, Steffen & Michalowski (1970) have estimated for human lymphocytes that a maximum of 45% of the cells surviving in culture at day 4 had responded to PHA, whereas in an antigenic specific system such as the rat MLR, Wilson, Blyth & Nowell (1968) consider that 1–3% of the initial population were stimulated into division. If cells from human lymphocyte cultures are collected by colchicine arrest at the maximum response to PHA about one per cent of the cells are found in metaphase in 1 h (McKinney, 1964).

To determine whether cells are capable of more than one division, Bender & Prescott (1962) labelled stimulated human lymphocytes with [³H]thymidine at 48 h and then examined the chromosome preparations of cells by autoradiography. At 72 h only one of 75 divisions was a second mitosis (the majority being first divisions). By 84 h approximately two-thirds of the cells were in second division and at 108 h one-third were in their third division (i.e. labelled in only half the chromosomes of the complement). In the 120 h sample there were some cells which must have gone through four divisions. Subsequent samples gave mixtures of all four labelling patterns but it is of interest that cells in S at 48 h were entering divisions as late as 156 h.

Direct observation of lymphocyte division has been made by a micro-cinematographic technique. A human lymphocyte in a PHA-treated culture was observed to enlarge and divide 67 h after setting up the culture. The daughter cells, which were indistinguishable from small lymphocytes, enlarged over the next 23 h and then divided (Marshall & Roberts, 1965). For another cell studied the first division occurred at 118 h and a daughter divided 38½ h later. In further studies using specific antigens as stimulants it was shown that an activated lymphocyte could divide and redivide to produce clones of 64 cells (six divisions) or more. The generation times (following the first division) of the majority of the 301 cells studied were between 8 and 13 h but the range was 7.5 to 38 h (Marshall, Valentine & Lawrence, 1969). There appeared to be no difference between the generation times of human lymphocytes stimulated with tuberculin, strepto-kinase–streptodornase, pokeweed or allogeneic lymphocytes. However, it should be stressed that the times referred to were for second and subsequent divisions. The times taken for activated cells to reach mitosis after setting up the cultures, as stated, are much longer (minimum 40 h) and may well depend on the type of stimulant employed and it is certainly affected by the dose (see below).

In the rat MLR system, Wilson, Blyth & Nowell (1968) observed that, following a lag time of 40 h from stimulation to mitosis (the same as that mentioned above for human cells), there was an exponential proliferation with a doubling time of 9–10 h (again similar to human lymphocytes as reported by Marshall *et al.* 1969). This phase lasted about 100 h. Cells which entered mitosis generally went through a series of divisions – few, if any, dropping out of the mitotic cycle. In addition to the cells which entered mitosis at the beginning of the proliferation phase, significant numbers of new, previously non-dividing cells, continued to enter the mitotic cycle during the whole of the exponential growth phase. Thus both with rat lymphocytes and in a human system there is obviously broad heterogeneity of the responding population.

Estimates of the time taken by stimulated human lymphocytes to pass through the different phases of the cell cycle differ somewhat from author to author. Bender & Prescott (1962) reported that in the first wave of mitosis following PHA stimulation the minimum lengths of G_1 and S were 24 h and 12 h and the maximum length of G_2 was 6 h. German (1964) observed that the G_2 period was 3–4 h. Kikuchi & Sandberg (1964) noted that the S period began after 16–20 h and that the G_2 period started 3 h before metaphase. Cave (1966) cultured normal human lymphocytes in the presence of PHA for 60 h and then measured the duration of the cell cycle by adding a pulse of [³H]thymidine and counting the subsequent labelled mitoses. The mean generation time for the dividing cells was found to be 17.7 h and the average G_1, S and G_2 were 4.6 h, 9.6 h and 3.5 h respectively. In a further study on the lymphocytes from 17 patients with Down's syndrome, Cave & Levitsky (1966) reported the mean G_2 to be 4.6 h and the mean S 8.4 h. In all the above studies with lymphocytes stimulated by PHA no attempts were made to synchronise the populations of dividing cells. Steffen & Stolzmann (1969) made a very detailed study of the cell proliferation kinetics of stimulated lymphocytes released from a block at the first G_1/S or early S phase in the cell cycle. The block could be maintained for up to 96 h, under the appropriate conditions, by the addition of amethopterin plus adenosine, fluorodeoxyuridine or high concentrations of thymidine. Following the reverse of the block at 70 h little or no mitotic activity was seen for 6 to 9 h. The highest fraction of cells in mitosis was normally found in cultures harvested 15 h after block release and in some experiments a second mitotic peak could be seen at 21 h. Some cells were entering mitosis 24 h after the block was removed. Thus the minimal and median times for $S + G_2$ were found to be in the 6–9 h and 16–18 h range, respectively. From their experience with cells in a random population of second and subsequent generation cells the authors conclude that the duration of $S + G_2$ does not depend upon the number of generations in-vitro.

Careful analysis of the size distributions of PHA-stimulated lymphocytes after 71 h of culture led Michalowski & Jasinska (1968) to the conclusion

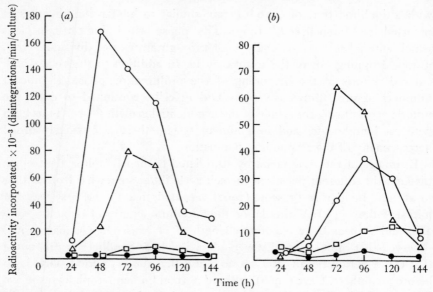

Fig. 4. Time–dose response curves for pig lymphocytes stimulated with PHA or SF. Pig blood lymphocytes were cultured at 2×10^6/ml in 20 % pig serum gelatin/Eagles MEM. 1 ml cultures were incubated in $3 \times \frac{1}{2}$ inch tubes in an atmosphere of 5 % carbon dioxide in air. Stimulant was added at time 0. (a) PHA (Burroughs Well-come, crude); ●, control; ○, 100 μg/ml; △, 10 μg/ml; □, 1 μg/ml; (b) SF; ●, control; △, final concentrations of 1 in 10; ○, 1 in 100; □, 1 in 1000. [³H]thy-midine (0.5 μCi; 150 mCi/mmole) was added 24 h before harvesting. Radioactivity in the TCA-insoluble material was estimated by a liquid scintillation technique. Values are means of triplicate cultures.

that there are at least two sub-populations of lymphocytes capable of responding to PHA. Although the pattern of response for the two cell types is somewhat different it appears that the transformed cells produced are homogeneous. The functional significance of the two sub-populations is not yet clear. It is possible that the differences in time taken by the different cells in the population to pass through the cell cycle is a reflection of these sub-populations of cells (Jasinska, Steffen & Michalowski, 1971).

There is evidence that the mean cell cycle time in activated lymphocytes varies with the dose of stimulant used. The observations relating to this topic will be considered in the following section.

(b) Changes of cell cycle time with dose of stimulant

In 1967 we observed that human lymphocytes stimulated with suboptimal doses of PHA or SF gave a later and often more sustained maximal response than that of cells treated with the optimal dose (Ling & Holt, 1967).

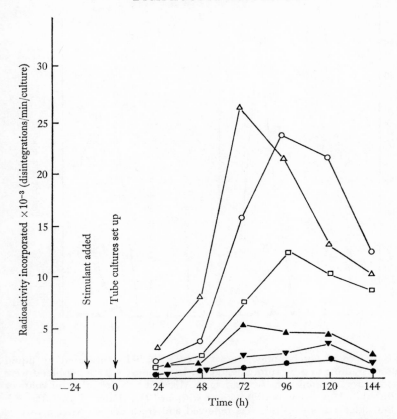

Fig. 5. Typical response curves for pig lymphocytes stimulated with 'pulse' doses of SF. Pig lymphocytes were separated from blood by the gelatin technique and suspended at 5×10^6 cells/ml in 20 % pig serum gelatin/Eagles MEM. 7 ml volumes were incubated in McCartney bottles with or without stimulant and cultured for 16 h. 6 ml of medium was then carefully removed without disturbing the cells and fresh added, a cell count performed, and the cell concentration adjusted to 2×10^6/ml. 1 ml cultures were incubated in $3 \times \frac{1}{2}$ inch capped tubes. ●, Control; △, SF 1 in 20; ○, SF 1 in 200; □, SF 1 in 1000; ▲, SF 1 in 2000; ▼, SF 1 in 10000. Final concentrations of SF refer to those in the McCartney bottles. [³H]thymidine was added 24 h before harvesting. Values are means of triplicate cultures. Viability counts performed throughout the culture period showed no significant differences between control and stimulated cultures.

Further studies with pig lymphocytes have confirmed and extended these findings. Lymphocytes were separated from defibrinated blood by a gelatin sedimentation method (Coulson & Chalmers, 1964) and cultured in Eagles medium supplemented with 20 % pig serum. Full details are given in Hardy (1969). Stimulant (PHA or SF) was added at various doses and the response measured by the uptake of [³H]methyl-thymidine or [³H]-5'-

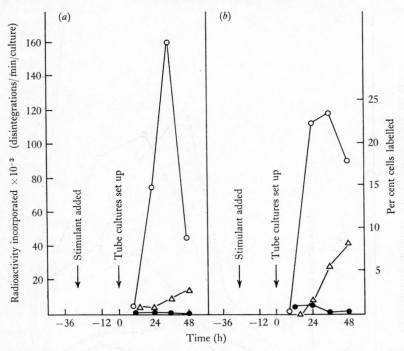

Fig. 6. Response of pig lymphocytes activated with PHA measured by liquid scintillation counting and by autoradiography. Lymphocytes from pig blood were suspended in 20 % homologous serum/Eagles MEM at 20×10^6/ml. 7 ml volumes were cultured in McCartney bottles with or without PHA (Batch X 5). After 18 h the medium (apart from 0.5 ml) was removed and fresh added. After a further 10 h incubation viability counts were performed, the cell concentration adjusted to 2×10^6/ml and tube cultures set up. [^{3}H]thymidine was added 12 h before harvesting. Scintillation counting was performed in the usual way (Fig. 6a). Smears were prepared for autoradiography by the stripping film technique and fixed and stained after exposure for 1 week. The percentage of labelled cells was calculated after counting at least 200 cells on smears from duplicate cultures (Fig. 6b). ●, Control; ○, PHA 100 μg/ml; △, PHA 10 μg/ml.

uridine into the trichloroacetic acid (TCA) insoluble material. In all, 14 experiments were performed. In some of these stimulant was present throughout the culture period (Fig. 4). In others the cells were 'pulsed' with stimulant for 16 h (Fig. 5), i.e. the medium was changed after this time and replaced by fresh which did not contain stimulant. No attempt was made, however, to wash the cells free of stimulant.

Preliminary autoradiographic studies have revealed that the percentage of cells labelled follows the same pattern as the gross thymidine uptake (Fig. 6). This indicates that the earlier peak with the higher dose of PHA and the delayed peak with the lower dose is a reflection of the numbers of

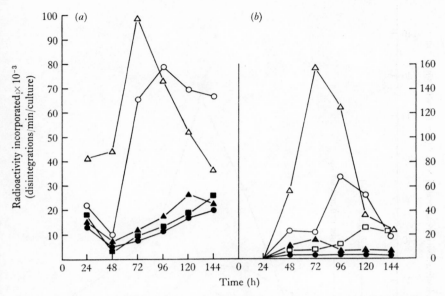

Fig. 7. Activation of pig lymphocytes by different doses of PHA measured by [³H]uridine or [³H]thymidine incorporation. Pig blood lymphocytes were cultured in 20 % pig serum/Eagles MEM in $3 \times \frac{1}{2}$ inch capped test tubes in an atmosphere of 5 % carbon dioxide in air. PHA (Burroughs Wellcome, Batch X 5) was added at the following final concentrations. ●, Control (no PHA); ■, 0.1 μg/ml; ▲, 0.2 μg/ml; □, 0.4 μg/ml; ○, 1 μg/ml or, △, 2 μg/ml. Stimulant was present throughout the culture period. [³H]uridine (1 μCi; 5000 mCi/mmole) (Fig. 7a) or [³H]thymidine (0.5 μCi; 150 mCi/mmole) (Fig. 7b) were added 24 h before harvesting. Cells were harvested at the time shown and the radioactivity in the TCA insoluble material counted. Values are means of triplicate cultures.

cells in S and not some artefact due to the high dose of stimulant causing higher thymidine uptake by the same cells.

The results of thirteen experiments in all showed the delayed peak with lower dose of stimulant. A similar effect was observed whether thymidine or uridine incorporation was measured (Fig. 7). In one experiment, however, the delayed peak effect was not observed when PHA was used as stimulant (Fig. 8) but cells from the same pig did show the effect with SF (Fig. 5). The reason for this was not clear. In some experiments a low, early peak was observed with low doses of stimulant, e.g. in Fig. 5 and Fig. 7b. This has also been observed by Jasinska, Steffen & Michalowski (1970) with synchronised human lymphocytes. Although the significance of this early peak with low doses of stimulant is not yet known it may reflect the effect of the mitogen on different sub-populations.

The results taken as a whole show that the dose of stimulant affects the

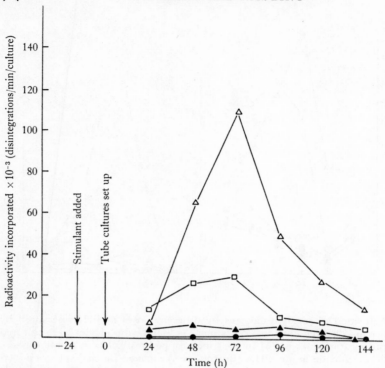

Fig. 8. Atypical response curves for pig lymphocytes stimulated with 'pulse' doses of PHA. Details as for Fig. 5. ●, Control; △, PHA (batch X 5) 1 μg/ml; □, PHA 0.5 μg/ml; ▲, PHA 0.2 μg/ml. Concentrations refer to those in the McCartney bottles for the 16 h 'pulse'. Viability counts showed no significant differences between control and stimulated cultures.

time at which the optimal response is obtained. Before this can be interpreted in terms of the action of stimulant on the cells three possible criticisms must be discussed. These are: the toxicity of stimulant, exhaustion of nutrients in the medium, or effects due to contaminating leucocytes or red cells in the lymphocyte preparation.

If the preparations of the stimulants used were toxic then the observed decline from the optimum nucleoside incorporation could be argued to be due to the cells dying. This would be expected to occur earlier and more rapidly with the lymphocytes exposed to the higher, and thus more toxic, doses of stimulant, and would thus be a possible explanation for the shapes of the curves obtained. To assess whether the PHA or SF were toxic to lymphocytes viability counts were performed on the cultures using a trypan blue dye exclusion test. At the doses of stimulant used no significant differences in viability were noted between the activated and the control cells throughout the culture period – indicating that the delayed optimum effect is not an artefact due to toxic material in the stimulant. Indeed, the observation that the cells first stimulated with optimal doses could respond to a higher degree to a subsequent dose of stimulant than sub-optimally or non-stimulated cells (see Section 2c) strongly argues in favour of the dose effect being genuine.

An explanation for the shapes of the curves, similar to that predicted if the stimulant was toxic, would apply if there was exhaustion from the medium of nutrients essential for proliferation in the cultures treated with the high dose of stimulant. Such an explanation, however, is extremely unlikely for it was found in preliminary experiments that lymphocytes stimulated with optimal doses of PHA showed no difference in their response pattern, whether the medium was changed at intervals through the culture period or not.

It is not known what effect, if any, the different doses of mitogens had on cells in the cultures other than lymphocytes. In the experiments described the pig blood lymphoctyes were contaminated with about 20% other leucocytes at the time of setting up the cultures and with red cells. However, this cannot be an important factor because, nearly 'pure' lymphocyte preparations from pig thymus also gave a similar delayed optimum with lower doses of stimulant, but because of the poorer viability of thymus cells this was not maintained (Fig. 9).

Thus we conclude that the rate at which the activated lymphocytes pass through the proliferation cycle is dependent upon the dose of stimulant employed. This observation would tend to indicate that lymphocytes are 'pushed' rather than 'triggered' into proliferation (Ling & Holt, 1967; Hardy & Ling, 1968). By a 'trigger' mechanism we mean that the reactive cells respond in an 'all-or-none' fashion to stimulant and would imply inherent characteristics of the kinetics of growth in the cell whereas a 'push' hypothesis would assume that there was a continuous requirement for stimulant molecule (or its effective agent) which would be responsible for the rate of proliferation.

Our studies showing differences in the rate of response to different doses of mitogen by bulk DNA or RNA synthesis are open to a number of interpretations. It is possible that the dose-dependent process is either in the time taken for the initiation of the response or on the number of divisions that the cells go through. A third possibility is that the time a cell takes to pass through the cell cycle is dose dependent. As pointed out by Jasinska, Steffen & Michalowski (1970), it is not possible to distinguish between these three possibilities in asynchronously dividing lymphocyte cultures by the technique we used. However, our experiments do provide

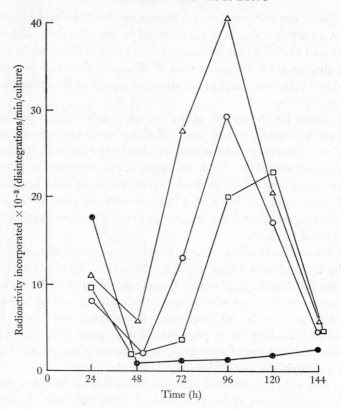

Fig. 9. Activation of pig thymus cells by SF. A suspension of pig thymus cells (5×10^6 viable lymphocytes/ml) was prepared by chopping a thymus with scissors. Debris was removed by filtering the preparation through nylon and washing the cells in medium. The cells (1 ml volumes) were cultured in 20% pig serum gelatin/Eagles MEM in $3 \times \frac{1}{2}$ inch capped tubes. Stimulant was added at time 0. ●, No stimulant; △, SF 1 in 10; ○, SF 1 in 100; □, SF 1 in 1000. [³H]thymidine was added 24 h prior to harvesting. Radioactivity in the TCA insoluble material was estimated by liquid scintillation counting. Cell counts performed on parallel cultures showed a rapid decline in viability after 48 h so that at 144 h only about 5% of the initial population remained alive.

data that can be explained on the hypothesis that the time taken for lymphocytes to pass through the cell cycle is dose dependent.

Michalowski and his co-workers have carried out more detailed studies to determine the effect of stimulant dose on the rate at which lymphocytes pass through the cell cycle. Using populations of stimulated lymphocytes released from a block at the G_1/S point of the cell cycle (Steffen & Stolzmann, 1969), Jasinska, Steffen & Michalowski (1970) demonstrated that the period of induction of DNA synthesis varied markedly from cell to cell, the extreme values being separated by as much as 50 h, but that this variability did not depend on the time of exposure to, or the availability of, PHA in the medium. However, the curves showing the rate of accumulation of cells at the G_1/S point of the cell cycle indicate that the arrival of cells at that point was delayed in cultures stimulated with sub-optimal concentrations of PHA. A biphasic response was also observed, suggesting the presence of two PHA-reactive lymphocyte sub-populations.

In a subsequent study, using the same reversible G_1/S blockage technique, Michalowski (1972) measured the modal duration of $S + G_2$ phases of lymphocytes activated with optimal or sub-optimal doses of PHA as given by the peak of the mitotic wave following reversal of the block of DNA synthesis. It was found that the mitotic activity of the optimally stimulated cultures generally peaked between 11 and 14 h (mean $\pm$ S.E. $= 12.75 \pm 0.56$h) whereas in sub-optimally stimulated cultures the time was significantly longer – between 14 and 17 h (mean $\pm$ S.E. $= 14.83 \pm 0.54$ h). Thus variations in the dose of PHA, as well as influencing the percentage of cells which respond, affect the duration of the induction time of DNA synthesis and the timing of the cell cycle.

As Michalowski (1972) points out, the question of whether or not PHA dose influences the duration of the induction phase of DNA synthesis and the cell cycle time at the individual cell level is still open. The experiments performed do not formally exclude the possibility that PHA at different doses merely selects from the population of potential responders those cells which differ in their intrinsically fixed duration for the cell cycle. However, it is not easy to see how such a mechanism could possibly work. It seems far more likely that individual lymphocytes are capable of adjusting the rate of their response to the intensity of the mitogenic stimulus. It is of great significance that the $S + G_2$ phase of the cell cycle was found to be affected, for it is generally supposed that the extension or contraction of the cell cycle time is due to variations in the length of G_1 (e.g. Gamow & Prescott, 1970).

Thus it would seem that the rate of response of lymphocytes to mitogen is dependent upon the dose employed. The mechanism of this dose-dependent response to stimulant is as yet unknown, but it provides further evidence that the lymphocytes are 'pushed' through the proliferation cycle by the continual presence of mitogen (or its effective agent), rather than the cells being simply 'triggered' into division by a once-and-for-all action of the stimulant. Such an interpretation would seem to us to imply control of division by a means other than by the direct 'turning on' of the genome.

(c) *Reactivation of stimulated lymphocytes*

Ling & Holt (1967) demonstrated that human lymphocytes that had been stimulated with a 16 h pulse of PHA or SF would respond to a second dose of either stimulant after the effect of the first dose had died away. The pre-stimulated cells were indistinguishable morphologically from the untreated controls at the time of re-stimulation. The response to the second dose of stimulant was more rapid than with control cells which had not been pre-stimulated. It was non-specific, i.e. cells 'pulsed' with a primary dose of PHA or SF would respond more rapidly than controls to either stimulant or to antigens such as PPD or in a MLR.

Hausen & Stein (1968) observed that human lymphocytes 'stopped' during activation by PHA with the addition of anti-PHA, would respond to a greater degree to a second dose of PHA after the antiserum had been removed. Polgar, Kibrick & Foster (1968) also report that cells which have reverted after transformation respond faster when re-stimulated with PHA.

Studies on pig lymphocytes (Hardy & Ling, 1968) have confirmed and extended our early findings with human cells. Seven experiments so far have been performed with PHA or SF as mitogens. In four of these clear evidence for reactivation was obtained. Thus pig lymphocytes 'pulse' stimulated with different doses of PHA showed a clear gradation in their capability to be reactivated (Fig. 10). With increasing primary doses of PHA the cells showed greater and quicker responses to a second dose of PHA or of SF although there was no difference in the incorporation of thymidine by the cells of the different batches not given a secondary stimulation. A further experiment is illustrated in Fig. 11.

A very low dose of SF (used at a concentration of 1 in 2400 instead of the optimal dose of about 1 in 10) was capable of 'priming' the cells so that they responded more rapidly and to a greater extent to a secondary dose. Again the control levels of thymidine incorporation of non-secondarily stimulated cells were indistinguishable whether the cells had been pre-stimulated or not. Higher primary doses of SF (1 in 240 or 1 in 24) also 'primed' the cells to respond quicker and more intensely to a secondary dose (curves not shown) but in these cases the thymidine uptake in cells not reactivated was significantly higher than control cells which were not reactivated. Studies on [³H]uridine incorporation showed increased uptake of label by the primed cells compared to unprimed controls at the time of reactivation even if no increase in thymidine uptake was detectable. This suggests that primed cells were, in fact, metabolically different from un-primed small lymphocytes even if they were indistinguishable from them morphologically or by their rate of DNA synthesis. This is further sup-

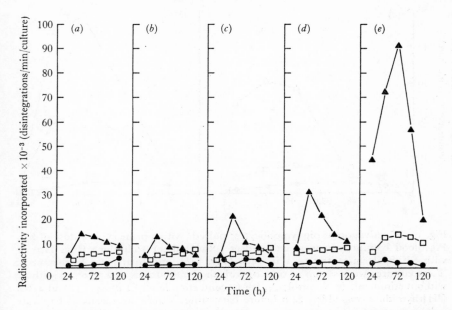

Fig. 10. Re-stimulation of pig lymphocytes 'pulsed' with different doses of PHA. Pig lymphocytes (5×10^6/ml; 60 ml) were cultured in medical flat bottles in 20 % (v/v) pig serum/Eagles MEM. PHA (Burroughs Wellcome, batch X 5) was added at a final concentration of (a) 0; (b) 0.02 μg/ml; (c) 0.1 μg/ml; (d) 0.5 μg/ml; (e) 2.0 μg/ml. After 16 h 55 ml of the medium was removed without disturbing the cells and fresh added. After a further 4 days the medium change was repeated and viability counts performed. No significant difference was observed between the five preparations of cells. The concentration of viable cells was adjusted to 2×10^6 ml and 1 ml volumes dispensed into tubes. Stimulant was added as follows. ●, 0; ▲, PHA 2 μg/ml; □, SF 1 in 100. Cells were cultured for the time shown. [³H]thymidine (0.5 μCi; 150 mCi/mmole) was added 24 h before harvesting. Radioactivity in the TCA insoluble material was estimated by scintillation counting. Viability counts performed at 120 h following secondary stimulation showed approximately 30 % of the cells viable. There was no significant difference between stimulated and non-stimulated cultures.

ported by the observation of Steffen & Soren (1968) who found in PHA-stimulated cultures at day 11, when the DNA synthesis had fallen to a very low level, that the dry mass of the cells was about twice that of the starting lymphocyte population. Also Polgar, Rutenburg, Kibrick & Kim (1971) observed increased enzyme levels in cells which had apparently reverted to small lymphocytes following activation (see Section 2d).

The reason why re-stimulation was not observed in three of the experiments was not clear. It may be due to technical difficulties or may perhaps reflect some form of 'tolerance' developed by the population of lymphocytes to the stimulants.

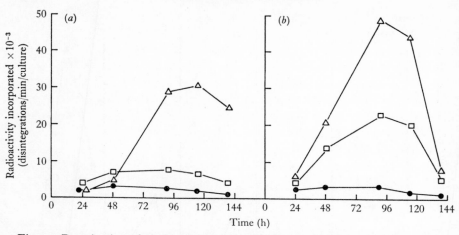

Fig. 11. Reactivation of pig lymphocytes 'pulsed' with a primary, low dose of SF. Pig blood lymphocytes were cultured at 5×10^6/ml for 16 h alone (Fig. 11a) or with SF–final concentration 1 in 2400 (Fig. 11b), and then for a further 96 h. Tube cultures were set up at time o, with the lymphocytes at 2×10^6/ml with or without stimulant. ●, Control; △, SF (second stimulation) 1 in 24; □, 1 in 2400. [³H]thymidine was added 24 h before harvesting. Values are means of triplicate cultures.

We have found recently that human cord blood lymphocytes stimulated with X-irradiated lymphoid cell line cells (Hardy, Knight & Ling, 1970) will respond, under suitable conditions, to a second dose of irradiated cells more rapidly and to a greater extent than unstimulated cells (unpublished data). Again, however, in certain of these experiments no secondary response was obtained. These experiments are open to the criticism that stimulant (X-irradiated lymphoid cell line cells) cannot be removed before a second dose is given, but the possibility of producing unresponsiveness in this culture system needs further investigating. Reports of experiments in which responding 'clones' of cells activated by specific antigen are claimed to be destroyed by very high specific activity thymidine (Salmon, Krakauer & Whitmore, 1971), or bromodeoxyuridine and light (Zoschke & Bach, 1970), if confirmed and if interpreted correctly, make reactivation an extremely interesting field for future research. The experiments reported here demonstrate clearly that reactivation is possible if the conditions are favourable.

(d) Reversibility of transformation

Closely connected with the idea of reactivation is the question of whether the stimulated lymphocytes are capable of reversing from a blast stage to a small lymphocyte again. Much confusion is likely to arise on this point if the terms are not adequately defined. Thus by reversibility we do not mean a 'stopping' of the cell at a particular point in the cell cycle nor do we take it to be the production of two small lymphocytes following the cell division of the lymphoblasts for there is evidence that the progeny are functionally not the same cells as non-activated small lymphocytes. We take true reversibility to be either the reversion of an activated blast cell to a small lymphocyte without it passing through cell division at all, or the reversion of the progeny of cell division to become true small lymphocytes.

At present there is no unequivocal evidence for this reversion from an activated phase to a resting small lymphocyte phase. Following a single PHA stimulus and passage of the cells through the stage of DNA synthesis, the number of blast cells has been shown to diminish by day 10, when 'small' lymphocytes were said to predominate (Polgar, Kibrick & Foster, 1968; Polgar & Kibrick, 1970). Using an acridine orange staining technique Yamomoto (1966) also obtained results which suggested that large blast cells reverted to smaller lymphocytes after a further period of incubation without stimulus. The morphological evidence for a reversion of the lymphocytes to a resting phase must be said, however, to be slight and no supporting evidence based on re-stimulation has been produced. Indeed, as noted earlier (Section 2c) Steffen & Soren (1968) found that the cell mass of PHA-stimulated lymphocytes in which DNA synthesis had returned to a very low level by day 11 was twice that of the initial small lymphocyte population. More recently Polgar, Rutenburg, Kibrick & Kim (1971) reported that the so-called small lymphocytes observed following activation are not identical with unstimulated cells in that the activated and reverted cells show elevated levels of at least one enzyme, namely acid phosphatase. The authors themselves conclude that despite the morphological similarity between lymphocytes which have not been stimulated and those which have reverted the reverted cells may be functionally different. It is an open question whether these functional differences reflect an altered gene expression of the cell or whether they are merely the expression of increased metabolic activity of the normal lymphocytes.

The question of true reversibility is a fundamental one for the understanding of the mechanism of lymphocyte activation. The fact that progression through the cell cycle in activated cells seems to be dependent upon the presence of stimulant makes the reversibility of cells receiving insufficient stimulation plausible, but this remains to be demonstrated conclusively.

(3) PROPERTIES OF ACTIVATED LYMPHOCYTES

Studies on the biochemical changes in lymphocytes soon after stimulation have so far failed to reveal any conclusive evidence for the activation of genes previously not expressed in the 'resting' cells. However, it may well be that gene derepression occurs as a later event as the activated cells pass through the S, G_2 or M phase of the cell cycle. If this were the case it may be possible to demonstrate the presence of enzymes or other proteins, or some functional difference, in the activated cells not observed in small lymphocytes. In this section we consider some of the characteristics of activated lymphocytes and look for evidence of changes in the genetic expression of the cells.

(a) *Changes in antigenic expression*

The expression of cellular antigens has been reported to vary as cells pass through the cell cycle (Bjaring, Klein & Popp, 1969; Cikes, 1970). Studies performed by Mrs Pauline Mackintosh and ourselves (Mackintosh, Hardy & Aviet, 1971; Mackintosh *et al.* 1972) have revealed that lymphocytes when cultured in-vitro show 'extra' reactions with certain HL-A alloantisera not found on the fresh cells of the donor. There are quantitative and qualitative differences when the cells are cultured in the presence of stimulant. These observations may indicate an antigenic change in the cell surface resulting from altered metabolism of the cell, perhaps itself the result of altered gene expression. However, it is not possible at this stage to establish that the 'extra' antigens found on stimulated cells are not present on resting lymphocytes in quantities too low to be detected by the cyto-toxicity test employed, or that the phenomenon is not due to some more trivial explanation (see Mackintosh *et al.* 1972 for discussion). Thus, although altered gene expression may be a possible explanation for the observed 'extra' antigens on stimulated lymphocytes further work is needed to establish the point with certainty.

(b) *Analysis of proteins synthesised in activated lymphocytes*

Although it has not been possible to demonstrate new species of messenger RNA in stimulated lymphocytes, altered gene expression could be said to occur if new proteins could be demonstrated in activated cells which were not found in the small lymphocyte. As has been mentioned (Section 1c) lymphocyte activation is accompanied by an increase in protein synthesis and we shall now consider the nature of the proteins produced.

Analysis of the soluble proteins by acrylamide electrophoresis has not revealed the synthesis of any new species of protein by activated lymphocytes not produced in smaller amounts in cultures of unstimulated lymphocytes (Neiman & MacDonnell, 1970). However, it must be remembered that such an observation does not exclude the possibility of the production of new proteins by the activated lymphocytes in quantities too small to be detected by the technique used. There is also the complication that populations of peripheral blood lymphocytes invariably contain a small percentage of activated lymphocytes so that even if lymphoblasts synthesised new proteins these would be produced in small amounts in the, so called, small lymphocyte preparations.

The synthesis of various enzymes by lymphocytes in relation to activation has also been studied. It has long been known that lymphocyte activation is accompanied by enhanced glycolysis (see Ling, 1968). Significant increases in the level of lactic dehydrogenase (LDH) occur during the first 24 h and analysis of the isoenzymes has shown a considerable increase in the proportion of muscle or M type enzyme compared with heart or H type (Rabinowitz, Lubrano, Wilhite & Dietz, 1967; Rabinowitz & Dietz, 1968). Although interpreted by the authors as a direct action of PHA on the genetic loci controlling M- and H-LDH the dependence of the isoenzyme ratio of cells on the oxygen tension of the system suggests simpler interpretations of the data (Hellung-Larsen & Anderson, 1969). Ornithine decarboxylase is another enzyme which increases in activity (about eight-fold) during the first 24 h after addition of PHA (Kay & Cooke, 1971). Stimulation also increases both DNA and RNA polymerase activity. The DNA polymerase activity of PHA activated cells has been found to increase about two-fold at 24 h to eight-fold at 48 h and ten-fold at 72 h (Rabinowitz, McCluskey, Wong & Wilhite, 1969). Sometimes PHA activation increases DNA polymerase activity as much as 30-fold (Loeb, Agarwal & Woodside, 1968). Rises in the level of DNA-dependent RNA polymerase have been found during the first 20 h of incubation with PHA (Handmaker & Graef, 1970). By 30 h both deoxyribonucleoprotein-bound and soluble RNA polymerases were enhanced by a factor of two (Hausen, Stein & Peters, 1969). An increase in RNA methylases occurs rather late, being observed at the S period or later, that is, after morphological transformation but prior to mitosis (Riddick & Gallo, 1971). These are a few examples of the presumably very large numbers of enzymes showing increased activity associated with the enhanced metabolic activity consequent upon activation. Studies have also been made of the levels of glucose 6-phosphate dehydrogenase (Monteleone, Nadler, Justice & Hsia, 1967) lysosomal enzymes (Nadler, Dowben &

Hsia, 1969), interferon (Friedman & Cooper, 1967) and nuclear proteins (Shapiro & Levina, 1967). Nobody has as yet described the production of a new enzyme as a result of lymphocyte activation. Even the thymidine metabolising enzymes are present prior to activation (Rabinowitz, Wong & Wilhite, 1971; Pegoraro & Benzio, 1971).

Whether the activated lymphocytes are capable of synthesising immunoglobulins is a question around which considerable confusion and controversy has arisen. There have been many claims to have shown increased synthesis of immunoglobulin or specific antibody by lymphocytes transformed in-vitro but others have failed to confirm these findings (see Greaves & Roitt, 1968 for references). The confusion seems to have arisen because of the different assay systems used, which may not always measure true immunoglobulin production, and because of the heterogeneity of the cell populations studied. Lymphocytes are functionally heterogeneous. They may be divided into two main types, namely thymus-derived or T lymphocytes and those lymphocytes (B lymphocytes) arising from bone marrow precursors which have not passed through a stage of maturation in the thymus (see review by Roitt, 1971). The T lymphocytes constitute the major sub-population ($>70\%$) of peripheral lymphocytes of humans and mice and probably most species. The B lymphocytes are distinguishable from T by the presence on them of immunoglobulin determinants detectable with fluorescein or radioiodine-labelled anti-immunoglobulin antisera. Some B lymphocytes are capable, after antigenic stimulation, of mitotic activation and differentiation into plasma cells producing immunoglobulin for export, whereas T cells, although antigen-reactive, do not possess this capacity.

Freshly isolated leucocytes would be expected to produce some immunoglobulin attributable to cells of the B lymphocyte sub-population and such synthesis has been shown to occur (Van Furth, Schuit & Hijmans, 1966). When peripheral human leucocytes are cultured for several days, the apparent immunoglobulin production (as determined by radioimmunoassay after incorporation of radio-amino acid) is increased in cultures containing PHA compared with unstimulated controls. This increase is approximately proportional to the increase in the synthesis of total protein (Gerhadt-Scheurlen, 1968; Smith, Lawton & Forbes, 1967). The three to five proteins synthesised in increased amounts by stimulated lymphocytes appeared to be identical, in serological tests, to normal serum globulins. Since many investigators have found that it is the T and not the B lymphocytes which are stimulated by PHA (Greaves, Roitt & Rose, 1968; Davies, 1969) a rise in immunoglobulin synthesis would not be expected unless the T lymphocytes were capable of some immunoglobulin synthesis or were able to enhance indirectly the production of immunoglobulin by B cells. One possible interpretation of the enhanced synthesis is that the IgG measured, representing only 0.1 % to 1.0 % of the total protein synthesised (Gerhadt-Scheurlen, 1968) may not be IgG but a protein of similar antigenic properties. This is unlikely. Another is that B lymphocytes may, after all, be directly stimulated by PHA. Recent reports suggest that B lymphocytes are stimulated by pokeweed mitogen (Parkhouse, Janossy & Greaves, 1972) and even by PHA, provided it is first coupled to an insoluble particle or surface (Greaves & Bauminger, 1972). When mixtures of T and B lymphocytes (e.g. from peripheral blood) are used it is probable that PHA attached to T lymphocytes will stimulate B lymphocytes.

It is certainly possible for lymphoid cells other than plasma cells to liberate immunoglobulin into the medium. This has been well shown with Burkitt lymphoma cells and cells from many other continuously dividing lymphoid cell lines which produce from 0.5–15 μg of specific immunoglobulin or heavy or light chains per 10^6 cells per day (Matsuoka, Moore, Yagi & Pressman, 1967). The immunoglobulin is produced at a rate proportional to the rate of cell division. Surface immunoglobulin on both normal lymphocytes and lymphoid cell line cells may be removed by trypsinisation but is resynthesised after a period of incubation (Pernis, Ferrarini, Forni & Amante, 1971).

There would not appear to be any dramatic change in the species of proteins produced when a lymphocyte is activated. It would seem more likely that when the lymphocyte is stimulated there is an acceleration of synthesis of proteins which are already being produced, albeit in small quantities, by non-activated lymphocytes. However, we must be careful to stress that, as yet, the question of the production of new proteins is still very much open. This is because: (a) it is very difficult to obtain small lymphocyte preparations completely free of lymphoblasts; (b) methods for the detection of new proteins which have been used: e.g. acrylamide electrophoresis, have been incapable of detecting trace proteins; (c) serological methods of detection have usually relied entirely upon antisera prepared against serum proteins, thus excluding the detection of proteins antigenically unrelated to serum proteins; (d) some proteins are membrane bound and difficult to detect.

(c) Cytotoxic properties

One clear indication that blast activation of lymphocytes is something more than merely preparation for cell division is the acquisition of a cytotoxic property of lymphocytes towards foreign cells of various types as a result of the activation (Perlmann & Holm, 1969). Cytotoxicity requiring direct cell–cell contact and not requiring the assistance of humoral antibody seems to be a distinctive property of thymus-derived lymphoblasts. It is not shown by leukaemic blast cells or Burkitt lymphoma cells or cells from continuously growing lymphoid lines or by activated thymocytes (Hardy, Ling, Wallin & Aviet, 1970). It is one of the few ways of distinguishing an immunocompetent T cell in-vitro from morphologically similar cells lacking this competence. This cytotoxic property could be part of an important defence mechanism for removing abnormal tissue cells, e.g. those transformed by virus in the body. Abnormal lymphocytes, such as those from human lymphoid cell lines, are killed by autochthonous normal lymphoblasts, which is suggestive of the operation of a surveillance mechanism (Hardy & Steel, 1971).

As measured by a ^{51}Cr-release technique the acquisition of cytotoxic capacity roughly follows the curve of thymidine incorporation as lymphocyte activation proceeds (Hardy, Wallin & Ling, 1970). However, DNA synthesis may be inhibited without cytotoxicity being affected (Perlmann & Holm, 1969). Unstimulated leucocyte preparations also show some cytotoxicity but part of this is due to non-lymphocyte leucocytes and disappears when purified lymphocytes are used. The slight activity remaining may be due to the presence of a small number of blast cells or alternatively to a low degree of cytotoxic activity of the small lymphocytes. Certain types of cells, e.g. Chang liver cells, L-strain fibroblasts, are more easily lysed than others, e.g. human or sheep red cells, and both antigen-specific and non-specific killing effects have been observed. Some of the cytotoxic effect may be explained by the release of a cytotoxic lymphokine into the medium but the major part of the effect requires cell–cell contact and may well depend upon membrane–membrane interactions. The extremely marked amoeboid activity of the lymphoblasts may play an essential part in the cytotoxic event by bringing the two surfaces into close apposition. The mechanism by which the cytotoxic activity is effected is unknown but there is no reason to believe that gene activation is a pre-requisite.

(d) Other biological activities of activated lymphocytes

It is likely that future research will reveal that blood and tissue lymphocytes consist of many sub-populations of lymphocytes with distinctive biological properties. Of the two types of lymphocytes at present recognised, namely T and B lymphocytes, most studies in-vitro have dealt, knowingly or unknowingly, with the T component. Activated B lymphocytes survive poorly under the usual culture conditions. In addition to their direct cytotoxic action on many cells, activated T lymphocytes also appear to be able to activate macrophages, either directly or by the production of soluble mediators (Nathan, Karnovsky & David, 1971; Zembala, 1972). They also have an important indirect role in antibody formation and in the carriage of immunological memory. In the reaction of mouse lymphoid cells to sheep red cells, which has been most studied, it is the T lymphocytes which are activated by antigen and initiate the events which culminate in haemo-lysin production by the B cells (Miller *et al.* 1971; Shearer & Cudkowicz, 1969). Although the mechanism of the co-operation between T and B cells is at present unknown, Miller *et al.* (1971) consider that there is a 'require-ment for differentiation' of the T cell which is induced by antigen. This conclusion is based on experiments in which lymphoid cells from tolerant animals, when coated with the antigen to which they were tolerant, failed to induce antibody production in bone marrow reconstituted, X-irradiated, thymectomised mice. Similarly treated normal T lymphocytes were capable of inducing antibody formation. However, even if it is assumed that T cell activation was taking place, as suggested by experiments with mitomycin-treated cells, there is no real evidence to distinguish whether differentia-tion, in the sense of multi-potent cells becoming functionally specialised to react to antigen, is occurring or whether cells are being accelerated in their pre-committed metabolic functions by the presence of antigen. Thus these experiments do not prove or disprove T lymphocyte differentiation but only an essential T lymphocyte involvement in the initiation of anti-body formation to sheep red blood cells. It may well be that T lymphocyte differentiation can take place in response to antigen but further experiments are needed to investigate this possibility.

CONCLUDING REMARKS

Small lymphocytes when cultured in-vitro do not synthesise DNA or pass through the cell cycle. The addition of stimulants such as PHA can induce them to do so. It is well documented that cells once activated are capable of many divisions. The exact manner in which the resting lymphocyte is pushed through the proliferation stage is not known. However, it seems to be related to the presence of stimulant or an active fragment, almost certainly at the cytoplasmic level. A fuller knowledge of the process may well help in our understanding of the fundamental problem of the control of cell division.

The question of whether the stimulation of a lymphocyte is associated with gene activation and differentiation of the cell is still open. However, we feel that at present the evidence is more in favour of the hypothesis that the activated cell is expressing the same genetic material as the small lymphocyte. There are obviously changes in the rate of cellular activity and there may well be changes in the proportions of the different macromolecules produced but as yet there is no good evidence for the expression of new genetic material. The fact that no new species of messenger RNA or of protein have been detected suggests that the lymphocyte, in passing from a resting to an actively dividing state, is utilising only genes that are already functioning. However, such a conclusion is based on experimental data which would not detect the presence of small quantities of products produced by the activation of new genes. It is clear that more refined techniques are required to determine whether previously unexpressed genes are activated or not during the course of stimulation.

The other possibility, which is still very much open, is that the small lymphocyte is a multipotent cell and that the activation process results in the 'shutting down' of certain genes which were previously expressed. However, to show such a phenomenon would require techniques sensitive enough to detect what could only be very minute amounts of species of messenger RNA or of protein that would be present in the small lymphocytes but not in the activated cells. There is no indication that such techniques will be available for some time to come.

The activation of lymphocytes in-vitro provides a very useful means of studying the control mechanisms of cellular processes involved in DNA synthesis and cell division. However, we feel that the many claims for lymphocyte stimulation being a system for studying the process of gene activation and cell differentiation have been overstated. Clearly the cell membrane and/or the cytoplasm play a major role in controlling the metabolic state of the lymphocyte.

The experimental work described in this paper was financed by grants from The Wellcome Trust, The Medical Research Council and The Ernest and Minnie Dawson Cancer Trust to whom we are most grateful. We also thank Mrs J. Wallin and Miss B. Newey for excellent technical assistance and Mrs F. O'Reilly for typing the manuscript.

REFERENCES

ALLWOOD, G., ASHERSON, G. L., DAVEY, M. J. & GOODFORD, P. J. (1971). The early uptake of radioactive calcium by human lymphocytes treated with phytohaemagglutinin. *Immunology*, **21**, 509–16.

AUER, G., ZETTERBERG, A. & KILLANDER, D. (1970). Changes in binding between DNA and arginine residues in histone induced by cell crowding. *Experimental Cell Research*, **62**, 32–8.

BAIN, B., VAS, M. R. & LOWENSTEIN, L. (1964). The development of large immature mononuclear cells in mixed leukocyte cultures. *Blood*, **23**, 108–16.

BENDER, M. A. & PRESCOTT, D. M. (1962). DNA synthesis and mitosis in cultures of human peripheral leukocytes. *Experimental Cell Research*, **27**, 221–9.

BJARING, B., KLEIN, G. & POPP, I. (1969). Cyclic variations in the H-2 isoantigenic expression of mouse lymphoma cells *in vitro*. *Transplantation*, **8**, 38–43.

BOLUND, L., DARŻYNKIEWICZ, Z. & RINGERTZ, N. R. (1970). Cell concentration and the staining properties of nuclear deoxyribonucleoprotein. *Experimental Cell Research*, **62**, 76–89.

BYRD, W. J., HARE, K., FINLEY, W. H. & FINLEY, S. C. (1967). Inhibition of the mitogenic factor in phytohaemagglutinin by an antiserum. *Nature, London*, **213**, 622–4.

CAVE, M. D. (1966). Incorporation of tritium-labeled thymidine and lysine into chromosomes of cultured human leukocytes. *Journal of Cell Biology*, **29**, 209–22.

CAVE, M. D. & LEVITSKY, J. M. (1966). Tritiated thymidine uptake by group G chromosomes of female individuals with Down's syndrome. *Experimental Cell Research*, **43**, 210–13.

CIKES, M. (1970). Antigenic expression of a murine lymphoma during growth *in vitro*. *Nature, London*, **225**, 645–7.

CONARD, R. A. (1967). Autoradiography of leucocytes cultured with tritiated bean extract. *Nature, London*, **214**, 709–10.

COOPER, H. L. (1969a). Ribosomal ribonucleic acid production and growth regulation in human lymphocytes. *Journal of Biological Chemistry*, **244**, 1946–52.

COOPER, H. L. (1969b). Ribosomal ribonucleic acid wastage in resting and growing lymphocytes. *Journal of Biological Chemistry*, **244**, 5590–6.

COOPER, H. L. (1970a). Early biochemical events in lymphocyte transformation and their possible relationship to growth regulation. *Proceedings of the Fifth Leukocyte Culture Conference*, ed. J. E. Harris, pp. 15–30. New York: Academic Press.

COOPER, H. L. (1970b). Control of synthesis and wastage of ribosomal RNA in lymphocytes. *Nature, London*, **227**, 1105–7.

COOPER, H. L. & RUBIN, A. D. (1965). RNA metabolism in lymphocytes stimulated by phytohemagglutinin: Initial responses to phytohemagglutinin. *Blood*, **25**, 1014–27.

COULSON, A. S. & CHALMERS, D. G. (1964). Separation of viable lymphocytes from human blood. *Lancet*, **i**, 468–9.

CROSS, M. E. & ORD, M. G. (1970). Changes in the phosphorylation and thiol content of histones in phytohaemagglutinin-stimulated lymphocytes. *Biochemical Journal*, **118**, 191–3.

CROSS, M. E. & ORD, M. G. (1971). Changes in histone phosphorylation and associated early metabolic events in pig lymphocyte cultures transformed by phytohaemagglutinin or 6-N,2'-O-dibutyryladenosine 3':5'-cyclic monophosphate. *Biochemical Journal*, **124**, 241–8.

DARŻYNKIEWICZ, Z., BOLUND, L. & RINGERTZ, N. R. (1969). Nucleoprotein changes and initiation of RNA synthesis in PHA stimulated lymphocytes. *Experimental Cell Research*, **56**, 418–24.

DAVIES, A. J. S. (1969). The thymus and the cellular basis of immunity. *Transplantation Reviews*, **1**, 43–91.

DENT, P. B. (1971). Inhibition by phytohemagglutinin of DNA synthesis in cultured mouse lymphomas. *Journal of the National Cancer Institute*, **46**, 763–73.

FISHER, D. B. & MUELLER, G. C. (1968). An early alteration in the phospholipid metabolism of lymphocytes by phytohemagglutinin. *Proceedings of the National Academy of Sciences, U.S.A.* **60**, 1396–402.

FISHER, D. B. & MUELLER, G. C. (1969). The stepwise acceleration of phosphatidyl choline synthesis in phytohemagglutinin-treated lymphocytes. *Biochemica et Biophysica Acta*, **176**, 316–23.

FISHER, D. B. & MUELLER, G. C. (1971). Rapid activation of phospholipid metabolism by PHA. *Proceedings of the Fourth Annual Leucocyte Culture Conference*, ed. O. R. McIntyre, pp. 89–95. New York: Appleton-Century-Crofts.

FRIEDMAN, R. M. & COOPER, H. L. (1967). Stimulation of interferon production in human lymphocytes by mitogens. *Proceedings of the Society for Experimental Biology and Medicine*, **125**, 901–5.

GAMOW, E. I. & PRESCOTT, D. M. (1970). The cell life cycle during early embryogenesis of the mouse. *Experimental Cell Research*, **59**, 117–23.

GERHADT-SCHEURLEN, P. (1968). Synthesis of some serum proteins in lymphocytes stimulated by PHA. *Nature, London*, **217**, 1267–8.

GERMAN, J. L. (1964). The pattern of DNA synthesis in the chromosomes of human blood cells. *Journal of Cell Biology*, **20**, 37–55.

GRASBECK, R., NORDMAN, C. T. & DE LA CHAPELLE, A. (1964). The leucocyte mitogenic effect of serum from rabbits immunised with human leucocytes. *Acta Medica Scandinavia*, **175**, supplement 412, 39–47.

GREAVES, M. F. & BAUMINGER, S. (1972). Activation of T and B lymphocytes by insoluble phytomitogens. *Nature, New Biology*, **235**, 67–70.

GREAVES, M. F. & ROITT, I. M. (1968). The effect of phytohaemagglutinin and other lymphocyte mitogens on immunoglobulin synthesis by human peripheral blood lymphocytes in vitro. *Clinical and Experimental Immunology*, **3**, 393–412.

GREAVES, M., ROITT, I. M. & ROSE, M. E. (1968). Effect of bursectomy and thymectomy on the responses of chicken peripheral blood lymphocytes to phytohaemagglutinin. *Nature, London*, **220**, 293–5.

HANDMAKER, S. D. & GRAEF, J. W. (1970). The effect of phytohemagglutinin on the DNA-dependent RNA polymerase activity of nuclei isolated from human lymphocytes. *Biochemica et Biophysica Acta*, **199**, 95–102.

HARDY, D. A. (1969). The activation of lymphocytes by chemical and cellular agents. Ph.D. Thesis, University of Birmingham.

HARDY, D. A., KNIGHT, S. & LING, N. R. (1970). The interaction of normal lymphocytes and cells from lymphoid cell lines. I. The nature of the activation process. *Immunology*, **19**, 329–42.

HARDY, D. A. & LING, N. R. (1968). Nucleic acid synthesis in stimulated pig lymphocytes in relation to time and dose of stimulant. *Biochemical Journal*, **108**, 23–4 P.

HARDY, D. A. & LING, N. R. (1969). Effects of some cellular antigens on lymphocytes and the nature of the mixed lymphocyte reaction. *Nature, London*, **221**, 545–8.

HARDY, D. A., LING, N. R., WALLIN, J. & AVIET, T. (1970). Destruction of lymphoid cells by activated human lymphocytes. *Nature, London*, **227**, 723–5.

HARDY, D. A. & STEEL, C. M. (1971). Cytotoxic potential of lymphocytes stimulated with autochthonous lymphoid cell line cells. *Experientia*, **27**, 1336–8.

HARDY, D. A., WALLIN, J. M. & LING, N. R. (1970). Cytotoxic activity of human lymphocytes stimulated with X-irradiated cells from lymphoid cell lines. *Proceedings of the Fifth Leukocyte Culture Conference*, ed. J. E. Harris, pp. 287–97. New York: Academic Press.

HAUSEN, P. & STEIN, H. (1968). On the synthesis of RNA in lymphocytes stimulated by phytohemagglutinin. I. Induction of uridine-kinase and the conversion of uridine to UTP. *European Journal of Biochemistry*, **4**, 401–6.

HAUSEN, P., STEIN, H. & PETERS, H. (1969). On the synthesis of RNA in lymphocytes stimulated by phytohemagglutinin. The activity of deoxyribonucleoprotein-bound and soluble RNA polymerase. *European Journal of Biochemistry*, 9, 542–9.

HELLUNG-LARSEN, P. & ANDERSON, V. (1969). Kinetics of oxygen-induced changes in lactate dehydrogenase isoenzymes of human lymphocytes in culture. *Experimental Cell Research*, 54, 201–4.

HIRSCHHORN, R., BRITTINGER, G., HIRSCHHORN, K. & WEISSMAN, G. (1968). Studies on lysosomes. XII. Redistribution of acid hydrolases in human lymphocytes stimulated by phytohemagglutinin. *Journal of Cell Biology*, 37, 412–23.

HOKIN, L. E. & HOKIN, M. R. (1965). Changes in phospholipid metabolism on stimulation of protein secretion in pancreas slices. *Journal of Histochemistry and Cytochemistry*, 13, 113–16.

HUBER, H., STRIEDER, N., WINNLER, H., REISER, G. & KOPPELSTAETTER, K. (1968). Studies on the incorporation of ^{14}C-sodium acetate into the phospholipids of phytohemagglutinin-stimulated and unstimulated lymphocytes. *British Journal of Haematology*, 15, 203–9.

JASINSKA, J., STEFFEN, J. A. & MICHALOWSKI, A. (1970). Studies on in vitro lymphocyte proliferation in cultures synchronized by the inhibition of DNA synthesis. II. Kinetics of the initiation of the proliferative response. *Experimental Cell Research*, 61, 333–41.

JASINSKA, J., STEFFEN, J. & MICHALOWSKI, A. (1971). Human lymphocyte proliferation kinetics following stimulation with different phytohemagglutinin (PHA) concentrations in cultures synchronized by reversible DNA synthesis inhibition. *Bulletin de l' Académie Polonaise des Sciences, Série des Sciences Biologiques*, 19, 37–42.

KAY, J. E. (1968a). Phytohaemagglutinin: an early effect on lymphocyte lipid metabolism. *Nature, London*, 219, 172–3.

KAY, J. E. (1968b). Early effects of phytohaemagglutinin on lymphocyte RNA synthesis. *European Journal of Biochemistry*, 4, 225–32.

KAY, J. E. (1969). The role of the stimulant in the activation of lymphocytes by PHA. *Experimental Cell Research*, 58, 185–8.

KAY, J. E. (1971). The binding of PHA to lymphocytes. *Proceedings of the Fourth Annual Leucocyte Culture Conference*, ed. O. R. McIntyre, pp. 21–35. New York: Appleton-Century-Crofts.

KAY, J. E. (1972). Lymphocyte stimulation by phytohaemagglutinin: Role of the early stimulation of potassium uptake. *Experimental Cell Research*, 71, 245–7.

KAY, J. E., AHERN, T. & ATKINS, M. (1971). Control of protein synthesis during the activation of lymphocytes by phytohaemagglutinin. *Biochimica et Biophysica Acta*, 247, 322–34.

KAY, J. E. & COOKE, A. (1971). Ornithine decarboxylase and ribosomal RNA synthesis during the stimulation of lymphocytes by phytohaemagglutinin. *FEBS Letters*, 16, 9–12.

KAY, J. E. & HANDMAKER, S. D. (1970). Uridine incorporation and RNA synthesis during stimulation of lymphocytes by PHA. *Experimental Cell Research*, 63, 411–21.

KAY, J. E. & KORNER, A. (1966). Effect of cycloheximide on protein and ribonucleic acid synthesis in cultured human lymphocytes. *Biochemical Journal*, 100, 815–22.

KAY, J. E. & OPPENHEIM, J. J. (1971). The specificity of antisera to PHA. *Proceedings of the Fourth Annual Leucocyte Culture Conference*, ed. O. R. McIntyre, pp. 59–67. New York: Appleton-Century-Crofts.

KIKUCHI, Y. & SANDBERG, A. A. (1964). Chronology and pattern of human chromosome replication. I. Blood leukocytes of normal subjects. *Journal of the National Cancer Institute*, 32, 1109–43.

KILLANDER, D. & RIGLER, R. (1965). Initial changes of deoxyribonucleoprotein and synthesis of nucleic acid in phytohemagglutinin-stimulated human leucocytes in vitro. *Experimental Cell Research*, **39**, 701–4.

KILLANDER, D. & RIGLER, R. (1969). Activation of deoxyribonucleoprotein in human leucocytes stimulated by phytohemagglutinin. I. Kinetics of the binding of acridine orange to deoxyribonucleoprotein. *Experimental Cell Research*, **54**, 163–70.

KLEINSMITH, L. J., ALLFREY, V. G. & MIRSKY, A. E. (1966). Phosphorylation of nuclear protein early in the course of gene activation in lymphocytes. *Science, New York*, **154**, 780–1.

KNIGHT, S. C., WALKER, D. & LING, N. R. (1971). Factors affecting the activation of rabbit peripheral, spleen, thymus and bone marrow cells by homologous lymphoid cells and heterologous erythrocytes. *Cytobios*, **3**, 65–76.

LINDAHL-KIESSLING, K. (1972). Mechanism of phytohemagglutinin (PHA) action. V. PHA compared with concanavalin A (con A). *Experimental Cell Research*, **70**, 17–26.

LINDAHL-KIESSLING, K. & MATTSON, A. (1971). Mechanism of phytohemagglutinin (PHA) action. IV. Effect of some metabolic inhibitors on binding of PHA to lymphocytes and the stimulatory potential of PHA-pretreated cells. *Experimental Cell Research*, **65**, 307–12.

LINDAHL-KIESSLING, K. & PETERSON, R. D. A. (1969a). The mechanism of phytohemagglutinin (PHA) action. II. The effect of certain enzymes and sugars. *Experimental Cell Research*, **55**, 81–4.

LINDAHL-KIESSLING, K. & PETERSON, R. D. A. (1969b). The mechanism of phytohemagglutinin (PHA) action. III. Stimulation of lymphocytes by allogeneic lymphocytes and phytohemagglutinin. *Experimental Cell Research*, **55**, 85–7.

LING, N. R. (1968). *Lymphocyte Stimulation*. Amsterdam: North-Holland Publishing Co.

LING, N. R. & HOLT, P. J. L. (1967). The activation and reactivation of peripheral lymphocytes in culture. *Journal of Cell Science*, **2**, 57–70.

LOEB, L. A., AGARWAL, S. S. & WOODSIDE, A. M. (1968). Induction of DNA polymerase in human lymphocytes by phytohemagglutinin. *Proceedings of the National Academy of Sciences, U.S.A.* **61**, 827–34.

LUCAS, D. O. (1970). Non-uniformity of phospholipid changes occurring with antigenic stimulation of lymphocytes in culture. *Proceedings of the Fifth Leukocyte Culture Conference*, ed. J. E. Harris, pp. 43–50. New York: Academic Press.

LUCAS, D. O., SHOHET, S. B. & MERLER, E. (1971). Changes in phospholipid metabolism which occur as a consequence of mitogenic stimulation of lymphocytes. *Journal of Immunology*, **106**, 768–72.

LUCAS, Z. J. (1967). Pyrimidine nucleotide synthesis: regulatory control during transformation of lymphocytes in vitro. *Science, New York*, **156**, 1237–40.

MACGILLIVRAY, A. J. & MONJARDINO, J. P. P. V. (1968). Phytohaemagglutinin and the acetylation of lymphocyte histones. *Biochemical Journal*, **108**, 22–3 P.

MACKINNEY, A. A. (1964). Dose response curve of phytohaemagglutinin in tissue culture of normal human leucocytes. *Nature, London*, **204**, 1002–3.

MACKINTOSH, P., HARDY, D. A. & AVIET, T. (1971). Lymphocyte-typing changes after short-term culture. *Lancet*, **i**, 1019.

MACKINTOSH, P., WALLIN, J., HARDY, D. A., LING, N. R. & STEEL, C. M. (1972). The interaction of normal lymphocytes and cells from lymphoid cell lines. IV. HL-A typing of the cell line cells. *Immunology*, in press.

MARSHALL, W. H. & ROBERTS, K. B. (1965). Continuous cinematography of human lymphocytes cultured with a phytohaemagglutinin including observations on cell division and interphase. *Quarterly Journal of Experimental Physiology*, **50**, 361–74.

MARSHALL, W. H., VALENTINE, F. T. & LAWRENCE, H. S. (1969). Cellular immunity *in vitro*. Clonal proliferation of antigen-stimulated lymphocytes. *Journal of Experimental Medicine*, **130**, 327–43.

MATSUOKA, Y., MOORE, G. E., YAGI, Y. & PRESSMAN, D. (1967). Production of free light chains of immunoglobulin by a hematopoietic cell line derived from a patient with multiple myeloma. *Proceedings of the Society of Experimental Biology & Medicine*, **125**, 1246–50.

MENDELSOHN, J., SKINNER, A. & KORNFELD, S. (1971). The rapid induction by phytohaemagglutinin of increased α-aminoisobutyric acid uptake by lymphocytes. *Journal of Clinical Investigation*, **50**, 818–26.

MICHALOWSKI, A. (1972). Duration of the lymphocyte reproductive cycle *in vitro* is phytohaemagglutinin dose-dependent. *Personal communication*.

MICHALOWSKI, A. & JASINSKA, J. (1968). Heterogeneity of circulating lymphocytes and their unlike behaviour in phytohaemagglutinin-treated cultures. *Bulletin de l'Académie Polonaise des Sciences Série des Sciences Biologiques*, **16**, 53–61.

MICHALOWSKI, A., JASINSKA, J., BRZOSKO, W. J. & NOWOSLAWSKI, A. (1964). Cellular localization of the mitogenic principle of phytohaemagglutinin in leucocyte cultures. *Experimental Cell Research*, **34**, 417–19.

MICHALOWSKI, A., JASINSKA, J., BRZOSKO, W. J. & NOWOSLAWSKI, A. (1965). Studies on phytohemagglutinin. II. Cellular localization of phytohemagglutinin in white blood cell cultures with immunohistochemical methods. *Experimental Medicine and Microbiology*, **17**, 197–203.

MILLER, J. F. A. P., SPRENT, J., BASTEN, A., WARNER, N. L., BREITNER, J. C. S., ROWLAND, G., HAMILTON, J., SILVER, H. & MARTIN, W. J. (1971). Cell to cell interaction in the immune response. VII. Requirement for differentiation to thymus-derived cells. *Journal of Experimental Medicine*, **134**, 1266–84.

MONTELEONE, P. L., NADLER, H. C., JUSTICE, P. & HSIA, D. Y. Y. (1967). Electrophoretic mobility of glucose-6-phosphate dehydrogenase in lymphocytes and granulocytes. *Clinica Chimica Acta*, **18**, 275–7.

MUKHERJEE, A. B. & COHEN, M. M. (1969). Histone acetylation: cytological evidence in human lymphocytes. *Experimental Cell Research*, **54**, 257–60.

NADLER, H. L., DOWBEN, R. M. & HSIA, D. Y. Y. (1969). Enzyme changes and polysome profiles in phytohemagglutinin stimulated lymphocytes. *Blood*, **34**, 52–62.

NATHAN, C. F., KARNOVSKY, M. L. & DAVID, J. R. (1971). Alterations of macrophage functions by mediators from lymphocytes. *Journal of Experimental Medicine*, **133**, 1356–76.

NEIMAN, P. E. & HENRY, P. H. (1971). An analysis of the rapidly synthesized ribonucleic acid of the normal human lymphocyte by agarose-polyacrylamide gel electrophoresis. *Biochemistry*, **10**, 1733–40.

NEIMAN, P. E. & MACDONNELL, D. M. (1970). Studies on the mechanism of increased protein synthesis in human phytohemagglutinin stimulated lymphocytes. *Proceedings of the Fifth Leukocyte Culture Conference*, ed. J. E. Harris, pp. 61–74. Academic Press: New York.

NOWELL, P. C. (1960). Phytohemagglutinin: an initiator of mitosis in cultures of normal human leukocytes. *Cancer Research*, **20**, 462–6.

PARKHOUSE, R. M. E., JANOSSY, G. & GREAVES, M. F. (1972). Selective stimulation of IgM synthesis in mouse B lymphocytes by pokeweed mitogen. *Nature, New Biology, London*, **235**, 21–3.

PEGORARO, L. & BENZIO, G. (1971). Effect of methotrexate on DNA synthesis and thymidine kinase activity of human lymphocytes stimulated with phyto-haemagglutinin. *Experientia*, **27**, 33–4.

PERLMANN, P. & HOLM, G. (1969). Cytotoxic effects of lymphoid cells *in vitro*. *Advances in Immunology*, **11**, 117–93.

PERNIS, B., FERRARINI, M., FORNI, L. & AMANTE, L. (1971). Immunoglobulins on lymphocyte membranes. *Progress in Immunology*, ed. B. Amos, pp. 95–106. New York: Academic Press.

PETERS, J. H. & HAUSEN, P. (1971a). Effect of phytohemagglutinin on lymphocyte membrane transport. I. Stimulation of uridine uptake. *European Journal of Biochemistry*, **19**, 502–8.

PETERS, J. H. & HAUSEN, P. (1971b). Effect of phytohemagglutinin on lymphocyte membrane transport. II. Stimulation of 'facilitated diffusion' of 3-O-methyl-glucose. *European Journal of Biochemistry*, **19**, 509–13.

POGO, B. G. T., ALLFREY, V. G. & MIRSKY, A. E. (1966). RNA synthesis and histone acetylation during the course of gene activation in lymphocytes. *Proceedings of the National Academy of Sciences, U.S.A.* **55**, 805–12.

POGO, B. G. T., ALLFREY, V. G. & MIRSKY, A. E. (1967). The effect of phyto-hemagglutinin on ribonucleic acid synthesis and histone acetylation in equine leukocytes. *Journal of Cell Biology*, **35**, 477–82.

POLGAR, P. R. & KIBRICK, S. (1970). Origin of small lymphocytes following blasto-genesis induced by short-term PHA stimulation. *Nature, London*, **225**, 857–8.

POLGAR, P. R., KIBRICK, S. & FOSTER, J. M. (1968). Reversal of PHA-induced blastogenesis in human lymphocyte cultures. *Nature, London*, **218**, 596–7.

POLGAR, P. R., RUTENBURG, A. M., KIBRICK, S. & KIM, H. (1971). Response of human lymphocytes to short-term PHA stimulation. *Proceedings of the Fourth Annual Leucocyte Culture Conference*, ed. O. R. McIntyre, pp. 367–73. New York: Appleton-Century-Crofts.

POWELL, A. E. & LEON, M. A. (1970). Reversible interaction of human lymphocytes with the mitogen concanavalin A. *Experimental Cell Research*, **62**, 315–25.

QUASTEL, M. R., DOW, D. S. & KAPLAN, J. G. (1970). Stimulation of K^{42} uptake into lymphocytes by phytohemagglutinin and role of intracellular K^+ in lymphocyte transformation. *Proceedings of the Fifth Leukocyte Culture Conference*, ed. J. E. Harris, pp. 97–123. New York: Academic Press.

QUASTEL, M. R. & KAPLAN, J. G. (1968). Inhibition by ouabain of human lympho-cyte transformation induced by phytohemagglutinin *in vitro*. *Nature, London*, **219**, 198–200.

QUASTEL, M. R. & KAPLAN, J. G. (1970). Lymphocyte stimulation: the effect of ouabain on nucleic acid and protein synthesis. *Experimental Cell Research*, **62**, 407–20.

RABINOWITZ, Y. & DIETZ, A. A. (1968). Effect of phytohemagglutinin in cultures on the lactate dehydrogenases of lymphocytes from chronic lymphatic leukae-mia. *Blood*, **31**, 166–74.

RABINOWITZ, Y., LUBRANO, T., WILHITE, B. A. & DIETZ, A. A. (1967). Lactic dehydrogenase of cultured lymphocytes; response to environmental conditions. *Experimental Cell Research*, **48**, 675–8.

RABINOWITZ, Y., McCLUSKEY, I. S., WONG, P. & WILHITE, B. A. (1969). DNA polymerase activity of cultured normal and leukaemic lymphocytes. Response to phytohemagglutinin. *Experimental Cell Research*, **57**, 257–62.

RABINOWITZ, Y., WONG, P. & WILHITE, B. (1971). Thymidine metabolizing enzymes of normal lymphocytes cultured with phytohemagglutinin. *Proceed-ings of the Fourth Annual Leucocyte Culture Conference*, ed. O. R. McIntyre, pp. 81–7. New York: Appleton-Century-Crofts.

RAZAVI, L. (1966). Cytoplasmic localization of phytohaemagglutinin in peripheral white cells. *Nature, London,* **210**, 444–5.

RICHTER, M., NASPITZ, C., SINGHAL, K., NOWICKI, S. & ROSE, B. (1966). The action of phytohemagglutinin *in vivo* and *in vitro*. (Abstract.) *Federation Proceedings,* **25**, 298.

RIDDICK, D. H. & GALLO, R. C. (1971). The transfer RNA methylases of human lymphocytes. I. Induction by PHA in normal lymphocytes. *Blood,* **37**, 282–92.

RIEKE, W. O. (1965). Radioautographic studies on the localization of phytohemagglutinin-^{125}I and of the lymphocytes which respond to it. Presentation at Third Lymphocyte Transformation Workshop, Washington D.C.

RIGLER, R. & KILLANDER, D. (1969). Activation of deoxyribonucleoprotein in human leucocytes stimulated by phytohemagglutinin. II. Structural changes of deoxyribonucleoprotein and synthesis of RNA. *Experimental Cell Research,* **54**, 171–80.

RIGLER, R., KILLANDER, D., BOLUND, L. & RINGERTZ, N. R. (1969). Cytochemical characterization of deoxyribonucleoprotein in individual cell nuclei. Techniques for obtaining heat denaturation curves with the aid of acridine orange microfluorimetry and ultraviolet microspectrophotometry. *Experimental Cell Research,* **55**, 215–24.

RINGERTZ, N. R., DARŻYNKIEWICZ, Z. & BOLUND, L. (1969). Actinomycin binding properties of stimulated human lymphocytes. *Experimental Cell Research,* **56**, 411–17.

ROBINEAUX, R., BONA, C., ANTEUNIO, A. & ORME-ROSSELLI, L. (1969). La capacité endocytaire des lymphocytes ganglionnaires du cobaye, normaux et transformes *in vitro*. *Annales de l'Institut Pasteur,* **117**, 790–5.

ROITT, I. M. (1971). In *Essential Immunology*, pp. 35–50. Oxford: Blackwell.

SALMON, S. E., KRAKAUER, R. S. & WHITMORE, W. F. (1971). Lymphocyte stimulation: selective destruction of cells during blastogenic response to transplantation antigens. *Science, New York,* **172**, 490–2.

SELL, S., ROWE, D. S. & GELL, P. G. H. (1965). Studies on rabbit lymphocytes *in vitro*. III. Protein, RNA and DNA synthesis by lymphocyte cultures after stimulation with phytohaemagglutinin, with staphylococcal filtrate, with antiallotype serum and with heterologous antiserum to rabbit whole serum. *Journal of Experimental Medicine,* **122**, 823–39.

SHAPIRO, I. M. & LEVINA, L. Y. (1967). Autoradiographical study on the time of protein synthesis in human leucocyte blood culture. *Experimental Cell Research,* **47**, 75–85.

SHEARER, G. M. & CUDKOWICZ, G. (1969). Distinct events in the immune response elicited by transferred marrow and thymus cells. I. Antigen requirements and proliferation of thymic antigen-reactive cells. *Journal of Experimental Medicine,* **130**, 1243–61.

SMITH, J. L., LAWTON, J. W. M. & FORBES, I. J. (1967). Characteristics of protein synthesis *in vitro* by lymphocytes from human peripheral blood. *Australian Journal of Experimental Biology and Medical Science,* **45**, 629–43.

SMITH, J. W., STEINER, A. L., NEWBERRY, W. M. & PARKER, C. W. (1971). Cyclic adenosine 3',5'-monophosphate in human lymphocytes. Alterations after phytohemagglutinin stimulation. *Journal of Clinical Investigation,* **50**, 432–41.

STANLEY, D. A., FRENSTER, J. H. & RIGAS, D. A. (1971). Localization of H^3-phytohemagglutinin within human lymphocytes and monocytes. *Proceedings of the Fourth Annual Leucocyte Culture Conference*, ed. O. R. McIntyre, pp. 1–9. New York: Appleton-Century-Crofts.

STEFFEN, J. & SOREN, L. (1968). Changes in dry mass of PHA stimulated human lymphocytes during blast transformation. *Experimental Cell Research*, **53**, 652–9.

STEFFEN, J. A. & STOLZMANN, W. M. (1969). Studies on *in vitro* lymphocyte proliferation in cultures synchronized by the inhibition of DNA synthesis. I. Variability of S plus G_2 periods of first generation cells. *Experimental Cell Research*, **56**, 453–60.

TORELLI, V. L., HENRY, P. H. & WEISSMAN, S. M. (1968). Characteristics of the RNA synthesized *in vitro* by the normal human small lymphocyte and the changes induced by phytohemagglutinin stimulation. *Journal of Clinical Investigation*, **47**, 1083–95.

TORMEY, D. C. & MUELLER, G. C. (1965). An assay for the mitogenic activity of phytohemagglutinin preparations. *Blood*, **26**, 569–78.

VAN FURTH, R., SCHUIT, H. R. E. & HIJMANS, W. (1966). The formation of immunoglobulins by human tissues *in vitro*. IV. Circulating lymphocytes in normal and pathological conditions. *Immunology*, **11**, 29–40.

WILSON, D. B., BLYTH, J. L. & NOWELL, P. C. (1968). Quantitative studies on the mixed lymphocyte interaction in rats. III. Kinetics of the response. *Journal of Experimental Medicine*, **128**, 1157–81.

YAMAMOTO, H. (1966). Reversible transformation of lymphocytes in human leucocyte cultures. *Nature, London*, **212**, 997–8.

ZEMBALA, M. (1972). Role of macrophages in delayed-type hypersensitivity. Paper presented at the *Fourth Immunology Meeting of cell mediated immunity and blastic transformation*, Poznań. To be published in *Annals of Immunology*.

ZETTERBERG, A. & AUER, G. (1969). Early changes in the binding between DNA and histone in human leucocytes exposed to phytohemagglutinin. *Experimental Cell Research*, **56**, 122–6.

ZOSCHKE, D. C. & BACH, F. H. (1970). Lymphocyte reactivity *in vitro*. IV. Specificity in the recognition of antigens in leukocyte cultures. *Proceedings of the Fifth Leukocyte Culture Conference*. ed. J. E. Harris, pp. 487–99. New York: Academic Press.

DNA METABOLISM IN THE MATURATION
OF ANTIBODY-PRODUCING CELLS

By G. HARRIS* and I. OLSEN*

Department of Immunology,
St Mary's Hospital Medical School, London w.2

INTRODUCTION

After injection of an antigen into an intact individual, a complex sequence of cellular events occurs in the responding lymphoid tissues, resulting in the development of cells capable of the synthesis and production of specific antibodies which are secreted into the blood and tissue fluids. Proliferation of cells which develop into antibody-producing cells has been shown to occur both in-vivo (Hanna, 1964) and in-vitro (Dutton, 1967), using radio-active-labelled thymidine as a specific precursor for DNA synthesis, and has indicated that synthesis of deoxyribonucleic acid (DNA) is intimately involved in immune responses.

Most investigators of this aspect of the immune response have assumed that all cells synthesising DNA were doing so in preparation for mitotic division, but more recent studies of the spleens of intact mice responding to antigenic challenge with sheep erythrocytes (SRC) (Pelc, Harris & Caldwell, 1972) have indicated that many cells synthesising DNA were not doing so for this purpose. Synthesis of DNA by non-dividing cells is now well established (Pelc, 1972) and was termed metabolic DNA (Pelc, 1968), since loss of DNA has also been observed to occur (Pelc & Viola-Magni, 1969). From the studies of intact mouse spleen (Harris & Pelc, 1970; Pelc, Harris & Caldwell, 1972) it was concluded that metabolic turnover of DNA was involved in the development of the capacity of cells to synthesise specific antibodies.

The present experiments were concerned with DNA metabolism in organ explants of the spleens of rabbits previously immunised to SRC, cultured in the presence of this antigen. Attempts have been made to correlate DNA metabolism with proliferation of cells and the development of antibody-producing cells in these cultures in order to assess the relationship between these processes.

* Present address: Experimental Pathology Division, The Mathilda and Terence Kennedy Institute of Rheumatology, Hammersmith, London, w6 7DW.

MATERIALS AND METHODS

Preparation of tissues for culture

New Zealand white rabbits of either sex were immunised with six intra-venous injections of 10^8 washed, sheep red cells (SRC) given twice weekly for three weeks. Spleens were removed under sterile conditions at least two months after the final injection of SRC after killing the rabbit by a blow on the back of the head. After removal of fat, the spleen was rinsed in ice-cold phosphate-buffered saline (PBS) and rapidly cut into fragments (explants) of at least 2 mm diameter using sharp scissors.

Six to twenty explants were cultured in sterile disposable plastic petri dishes of 1 inch diameter containing 1 ml of Eagle's minimum essential medium (MEM) supplemented with 10–15 % foetal calf serum, inactivated by heating at 56 °C for 30 min. Glutamine 292 μg/ml, folic acid 10 μg/ml, hydrocortisone 5 μg/ml, and the following antibiotics: mycostatin 100 u/ml, streptomycin 100 μg/ml, and penicillin 100 u/ml, SRC, washed three times with sterile PBS and suspended in MEM, were added to the culture medium at the required concentrations. The explants were cultured at 37 °C in an atmosphere of 5 % CO_2 in air and were rocked continuously at six movements per minute.

Radioactive-labelled thymidine

[³H]thymidine at 15.6 Ci/mmole and [¹⁴C]thymidine at 60.6 mCi/mmole were obtained from the Radiochemical Centre, Amersham, England. [³H]thymidine was used at 5 μCi/ml and [¹⁴C]thymidine at 0.02 to 0.1 μCi/ml in these experiments.

Estimation of uptake of radioactive thymidine

This was carried out by measuring the incorporation of radioactive-labelled thymidine into the trichloracetic acid (TCA) precipitable fraction of the cells by methods described in detail elsewhere (Pelc, Harris & Caldwell, 1972).

Isolation of DNA

DNA was isolated by the method of Marmur (1961).

Chromatography of DNA

This was carried out using paper pulp prepared from DEAE–cellulose paper, D.E. 81 and centrifugal elution (Davila, Charles & Ledoux, 1965) with only minor modifications from the original method.

Estimation of antibody-producing cells (PFC)

The specific haemolytic plaque assay to estimate the number of PFC was carried out according to the original method (Jerne & Nordin, 1963) as described in detail (Harris & Littleton, 1966).

Autoradiography

Autoradiographs of smears of cells were prepared as described in detail (Harris & Littleton, 1966). The preparation of autoradiographs of PFC was carried out using [^{14}C]thymidine to label these cells (Harris, 1968). Labelling indices of PFC were based on counts of at least 100 cells. Nuclear K 5 emulsion (Ilford Ltd.) was used routinely, but for grain count analysis, the stripping-film method was used.

Mitotic counts

Wet fixed cells and haemocytometer counting was used (Schindler, 1962).

RESULTS

The immune response

The development of PFC in the cultures was directly proportional to the concentration of SRC in the medium as shown in Table 1, where it can also be seen that the majority of these cells occurred among the migrating cells (outgrowths). The increase of PFC in the outgrowths with time in culture in relation to the concentration of SRC in the medium is shown in Fig. 1, the peak development having occurred on day three.

Table 1. *Distribution of PFC in cultures*

SRC/ml	Outgrowths		Explants	
	Cell count × 10^6	PFC/10^6	Cell count × 10^6	PFC/10^6
0	9.6	42	13.2	3
2 × 10^6	12.1	503	12.9	2.8
2 × 10^7	11.4	913	13.7	1
2 × 10^8	11.5	1496	8.5	4

Explants were cultured in triplicate for 72 h in the presence of varying concentrations of SRC. Outgrowths and explants were harvested separately and the PFC and cell numbers estimated. The results are the means of 3 separate cultures which did not vary by more than 30 %.

The migration of cells from the explants during the first three days of culture is shown in Table 2. The greatest number of cells entered the

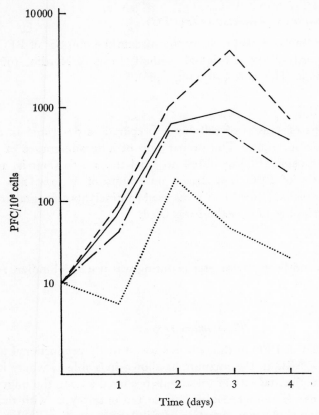

Fig. 1. The development of PFC in outgrowing cells from explants in culture. The effect of concentration of SRC. ......, o; $-\cdot-$, 2×10^6; ———, 2×10^7; $--$, 2×10^8/ml.

Table 2. *The analysis of cells migrating (outgrowths) from explants at different periods of culture*

Period culture (h)	Cell count ($\times 10^6$)		Differential count (%)				PFC/10^6	[³H]TdR (cpm/10^6)	LI%
	Outgr.	Expl.	SL	L	M	P			
0–24	29.8	21.7	63.5	13.5	5.5	17.5	36	4412	6.3
24–48	1.52	14	5	65	29.5	0.5	2834	19638	42
48–72	1.41	15.7	3	71	26	0	3293	19503	51

Explants were cultured in triplicate in the presence of 5×10^7 SRC per ml of culture medium. The number of cells migrating from the explants at the times shown were estimated by rinsing the tissues as required in ice-cold medium and then continuing their culture for the subsequent 24 h. Differential counts were performed by counting at least 300 cells in giemsa-stained smears. Separate identical cultures were used to estimate PFC and incorporation of [³H]thymidine ([³H]TdR). SL = small lymphocytes, L = lymphoblasts, M = macrophages, P = polymorphonuclear leucocytes, LI = labelling index.

medium during the first 24 h, but these cells consisted mainly of small lymphocytes and were much less active with respect to numbers of PFC and DNA synthesis, when compared to cells migrating from the explants at later times. All further experiments described here have been confined to cells migrating from explants during the first three days of culture, over relatively short periods of time.

Table 3. *The acute development of PFC in the outgrowths from explant cultures*

Antigen SRC/ml	PFC/10^6 time (h)			
	0	2	4	6
0	14	63	366	532
5×10^5	—	183	1100	686
5×10^6	—	166	755	1473
5×10^7	—	99	1379	1496

Explants were rinsed in ice-cold medium after 72 h incubation in the presence of SRC, 5×10^7/ml, and then cultured in triplicate in the presence of varying concentrations of SRC in the subsequent 6 h. The outgrowths from different cultures were harvested at the times shown and the numbers of PFC estimated.

Table 4. *The absolute increase of PFC in culture*

Time (h)	Explant	Total PFC outgrowth	Explant and outgrowth
0	140	0	140
2	80	62	142
4	80	1379	1459
6	110	1496	1606

Explants were treated as in Table 3, but the results have been expressed quantitatively on the basis that 10^7 explant cells will produce 10^6 cells in the outgrowth during the periods of culture shown.

The acute development of PFC in the cells migrating from explants between 72 and 78 h of culture is shown in Table 3. It can be seen that the presence of SRC in the medium was important for their maximal occurrence. In Table 4 the development of PFC in the presence of SRC has been expressed in absolute terms. Since the amount of explanted tissue varied in the different cultures used, the total number of cells was not uniform. From many observations it was established that explants containing 10^7 cells would produce approximately 10^6 cells in their outgrowths at the times studied here. The increase of migrating cells between 2 and 6 h was usually not significant and therefore the same figure of 10^6 was used to calculate the total PFC developing from explants of 10^7 cells at the times of culture shown. The absolute increase of PFC in this experiment was by a factor of 11.4 in only 6 h of culture, the majority of these cells being found in the outgrowths.

DNA synthesis

As shown in Table 5 many PFC can incorporate [^{14}C]thymidine. The proportion of labelled PFC increased with time but a long pulse of 24 h was required for maximum labelling in all experiments. This is emphasised in Table 6 which shows fewer PFC were labelled when pulsed only for one hour than when similar cells were pulsed for twenty-four hours before harvesting, indicating that the two processes were synchronous in only a proportion of PFC. The proportion of all cells labelling with [^{14}C]thymidine in the outgrowths which were PFC was found to be very low as illustrated in Table 7.

Table 5. *Labelling of PFC in outgrowths with [^{14}C]thymidine*

Pulse (h)	PFC/10^6	Labelled PFC (%)	Total PFC labelled/10^6
2	164	31.8	52
4	777	27.0	220
6	681	26.0	177
8	790	20.0	158
24	1246	69.3	863

Explants were cultured in the presence of SRC at a concentration of 5×10^7/ml. At 48 h, the explants were rinsed and culture continued in the presence of [^{14}C]thymidine, 0.02 μCi/ml. The numbers of PFC in the outgrowths were estimated at the times shown. The proportion of labelled PFC was also estimated in autoradiographs and the total number of PFC calculated as indicated.

Table 6. *Comparison of labelling of PFC with [^{14}C]thymidine with different pulse times*

Pulse period (h of culture)	PFC, labelling index (%)
48–72	69.3
71–72	23.0
72–96	78.0
95–96	8.4

Explants were incubated in the presence of [^{14}C]thymidine, 0.02 μCi/ml. and SRC, 5×10^7/ml after rinsing at 48 h or 72 h of culture. The labelled thymidine was added either 24 h or 1 h before harvesting the outgrowths to estimate the labelling index of PFC.

It was therefore of interest to investigate whether PFC incorporated labelled thymidine before or after migration from the explants. Explants

Table 7. *The proportion of DNA-synthesising cells in the outgrowths which are PFC*

Time (h)	PFC/10^6	All cells labelled with [^{14}C]thymidine		PFC labelled with [^{14}C]thymidine		Proportion of labelled cells which are PFC (%)
		%	Total/10^6	%	Total/10^6	
2	375	5.1	5.1×10^4	32	120	0.23
4	1015	9.1	9.1×10^4	57	578	0.63
6	1274	10.5	10.5×10^4	65	828	0.78

Explants cultured in the presence of SRC, 5×10^7/ml, were rinsed at 48 h and culture continued in fresh medium and SRC with [^{14}C]thymidine, 0.02 μCi/ml. Outgrowths were harvested at the times shown and the PFC estimated. The proportions of all cells and PFC labelled with [^{14}C]thymidine were counted in autoradiographs and from this the proportion of labelled cells which were PFC was estimated.

were labelled for 24 h with either ^{3}H- or ^{14}C-labelled thymidine and after rinsing in three changes of ice-cold medium were cultured for further periods in the absence of labelled material. It was found that virtually all the counts incorporated into the tissues under these conditions were in the DNA. A chase with cold thymidine was not employed in the post-labelling period since it could interfere with the development of PFC at concentrations of 10^{-4} M, while other experiments showed its presence in the medium did not influence the labelling of cells in the outgrowths at any time in the post-labelling period.

Table 8. *Labelling of PFC in outgrowths with [^{14}C]thymidine (prelabelling experiment)*

Post-labelling period (h)	PFC/10^6	Labelled PFC (%)	Total labelled PFC/10^6	[^{3}H]thymidine uptake (cpm/10^6 cells)	[^{3}H]thymidine labelled cells (%)
2	252	73	184	2872	10.3
4	565	63	356	2191	37
6	659	85	590	3675	58
24	4750	100	4750	3864	87

Explants were cultured in the presence of [^{14}C]thymidine, 0.1 μCi/ml and SRC 5×10^7/ml between 24 and 48 h. After 3 rinses in ice-cold medium, the tissues were transferred to fresh dishes in the absence only of [^{14}C]thymidine. Outgrowths were harvested at the times shown and the PFC estimated. The proportion of PFC labelled with [^{14}C]thymidine was estimated in autoradiographs, and from this the total labelled PFC calculated. Similar explants were labelled in the same way with [^{3}H]thymidine, 5 μCi/ml, and its uptake by cells in the outgrowths estimated by scintillation counting and the proportion of labelled cells in the autoradiographs of smears.

The results of Table 8 show that the number of labelled PFC developing in the outgrowths with time in the post-labelling period increased until all were found to be labelled at twenty-four hours. The proportion of all cells labelled with [³H]thymidine was also estimated in outgrowths at various times in the post-labelling period. The labelling index increased by a factor of 8.4 between two and twenty-four hours while the total counts, estimated by scintillation counting, increased only by a factor of 1.3.

Table 9. *Comparison of labelling with [¹⁴C]thymidine during pulse and post-labelling period*

Pulse (h)	Labelling index (%)		[¹⁴C]thy-midine cpm/10⁶ cells	Post label (h)	Labelling index (%)		[¹⁴C]thy-midine cpm/10⁶ cells
	PFC	All cells			PFC	All cells	
0–24	98.2	15.1	2572	0	98.2	15.1	2572
24–48	97.0	18.9	7534	24	100	40.0	6805
48–72	87.4	34.3	9741	48	89.0	37.6	2313
72–96	65.0	25.0	4781	72	76.5	32.3	354

Triplicate cultures of explants were pulsed with [¹⁴C]thymidine, 0.1 μCi/ml for 24 h periods in the presence of SRC, 5 × 10⁷/ml. Similar cultures were incubated with [¹⁴C]thymidine, 0.1 μCi/ml, for the first 24 h of culture. After rinsing the tissues were cultured for the subsequent 24 to 96 h period of culture with rinsing and transfer every 24 h. The outgrowths from the triplicate cultures were harvested and pooled at the times shown and the proportion of PFC and of all cells which were labelled as well as the uptake of [¹⁴C]thymidine was estimated.

In another experiment (Table 9) explants were either pre-labelled for 0 to 24 h of culture with [¹⁴C]thymidine and then studied for the subsequent 72 h or similar cultures were pulsed with [¹⁴C]thymidine for 24 h periods. It can be seen that with time in the post-labelling period the specific activity of migrating cells from pre-labelled explants dropped and was always lower than similar cells from non-labelled explants which were pulsed for the 24 h periods immediately before harvesting. However, the labelling index of all cells and PFC was better in the outgrowths of pre-labelled tissues than when [¹⁴C]thymidine was added for the appropriate pulse periods. The distribution of [³H]thymidine in migrating cells during the post-labelling period is shown in Fig. 2. Between 2 h and 24 h the labelling index increased by a factor of 5.4, while the uptake of [³H]thymidine had increased by only a factor of 1.8 in the same time. This was reflected in the grain count analysis which showed a greater proportion of lightly labelled nuclei at 24 h as compared to 2 h of culture. Similar explants were transferred to fresh dishes after rinsing, and the outgrowing cells collected between 9 and 24 h of the post-labelling period. The labelling

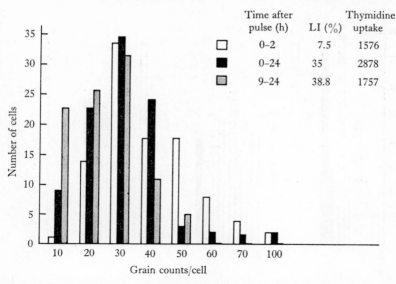

Fig. 2. The distribution of [³H]thymidine in outgrowth cells. Explants were incubated in [³H]thymidine, 5 μCi/ml, for 24 to 48 h of culture, then rinsed and culture continued without [³H]thymidine. Cells migrating from triplicate cultures of these explants between 0 to 2 h, 0 to 24 h, and 9 to 24 h after removal of the tissues from the presence of [³H]thymidine were harvested and pooled. Stripping film autoradiographs of smears, exposed for 3 weeks before developing were prepared. Uptake of [³H]thymidine, proportion of labelled cells, and the distribution of grains/100 cells was estimated.

index of these cells was very similar to the 0 to 24 h outgrowths but the [³H]thymidine uptake was lower. This was confirmed by the grain count analysis which showed that no cells migrating from explants after 9 h in the post-labelling period had a grain count > 50. Thus cells which migrated early in the post-labelling period were more heavily labelled than those migrating at later times.

In order to try and correlate cell division with the increase of PFC in these cultures the effect of colcemid on PFC and mitosis was studied as shown in Fig. 3. It can be seen that varying concentrations of this inhibitor of mitosis produced significant reduction of PFC as compared to controls after 2 h only at the highest concentrations used. Lower concentrations which were effective in arresting cells in mitosis failed to significantly prevent the development of PFC and only produced some inhibition of these cells after 8 h exposure. This strongly suggested that the acute increase of PFC was not dependent on cell division.

In contrast to the effects of colcemid, the development of PFC was highly sensitive to X-irradiation as shown in Fig. 4. Doses as low as 50 rad

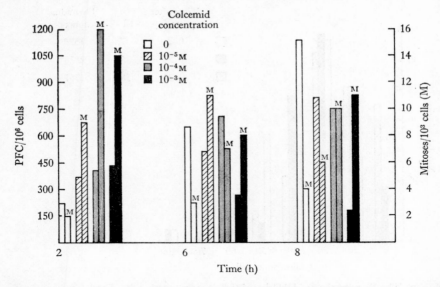

Fig. 3. The effect of varying concentrations of colcemid added to cultures of explants cultured in the presence of SRC, 5×10^7/ml. The explants were rinsed at 48 h and culture continued in triplicate in the presence of colcemid. Outgrowths were harvested at the times shown and the number of PFC and mitoses counted.

had severe inhibitory effects on both the incorporation of [^{3}H]thymidine and the development of PFC. The radiosensitivity shown here suggested very strongly that the development of plaque-forming capacity was a DNA-dependent process which was in contrast to the resistance to colcemid found during the same period in these studies.

Experiments to investigate the synthesis and release of [^{3}H]thymidine-labelled material from cells in these explant cultures were therefore carried out, and characterisation indicated it to be DNA of varying molecular weight. Table 10 shows the chromatographic behaviour of isolated DNA on centrifugal paper pulp chromatography and the molecular weight of each fraction.

Material labelled with [^{3}H]thymidine and released from explants into culture medium has been analysed chromatographically as in Table 11. It can be seen that all the fractions were non-dialysable but the low molecular weight materials were water-soluble as compared to those of higher molecular weight. The chromatographic characteristics of thymidine and calf thymus DNA have been shown for comparison.

The effect of enzymes on the various chromatographic fractions of DNA released into the culture medium as well as DNA isolated from explant tissue itself are shown in Table 12. DNAase had no effect on fraction II,

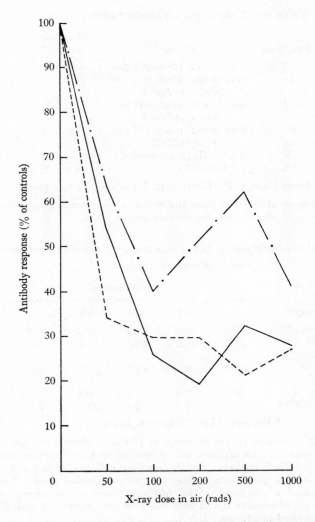

Fig. 4. The effect of X-radiation on the response of explants to SRC. Explants were cultured with daily rinsing and transfer to fresh culture dishes in the presence of SRC, 5×10^7/ml. After irradiation at room temperature from above at 10 MA, 240 kV, with a 1.00 mm Cu filter, tube distance 20 cm, and a dose-rate in air of 324 rad per minute, explants were cultured in triplicate for 4 h. The outgrowths from separate triplicate cultures were harvested, pooled, and used to estimate the uptake of [³H]thymidine, 5 μCi/ml, or PFC. The results have been recorded as per cent of unirradiated controls. —·—, [³H]thymidine uptake, day 3; ——, PFC, day 2; -----, PFC, day 3.

Table 10. *Centrifugal chromatography (CPP)*

Fraction	Buffer	Molecular weight*
I	0.01 M sod. phos. pH 7.0	500
II	0.01 M sod. phos. pH 7.0 +0.14 M-NaCl	2000
III	0.01 M sod. phos. pH 7.0 +0.50 M-NaCl	50000
IV	0.01 M sod. phos. pH 7.0 +1.0 M-NaCl	500000
V	0.2 M-NH_3+2.0 M-NaCl	10^6
VI	1.0 M-NaOH	5×10^6

* From Davilla, C., Charles, P. & Ledoux, L. (1965).

Chromatography of DNA to show the molecular weights of the fractions eluted with the various buffers used.

Table 11. *Effects of dialysis of DNA released into the culture medium from rabbit spleen explants*

Treatment sample	Fraction					
	I	II	III	IV	V	VI
Buffer dialysed	26.0	12.0	29.5	27.5	2.7	1.9
Water-soluble (supernatant)	33.5	32.5	13.5	13.0	5.0	6.5
Water-insoluble (precipitate)	19.0	0	45.0	26.0	4.0	6.0
[^{3}H]thymidine	89.3	0.4	2.5	0.4	7.2	0.2
Calf thymus DNA*	4.4	6.3	5.1	21.8	24.6	37.8

* Estimated by UV spectrophotometry

Explants were incubated in the presence of [^{3}H]thymidine, 5 μCi/ml between 24 and 48 h of culture. The explants were rinsed at 48 h and culture continued for a further 24 h in fresh medium. The medium was collected at 72 h and dialysed against 0.01 M sodium phosphate buffer, pH 7.0 or distilled water. The latter treatment produced a precipitate. The materials were chromatographed after dialysis and the distribution of the various fractions were compared with [^{3}H]thymidine and unlabelled calf thymus DNA.

but rendered the other fractions soluble in TCA with an effect increasing with the molecular weight of the material. Heat denatured DNAase was inactive in this respect. In contrast nuclease significantly affected only the high molecular weight components. While RNAase had no effect on these materials tested, the effect of protease was of interest since it was found to render low molecular weight fraction II TCA-soluble. It was reported that after chromatography of these fractions of DNA (Davila *et al.* 1965) that fraction II, consisting of thymidine di- and triphosphates was dialysable.

Table 12. *The effect of enzymes on DNA fractions*

CPP fractions (CP 10 min)

Enzyme treatment	II		III		IV		V		VI		FJ34*	
	−TCA	+TCA	−TCA	+TCA	−TCA	+TCA	−TCA	+TCA	−TCA	+TCA	−TCA	+TCA
Buffer (Acetate)												
Magnesium	1606	973	1427	1126	3965	3757	2037	1922	1462	1156	3043	2878
DNAase	1552	997	735	408	1616	324	584	196	510	252	981	425
Heat-denatured DNAase	—	—	1369	1425	3641	4622	1922	1798	—	—	—	—
Buffer (Tris calcium)	1446	1101	1523	1199	7910	6679	2333	2371	1242	1615	3904	3365
Nuclease	1518	1045	1084	748	4593	3259	1478	345	475	128	3041	1499
SSC	1436	1036	1045	848	3818	3766	1921	1763	849	831	2301	2349
RNAase	1497	1134	1124	893	4000	3939	1300	1457	1361	1079	2547	2311
Protease	994	256	752	677	3724	3372	1699	1492	568	415	2107	2130
1 M-NaOH	918	214	709	658	2820	3274	1352	1280	863	1203	2089	1998

Pools of chromatographic fractions of the DNA from rabbit spleen explants labelled with [^{3}H]thymidine in culture were treated with DNAase I, 75 μg/ml at room temperature, micrococcal nuclease, 150 μg/ml at 30 °C, RNAase IIIA, 150 μg/ml at 37 °C, or fungal protease, 500 μg/ml at 37 °C. All incubations were overnight and controls using the appropriate buffers were included. Aliquots of the fractions were placed on Whatman paper discs for incubation and then duplicates treated with or without 5 % TCA before processing for counting.

* Rabbit spleen DNA.

This contrasts to the observation here that before chromatography it was not. In view of the effect of protease shown here, fraction II would appear to be bound to protein and on chromatography was eluted as fraction II in unbound form and thus rendered dialysable. The results of this table emphasise the DNA nature of fractions III and IV in the material released from the cells in these cultures with respect to enzyme sensitivity and resistance to alkali. This is in contrast to fraction II which was susceptible to 1 M-NaOH.

Table 13. *Comparison of chromatographic distribution of DNA isolated from spleen of intact animals and DNA labelled during culture*

	Labelled with [³H]thy-midine	CPP fractions (%)					
		I	II	III	IV	V	VI
Mouse spleen	In-vivo (6 h)	0	0	2.0	17.5	19.5	61.0
Mouse spleen (germ free)	In-vivo (6 h)	0	0	0	18.6	21.1	60.3
Rat spleen*	Unlabelled	5.8	6.7	9.7	10.8	14.5	52.5
Rabbit spleen	In-vitro (0–24 h)	2.79	0.91	13.98	37.90	30.39	14.01
	In-vitro (24–48 h)	1.56	0.77	13.26	57.95	15.94	10.53
	In-vitro (48–72 h)	2.39	1.25	14.89	48.34	17.89	15.23

* Measured spectrophotometrically.

DNA was isolated from mouse and rat spleens by the method of Marmur. The mice had been given [³H]thymidine, 3 μCi/gm body weight, in 3 equal 2-hourly injections and killed 2 h after the last. Explants of rabbit spleen were cultured with [³H]thymidine, 5 μCi/ml for the times shown and their DNA extracted in the same way. Chromatography was then carried out.

A chromatographic comparison of the DNA synthesised by mouse spleen in-vivo as well as unlabelled rat spleen DNA with DNA synthesised by rabbit spleen explants in-vitro has been made in Table 13. It can be seen that the in-vitro labelled DNA was composed of a greater proportion of fraction IV than DNA in-vivo.

The results of Table 14 show the release of labelled DNA from rabbit spleen explants after pulsing with [³H]thymidine for varying periods of time. During the post-labelling periods there was a progressive accumulation of all fractions in the medium but III and IV formed the major components. Preliminary experiments involving killing or damage of cells prelabelled with [³H]thymidine have shown that the bulk of labelled material

Table 14. *Release of labelled material from explants in culture*

Pulse period (h)	Post-label period (h)	CPP fractions (c.p.m.)						Total counts/ml
		I	II	III	IV	V	VI	
0–24	4	309	84	1047	2014	137	25	3616
—	8	191	89	1534	2166	108	0	4097
—	24	428	359	7235	6791	445	66	15322
0–48	4	50	104	168	1060	124	16	1522
—	8	41	225	170	1123	168	23	1750
—	24	1075	357	15438	11464	768	249	29351
24–48	4	44	105	61	656	119	23	1008
—	8	116	126	2116	2595	385	79	5417
—	24	748	306	8025	6814	356	206	16455

Explants were cultured in triplicate in the presence of SRC 5×10^7/ml and after rinsing and transfer to fresh medium incubated with [³H]thymidine 5 μCi/ml, for the times shown. For the 0 to 48 h pulse the explants were rinsed at 24 h and [³H]thymidine was added once more after transfer to fresh medium. At the end of each pulse period, the explants were rinsed and culture continued, triplicate cultures being harvested at the times shown. The medium collected at each time was analysed chromatographically.

released was of low molecular weight and dialysable. While further studies are needed, the occurrence in DNA-synthesising cells in these cultures of low molecular weight fractions of DNA which was released into the medium would appear to depend on the metabolic activity of cells and was not simply the release of breakdown products of DNA due to cell death. The amount of DNA material released in the 24 h period following a 24 h pulse with [³H]thymidine has still to be determined, but in terms of incorporated [³H]thymidine, about 25 % of the counts were released into the medium in the subsequent 24 h after pulsing for the same time.

DISCUSSION

Cultured explants of rabbit spleens responded to stimulation by SRC, the antigen used to immunise the animals at least 2 months before killing. This immune response was characterised by the appearance of cells capable of producing specific antibodies (PFC) to the SRC, the majority of PFC being found among those cells which had migrated from the explanted tissues into the culture medium. The concentration of SRC in the culture medium was important for optimal development of PFC in these cultures. This strongly suggested that interaction must occur between the antigen and those cells destined to produce antibodies and led to the idea (Harris, 1968) of antigen acting like a hormone to induce the synthesis of a specific protein by the responsive cells, viz. antibody.

Autoradiography showed that the majority of PFC were capable of DNA synthesis as manifest by the incorporation of [¹⁴C]thymidine into their nuclear DNA. Experiments, not shown here, revealed that virtually all the PFC developing during the first three days of culture could be so-labelled but, as indicated here, long pulses with [¹⁴C]thymidine were required for this purpose. When pulsed in the same way at 96 h of culture very few PFC could be labelled. Only a small proportion of PFC, when plated for assay immediately after a 1 h pulse, were labelled as compared to similar cells labelled during 24 h exposure to [¹⁴C]thymidine before plating. This indicated that DNA synthesis and plaque-forming capacity were not synchronous in the majority of PFC but the former preceded the development of this latter capacity. The question thus arose whether cell labelled with [¹⁴C]thymidine were in S phase preparatory to division and becoming PFC, or whether some other explanation could be entertained such as the turnover of DNA by non-dividing cells as a preliminary to the development of antibody-producing capacity.

The estimation of the rate of increase of PFC during short periods of culture indicated this was too rapid to be accounted for by cell division alone. At least three to four mitoses would have been needed to give the rise of a factor of 10 seen in 6 h. This was supported by the finding that colcemid in concentrations capable of arresting mitosis did not significantly prevent this increase of PFC.

If cell division was not responsible for the acute increase of PFC seen in these experiments then DNA synthesis must have been playing some other role. The sensitivity to irradiation of the PFC seen here would indicate that DNA-dependent processes were involved in the development of plaque-forming capacity by PFC since RNA and protein synthesis are both considered to be highly radio-resistant processes in higher cells.

Estimation of mitosis after treatment of cultures with colcemid gave a mitotic rate of 1 % in 2 h among cells migrating from explants between 48 and 72 h of culture. Therefore 12 % of those cells would divide once in 24 h. In contrast 40 % of similar cells, untreated with colcemid, were labelled after 24 h incubation with [³H]thymidine, indicating that many more cells were labelled than were capable of division. Similar discrepancies between mitosis and labelling with [³H]thymidine have been found in the spleens of intact mice (Harris, Pelc & Blackmore, *Eur. J. Immunology*, in press).

Experiments designed to investigate whether PFC were synthesising DNA before or after migration from explants gave unexpected results. When explants were pre-labelled with ¹⁴C- or ³H-labelled thymidine, the cells which migrated from them into the medium showed an increase of their labelling index with time after removal from the presence of the labelled material. All the PFC estimated after 24 h in the post-labelling period were found to be labelled with [¹⁴C]thymidine while the labelling index of all migrating cells, using [³H]thymidine to pre-label the explants, increased by a factor of 8 between 2 and 24 h of the same period. In contrast the total amount of [³H]thymidine, estimated by scintillation counting, had increased only by a factor of 1.3. Grain count analysis of autoradiographs of these cells showed that the first cells to leave the explants were more heavily labelled as compared to those cells which entered the medium at later times and explained these findings.

One possibility was that pre-labelled cells which had migrated early in the post-labelling period had then divided. While some nuclei, having divided, would become less strongly labelled, a reduction in grain count resulting solely from division could not explain these results. The increased labelling index must have been due to cells with less label migrating from the explants at later times, cells losing label by means other than division, or new nuclei becoming labelled after cells had left the explants.

If cell division was solely responsible for the increase of labelling index with time in the post-labelling period, at least three divisions would have to be assumed, giving a cycle time of 7 h which is short for most estimates of 12 to 24 h for mammalian cells. In addition to this the total number of cells in the outgrowths should have doubled in order to produce the increased labelling found here even if cell migration had stopped after the first 2 h. In fact, cells continued to migrate after 2 h and any increase of cells in the outgrowths by 24 h could be accounted for by this process as much as actual cell division. Therefore since the numbers of unlabelled cells would also increase in these cultures a much shorter cycle time than 7 h would have been required to account for this increase of labelled cells, if due to division alone. As shown by the use of colcemid the actual mitotic rate in these cultures was too low for cell division alone to explain these results.

Another possibility to be considered was the transfer of labelled material to previously unlabelled cells. The present studies have shown that material of varying molecular weight, and with the properties characteristic of DNA in respect to susceptibility to enzymes was being synthesised in the cells of these tissues and was being released into the culture medium.

It was therefore possible to explain the present data as due to a combination of all the processes outlined here:

(a) Pre-labelled cells migrating from explants and then dividing.

(b) A metabolic loss of DNA from pre-labelled cells as they migrated from the explants into the culture medium.

(c) Transfer to and utilisation of DNA by cells metabolically active in this respect.

Chromatographic characterisation of DNA, newly synthesised by cells in these explants, showed that a significant proportion of this material could be isolated in a size of 200000–500000 molecular weight. Identical material could be also found to form the bulk of DNA released into the culture medium. Therefore the loss of labelled DNA from these explants could be accounted for by migration of labelled cells and the release of DNA which was available for re-utilisation by other, non-labelled migrating cells. Further studies are now in progress to elucidate how relevant this metabolic turnover is to the development of antibody-producing capacity by cells.

REFERENCES

DAVILA, C., CHARLES, P. & LEDOUX, L. (1965). The chromatography of nucleic acid preparations on DEAE–cellulose paper. I. Fractionation of deoxyribonucleic acid on paper strips or on centrifuged paper pulp. *Journal of Chromatography*, **19**, 382–95.

DUTTON, R. W. (1967). *In vitro* studies of immunological responses of lymphoid cells. *Advances in Immunology*, **6**, 253–336.

HANNA, M. G. (1964). An autoradiographic study of the germinal centre in spleen white pulp during early intervals of the immune response. *Laboratory Investigation*, **13**, 95–104.

HARRIS, G. (1968). Antibody production *in vitro*. I. Single cell studies of the secondary response to sheep erythrocytes. *Journal of Experimental Medicine*, **127**, 661–74.

HARRIS, G. & LITTLETON, R. J. (1966). The effects of antigen and phytohaemagglutinin on rabbit spleen cell suspensions. *Journal of Experimental Medicine*, **124**, 621–34.

HARRIS, G. & PELC, S. R. (1970). Incorporation of [^{3}H]thymidine into the spleen of intact mice during the immune response to sheep erythrocytes. *Immunology* **19**, 865–78.

JERNE, N. K. & NORDIN, A. A. (1963). Plaque formation in agar by single antibody-producing cells. *Science, New York*, **140**, 465–7.

MARMUR, J. (1961). A procedure for the isolation of deoxyribonucleic acid from micro-organisms. *Journal of Molecular Biology*, **3**, 208–18.

PELC, S. R. (1968). Turnover of DNA and function. *Nature, London*, **219**, 163–4.

PELC, S. R. (1972). Metabolic DNA in ciliated protozoa, salivary gland chromosomes and mammalian cells. *International Review of Cytology*, **32**, 327–55.

PELC, S. R., HARRIS, G. & CALDWELL, I. (1972). The relationship between antibody formation and deoxyribonucleic acid synthesis in mouse spleens during primary and secondary response to sheep erythrocytes. *Immunology*, **23**, 183.

PELC, S. R. & VIOLA-MAGNI, M. D. (1969). Decrease of labelled DNA in cells of the adrenal medulla after intermittent exposure to cold. *Journal of Cellular Biology*, **42**, 460–68.

SCHINDLER, R. (1962). Desacetylamino-colchicine: a derivative of colchicine with increased cytotoxic activity in mammalian cell cultures. *Nature, London*, **196**, 73–4.

CONTROL OF CELL DIVISION
IN MAMMALIAN CELLS

By T. RYTÖMAA*

Second Department of Pathology, University of Helsinki,
Haartmaninkatu 3, 00290 Helsinki 29, Finland

INTRODUCTION

Cells have the inherent tendency to proliferate. Yet, in a typical adult mammalian organism cell division and cell loss are balanced, i.e. within certain time limits the numbers of cells within a population remain constant. In unicellular organisms it is mostly the quantity and quality of food available which limits growth; in multi-cellular organisms specific control mechanisms are needed. It is now clear that one basic component in this mechanism is a cell-line specific chalone, a feed-back inhibitor which is produced within the same cell population on which it acts.

Chalone theory is basically simple; it is also intellectually satisfying. However, it is clear that this mechanism alone cannot explain all the mitotic phenomena found in the various tissues of a normal adult mammalian organism. Thus, for instance, in haematopoietic tissues the point of balance may change from time to time, according to the functional demand imposed on the tissue, and in certain target tissues cell division and cell loss are influenced by specific 'mitogenic' hormones. It is possible, of course, that these and other similar regulatory mechanisms act through the chalone system, e.g. by inactivating the inhibitor directly, or by interacting with the transfer of the inhibitory message from chalone to the responding 'mitotic gene(s)'.

All these complicating features make it difficult at present to visualise the control of cell division adequately; however, for practical purposes it is perhaps wise to stick to the assumption that the underlying specific system is simple. Chalone mechanism may be the unique basic parameter of this system.

GENERAL PROPERTIES OF CHALONES

Chalones can be defined in many ways. Evidently the leading definition is that of Bullough (1967): a chalone is 'an internal secretion produced by a tissue for the purpose of controlling by inhibition the mitotic activity of

* Supported by grants from the Lady Tata Memorial Trust, London, and from the Sigrid Jusélius Foundation, Helsinki.

[457]

that same tissue'. There are also other definitions in the literature, which are commonly related to the manner in which the chalone problem has been approached (Laurence, Randers Hansen, Christophers & Rytömaa, 1972); even a list of certain biological properties of chalones may serve as a useful definition.

The four most important biological properties of chalones are: they inhibit cell proliferation in-vivo; they are produced by the same tissue on which they act; their action is reversible; and their action is cell-line specific. In my opinion any substance which fulfils these criteria deserves the name chalone.

Some 12–13 different chalone systems have been reported in the literature. These are from the epidermis (Bullough & Laurence, 1964; Iversen & Elgjo, 1967; Voaden, 1968; Frankfurt, 1971; Marrs & Vorhees, 1971; Marks, 1971), granulocytes (Rytömaa & Kiviniemi, 1968a, b, c; Paukovits, 1971, 1973; Benestad, Rytömaa & Kiviniemi, 1973; Schütt & Langen, 1973), erythrocytes (Kivilaakso & Rytömaa, 1971), lymphocytes (Moorhead, Paraskova-Tchernozemska, Pirrie & Hayes, 1969; Bullough & Laurence, 1970a; Garcia-Giralt, Lasalvia, Florentin & Mathé, 1970; Houck, Irasquin & Leikin, 1971; Chung, 1973), sebaceous glands (Bullough & Laurence, 1970b), melanocytes (Bullough & Laurence, 1968b; Dewey, 1973), kidney (Saetren, 1956; Chopra & Simnett, 1970), liver (Saetren, 1956; Volm, Kinzel & Süss, 1969; Scaife, 1970; Verly, Deschamps, Pushpathadam & Desrosiers, 1971), lung (Simnett, Fisher & Heppleston, 1969), fibroblasts (Houck, 1973), certain ascites tumour cells (Bichel, 1971, 1972), endometrial tissue (Volm, Wayss & Hinderer, 1969), and thyroid gland (Garry & Hall, 1970). If the experimental evidence for the existence of these 'chalones' is evaluated in terms of the four criteria mentioned above, it is clear that most of the substances included in the list have been studied very superficially. Indeed, it appears from the survey of the available literature that so far only three of the chalones have been studied in any detail: these are the epidermal, the granulocytic, and the lymphocytic chalones.

From the chemical point of view chalones may appear a little dull: as far as characterised, they are proteins or polypeptides with no striking peculiarities (Hondius Boldingh & Laurence, 1968; Marks, 1973; Lasalvia, Garcia-Giralt & Macieira-Coelho, 1970; Houck et al. 1971; Paukovits, 1973). With respect to the molecular weights, the substances purified tend to fall into two distinct groups: one group, including, for example, the granulocytic chalone (see Rytömaa & Kiviniemi, 1968a, c; Paukovits, 1971, 1973), consists of polypeptides with molecular weights below 5000 daltons, and the other group, including, for example, the lymphocytic chalone

(Lasalvia *et al.* 1970; Houck *et al.* 1971), consists of proteins with molecular weights somewhere in the region of 40000 daltons. So far none of the chalones has been tested for biological activity in a chemically pure form, but some of the preparations may not be very far from this stage (e.g. the granulocytic chalone; see Paukovits, 1973).

The target for a chalone action in the cell cycle is not clear at present. There is unequivocal evidence that an epidermal inhibitor acts in the G_2 phase (Bullough & Laurence, 1964; Iversen & Elgjo, 1967; Marrs & Vorhees, 1971), and there is also good experimental evidence for an action in the G_1 phase (Hennings, Elgjo & Iversen, 1969; Frankfurt, 1971; Marks, 1971). It has been claimed, however, that there are actually two different tissue specific factors which control mitosis (G_2 action) and DNA synthesis (G_1 action), respectively, in mouse epidermis. In addition to the different points of action in the cell cycle, these two factors differ in regard to their origin (basal and differentiating cells, respectively) and also to their chemistry (Elgjo & Hennings, 1971*a*; Elgjo, 1973; Marks, 1971, 1973).

Most of the other chalones have been studied with respect to their effect on the transition $G_1 \rightarrow S$ only, presumably because it is difficult to explain the control of cell proliferation on the basis of the $G_2 \rightarrow M$ blocking alone. However, there is no *a priori* reason to assume that chalone action is exerted only in a specific point of the cell cycle; it is possible, at least in theory, that the inhibitory action extends over a large part of the cell cycle (see Bullough, 1973).

SPECIFICITY OF CHALONE ACTION

Cell-line specificity of the inhibitory action of the chalones is perhaps their most important biological characteristic; unfortunately, however, extensive evidence of this property is not easy to obtain. With respect to the epidermal chalone, there is plenty of evidence that crude extracts obtained from a large number of non-epidermal tissues do not inhibit cell division in epidermis either in-vivo or in-vitro, and that epidermal extracts do not inhibit cell division in a number of non-epidermal cells (Bullough & Laurence, 1964; Rytömaa & Kiviniemi, 1967; Bullough, Laurence, Iversen & Elgjo, 1967; Hennings *et al.* 1969; Frankfurt, 1971; Marks, 1971; Marrs & Vorhees, 1971).

Evidence for the cell-line specific action of the granulocytic chalone is equally strong. Thus it has been demonstrated by means of autoradiographic analysis of rat bone marrow cells, labelled with [³H]thymidine in a short-term in-vitro culture, that the number of labelled granulocytic cells was significantly decreased in the presence of an unpurified preparation of the granulocytic chalone, but that no change could be observed in the labelling index of the other cell types of the bone marrow (Rytömaa & Kiviniemi, 1968a, c). An essentially similar result has also been obtained with a partially purified preparation of the granulocytic chalone; guinea-pig bone-marrow cells were separated into three different subpopulations by means of Ficoll density step centrifugation and the effect of the test preparation was measured in terms of [^{3}H]thymidine incorporation into the cells; only the top layer, containing up to 80% of mitotically competent granulocytic precursor cells, was inhibited by the test preparation (Pauko-vits, 1973).

A third type of evidence for the cell-line specific action of the granulo-cytic chalone has been obtained by means of a closed culture in-vivo utilis-ing diffusion chambers implanted in the peritoneal cavity of mice (Benestad et al. 1972). In these experiments each mouse carried two different kinds of chamber cultures in the peritoneal cavity: one chamber contained bone-marrow cells, i.e. proliferating granulocytes and macrophages (see Benestad, 1970), and the other chamber mononuclear cells isolated from rat blood, i.e. proliferating immunoblasts and macrophages (see Benestad, Iversen & Rolstad, 1971). The chalone and control preparations ('extract' of liver) were then injected into the mice, and cell proliferation was measured in terms of [^{3}H]thymidine incorporation into the chamber cells; the results showed that the chalone preparation inhibited DNA synthesis in the granulocytic precursor cells only.

Further evidence for the cell-line specificity of the granulocytic chalone has been obtained by different means; some of this has been briefly re-viewed recently (Rytömaa, 1973). With respect to the other chalones reported in the literature (see p. 458), the strength of the evidence for a tissue-specific action varies from case to case; besides the epidermal and granulocytic chalones, the most extensive evidence has so far been obtained of the lymphocytic chalone (Garcia-Giralt et al. 1970; Houck et al. 1971). It may be mentioned, however, that a lymphoid cell extract, presumably containing the lymphocytic chalone, has also been reported to have suppressing action which is not only confined to lymphocyte DNA synthesis (Jones, Paraskova-Tchernozemska & Moorhead, 1970). Similarly, an extract ('chalone') prepared from human endometrium has been found to inhibit DNA synthesis also in embryonic human liver and kidney cells (Volm & Wayss, 1971). It must be realised, however, that these non-specific inhibitory effects, produced by crude tissue extracts, do not provide meaningful experimental evidence for concluding that the chalone action may lack tissue specificity.

It is clear from the foregoing that cell-line specificity must be considered as one of the most important biological characteristics of chalones. In spite of this, it may be worth emphasising that nevertheless the specificity need not be absolute; a certain amount of non-specificity may even be expected. Thus, for instance, highly undifferentiated (embryonic) and pathological (malignant) cells may perhaps fail to recognise the specific chalone with an absolute precision; they could thus well respond even to a number of closely related chalones. This is especially so, because it is certainly not impossible that all chalones share the same active group in their molecule (see Bullough, 1967, 1973). Furthermore, it is also thinkable that a chalone exists which is effectively retained within a short range in-vivo; this could be the only mechanism which prevents its action on other potentially responsive cell populations.

Like many hormones chalones are not species-specific in their action (e.g. Bullough *et al.* 1967; Rytömaa & Kiviniemi, 1970). This does not mean, of course, that the same chalone is chemically identical in all species. Consequently, antisera can be produced against chalones (see Chopra & Simnett, 1970; Garry & Hall, 1970; Chopra, 1973).

MODE OF ACTION OF CHALONES

It has been shown that the action of the epidermal chalone in-vitro is strengthened by adrenalin (Bullough & Laurence, 1964; Marrs & Vorhees, 1971); adrenalin, in turn, is known to act via the cyclic AMP system, and cyclic AMP may be involved in the control of cell proliferation (Heidrich & Ryan, 1970). It is therefore suggested (Iversen, 1969) that the epidermal chalone–adrenalin 'complex' exerts its inhibitory action by activating the epidermal adenyl cyclase, which would then mean that cyclic AMP is a 'secondary messenger' also in the chalone action. There is some experimental evidence which supports this hypothesis in epidermis (Powell, Duell & Vorhees, 1971; Brønstad, Elgjo & Øye, 1971; Marks & Rebien, 1972), even if the original idea of an active chalone–adrenalin 'complex' seems to be wrong (Laurence *et al.* 1972).

The possible role of cyclic AMP in other chalone systems is even more obscure at present. It may be noted, however, that there are chalone systems, e.g. the granulocytic chalone, in which adrenalin does not play any role (Rytömaa & Kiviniemi, 1969a), and that there are cell systems, e.g. the haematopoietic stem cells, in which cyclic AMP seems to exert a stimulatory action on cell proliferation (Byron, 1971).

It follows from the foregoing that although cyclic AMP is now generally recognised as a versatile regulatory agent in numerous cellular processes, its possible role in the chalone action remains to be solved.

There is also another theoretically interesting possibility to explain chalone action by linking it to the '*retine/promine*' idea of Szent-Györgyi (1968). Many years ago Szent-Györgyi and his associates reported that some tissue extracts retarded ('*retine*'), whilst others promoted ('*promine*') growth (see Szent-Györgyi, Együd & McLaughlin, 1967). Arduous efforts to isolate retine failed, but the experiments indicated that it may be a ketoaldehyde (Együd, 1965), probably methylglyoxal (see Szent-Györgyi, 1968).

It is now well known that there is a relation between SH groups and cell division (Harrington, 1967; Knock, 1967). Methylglyoxal, in turn, readily interacts with SH groups and thus exerts an inhibitory action on cell division; this inhibition is completely abolished, e.g. by cysteine added to the cell suspension in quantities equimolar to the methylglyoxal present (see Szent-Györgyi, 1968).

All living cells contain a most active enzymic system, glyoxalase, which transforms methylglyoxal into lactic acid. It is, of course, tempting to assume that methylglyoxal is the 'missing' substrate of the glyoxalase system. This would also make it easy to explain why methylglyoxal ('retine') could not be isolated: it readily interacts with SH and amines, and can thus be bound by the various cell constituents, if not decomposed by glyoxalase ('promine'). Indeed, with the high biological activity of methylglyoxal it can be present in the normal cell only in trace amounts (see Szent-Györgyi, 1968).

If it is actually methylglyoxal which inhibits cell division, by interacting with SH groups, then it is necessary also to assume that the glyoxalase is kept, in the inhibited cell, separated from the methylglyoxal (see Szent-Györgyi, 1968). In the light of this theory, chalone action could increase the intracellular amount of methylglyoxal and, therefore, chalone may act by binding and inactivating glyoxalase or, alternatively, by increasing the formation of methylglyoxal from trioses. This theory can be tested experimentally.

To this end, the actions of methylglyoxal, cysteine, and granulocytic chalone were tested singly and in various combinations on DNA synthesis in normal rat bone-marrow cells in-vitro, using a short-term assay system which has been described in detail before (Rytömaa & Kiviniemi, 1967, 1968a; Rytömaa, 1969). The results are shown in Table 1.

It is apparent from these results that qualitative effects of the test factors were essentially the same in both experiments. Quantitatively a difference exists in one respect: the response of the target cells to the inhibitory actions of chalone and methylglyoxal was excessive in the second experiment. The reason for this difference is not known.

Table 1. *Actions of chalone, methylglyoxal and cysteine on the cumulative incorporation of [^3H]thymidine in normal rat bone marrow cells in-vitro*

	Incorporation of [^3H]thy-midine in cells (relative units)	
Test factors	Exp. 32/72	Exp. 42/72
None	100 ± 6	100 ± 5
Granulocytic chalone*	79 ± 6	34 ± 4
Methylglyoxal†	68 ± 8	40 ± 6
Cysteine‡	114 ± 2	—
Chalone* + cysteine‡	71 ± 5	37 ± 5
Methylglyoxal† + cysteine‡	119 ± 5	108 ± 3
Chalone* + methylglyoxal†	39 ± 4	26 ± 3
Chalone* + methylglyoxal† + cysteine‡	—	38 ± 6

The actions of the test factors are expressed in relative units by giving the value 100 to the mean radioactivities incorporated in control cultures within each culture time group. The values shown give $\bar{x} \pm$ s.e., where $N = 4$ (number of culture time groups used; the number of individual cultures per test group was 14 in both experiments).

* Crude preparations obtained from rat bone marrow cells (experiment 32/72) and from ascites granulocytes (experiment 42/72).

† 10^{-4} M.

‡ 10^{-3} M.

Besides the finding that both the chalone and the methyglyoxal inhibited the incorporation of [^3H]thymidine in the target cells, the results demonstrate that the methylglyoxal inhibition was completely abolished by cysteine, but that cysteine was totally inactive in releasing the chalone inhibition. It thus follows that chalone action cannot be mediated by methylglyoxal ('retine').

OTHER REGULATOR SUBSTANCES OF CELL DIVISION

Factors which may interact with chalone mechanism

The number of (physiological) substances which have been suggested in the literature to control cell division is extremely large; indeed, the situation may appear chaotic. However, if it is assumed, rightly or wrongly, that the basic control mechanism is simple and that the unique parameter of this system is chalone, then the situation is perhaps a little easier to comprehend.

In the light of the chalone theory a group of regulator substances may be expected to exist which modify the chalone mechanism in some specific manner, e.g. by interacting with chalone production, transport, potency, binding, or possible triggering and transfer of secondary messenger(s). Some such substances have indeed been described in the literature. These at least include adrenalin, hydrocortisone, 'antichalone/chalone antagonists', and 'mitogenic' hormones.

It has been shown that the action of the epidermal and some other chalones require adrenalin as a cofactor, at least for blocking the transition $G_2 \to M$ in-vitro (Bullough & Laurence, 1964, 1968a, b, 1970b; Marrs & Vorhees, 1971), but it has never been really understood how this action takes place. However, it is now clear that adrenalin has no direct action on the epidermal chalone, e.g. because epidermal chalone causes the same mitotic depression in adrenalectomised animals as in normal intact mice (Laurence & Randers Hansen, 1971). On the basis of this and other evidence, it has recently been suggested that adrenalin destroys or blocks the action of an epidermal chalone antagonist, probably mediated by the dermis (Laurence et al. 1972).

It has also been shown that the inhibitory power of the epidermal chalone (plus adrenalin) is augmented by hydrocortisone, or rather that the hydrocortisone prolongs the mitotic depression induced by the chalone–adrenalin 'complex' (Bullough & Laurence, 1968c). Re-evaluation of the available evidence has led to the conclusion that hydrocortisone may suppress the production or action of the chalone antagonist (Laurence et al. 1972).

Evidence for the existence in the body of substances which antagonise the action of chalones has been found, for example for granulocytes (Rytömaa & Kiviniemi, 1967, 1968a, b; Rytömaa, 1969), erythrocytes (Kivilaakso, 1971), fibroblasts (Houck, 1973), macrophages (Mauel & Defendi, 1971), and epidermis (Laurence et al. 1972); some of these substances have also been characterised to some extent. Thus, for instance, in conditions of acute functional demand for granulocytes a tissue-specific stimulator, the 'granulocytic antichalone', replaces chalone in rat serum; on the basis of elution parameters of this substance on Sephadex G-75 and on G-200, the 'granulocytic antichalone' appears to have a molecular weight of about 30000 daltons (Rytömaa & Kiviniemi, 1967, 1968a). It has also been observed that the inhibition of DNA synthesis, caused by the granulocytic chalone in normal rat bone-marrow cells in-vitro, can be abolished by the 'granulocytic antichalone' (Rytömaa & Kiviniemi, unpublished results). However, direct interaction between these two substances has not been demonstrated.

Another group of factors which stimulate cell proliferation in a tissue-specific manner are the 'mitogenic' hormones, such as androgens and oestrogens. It has been suggested that these hormones act by neutralising the chalone of the target tissue (Bullough, 1965, 1967), but direct experimental evidence supporting this suggestion is not available. More details of the hormonal regulation of cell proliferation can be found in a recent review of the effects of hormones on the cell cycle (Epifanova, 1971).

Examples of factors which may not interact with chalone mechanism

In addition to the heterogeneous group of regulator substances which are known or may be expected to modify the chalone action, there is another and even more heterogeneous group of factors which evidently do not exert their main effect on cell proliferation through the chalone mechanism. Only some of these factors are presented below.

A large number of essentially non-specific factors, such as deficient nutrient supply, accumulation of toxic metabolites, oxygen deficiency, or changes in pH, do not usually play an important role in the regulation of cell division in multicellular organisms. This is demonstrated, for example, by the discovery that, when two different ascites tumours, both reaching a plateau before the death of the host animal, were inoculated in the same mouse, both tumours developed independently (Brown, 1970; Bichel, 1972). It is thus clear that the plateau phenomenon in mice bearing one ascites tumour is not explicable in terms of trivial environmental factors, such as nutrient exhaustion or accumulation of toxic metabolites; in fact, it has been shown that the decrease in the growth rate is based on a specific feed-back inhibition, i.e. on a chalone-like mechanism (Bichel, 1971, 1972).

There are some cell systems in the mammalian organism in which cell recruitment is different from 'ordinary' tissues; these include, for example, the granulocyte and erythrocyte systems. One characteristic feature of these systems is that they form a series of transit populations which are maintained only by the continuous feeding of new cells from a pluripotent stem cell compartment (see Lajtha, 1967, 1973), i.e. from a pool of progenitor cells which may differentiate in one of several possible directions. It is improbable that chalones would control this type of induction process, and indeed the activation of 'chalone gene(s)' (Bullough, 1965), directing the synthesis of a specific chalone, may itself be the consequence of the induction. What, then, determines the differentiation path which a pluripotent haematopoietic stem cell takes? A number of possibilities have been suggested in the literature, but perhaps the two most plausible ones are that the 'differentiation choice' depends on stromal effects (Trentin, 1971), and/or that the choice depends on a short range cell-to-cell interaction with other haematopoietic cells (McCulloch & Till, 1970).

When a pluripotent haematopoietic stem cell differentiates, it becomes a unipotent (committed) stem cell; from this second stem cell compartment then originate the morphologically recognisable haemic cells (see Lajtha, 1973). Feeding of cells from the unipotent progenitor pool into the final system, which consists of a series of transit populations, can be stimulated by tissue-specific humoral factors. One of these factors, erythropoietin, is now generally known; indeed, for a long time erythropoietin was the only firmly established tissue specific humoral regulator of cell proliferation.

It seems clear that the primary action of erythropoietin is indeed on the committed (unipotent) stem cells, i.e. it is an inducer of haemoglobin synthesis (Alpen, Cranmore & Johnston, 1962; Stohlman et al. 1968). Whether erythropoietin also acts on the precursor cells beyond the stem cell stage, i.e. on the haemoglobin-synthesising cells within the erythron, is not yet unambiguously established. It may be mentioned in this context that an erythropoiesis-inhibiting factor has been extracted from urine (Lindeman, 1971), which in some aspects resembles the erythrocytic chalone (see Seip, 1971); however, judging from the assay technique used (see Lindeman, 1971), the factor looks more like a direct inhibitor of erythropoietin.

The existence of other poietins, especially that of granulopoietin, has not been experimentally proven; the main reason is perhaps the complicated experimental situation in granulopioesis in-vivo. It has been suggested, however, that the colony-stimulating factor, required for the development of granulocyte colonies in the agar culture technique, may be the granulo-poietin (Metcalf & Moore, 1971).

Another substance worth mentioning here which can stimulate cell proliferation in a cell-line specific manner is antigen; it stimulates some small lymphocytes in-vitro and in-vivo roughly in the same way as does the plant mitogen phytohaemagglutinin in-vitro (see Daguillard & Richter, 1970; Benestad et al. 1971). Basically this stimulation resembles the poietin action: both types of substances induce a specific and an irreversible trans-formation of the target cells, and this transformation is then usually followed by mitotic activity (see also Bullough, 1965). It seems that both the antigens and the poietins may have little, if any, direct effect on the proliferation rate of the transformed cells; this rate is controlled, for example, by the chalone mechanism.

CONTROL OF CELL DIVISION AND CANCER

The difference between cancer and normal cells is not that malignant cells proliferate at an extremely fast rate; many normal cells proliferate faster than malignant cells. The difference is that in malignant tumours cell proliferation permanently exceeds cell loss. However, even this difference is not absolute: for instance, in ascites tumours the growth rate decreases steadily, and usually the tumour cells reach a plateau before the death of the host (Brown, 1970; Bichel, 1971, 1972).

It has been obvious for a long time (see Bayne-Jones *et al.* 1938) that the outstanding feature of malignant growth is a break in the cellular control mechanisms, i.e. in the normal balance between cell production and cell loss. However, for example, as indicated by the ascites tumours, this break is not necessarily complete; indeed, malignant cells may always have some point of balance. The implication then is that most, if not all, cancers have retained some ability to respond to the specific control mechanisms, especially to the chalones. This has actually been shown experimentally in a number of different tumour cells, including some spontaneous human tumours (Bullough & Laurence, 1968a, b, 1970a; Bullough & Deol, 1971; Rytömaa & Kiviniemi, 1968c, d, 1969b, 1970; Bichel, 1971, 1972; Elgjo & Hennings, 1971a; Houck *et al.* 1971).

Another important aspect in malignant growth is the fact that cell loss from the tumours is not zero. Indeed, from spontaneous human tumours cell loss may always be more than 50 % of the rate at which cells are being added by mitosis (Steel, 1967), and it may sometimes be as much as 95–99 % (Refsum & Berdal, 1967; Iversen, 1967). The biological mechanism of this cell loss is not quite clear, but there are good reasons to believe that the most important factor is the immunological mechanism (see Mathé, 1969).

The ability of malignant cells to respond by mitotic inhibition to the chalone of their tissue of origin, and the high rate of cell loss from malignant tumours, are two very interesting features; they make a new therapeutic approach possible. Thus, if the inhibitory control mechanism is strengthened by artificially raising the chalone content, without simultaneously reducing the rate of cell loss, it is clearly possible to achieve a situation in which cell loss exceeds cell proliferation. If this situation can be maintained for long enough, all tumour cells would of course disappear from the body (note that the corresponding normal tissue is not necessarily destroyed by the treatment; for explanation, see Rytömaa & Kiviniemi, 1970; Rytömaa, 1970). It has been demonstrated that all this is indeed possible in practice, at least in the case of some experimental (transplantable) tumours (Mohr *et al.* 1968; Rytömaa & Kiviniemi, 1969b, 1970).

REFERENCES

ALPEN, E. L., CRANMORE, D. & JOHNSTON, M. E. (1962). Early observations on the effects of blood loss. In *Erythropoiesis*, ed. L. O. Jacobson & M. Doyle, pp. 184–8. New York & London: Grune & Stratton.

BAYNE-JONES, S., HARRISON, R. G., LITTLE, C. C., NORTHROP, J. & MURPHY, J. B. (1938). Fundamental cancer research. *Public Health Reports*, **53**, 2121–30.

BENESTAD, H. B. (1970). Formation of granulocytes and macrophages in diffusion chamber cultures of mouse blood leucocytes. *Scandinavian Journal of Haematology*, **7**, 279–88.

BENESTAD, H. B., IVERSEN, J.-G. & ROLSTAD, B. (1971). Immunoblast formation by recirculating and non-recirculating rat lymphocytes cultured in diffusion chambers. *Scandinavian Journal of Haematology*, **8**, 32–43.

BENESTAD, H. B., RYTÖMAA, T. & KIVINIEMI, K. (1973). The cell-specific effect of the granulocyte chalone demonstrated with the diffusion chamber technique. *Cell and Tissue Kinetics*, in press.

BICHEL, P. (1971). Autoregulation of ascites tumour growth by inhibition of the G-1 and the G-2 phase. *European Journal of Cancer*, **7**, 349–55.

BICHEL, P. (1972). Specific growth regulation in three ascitic tumours. *European Journal of Cancer*, **8**, 167–73.

BØRNSTAD, G. O., ELGJO, K. & ØYE, I. (1971). Adrenalin increases cyclic $3',5'$-AMP formation in hamster epidermis. *Nature, New Biology, London*, **233**, 78–9.

BROWN, H. R. (1970). The growth of Ehrlich carcinoma and Crocker sarcoma ascitic cells in the same host. *Anatomical Record*, **166**, 283.

BULLOUGH, W. S. (1965). Mitotic and functional homeostasis. A speculative review. *Cancer Research*, **25**, 1683–727.

BULLOUGH, W. S. (1967). *The Evolution of Differentiation*. London & New York: Academic Press.

BULLOUGH, W. S. (1972). Epidermal chalone mechanism. In *Proceedings of the First International Chalone Conference*, ed. B. Forscher. *National Cancer Institute Monograph*. (In press.)

BULLOUGH, W. S. & DEOL, J. U. R. (1971). The pattern of tumour growth. In *Control Mechanisms of Growth and Differentiation, Symposia of the Society for Experimental Biology*, **25**, 255–75.

BULLOUGH, W. S. & LAURENCE, E. B. (1964). Mitotic control by internal secretion: the role of the chalone–adrenalin complex. *Experimental Cell Research*, **33**, 176–94.

BULLOUGH, W. S. & LAURENCE, E. B. (1968a). Control of mitosis in rabbit V × 2 epidermal tumours by means of the epidermal chalone. *European Journal of Cancer*, **4**, 587–94.

BULLOUGH, W. S. & LAURENCE, E. B. (1968b). Control of mitosis in mouse and hamster melanomata by means of the melanocyte chalone. *European Journal of Cancer*, **4**, 607–15.

BULLOUGH, W. S. & LAURENCE, E. B. (1968c). The role of glucocorticoid hormones in the control of epidermal mitosis. *Cell and Tissue Kinetics*, **1**, 5–10.

BULLOUGH, W. S. & LAURENCE, E. B. (1970a). The lymphocytic chalone and its antimitotic action on mouse lymphoma *in vitro*. *European Journal of Cancer*, **6**, 525–31.

BULLOUGH, W. S. & LAURENCE, E. B. (1970b). Chalone control of mitotic activity in sebaceous glands. *Cell and Tissue Kinetics*, **3**, 291–300.

BULLOUGH, W. S., LAURENCE, E. B., IVERSEN, O. H. & ELGJO, K. (1967). The vertebrate epidermal chalone. *Nature, London*, **214**, 578–80.

BYRON, J. W. (1971). Effect of steroids and dibutyryl cyclic AMP on the sensitivity of haemopoietic stem cells to ³H-thymidine *in vitro*. *Nature, London*, **234**, 39–40.

CHOPRA, D. P. (1973). The lung and kidney chalone. In *Proceedings of the First International Chalone Conference*, ed. B. Forscher. *National Cancer Institute Monograph*. (In press.)

CHOPRA, D. P. & SIMNETT, J. D. (1970). Stimulation of mitosis in amphibian kidney by organ specific antiserum. *Nature, London*, **225**, 657–8.

CHUNG, A. (1973). Some *in vivo* effects of chalone (mitotic inhibitor) obtained from lymphoid tissues. In *Proceedings of the First International Chalone Conference*, ed. B. Forscher. *National Cancer Institute Monograph*. (In press.)

DAGUILLARD, F. & RICHTER, M. (1970). Cells involved in the immune response. XVI. The response of immune rabbit cells to phytohemagglutinin, antigen, and goat anti-rabbit immunoglobulin antiserum. *Journal of Experimental Medicine*, **131**, 119–31.

DEWEY, D. L. (1973). Development of an assay system for the melanocyte chalone *in vitro*. In *Proceedings of the First International Chalone Conference*, ed. B. Forscher. *National Cancer Institute Monograph*. (In press.)

EGYÜD, L. G. (1965). Studies on autobiotics: chemical nature of retine. *Proceedings of the National Academy of Sciences, U.S.A.* **54**, 200–2.

ELGJO, K. (1973). Cell cycle specificity of two epidermal growth inhibitors. In *Proceedings of the First International Chalone Conference*, ed. B. Forscher. *National Cancer Institute Monograph*. (In press.)

ELGJO, K. & HENNINGS, H. (1971a). Epidermal chalone and cell proliferation in a transplantable squamous cell carcinoma in hamsters. I. *In vivo* results. *Virchows Archiv, Abteilung B, Zellpathologie*, **7**, 1–7.

ELGJO, K. & HENNINGS, H. (1971b). Epidermal mitotic rate and DNA synthesis after injection of water extracts made from mouse skin treated with actinomycin D: Two or more growth-regulating substances? *Virchows Archiv, Abteilung B, Zellpathologie*, **7**, 342–7.

EPIFANOVA, O. I. (1971). Effects of hormones on the cell cycle. In *The Cell Cycle and Cancer*, ed. R. Baserga, chapt. 6, pp. 145–90. New York: Marcel Dekker.

FRANKFURT, O. S. (1971). Epidermal chalone. Effect on cell cycle and on development of hyperplasia. *Experimental Cell Research*, **64**, 140–4.

GARCIA-GIRALT, E., LASALVIA, E., FLORENTIN, I. & MATHÉ, G. (1970). Evidence for a lymphocytic chalone. *European Journal of Clinical and Biological Research*, **15**, 1012–15.

GARRY, R. & HALL, R. (1970). Stimulation of mitosis in rat thyroid by long-acting thyroid stimulator. *Lancet*, **ii**, 693–5.

HARRINGTON, J. S. (1967). The sulfhydryl group and carcinogenesis. In *Advances in Cancer Research*, vol. 10, ed. A. Haddow & S. Weinhouse, pp. 247–309. New York & London: Academic Press.

HEIDRICH, M. L. & RYAN, W. R. (1970). Cyclic nucleotides on cell growth *in vitro*. *Cancer Research*, **30**, 376–8.

HENNINGS, H., ELGJO, K. & IVERSEN, O. H. (1969). Delayed inhibition of epidermal DNA synthesis after injection of aqueous skin extract (chalone). *Virchows Archiv, Abteilung B, Zellpathologie*, **4**, 45–53.

HONDIUS BOLDINGH, W. & LAURENCE, E. B. (1968). Extraction, purification and preliminary characterisation of the epidermal chalone: a tissue specific inhibitor obtained from vertebrate skin. *European Journal of Biochemistry*, **5**, 191–8.

HOUCK, J. C. (1973). The fibroblast chalone. In *Proceedings of the First International Chalone Conference*, ed. B. Forscher. *National Cancer Institute Monograph*. (In press.)

470 T. RYTÖMAA

HOUCK, J. C., IRASQUIN, H. & LEIKIN, S. (1971). Lymphocyte DNA synthesis inhibition. *Science, New York*, **173**, 1139–41.

IVERSEN, O. H. (1967). Kinetics of cellular proliferation and cell loss in human carcinomas. A discussion of methods available for *in vivo* studies. *European Journal of Cancer*, **3**, 389–94.

IVERSEN, O. H. (1969). Chalones of the skin. In *Ciba Foundation Symposium on Homeostatic Regulators*, ed. G. E. W. Wolstenholme & J. Knight, pp. 29–53. London: Churchill.

IVERSEN, O. H. & ELGJO, K. (1967). The effect of chalone on the mitotic rate and on the mitotic duration in hairless mouse epidermis. In *Control of Cellular Growth in Adult Organisms*, ed. H. Teir & T. Rytömaa, pp. 83–91. London & New York: Academic Press.

JONES, J., PARASKOVA-TCHERNOZEMSKA, E. & MOORHEAD, J. F. (1970). In-vitro inhibition of D.N.A. synthesis in human leukaemic cells by a lymphoid cell extract. *Lancet*, **i**, 654–5.

KIVILAAKSO, E. (1970). In-vitro assay of the stimulatory effect of hypoxic serum on cellular proliferation in the erythron. *Acta physiologica scandinavica*, **80**, 412–19.

KIVILAAKSO, E. & RYTÖMAA, T. (1971). Erythrocytic chalone, a tissue-specific inhibitor of cell proliferation in the erythron. *Cell and Tissue Kinetics*, **4**, 1–9.

KNOCK, F. E. (1967). *Anticancer Agents*. Illinois: Thomas, Springfield.

LAJTHA, L. G. (1967). Proliferation kinetics of steady state cell populations. In *Control of Cellular Growth in Adult Organisms*, ed. H. Teir & T. Rytömaa, pp. 97–105. London & New York: Academic Press.

LAJTHA, L. G. (1973). Review of leukocytes. In *Proceedings of the First International Chalone Conference*, ed. B. Forscher. *National Cancer Institute Monograph.* (In press.)

LASALVIA, E., GARCIA-GIRALT, E. & MACIEIRA-COELHO, A. (1970). Extraction of an inhibitor of DNA synthesis from human peripheral blood lymphocytes and bovine spleen. *European Journal of Clinical and Biological Research*, **15**, 789–92.

LAURENCE, E. B. & RANDERS HANSEN, E. (1971). An *in vivo* study of epidermal chalone and stress hormones on mitosis in tongue epithelium and ear epidermis of the mouse. *Virchows Archiv, Abteilung B, Zellpathologie*, **9**, 271–9.

LAURENCE, E. B., RANDERS HANSEN, E., CHRISTOPHERS, E. & RYTÖMAA, T. (1972). Systemic factors influencing epidermal mitosis. *European Journal of Clinical and Biological Research*, **17**, 133–9.

LINDEMAN, R. (1971). Erythropoiesis inhibiting factor (EIF). I. Fractionation and demonstration of urinary EIF. *British Journal of Haematology*, **21**, 623–31.

McCULLOCH, E. A. & TILL, J. E. (1970). Cellular interactions in the control of hemopoiesis. In *Hemopoietic Cellular Proliferation*, ed. F. Stohlman, pp. 15–25. New York: Grune & Stratton.

MARKS, F. (1971). Direct evidence of two tissue-specific chalone-like factors regulating mitosis and DNA synthesis in mouse epidermis. *Hoppe-Seyler's Zeitschrift für Physiogische Chemie*, **352**, 1273–4.

MARKS, F. (1973). A tissue-specific inhibitor of epidermal DNA synthesis. In *Proceedings of the First International Chalone Conference*, ed. B. Forscher. *National Cancer Institute Monograph.* (In press.)

MARKS, F. & REBIEN, W. (1972). Cyclic 3′,5′-AMP and theophylline inhibit epidermal mitosis in G_2 phase. *Naturwissenschaften*, **59**, 41–2.

MARRS, J. M. & VORHEES, J. J. (1971). Preliminary characterization of an epidermal chalone-like inhibitor. *Journal of Investigative Dermatology*, **56**, 353–8.

MATHÉ, G. (1969). Approaches to the immunological treatment of cancer in man. *British Medical Journal*, **4**, 7–10.

MAUEL, J. & DEFENDI, V. (1971). Regulation of DNA synthesis in mouse macrophages. II. Studies on mechanism of action of the mouse macrophage growth factor. *Experimental Cell Research*, **65**, 377–85.

METCALF, D. & MOORE, M. A. S. (1971). *Haemopoietic Cells*. Amsterdam & London: North-Holland Publishing Co.

MOHR, U., ALTHOFF, J., KINZEL, V., SÜSS, R. & VOLM, M. (1968). Melanoma regression induced by 'chalone': a new tumour inhibiting principle acting *in vivo*. *Nature, London*, **220**, 138–9.

MOORHEAD, J. F., PARASKOVA-TCHERNOZEMSKA, E., PIRRIE, A. J. & HAYES, C. (1969). Lymphoid inhibitor of human lymphocyte DNA synthesis and mitosis *in vitro*. *Nature, London*, **224**, 1207–8.

PAUKOVITS, W. R. (1971). Control of granulocyte production: separation and chemical identification of a specific inhibitor (chalone). *Cell and Tissue Kinetics*, **4**, 539–47.

PAUKOVITS, W. R. (1973). Granulopoiesis inhibiting factor (GIF): demonstration and preliminary chemical–biological characterization of a specific polypeptide (chalone). In *Proceedings of the First International Chalone Conference*, ed. B. Forscher. *National Cancer Institute Monograph*. (In press.)

POWELL, J. A., DUELL, E. A. & VORHEES, J. J. (1971). Beta adrenergic stimulation of endogenous epidermal cyclic AMP formation. *Archives of Dermatology*, **104**, 359–65.

REFSUM, S. B. & BERDAL, P. (1967). Cell loss in malignant tumours in man. *European Journal of Cancer*, **3**, 235–6.

RYTÖMAA, T. (1969). Granulocytic chalone and antichalone. In *Hemic Cells in vitro*, ed. P. Farnes, *In Vitro*, vol. 4, pp. 47–58. Baltimore: Williams & Wilkins.

RYTÖMAA, T. (1970). Regulation of cell production by chalones. *Annals of Clinical Research*, **2**, 94–5.

RYTÖMAA, T. (1972). Granulocytic chalone. In *Proceedings of the Vth International Symposium on Comparative Leukemia Research*. *Bibliotheca Haematologica*, no. 39. Basel & New York: Karger. (In press.)

RYTÖMAA, T. (1973). Chalone of the granulocyte system. In *Proceedings of the First International Chalone Conference*, ed. B. Forscher. *National Cancer Institute Monograph*. (In press.)

RYTÖMAA, T. & KIVINIEMI, K. (1967). Regulation system of blood cell production. In *Control of Cellular Growth in Adult Organisms*, ed. H. Teir & T. Rytömaa, pp. 106–38. London & New York: Academic Press.

RYTÖMAA, T. & KIVINIEMI, K. (1968a). Control of granulocyte production. I. Chalone and antichalone, two specific humoral regulators. *Cell and Tissue Kinetics*, **1**, 329–40.

RYTÖMAA, T. & KIVINIEMI, K. (1968b). Control of granulocyte production. II. Mode of action of chalone and antichalone. *Cell and Tissue Kinetics*, **1**, 341–50.

RYTÖMAA T. & KIVINIEMI, K. (1968c). Control of DNA duplication in rat chloroleukaemia by means of the granulocytic chalone. *European Journal of Cancer*, **4**, 595–606.

RYTÖMAA, T. & KIVINIEMI, K. (1968d). Control of cell production in rat chloroleukaemia by means of the granulocytic chalone. *Nature, London*, **220**, 136–7.

RYTÖMAA, T. & KIVINIEMI, K. (1969a). The role of stress hormones in the control of granulocyte production. *Cell and Tissue Kinetics*, **2**, 263–8.

RYTÖMAA, T. & KIVINIEMI, K. (1969b). Chloroma regression induced by the granulocytic chalone. *Nature, London*, **222**, 995–6.

RYTÖMAA, T. & KIVINIEMI, K. (1970). Regression of generalized leukaemia in rat induced by the granulocytic chalone. *European Journal of Cancer*, **6**, 401–10.

SAETREN, H. (1956). A principle of auto-regulation of growth. Production of organ specific mitose-inhibitors in kidney and liver. *Experimental Cell Research*, **11**, 229–32.

SCAIFE, J. F. (1970). Liver homeostasis: an *in vitro* evaluation of a possible specfic chalone. *Experientia*, **26**, 1071–2.

SCHÜTT, M. & LANGEN, P. (1972). Comments on granulocytic chalone action. In *Symposium on Active Control of Nucleic Acid Metabolism. Studia Biophysica.* (In press.)

SEIP, M. (1971). Erythropoietic inhibiting factor (EIF) – an erythrocytic chalone? *Annals of Clinical Research*, **3**, 3–4.

SIMNETT, J. D., FISHER, J. M. & HEPPLESTON, A. G. (1969). Tissue-specific inhibition of lung alveolar cell mitosis in organ culture. *Nature, London*, **223**, 944–6.

STEEL, G. G. (1967). Cell loss as a factor in the growth rate of human tumours. *European Journal of Cancer*, **3**, 381–7.

STOHLMAN, F., EBBE, S., MORSE, B., HOWARD, D. & DONOVAN, J. (1968). Regulation of erythropoiesis. XX. Kinetics of red cell production. *Annals of the New York Academy of Sciences*, **149**, 156–72.

SZENT-GYÖRGYI, A. (1968). *Bioelectronics. A Study in Cellular Regulations, Defense, and Cancer.* New York & London: Academic Press.

SZENT-GYÖRGYI, A., EGYÜD, L. G. & MCLAUGHLIN, J. A. (1967). Keto-aldehydes and cell division. *Science, New York*, **155**, 539–41.

TRENTIN, J. J. (1971). Determination of bone marrow stem cell differentiation by stromal hemopoietic inductive microenvironments (HIM). *Americal Journal of Pathology*, **65**, 621–8.

VERLY, W. G., DESCHAMPS. Y., PUSHPATHADAM, J. & DESROSIERS, M. (1971). The hepatic chalone. I. Assay method for the hormone and purification of the rabbit liver chalone. *Canadian Journal of Biochemistry*, **12**, 1376–83.

VOADEN, M. J. A. (1968). A chalone in the rabbit lens? *Experimental Eye Research*, **7**, 326–31.

VOLM, M., KINZEL, V., MOHR, U. & SÜSS, R. (1969). Inactivation of tissue-specific inhibitors by a carcinogen (diethylnitrosamine). *Experientia*, **25**, 68–9.

VOLM, M. & WAYSS, K. (1971). Zur Gewebsspezifität der Hemmfaktoren im Endometrium. *Naturwissenschaften*, **58**, 458.

VOLM, M., WAYSS, K. & HINDERER, H. (1969). A new model of growth regulation. *Naturwissenschaften*, **56**, 566–7.

INDEX

Acer pseudoplatanus, synchronisation of cell culture of, 186

acetate: labelled, for following sporopollenin synthesis in *Chlorella*, 73–4; respiratory rate of *Chlorella* in, 67

acid phosphatase: in activated and reverted lymphocytes, 421; in cell cultures of tubers, 188, 191

actidione, mutant of *Physarum* resistant to, 78

actinomycin D: and cell cycle length, 38; and DNA replication in *Amoeba*, 34; and enzymes in cell cultures of tubers, 191; and erythropoietin-induced syntheses, 344–5; and protein synthesis in sea urchin blastulae, 210; and RNA synthesis, 58–9, 90–1

adenyl cyclase: in activated lymphocytes, 404; epidermal chalone and, 461

adrenalin, enhances effect of epidermal chalone, 461, 464

age of cells, distribution of, 14–15

ageing, and cell cycle, 268–70, 274, 280; in ciliates, 51, 53–4, 59; and mitotic cycle in roots, 144–5

Alliaria petiolata, meiosis duration in, 113, 121

Allium cepa; cell cycle in root meristems of, 136, 137, 151, 154, 155; histones during cell differentiation in, 147; induction of cells with two nuclei in, 143; meiosis duration in, 112, 117, 121; polyploidy in ungerminated seeds of, 134

Allium sativum, histones during cell differentiation in, 146

α-amanitin, and RNA synthesis in nucleus and nucleolus, 96, 98

amethopterin + adenosine, growth inhibition by, 409

5-amino uracil, synchronisation of plant cells by, 185, 186

amnion, human: polyploidy in cells of, 304

Amoeba proteus: DNA synthesis in cell cycle of, 32–7; polyploidy in, 46; protein synthesis in, 44–6; RNA synthesis in, 37–44

AMP, cyclic, 461

amphibia, cell cycles in, 257; *see also individual species*

anaemia: erythroblasts in, 349, 350, 360; in foetus, 374; genetic, in mice, 283–91; mature erythroid cells in, 346; response of adult erythroid system to, 370–4

androgens, as mitogenic hormones, 337, 464

antibodies, DNA metabolism in maturation of cells producing, 437–54

antichalones, 464

anti-erythropoietin, 368, 382, 384

antigens: of activated lymphocytes, 422; activation of lymphocytes by, 397, 466

anti-immunoglobulin serum, 424

anti-lymphocyte serum (ALS), activates lymphocytes, 397, 398

antiphytohaemagglutinin (anti-PHA) serum, 399, 418

Antirrhinum majus, meiosis duration in, 112, 121

apical meristems, cell cycle of: in different regions, 168–72; lengths of phases of, 173–81; synthesis during, 172–3; in transition from vegetative to floral growth, 181–2; in whole apex, 167–8

arginase: synthesis of, in yeast, 6

ascites tumours, growth rate of, 465, 467

ascorbic oxidase, in cell cultures of tubers, 188, 190

ATP glucokinase, in cell cultures of tubers, 195, 196, 197, 198

auxins, effects of: on axillary buds, 180; on cell cultures of tubers, 187, 190, 191

avian myeloblastosis virus, reverse transcriptase of, 361

azetidine-2-carboxylic acid (proline analogue), and enzyme synthesis in *Chlorella*, 69

Bacillus subtilis, enzyme synthesis in, 6, 7

bacteria: gene dosage effect in, 193; inducible enzymes in, 7; short cycle cell of, 38

Beta vulgaris, meiosis duration in, 112

bilaterality, in developing molluscan egg, 218–19, 219–20, 222–3

blastocysts, mammalian: differential sensitivity to radiation of outer and inner cells of, 320–1; effects of tritiated thymidine on development of, 314–19; regions of, 311

blastomeres, in molluscan egg: qualitative difference of, associated with difference in cleavage rhythm, 217

Bombyx, cell cycle and response to X-rays in embryo of, 231

bone, cell cycle in mineralisation of, 325

[473]

cytochrome oxidase: stepwise synthesis of, in *Chlorella*, 63, 64, 66, 68

cytoplasm: distribution of RNA between nucleus and, in *Amoeba*, 41–3; and DNA synthesis, 3, 34, 35–6; radiation effects on, 240–1; and RNA synthesis, 43–4; RNAs of, in *Physarum*, 85–6

cytosine arabinoside, inhibits DNA synthesis and cell division in foetal liver cells 356, 357–8, 359

cytotoxins, in activated lymphocytes, 425–6

Datura, cell cycle in apical meristem of, 169, 170, 181

2,4-dichlorophenoxyacetic acid (2,4-D, auxin): and growth and enzyme activity in cell cultures of tubers, 190, 191; synchronisation of plant cell cultures by, 187, 191

Dictyostelium, enzyme synthesis in, 71

Diptera, cyclic DNA synthesis in polytene chromosomes of, 279–88

dipyridamole, inhibits uptake of nucleotides by activated lymphocytes, 401, 403

DNA: fractions of, released from spleen explants giving immune response, 447–52, 454; mitochondrial, 87, 90, 100, 101, 104; of *Physarum*, highly repetitious, 103; protein associated with, in activated lymphocytes, 405; satellite, 4, 91, 100, 101, 304; types of, in micro- and macronuclei of ciliates, 57–8, 59; ultraviolet light selectively absorbed by, 238, 244

DNA, cell content of: and cell cycle in plant meristems, (apical) 176–9, (root) 123, 150–4; in development of macronucleus of ciliates, 55, 56; and meiosis duration in plants, 114–16, 128; in pea apical and root meristems, 172; and resistance of embryo to X-rays, 232, 233–4; in ungerminated seeds, 133–4; *see also* polyploidy

DNA polymerase: in activated lymphocytes, 423; in cell cultures of tubers, 193, 194, 199; in germinating seeds, 135; and replication of nucleolar DNA, 287

DNA synthesis: in activated lymphocytes, 403, 410–17, (reactivated) 418–20; in antibody-producing cells, 442–52; in cell cycle of *Amoeba*, 32–7, 46, and of *Chlorella*, 61, 63, 64, 65; and enzyme synthesis in cell cultures of tubers, 193, 199; in erythroblasts of normal and genetically anaemic mice, 387; erythropoietin and, in cell cultures, 342–3, 344, 346–8, 351, 356, 390–1; genes for, in *Saccharomyces*, 105, 149; in germinating seeds, 135; in immune responses, 437;

inhibited by chalones, 460, 464, and by thymidine, 312; in inner cells of blastocyst, 321; in macronucleus of ciliates, 36, 53, 56; in mammalian embryo, 293–4, in meristems, (apical) 119, (root) 137–43, 146, 147, (after irradiation) 157, 159; metabolic, by non-dividing cells, 437; in molluscan development, 225–6; periodic, 3, 17, 156; periodic synthesis of enzymes involved in, 5; in *Physarum*, 79, 87; in polytene chromosomes of Diptera, 85, 279, 280, 283–8; in presumptive neurectoderm of *Xenopus*, 249–55; radiation effects on, 241, 321, 369; stathmokinetic effects on, in adult and foetal tissue, 332–4; in *Stylonychia*, 53; testosterone and, 337, 340; in trophoblasts, 300–4

DNAase: in cell cultures of tubers, 188, 189; in developing microspores of *Lilium*, 186; effect of, on DNA fractions from spleen explants giving immune response, 447–8, 449

Drosophila: cell cycle and radiation effects on, 231–2, 238, 242; DNA synthesis in polytene chromosomes of, 280–5; sulphydryl compounds and radiation effects on, 235, 243

Dryopteris, cell cycle in apical meristem of, 167, 168

ecdysone (moulting hormone), not directly involved in control of DNA synthesis in polytene chromosomes of Diptera, 281–2

Elaeis, cell cycle in apical meristem of, 167, 168

embryos: cell cycle during development of, 31, 293–306; cell cycle and radiation effects on, 229–45; differential effect of tritiated thymidine on two populations of, 311–21; specific action of stathmokinetic agents on, 331–4

endoplasmic reticulum: synthesis of, in cell cycle of *Chlorella*, 63

end-product repression, as possible cause of oscillations in enzyme synthesis, 5–6, 65, 69, 72

Endymion nonscriptus, meiosis duration in, 112, 114, 115, 120

environments: different, required for development of different types of blood cells, 367–8, 379, 391

enzymes: changes in activity of, during division cycle of plant cells in culture, 185–99; integration of, into respiratory metabolism of *Chlorella*, 66–8; periodic and continuous synthesis of, 5–9; synthesis of, in *Chlorella*, 63–5, 68–73; *see also individual enzymes*